Intensive Sugarcane Production:
Meeting the Challenge Beyond 2000

Intensive Sugarcane Production: Meeting the Challenge Beyond 2000

Proceedings of the Sugar2000 Symposium
Brisbane, Australia, 20–23 August 1996

Edited by

Brian Keating

and

John Wilson

CSIRO Tropical Agriculture
306 Carmody Road
St Lucia
Queensland 4067
Australia

CAB INTERNATIONAL

CAB INTERNATIONAL
Wallingford
Oxon OX10 8DE
UK

Tel: +44 (0)1491 832111
Fax: +44 (0)1491 833508
E-mail: cabi@cabi.org

CAB INTERNATIONAL
198 Madison Avenue
New York, NY 10016-4314
USA

Tel: +1 212 726 6490
Fax: +1 212 686 7993
E-mail: cabi-nao@cabi.org

A catalogue record for this book is available from the British Library, London, UK.

Library of Congress Cataloging-in-Publication Data
Intensive sugarcane production : meeting the challenges beyond 2000 /
edited by B.A. Keating and J. Wilson.
p. cm.
papers presented at the Sugar 2000 symposium held in Brisbane.
Australia, in August 1996.
Includes index.
ISBN 0-85199-193-9 (alk. paper)
1. Sugarcane--Congresses. I. Keatin, B. A. II. Wilson, J. R.
III. International Symposium "Sugar 2000" (1996 : Brisbane, Qld.)
SB226.2.I58 1997
338.1'7361'0994--dc21 97-15282
CIP

ISBN 0 85199 193 9

Printed and bound from copy supplied by the editors in the UK
by Biddles Ltd, Guildford and King's Lynn

Contents

THEME: ***OPPORTUNITIES FOR IMPROVED PLANTS AND MANAGEMENT SYSTEMS***

THEME: *SUMMARY ADDRESS*

Preface

As we approach the beginning of the 21st Century, agriculture is facing a set of challenges and opportunities that are simultaneously daunting and exhilarating. This is the case regardless of where we look, whether it be the subsistence food production systems of Africa or the intensive agro-industrial farming systems of Europe, North and South America and Oceania. The steadily increasing world population creates the underlying challenge. The need for production to keep pace with population growth, whilst preserving the soil and water resources for productivity in future generations is a driving force at the global scale. Erosion, soil fertility depletion, salinisation, acidification and soil physical degradation have taken their toll on the productive capacity of the world's food production lands. In many parts of the world, we are increasingly reminded that agriculture is not the only claimant on land and water resources. Expansion of urban communities, the needs of other industries such as mining and tourism, as well as the expectation that very special natural places should be preserved for their own intrinsic sake and for the sake of future generations, are all legitimate claimants for these resources. We all have to share the land with agriculture.

These forces are nowhere more obvious than in the intensive cropping industries of the world that are the prime focus of this book. The focus is largely on sugarcane, as an example of a cropping system that uses intensive land, labour, capital and other inputs in the search for high productivity and profitability. While the majority of chapters reflect upon the Australian sugar industry, the issues raised are relevant to other sugar producing regions of the world, and more generally to many intensive agricultural production systems facing the demands of international competitiveness and ecological sustainability. With 90% of Australian sugar exported and subject to the pressures of world market forces, with a 100-year history of research and development, and with a location amongst the splendid natural environment of wet tropical rainforests and the Great Barrier Reef, there are few better case studies of the past gains and future challenges facing an intensive agricultural industry.

While the challenges facing agriculture in the 21st century are large and varied, the opportunities through modern scientific advances for more efficient and effective agriculture are equally impressive. A number of contributions to this book consider the role of emerging technologies for genetic manipulation in areas as diverse as potential productivity, pest and disease resistance, nutrient utilisation and alternative economic products from the world's sugar industries. Other contributions focus on the role of crop and soil management in the achievement of productive and sustainable sugarcane cultivation. Several chapters make excellent contributions to the discussion of the balance between agricultural demands and protecting the environment. The progress to be made through more integrated approaches to system analysis and management, and to technology transfer, is an outstanding feature of a number of contributions. A third group of opportunities resides in the area of planning and policy beyond the farm, and contributors in this area have raised key contemporary issues.

This book comprises the 28 invited review papers presented to the International Symposium 'Sugar 2000: Challenges for Intensive Crop Production' held in Brisbane, Australia from 19-23 August 1996. This symposium was organised by CSIRO Tropical Agriculture with support from other research, development, grower and miller groups connected with the Australian sugar industry. These included the Bureau of Sugar Experiment Stations, CSR Technical Field Department, CANEGROWERS, INCITEC, John Deere and the Sugar Research and Development Corporation. Over 300 delegates attended this meeting from 21 different countries. The symposium represented the first comprehensive international conference focusing on the Australian sugar industry and the first international sugar conference where an attempt was made to compare and contrast sugarcane production with other intensive cropping industries.

Almost 100 contributed short research papers related to the various topics of the symposium have been separately published by CSIRO Tropical Agriculture in 1996 as a Poster book 'Sugarcane: Research Towards Efficient and Sustainable Production' (ed. J.R. Wilson *et al.*).

The Symposium covered a very broad range of issues, from biotechnology to agronomy, from resource management to policy and planning, from by-product utilisation to approaches to research and develop ent. While the scope of the subject matter was large by any measure, this simply reflects the range of issues that must be addressed by an intensive agricultural industry to prosper in a highly complex and competitive world.

A great many people contributed to the success of the symposium and in the subsequent preparation of the papers found in this book. The Symposium Organising Committee consisted of Russell Muchow (Chairman), James Campbell, Mac Hogarth, Kelso Greenword, MayLing Goode, Alan Garside, Andrew Johnson, Brian Keating, Marshall Mackay, John Wilson and Keith Weier. In addition, the editors wish to acknowledge the special contribution made by Professor Graham Wilson for his assistance in editorial matters and to Dr James Campbell, Ms Leesa Austin and Ms Barbara Philp for their efforts associated with the editing, reviewing and preparation of the manuscript, and the many readers who assisted in the refereeing process.

Brian Keating
and
John Wilson

Brisbane, Australia
June 1997.

Chapter one:

A Global Perspective of the Sugar Industry

J. Fry
Managing Director, LMC International, 14-16 George Street, Oxford, UK

Abstract

The world sugar and sweetener industry is reviewed in terms of main topics:

1. The changing structure of sugar production, consumption and trade.
2. The growing competition for sugar within the market for sweeteners.
3. The competitiveness of leading sugar producers.
4. The impact of external factors upon different sugar industries.
5. The competition between sugar crops and other crops at the farm level.
6. The likely trends early in the next century.

Sugar has lost ground to competing sweeteners, which are cheaper or offer fewer calories. If sugar had maintained its 1980 share of world sweetener demand, sugar sales would have been 10% higher in 1995 than they were.

Australia, India and Thailand have lowered their real field costs substantially since 1980. Then, Australia enjoyed the lowest field costs of this group. By 1995, India was in pole position. In milling costs, Australia has been overtaken by Thailand. Australia's special advantage is its low costs of getting sugar into export vessels. Australia's high yields in the field and good factory recoveries are targets for other producers. The gap between Australia's performance and that in other countries has shrunk. Production costs feedback to world prices. Low cost producers in Australia, Thailand, Brazil, India and Southern Africa will therefore continue to grow in importance, while the role of high cost producers in Europe, the US, the Caribbean and Eastern Asia will diminish.

In: Keating, B.A. and Wilson, J.R. (eds) *Intensive Sugarcane Production: Meeting the Challenges Beyond 2000*. CAB International, Wallingford, UK, 1997.

Beet factories lead the way in energy conservation; they buy all their fuel, whereas cane mills often take bagasse for granted. Sugar crops in both the EU and US generate higher margins per hectare than the other major crops.

Technical progress is essential, since the real sugar price will fall at 2% per year or so over time. Australia's willingness to invest a higher proportion of sugar revenues in R & D than any other country bodes well for the future. If wages are to continue to rise, successful sugar industries need yearly productivity increases of 3% to remain internationally competitive.

Changing Structure of the World Sugar Market

The best starting point in understanding the changes under way in the structure of the world sugar market is an assessment of the key shifts in sugar production, consumption and trade over the past 20 years, following the deep recession of the mid-1970s. In 1975, world sugar demand amounted to approximately 75 million tonnes, raw value (r.v.). Since then, it has risen by 50% to almost 114 million tonnes in 1994, implying average annual growth of 2.1%.

The regional composition of production has shifted considerably during this period. In the mid-1970s, Europe was the world's largest producing region, with 33% of world output, followed by Asia (20%), Central America (15%) and South America (14%). By 1994, Asia had overtaken Europe, with 30% of world production, while Europe had fallen back to 25%. Behind these two regions, South America had moved ahead of Central America, with 16% and 10%, respectively. Elsewhere, North America's share fell by 1% to 6.5% during the 20 years, while Africa remained static at 6.5% and Oceania rose from 3.9% to 5.3% of the market. The falling shares of Europe and North America were translated into a sharp drop in the beet share of world output, from 39% to 31%.

A checklist of the world's ten largest producers in 1994 reflects the regional picture. In millions of tonnes, r.v., Asia is represented by India (11.9), China (6.3) and Thailand (4.2); Europe by the EU (15.7) and the former Soviet Union (5.7, barely half its peak before the collapse of Communism); South America by Brazil (12.3); Central America by Cuba (4.0, another casualty of the collapse of Communism) and Mexico (4.0); while Oceania is dominated by Australia (5.2).

On the consumption side, the regional shifts have largely mirrored those in production. Between 1975 and 1994, Europe's share fell from 39% to 27%, while Asia's surged from 23% to 37%. In North America, competition from starch sweeteners actually led to an absolute fall in sugar demand, lowering the region's share of the world market from 13% to 8%, allowing it to be overtaken by South America, whose share was stable at 12%. Both Central America's and Africa's shares moved up by roughly 1% to 6% and 8%, respectively; while Oceania remained just ahead of 1% of the world total.

The proportion of world sugar output which enters foreign trade has altered little during the 20 years, rising from 26.1% to 27.2%. However, the composition of this trade has altered dramatically. White sugar increased its percentage of the volume of sugar exports from only 25% to 46%, led by the growth in EU exports. In 1994, the ten leading exporters, in millions of tonnes, r.v., were the EU (5.1); Australia (4.5); Brazil (3.6); Cuba (3.2); Thailand (2.7); the former Soviet Union (2.0, largely of exports from one Republic to another); China (1.0, mainly re-exports); Guatemala (0.8); Mauritius (0.6) and Fiji (0.5). It is significant that, of those exporting over one million tonnes, only Australia and Cuba are still seen primarily as raw sugar exporters.

Sugar's Place in the World of Sweeteners

There are many sweeteners competing for the world's sweet tooth. Among nutritive sweeteners, sugar competes with starch-based glucose and fructose, notably in the form of High Fructose Syrup and glucose syrup. High Fructose Syrup is widely known as HFCS, or High Fructose Corn Syrup, because virtually all of the world's HFS is made from maize - 'corn' to North Americans.

Where HFCS is cheaper than sugar, as in the US, the starch sweeteners have more or less eliminated sugar from all liquid markets. As yet, the solid forms of starch sweeteners, such as dextrose and crystalline fructose, have not made much headway in competition with sugar, both because they are expensive and because they cannot match sugar in its functionality.

Intense sweeteners include long established products, such as saccharin and cyclamates, and a recent generation which includes aspartame, acesulfame-K, sucralose and alitame.

The attractions of intense sweeteners are twofold: as a cheap alternative to sugar, and as a calorie-free sweetener for dieting consumers. Worldwide, cheapness is the more important. In China, more than the equivalent of four million tonnes of sugar is consumed as intense sweeteners.

Figures 1.1 and 1.2 present the trends since 1980 in the market shares of alternative sweeteners in Oceania, as well as in the world as a whole. Globally, and also in every continent, sugar has lost market share since 1980. High Fructose Syrup now has over 7% of global demand for sweeteners. The older intense sweeteners and the older starch sweeteners, glucose and dextrose, have each captured 5% to 6% of the world market, while the new intense sweeteners account for 2% of the total. As a result, sugar's proportion of world sweetener demand was below 80% in 1995, as against over 87% in 1980.

This means that, if sugar had been able to do no better than maintain its share of the world sweetener market since 1980, sugar demand in 1995 would have been 10% - or over 11 million tonnes - higher than it actually was.

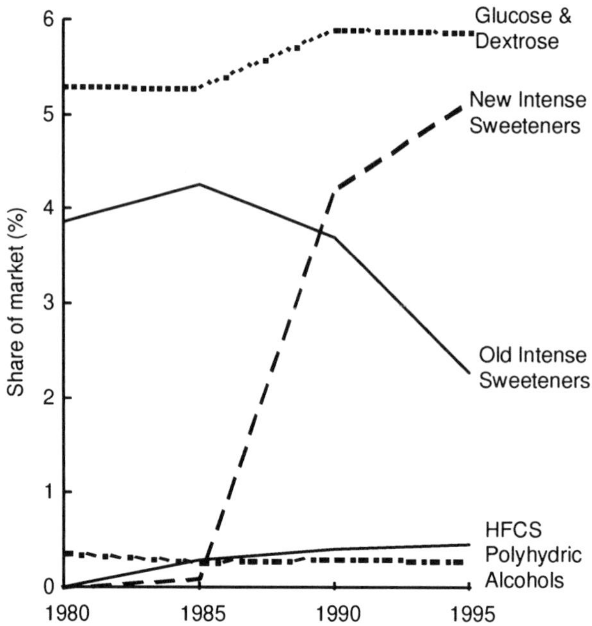

Fig. 1.1. Non-sugar share of sweetener markets — Oceania.

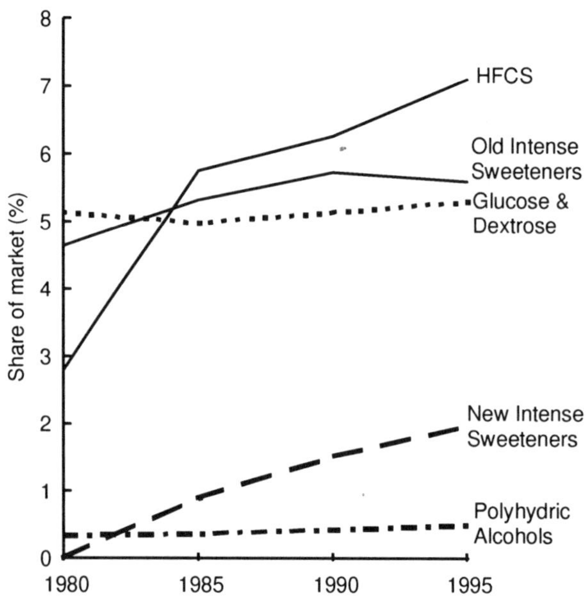

Fig. 1.2. Non-sugar share of sweetener markets — World.

Cost Competitiveness of Leading Sugar Producers

Ever since the late 1970s, my colleagues have made regular surveys of the costs of production of sugar around the world. Using these results, together with my own initial estimates for 1995, I have prepared a trio of figures (Figs. 1.3 to 1.5) depicting the trends in real (inflation-adjusted) ex-mill field, factory and total costs for leading producers in this region: Australia, India and Thailand. All three countries have lowered their real field costs substantially since the early 1980s. Within the group, the balance of advantage has shifted steadily. In the early 1980s, Australia enjoyed the lowest field costs. By the mid 1990s, India was in pole position, with Australia close behind, and Thailand not much further back in third place. All three were operating with costs well below the world average.

The relativities are not the same for factory costs. Once again, Australia has been overtaken by Thailand, but India remains some way behind both, hindered by its reliance upon small factories. When field and factory costs are combined, Thailand and Australia are the most competitive. India lags behind, but with a significant advantage over the world cane industry as a whole.

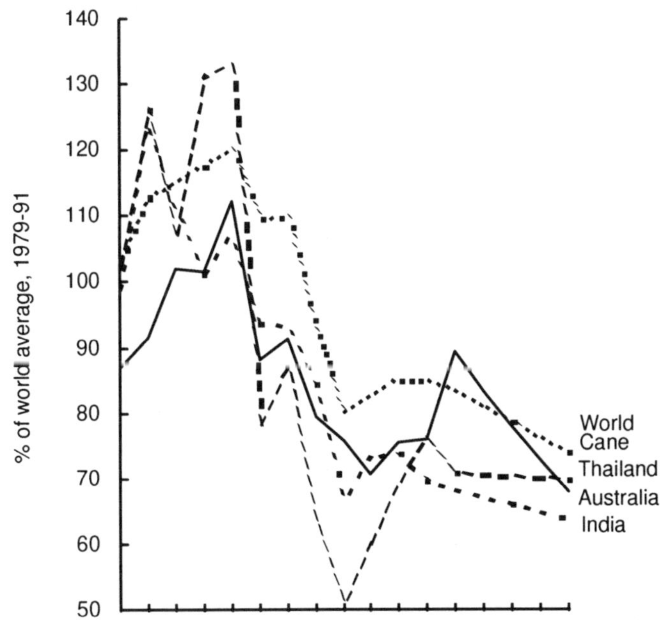

Fig. 1.3. Field costs for Australia, Thailand and India.

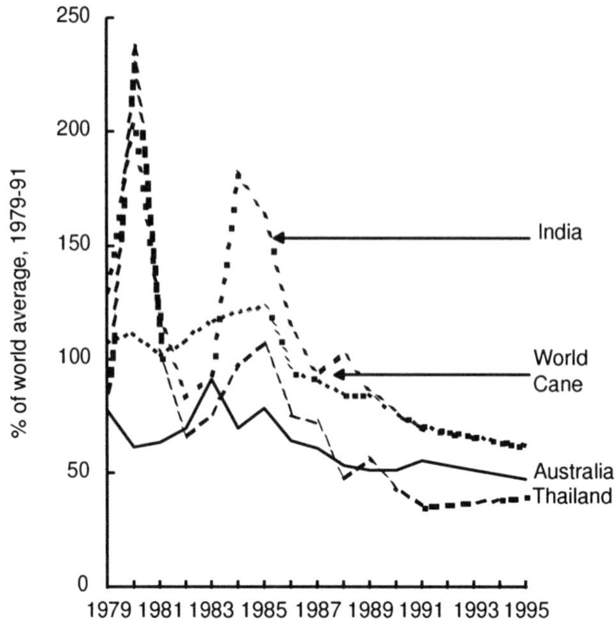

Fig. 1.4. Mill production costs for Australia, Thailand and India.

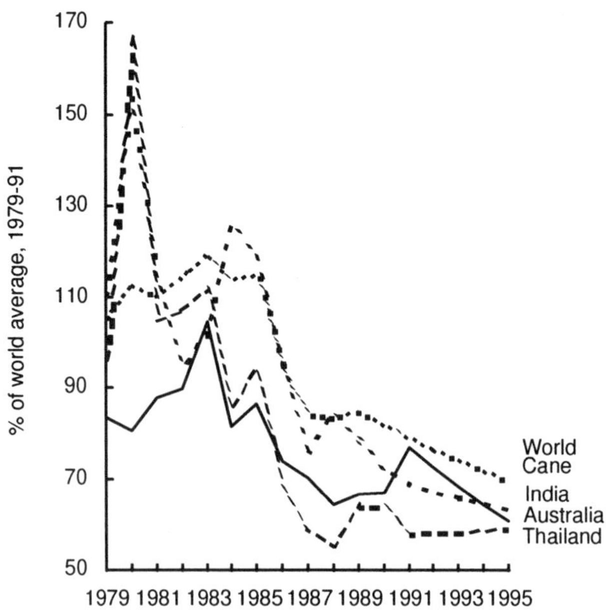

Fig. 1.5. Total production costs for Australia, Thailand and India.

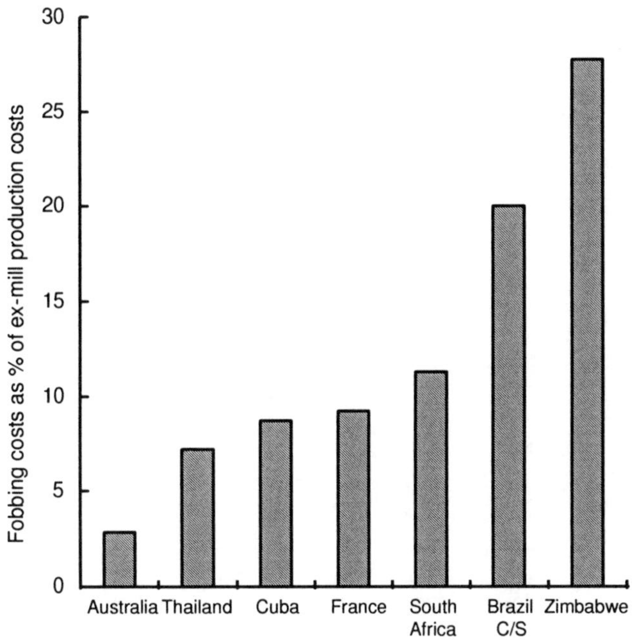

Fig. 1.6. Fobbing costs as a % of total costs.

These comparisons are for ex-mill costs. However, the world price is for sugar loaded onto a ship on a 'free on board and stowed' (f.o.b.s.) basis. Figure 1.6 reveals how the 'fobbing' costs of getting sugar from a mill onto a vessel vary among a cross-section of exporters. The comparison leaves little doubt about the importance of efficient export marketing in maintaining Australia's overall competitiveness.

Technical Competitiveness of Leading Sugar Producers

Many determinants of cost competitiveness, such as exchange rates, lie beyond the control of sugar producers. The only respect in which producers can influence their destiny is by improving the technical efficiencies of their operations. Figure 1.7 illustrates the trend in the yields of recovered sugar per hectare per year for some leading producers, and shows how other countries have narrowed the gap between their yields and those in Australia. The reasons for the narrowing of the gap are by no means all ones of which the Australian industry should be ashamed. Changes in farming practices do not necessarily improve the industry's technical performance; for example, the switch to 100% mechanical harvesting

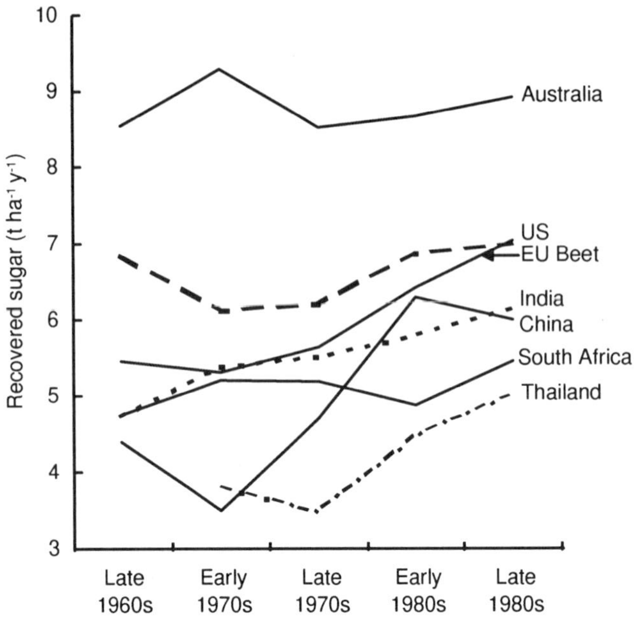

Fig. 1.7. Trend in yields of recovered sugar per hectare per year (sugarcane unless otherwise indicated).

reduced costs, even though, at least initially, it lowered recovered sugar output per hectare. Likewise, the trend towards a higher number of ratoon crops was justified on economic rather than technical grounds.

Australia cannot rest on its laurels. Its performance acts as a target to which other producers aspire. It is encouraging to note the success of efforts in the early 1990s (a period not covered in the figure) to break out of the flat yield trend of the past, and to note the commitment of resources to maintain the improvement. Australia now spends over 1.5% of its sugar revenues on R & D, while countries such as India and Thailand spend less than 0.25%.

The scope for yield improvements in the field is much greater than that for higher recoveries in the factory. Figure 1.8 contrasts Australian factory recoveries with those in other countries, and suggests that Australian mills are coming close to the realistic economic limits on recoveries. The Australian performance again acts as a target for other processors; and Australian machinery companies make a living selling foreigners the technologies first applied at home.

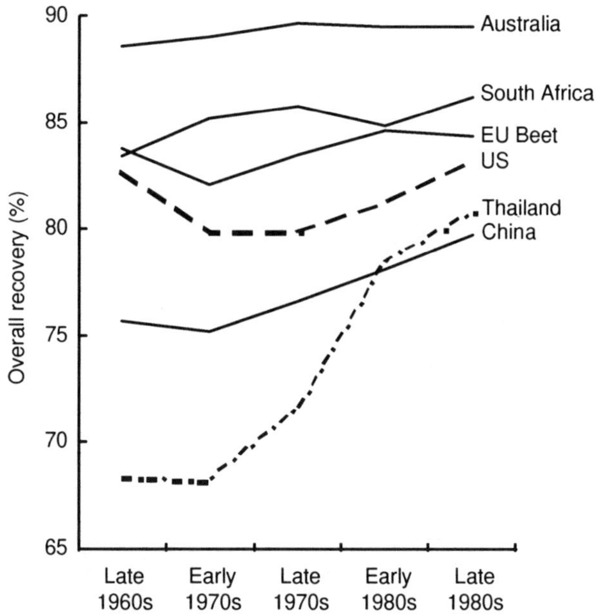

Fig. 1.8. Trend in overall factory recoveries of sugar (sugarcane unless otherwise indicated).

Energy Saving and Environmental Concerns

One very clear trend in technology has been the use of chemicals and mechanisation to boost labour productivity. In the field, environmental restrictions and the wider adoption of quality payment systems, rewarding sugar content rather than gross weight, have encouraged the decline in fertiliser (notably nitrogenous fertiliser) use. However, cane farming has also been at the expense of the environment, as in North East Thailand, where forest cover is disappearing fast, and in India, where heavily subsidised electricity has encouraged an excessive use of groundwater from great depths.

In the factory, environmental and cost pressures have had their main impact upon energy efficiencies. The trend has been one of reduced energy consumption and a better appreciation of the value of captive energy sources such as bagasse. Energy conservation is one area in which the beet industry has been to the fore.

Figure 1.9 contrasts the energy use per tonne of sugar in several countries. Within the cane group, Australia is more efficient than the other countries depicted in the figure; but Australia's average energy requirements are more than double those in leading beet industries, even though the beet factories are producing white, rather than raw sugar.

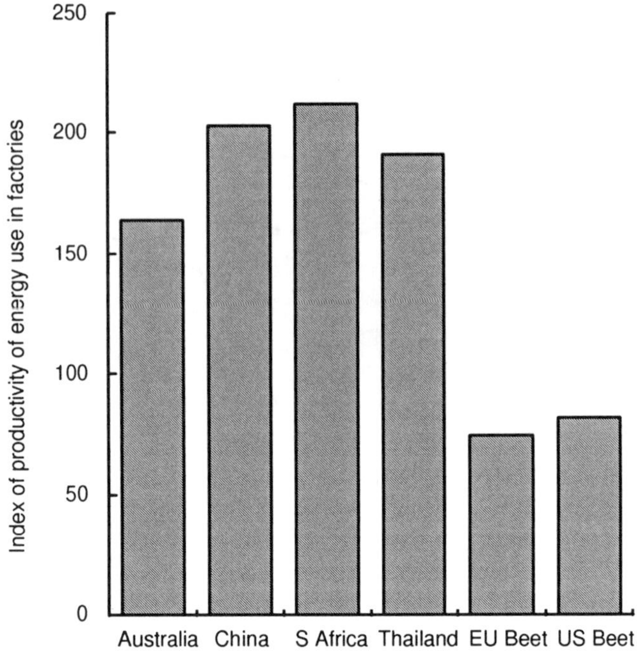

Fig. 1.9. Energy use/tonne of sugar output — late 1980s (sugarcane unless otherwise indicated).

The main reasons for this contrast are, first, that whereas the beet industry has to spend money every time it uses energy, the cane sector has traditionally treated bagasse as a free source of fuel; and second, that many of the cane millers who know that they can convert bagasse into cash by cogeneration find it difficult to secure a good price from electric power distribution companies. Unless a mill can exploit its own captive energy in the manufacture of energy-intensive products, it has to rely upon utility companies to offer worthwhile prices which would stimulate investment in cogeneration capacity.

Sensitivity of Cost Competitiveness to External Factors

The difference in the sensitivity of beet and cane processors to energy prices demonstrates very clearly the lack of uniformity across the sugar industry in its reaction to external developments.

This lack of uniformity applies also in other respects. In Fig. 1.10, I have taken Australia, France and India as examples. For each of them, I have examined how their overall sugar production costs would be affected under four separate simulations: first, if their crop yields were 10% higher; second, if their

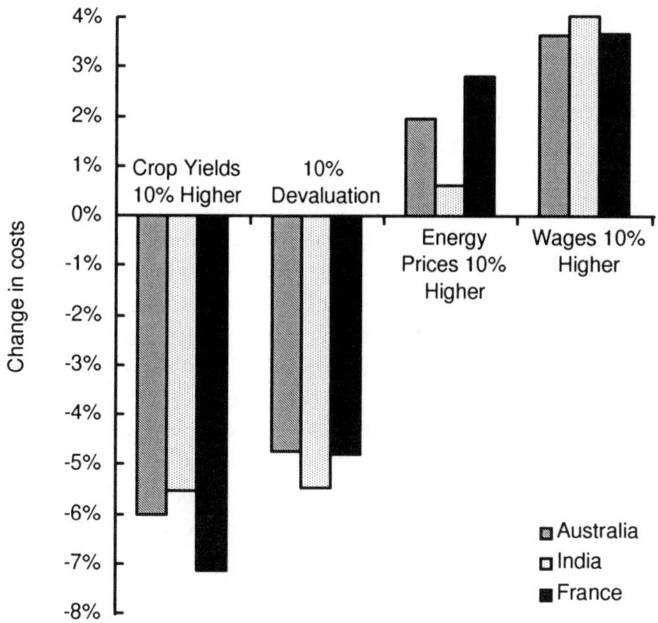

Fig. 1.10. Sensitivity of sugar production costs to yields, exchange rates and input costs.

currencies were devalued by 10%; third, if their energy prices, and the prices of energy-based products such as fertilisers and chemicals, were to rise by 10%; and fourth, if local wages increased by 10%.

Higher crop yields and a devaluation of the currency would both reduce costs expressed in US dollars per tonne. Figure 1.10 reveals that the beet sugar producer, France, would gain the most from better crop yields, because its fixed costs are relatively high. India, with low fixed costs, would enjoy the smallest boost from higher yields. In the case of a devaluation, India benefits more than either Australia or France. Local inputs, such as labour, weigh heavily in the composition of Indian production costs.

Turning to those changes which raise production costs, French beet sugar producers are the most sensitive to the higher energy prices. Because Indian cane farming is not mechanised and uses only limited amounts of fertiliser, it is barely affected by changes in energy prices. It is surprising to see that the sensitivity of production costs to changes in wages is similar in all three countries. The much lower labour productivity in India is more or less exactly offset by its lower wages.

The Competition Between Sugar and Alternative Crops

The competitiveness of sugarcane and beet crops in relation to alternatives is determined not only by technical and cost factors, but also by government pricing policies. Beet sugar is typically costlier to produce than cane sugar. Yet, the European beet industry survives.

In Fig. 1.11, we have contrasted the gross margins earned per hectare from grain, oilseed, protein and sugar beet crops in leading EU countries in the late 1980s. The attractions of sugar beet are clear cut. Beet generates larger profits than the other major crops. Similar conclusions apply to the US sugar sector.

Critics of EU and US agricultural policy will rightly emphasise the distortions caused by government intervention, and will point to the encouragement that such intervention provides to the cultivation of sugar crops. However, the same government policies often boost the returns from non-sugar crops, too. European farmers enjoy high returns from grain and oilseeds, as well as sugar beet. Therefore, a comprehensive scheme for reducing government assistance to EU farmers will hit the returns from alternative crops. In a free market economy, the natural response to lower official support prices for a broad range of crops will be a fall in land values, caused by lower gross margins per hectare. Exactly how it will affect the *relative profitability* of sugar and other crops is less certain. It is quite possible that European sugar production will fall only modestly if agricultural price reform affects all crops equally.

In contrast with beet areas, cane growing regions often experience only limited competition between cane and other crops. Many major cane areas, as in much of Queensland, are effectively cane monocultures, since cattle grazing is the only large scale alternative. However, in S. and S.E. Asia, in particular, cane often competes with rice for irrigated land, and rice receives official price support. Therefore, a broad policy of reduced government intervention would not necessarily harm cane's attractions to farmers.

A special form of competition is that found in Brazil between cane grown for sugar production and that grown for ethanol distillation. The major part of Brazil's cane is destined for the production of fuel alcohol, which has been prompted as a locally grown, renewable source of energy. In the past, the Brazilian Government gave special incentives to encourage the use of alcohol fuel. As the market is liberalised, ethanol is looking increasingly uncompetitive as an outlet for sugarcane, and there is a growing prospect of an acceleration in the shift of cane from distillation towards sugar production.

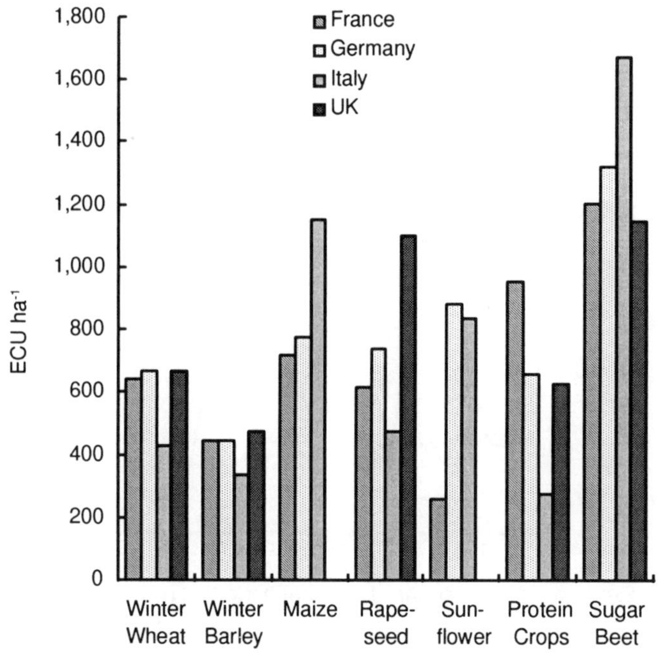

Fig. 1.11. Gross margins per hectare, European Union, 1986-90.

Prospects for the Sugar Industry Beyond 2000

After examining all these influences upon the world sugar and sweetener market, it is possible to draw some conclusions about the outlook for the sector as we enter the next millennium.

In order to understand the underlying factors which will help to determine my answer, I have prepared two last figures and one table. Figure 1.12 plots the long run *real* price (expressed in US\$ t^{-1}, at 1995 purchasing power) of world market raw sugar, alongside the long term trend in the real price. It is interesting to note that the trend value of the raw sugar price in 1995 was slightly under US\$300 t^{-1} (13 or so USc lb^{-1}).

In the Far East region, raw sugar typically commands a premium to the world price. The size of the premium has tended to be between US\$10 and US\$20 t^{-1}. Since Australian exports to the Far East benefit from this premium, one could argue that the long term trend price for raw sugar exports from Australia is currently of the order of 13 to 14 USc lb^{-1}.

It is sometimes argued that historical trends may have little relevance to the future course of the sugar market, when trade will be liberalised and the government protection much reduced. The same argument says that the feedback from world prices to individual producers has often been so poor in the past that producers' behaviour has been little affected by world price movements, and that little significance should be attached to historical patterns in these prices.

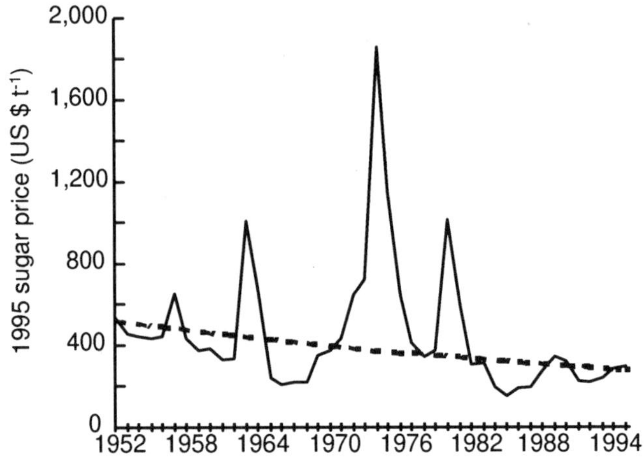

Fig. 1.12. Real world raw sugar price (continuous line) and long term trend (broken line).

With this in mind, Fig. 1.13 presents the trends for three variables from 1979 to 1991: the real world raw sugar price; the long term trend in this price (based upon data in Fig. 1.12); and the real average world-wide cane sugar production costs, derived from LMC's cost of production surveys.

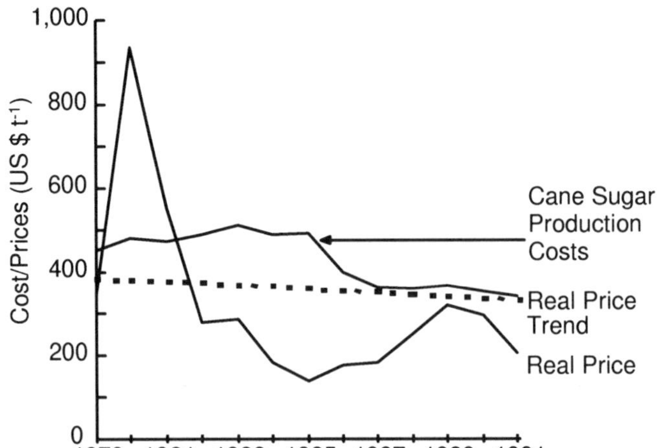

Fig. 1.13. World cane sugar production costs and raw sugar prices.

There is no doubt that the closeness of the production cost curve and the long run price trend is uncanny. At the least, it suggests that, in the very long run, there is a feedback from production costs to world prices. This is not as fanciful as it may seem. Any producer expanding sugar production has to be able to justify this decision. If the country is an exporter of sugar, the criterion used to assess the expansion must surely be a comparison of the costs of producing the extra sugar with the export revenue expected from its sale, which is given by the trend projected in the world sugar price.

Countries which expand their sugar output in order to reduce imports are in a different position, since they can hope for government protection. However, governments are not entirely out of touch with reality. At some stage they have to decide whether it is better to devote resources to enlarging the domestic sugar industry or to turn to imports.

In the very long term, those sugar industries producing sugar at costs below the world-wide average will tend to be profitable, and will expand, if suitable land is available. Industries with above average costs will tend to be unprofitable and will contract (at least as a share of world output) over time. The faster the implementation of agricultural liberalisation, the stronger will be the feedback from world prices to production trends in individual countries.

On balance, these conclusions imply that low cost sugar producers, such as Australia, Thailand, Brazil, India and the group of Southern African exporters, will continue to grow in importance, while the high cost producers of parts of Europe and the US, as well as some countries in the Caribbean and Eastern Asia, will see their role diminish. In Brazil and India, the expansion in sugar output will be accelerated by the shift of cane away from non-sugar uses, namely fuel alcohol and artisanal non-centrifugal gur, respectively.

My belief is that the pace of economic development in a few select areas around the Pacific Rim, notably Thailand, the Philippines, Chile and regions of China, will force up wages to the point where, unless there is massive mechanisation and dramatic changes in land tenure systems to permit the consolidation and reforming of sugar farms, their cost competitiveness will be undermined and sugar production will plateau.

A further important conclusion which I would draw from the preceding discussion is that technical progress is essential for success. Real prices of sugar are falling over time, and the beet industry has shown the cane industry that there is a great deal which can still be done to boost the competitiveness of sugar production by raising the productivity of sugar factories, such as in energy efficiency, and in benefiting from research in improved varieties. Therefore, the willingness of the Australian sugar industry to invest a higher proportion of its revenues in R & D than any other country must bode well for the future (even allowing for some export of equipment, technology or know-how).

This list of conclusions which are generally favourable to Australian sugar producers needs two major qualifications. The first is that, whatever the sugar industry does to improve its competitiveness, it is still at the mercy of external economic factors: for example, a strengthening of the exchange rate. The second qualification relates to the competition from alternative sweeteners, with which I began this paper. Starch sweeteners are the major competitor to sugar, and their production costs depend heavily upon the price of maize. Therefore, it is significant that the percentage annual decline in maize prices is significantly higher than that for sugar prices, as is revealed by the long run rates of decline in real commodity prices, detailed in Table 1.1. Indeed, every other major agricultural food commodity has recorded steeper rates of price decrease than sugar over the post-war period. If these trends continue, it is likely that starch sweeteners will become steadily more competitive with sugar. In addition, should the trends persist, farmers in a free market, unrestricted by distortions such as production quotas, are likely to move into sugar crops from the others, whose prices are falling faster. This should eventually force sugar prices down more steeply, as the market adjusts to keep a balance between supply and demand.

Table 1.1. Trend rates of long term decline in real commodity prices (%).

Commodity	Annual % decline in real prices
Sugar	1.5
Cocoa	1.9
Robusta Coffee	1.9
Soybean Meal	2.1
Arabica Coffee	2.1
Soybean	2.2
Wheat	2.2
Maize	2.6
Rice	2.7
Coconut Oil	2.9
Palm Oil	3.0
Copra	3.0
Soybean Oil	3.1
Rubber	3.5

As the behaviour of world sugar prices becomes more similar to that of other agricultural commodities, real sugar prices should fall at an underlying rate of well over 2% per annum. The rate of productivity improvement which this implies for farmers and millers will have to be considerably higher, if wages are to continue to rise. All in all, successful sugar industries are likely to have to sustain yearly productivity increases of 3% or more for all of their inputs, taken as a group, if they are to remain competitive internationally.

Chapter two:

Limits to Intensive Agricultural Systems and Opportunities for Sustainable Agricultural Development

M.J. Kropff[1,2], R. C. Muchow[3], R. Rabbinge[1] and P.K. Aggarwal[4]

1. *Department of Theoretical Production Ecology, Wageningen Agricultural University, Bornsesteeg 47, 6708 PD Wageningen, The Netherlands*
2. *DLO-Research Institute for Agrobiology and Soil Fertility (AB-DLO), P.O. Box 14, 6700 AA Wageningen, The Netherlands*
3. *CSIRO Tropical Agriculture and CRC for Sustainable Sugar Production, 306 Carmody Road, St Lucia, Qld 4067, Australia*
4. *International Rice Research Institute, P.O. Box 933, Manila, Philippines*

Abstract

World-wide, agricultural production systems have increased in productivity since the second world war. Improved varieties (higher harvest index and better resistance to pests and diseases), better water management (irrigation and drainage) and better nutrient management systems have contributed to increased productivity. However, world food production will have to increase further on less land, with less water and with less labour and must be sustainable. Concern about the environmental impact of intensive agricultural systems requires an improvement in production technologies to maximise water use efficiency and nutrient use efficiency, and better pest, disease and weed management. In this paper, opportunities for sustainable agricultural development are discussed using the production ecological framework to distinguish agricultural production situations: potential production, attainable production (limited by water and/or nutrients) and actual production (reduced by insects, diseases and weeds). After a historical overview of agricultural intensification, the current level of quantitative understanding of different production situations is discussed. Systems analysis and simulation modelling have reached the level at which yield potential in a wide range of environments can be predicted. The gap between potential,

In: Keating, B.A. and Wilson, J.R. (eds) *Intensive Sugarcane Production: Meeting the Challenges Beyond 2000*. CAB International, Wallingford, UK, 1997.

attainable and actual can be quantitatively analysed using these tools. Opportunities to increase the productivity and sustainability of agricultural systems are analysed and it is concluded that this can be achieved by increasing resource use efficiency. New methodologies to explore options for the development of integrated and ecological farming systems designed to achieve multiple objectives are discussed.

Introduction

Agricultural production has increased tremendously in the last 50 years, mainly as a result of intensification by increasing external inputs per unit of land. Therefore, the enormous increase in food demand as a result of population growth from 1.6 billion around 1900 to over 5.7 billion today could be met at a global scale. The supply of food, which once was a simple process of collecting and hunting, has evolved into a complex system of production activities, land use needs, industrial processes, trade, market and price mechanisms, and national and international policies. Despite success in the past, food production must be increased further in the coming decades, to meet the demand created by growing populations. This production increase has to be achieved on less land, with less labour, less water and less pesticides, and must be sustainable through conserving scarce natural resources. This is especially the case in Asia where the population pressure is the highest in the world (e.g. IRRI, 1993). To meet the challenges to increase food production, ways have to be found to improve the productivity and profitability of agricultural production systems.

The increased production per unit of land area by increased external inputs as fertilisers and biocides and often large scale reconstruction in regions with an agricultural function has led to irreversible changes in the landscape, soil degradation, diversity of the environment and the quality of the natural resource base by pollution (De Wit *et al.*, 1987). However, soil degradation as a result of over-exploitation in low input systems world-wide is an enormous threat as well. This has resulted in major questions world-wide in relation to the sustainability of intensive as well as extensive agricultural production systems.

The World Commission for Environment and Development (1987, the Brundtland Commission) defined the concept of sustainable development as one 'that meets the needs of the present without compromising the ability of future generations to meet their own needs'. In its report on 'Sustained risks, a lasting phenomenon', The Netherlands Scientific Council for Government Policy (WRR, 1995) demonstrated that decisions on what needs to be safeguarded for the present generation and coming generations cannot be made on objective, measurable ecological data alone, because normative choices are at issue as well. This can be translated in operational perspectives that demonstrate the possibilities to feed the world using various production technologies (WRR, 1995).

Thus, opportunities for sustainable agricultural development of intensive agricultural production systems are not just determined by biophysical factors, but also by socio-economic factors. Clear goals from different perspectives have to be set for agricultural development. Major questions that arise are: What are the biophysical limits to productivity; and What are the potentials for further increase of productivity and resource use efficiency? Is it possible to develop systems that can simultaneously achieve different goals with respect to productivity and sustainability? This paper aims to (i) present a historical perspective of agricultural intensification in relation to sustainability; (ii) discuss the biophysical, technical and socio-economic factors related to the development of intensive agricultural production systems; and (iii) to examine the role of resource use efficiency in the development of sustainable and productive agricultural systems and report on recent developments with respect to the optimisation of integrated and ecological agricultural systems.

Agricultural Intensification: a Historical Perspective

Until recently, the major activity for most of the population was to produce food and in many countries this is still the case. This situation has changed rapidly in industrialised countries since the second world war and more recently in developing countries. In The Netherlands and Australia, <5% of the working population is engaged in agriculture (Rabbinge, 1993; Wonder and Fisher, 1990). This has been made possible by the large increases in external inputs (machines, fertilisers, biocides) giving an enormous increase in productivity and, counter-intuitively, also in increased efficiency and efficacy of input use.

In the middle ages in Europe, farming systems were characterised by low yields and high labour requirements. French monks and their servants in Antver produced <800kg $ha^{-1}$of wheat, of which 200kg was used as seed for next year and 300kg for the draught animals (De Wit, 1972). The low yields were mainly the result of low nutrient availability. Because of the low mechanisation level, about 500 man hours per ha were required for cultivation, weeding and harvesting. Now, farmers produce more than 8t ha^{-1} of wheat with <15 hours of labour per ha. Up to 1900, yields increased at about 3-4kg ha^{-1} y^{-1} (De Wit, 1972). After 1900, annual yield increases in The Netherlands climbed to 15kg ha^{-1} y^{-1} until the second world war, after which productivity increased dramatically to 160kg ha^{-1} y^{-1} (De Wit *et al.*, 1987; Rabbinge, 1993). Similar trends in crop yield have been observed in other European countries and in the USA.

Until the 1970s, sugarcane yields in Australia also showed a continuous increase. During the 1970s and 1980s sugarcane yields in Australia appeared to plateau. However, in recent times yields have increased again. Many factors contributed to this trend including climate, varieties, and expansion on to new land. This issue is discussed by Garside *et al.* (1997).

The enormous increases in productivity are the result of simultaneous changes in several inputs. Nitrogen use changed from 30kg N uptake per crop without external inputs before 1900 to an external fertiliser input of N of c.200kg ha^{-1} today. At the same time inputs of P and K mineral fertilisers increased and chemical pesticides reduced yield losses by insects and diseases and labour inputs for weeding. Many other examples are given by Evans (1993) and De Wit (1992). This process of tremendous yield increase, in which increased productivity per ha was associated with a huge increase in labour productivity, is often referred to as the first green revolution, and took place in the industrialised world.

The second green revolution was initiated in the 1960s in developing countries. The Rockefeller and Ford Foundations established International Agricultural Research Institutions like the International Rice Research Institute to transfer the yield-increasing technologies to the developing world. The combination of semi-dwarf varieties and increased inputs of fertilisers and pesticides resulted in large yield increases that were even higher than in the first green revolution. This resulted for a long time in yield increases that exceeded population growth in Asia which had a major impact on food supply, although problems of hunger due to poverty and problems related to equity remained.

These evolutions resulted in increased food production, but also environmental and health problems due to intensification of land use and the use of fertilisers and pesticides. An extreme example is found in intensive grassland systems in The Netherlands. Manure and fertiliser inputs amounted to over 560kg N ha^{-1} on more than 1M ha whereas only 133kg N ha^{-1} was exported in the products of milk and meat (Meer, 1991). The rest (427kg N ha^{-1}) accumulated in the environment causing problems in nature reservations with vegetation typical of low fertile soils suffering as a result of increased N deposition (50kg N ha^{-1} y^{-1}) and problems with nitrate in groundwater. Examples with respect to pesticides impacts on the environment are numerous. Although the persistence and amount of active ingredient of newer pesticides are much smaller than those of the early products, many heavily toxic compounds are still used and found in ground, drinking and surface water. In many European countries legislation for pesticide use is becoming more and more restricted and plans to reduce pesticide use have been implemented.

Intensification of agricultural systems also resulted in other problems related to sustainability. For a long time, increased food demands were met by expanding the area under cultivation. Sometimes, sustainable systems were developed like the shifting cultivation systems in tropical rainforest areas and rainfed bunded rice systems in Asia. However, productivity demands generally were stronger than sustainability demands, which led to intensification and over-exploitation of soil resources. This resulted in vast areas of land which once were very suitable for agriculture, being lost or damaged by erosion, salinisation or exhaustion. Today soil degradation is one of the major threats to agriculture. In subsistence farming systems such as shifting cultivation in Asia and Africa, rotation cycles are being reduced with fallow periods becoming too short to restore soil fertility. Population pressure also leads to the cultivation of marginal fragile soils with increasing risks of soil degradation. The over-exploitation of the potential of agro-ecosystems is not confined to developing countries, but is also found in large scale agriculture in developed countries. Agricultural systems that are geared toward short term economic gains can have drastic effects on the environment. A good farmer ensures the continuity of his farm, but can only do so if the right preconditions are created and there is no net economic gain to be had from exhaustive cultivation (Rabbinge, 1993). The International Soil Reference and Information Centre (ISRIC) estimated that soil degradation reduces the potential agricultural land in the world by c.7% per year (Oldeman *et al.*, 1991). In most of the world, except Oceania and the Americas, >10% of the vegetated area is degraded according to the World Resources Institute (1992).

The Australian Sugar Industry also faces potential environmental and health problems associated with the use of fertilisers and pesticides that threaten its future competitiveness and sustainability. The coastal location of the Industry adjacent to heritage areas such as the Great Barrier Reef and Tropical Rainforests has heightened environmental awareness. The impact of N fertiliser practices on nutrient leaching and groundwater contamination is considered by Keating *et al.* (1997). Other papers in this book address the potential impacts of sugarcane production on riparian, freshwater and marine environments (Arthington *et al.*, 1997; Crossland *et al.*, 1997).

Production Situations

To compare different production systems, a distinction has to be made between production situations, technologies and production levels (Rabbinge, 1993). The production situation of a system at a specific site can be characterised biophysically by crop characteristics, and climate and soil characteristics like water holding capacity and soil fertility. Production technology encompasses all technologies used by the farmer ranging from tillage to fertiliser application and irrigation. The combination of the production situation with technologies results in a production level which is characterised by the amount of produce per ha. The production situation determines the inputs needed to obtain a specific production. Rabbinge (1993) defined three categories of growth factors determining the 'potential', 'attainable' and 'actual' production level: growth determining factors like radiation and temperature, growth limiting factors like water and nutrients and growth reducing factors like pests, diseases and weeds (Fig. 2.1). This approach can be used to assess the gap between actual, attainable and potential yields and what opportunities are available for yield increment in different agricultural systems. Potential production is achieved in protected cultivation like in greenhouse crops in The Netherlands, but is rarely achieved in field crops.

Generally only a fraction of calculated potential production is obtained, ranging from <5% up to 60% for country average yields (Oerke *et al.*, 1994).

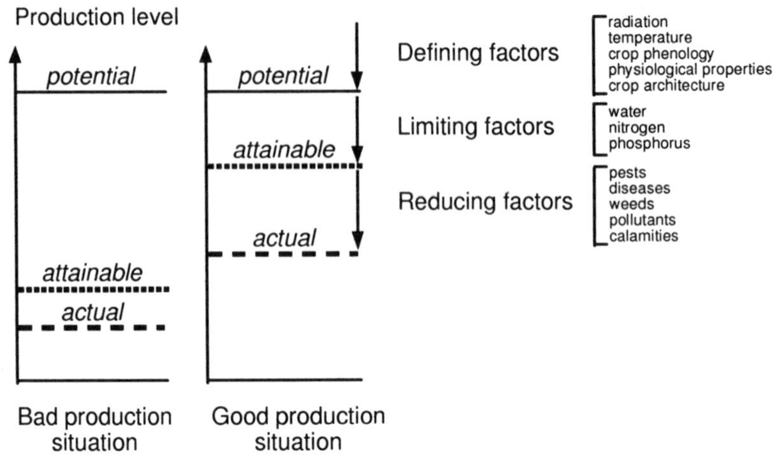

Fig. 2.1. Production situations, production levels and associated principal growth factors. [After Rabbinge *et al.* (1994)]

Potential production

Defined as the yield where only growth and yield determining factors play a role, potential production varies across agro-ecologies in response to solar radiation and temperature. Muchow and Kropff (1996) describe the role of field experimentation and crop simulation modelling in assessing potential production in different systems, and other papers in this book describe the limits to potential productivity in sugarcane (Muchow *et al.*, 1997) and grain crops (Sinclair, 1997).

Production of crops can be described in terms of the capture and use of solar radiation. For many crops grown under potential conditions, biological yield (biomass production) is linearly related to the amount of radiation intercepted by the crop canopy (Williams *et al.,* 1965; Monteith, 1977). The slope of the linear relationship, often termed the radiation use efficiency (RUE), is curvilinearly related to light-saturated leaf CO_2 assimilation rate, C_A. Muchow and Kropff (1996) from a review of literature concluded that there is little scope for improvement in maximum RUE. Of more importance in realising yield potential is the maintenance of high rates of C_A throughout the total growing season. This is particularly so during grain-filling, where a decline in C_A is often related to inadequate N supply (Kropff *et al.,* 1993; Muchow and Sinclair, 1994). Likewise, the prevention of abiotic and biotic crop growth-reducing factors such as pests, diseases, weeds and air pollution is important in raising yields towards the potential. The narrowing of the gap between actual, attainable and potential yields should be the focus of all crop improvement programs.

The proportion of biological yield present as economic yield is referred to as the harvest index. Evans (1993) reviewed the major changes in crops brought about by co-selection for high yield and suitability for modern, often high-input husbandry, and concluded that the increase in harvest index was one of the most significant advances in crop yield potential. In many crops, harvest index under potential growth conditions is >0.5, and there is unlikely to be much scope for further improvement (Muchow and Kropff, 1996; Sinclair, 1997).

Using experimental and simulation analysis, Muchow and Kropff (1996) and Sinclair (1997) demonstrate that both temperature and solar radiation influence the variation in benchmark potential yields in contrasting environments. Examples for maize, rice and wheat show a wide variation in potential yield, with low yields in tropical environments and high yields in temperate environments at higher latitude (and altitude), e.g. in rice, potential rice yield ranged from 6t ha^{-1} in tropical environments (wet season, Los Baños, Philippines) to 15t ha^{-1} at higher latitudes (Yanco, Australia) (Kropff *et al.*, 1995a) (Table 2.1). The primary influence of temperature was on growth duration, with lower temperature increasing the time that the crop can intercept radiation. Biomass accumulation is directly proportional to the amount of radiation intercepted, and for a given harvest index, grain yield is directly proportional to biomass. Consequently, high yield was associated with low temperature and high solar radiation within the range of environments studied. Furthermore, these analyses indicate that the solar radiation and temperature regime set a finite limit to potential yield of grain crops in a given environment.

In sugarcane, where vegetative growth comprises the economic yield, the duration of crop growth is set by management decisions relating to planting, ratooning and harvesting. Since radiation utilisation under potential growth conditions is depressed at supra-optimal and sub-optimal temperature, even for a specified growth duration, temperature and solar radiation receipt also set the yield limit in different sugarcane growing environments. Muchow *et al.* (1977) have determined benchmark annual potential sucrose yields for the Australian Sugar Industry and showed that potential productivity varied at similar and different latitudes due to variation in temperature and radiation.

Given that there appears to be relatively little scope for increasing maximum RUE or for increasing harvest index under potential growth conditions, high yield will be obtained where radiation capture is maximised and RUE is maintained at maximum values. Attainment of potential productivity requires appropriate inputs at the right time, i.e. water and nutrients. Frequently, crop demand for these resources cannot be met, resulting in attainable yields below the potential yield (Fig. 2.1).

Table 2.1. Observed and simulated crop characteristics for field experiments on yield potential in rice in different environments at IRRI (Philippines, WS = wet season, DS = dry season), Japan and Yanco (Australia) (standard errors in parentheses). [After Kropff *et al.*, (1995a)]

	IRRI WS	Japan	IRRI DS	Yanco
Variety	IR72	Nipponbare	IR72	YRL39
Biomass obs. (t ha^{-1})	11.5 (0.6)	12.8 (0.5)	17.8 (0.6)	28.9 (1.1)
Biomass sim. (t ha^{-1})	11.5	13.2	15.4	27.3
Yield obs. (t ha^{-1}, panicle d.w.)	5.7 (0.2)	7.8 (0.5)	9.8 (0.2)	14.7 (0.6)
Yield sim. (t ha^{-1}, panicle d.w.)	7.26	7.41	9.49	14.3
Crop duration (d)	109	153	110	159
Mean temperature				
- sowing-flowering (°C)	28.1	23.3	26.0	21.3
- flowering-maturity (°C)	27.7	25.6	27.9	22.9
Absorbed total radiation				
- sowing-flowering (MJ m^{-2})	189	195	278	413
- flowering-maturity	215	301	371	747

Attainable production

Water- or nutrient-limited yields range from near zero to near potential yields for different varieties depending on climatic, soil and management conditions. The availability of water and nutrients determines the capture of radiation by crops with stress reducing the development of early leaf area, maximum canopy size, and increasing rate of leaf senescence (Kropff *et al.*, 1994a; Muchow, 1994). RUE is also decreased by water and N shortage (Muchow 1989, 1994). Even under potential growth conditions, it is difficult to maintain leaf N values sufficiently high for the whole growth duration due to mobilisation of leaf N to the grain, with subsequent decrease in RUE during grainfilling (Kropff *et al.*, 1993; Muchow and Sinclair, 1994).

Climatic variability greatly influences attainable yield under rainfed agriculture. In the semiarid tropics and subtropics, rainfall is highly variable within and among seasons, and among locations, and hence it is difficult to determine benchmark attainable yields using conventional field experiments. Muchow *et al.* (1991) analysed maize and sorghum yields in semi-arid tropical Australia using growth simulation models coupled to historical climatic data. Yields were highly variable from year to year, and while on average maize yielded more than sorghum, the coefficient of variation was less for sorghum (Muchow *et al.*, 1991). The choice of crop to sow and the inputs to be used would depend on the level of risk aversion by the farmer. Knowledge of how far attainable yields under rainfed conditions are below the potential yield allows decisions to be made on the need for investment in irrigation where water resources are available. Crop simulation is a powerful tool to assess temporal and spatial variation in water requirement, as illustrated by Robertson *et al.* (1997) for sugarcane production.

In spite of half a century of intensive fertiliser research in many crops, nutrient supply during the growing season is still one of the major unknowns that determine attainable yields. In the past, many fertiliser trials were conducted without a systems approach. This provided general fertiliser recommendations which sometimes resulted in excessive fertiliser use (e.g. for grasslands in The Netherlands) or in suboptimal fertiliser use (e.g. in rice, resulting in yield levels of 7t ha^{-1} whereas more than 9t ha^{-1} were attainable with adequate N fertiliser) (Cassman *et al.*, 1993; Kropff *et al.*, 1994b,c).

Actual production

The difference between attainable and actual production is the result of growth reducing factors. Pests, diseases and weeds cause crop losses in spite of the intensive crop protection measures taken (Oerke *et al.*, 1994). World-wide this would add up to c.40% loss due to the three components (Table 2.2). In an earlier study, sugarcane was included as well and was found to be the most damaged crop with an overall loss to insects, diseases and weeds of 55% (Cramer, 1967).

Table 2.2. Estimated proportional (%) losses of crop yield due to diseases, weeds and animal pests for 1988-90. [After Cramer (1967) and Oerke *et al.* (1994) (in parentheses)]

Crop	Diseases	Animal pests	Weeds	Total
Rice	9 (15)	27 (21)	11 (16)	46 (51)
Wheat	9 (12)	5 (9)	10 (12)	24 (34)
Maize	9 (11)	27 (15)	11 (13)	46 (38)
Cotton	12 (11)	16 (15)	6 (12)	34 (38)
Sugarcane	19	21	16	55
Total (8 crops)	14 (13)	12 (16)	10 (13)	35 (42)

The approach of linking pest and crop models provides a tool to explore the dynamics of the interactions and to optimise pest management strategies (Kropff *et al.*, 1995b). Models at different levels of detail have been developed to quantify pest damage ranging from empirical functions to mechanistic simulation models. Empirical approaches are often used to quantify economic threshold levels in decision support systems for pest management. These empirical damage functions are generally derived by regression analysis relating pest severity at a given crop stage to yield loss. They have been useful because of their relative simplicity, but they ignore the dynamics of crop-pest interactions and their value is generally limited to the specific conditions for which measurements were taken. More insight can be obtained by linking pest-crop models (Kropff *et al.*,

1995b). Coupling points are located at the level of resource capture (radiation, water, nutrients), at the process level (photosynthesis, respiration, translocation) or at the state variable level (consumption of assimilates, biomass, leaf area). Different categories of damage mechanisms can be distinguished (Rabbinge *et al.*, 1994): competition for resources, plant killing, reduction of assimilation rate, effects on respiration, tissue necrosis and interception of radiation, tissue consumption, assimilate consumption, hampering of water uptake, induction of hormonal effects on stomatal regulation and deformations.

Other 'random non-controllable' factors can also result in actual yield being below attainable yield. These factors include waterlogging and lodging, and biomass loss and seed shattering that are difficult to control, even under potential growth conditions. In modern, semi-dwarf rice varieties, lodging is often a constraint to potential yield when N supply is high (Setter *et al.,* 1994). The breeding of semi-dwarf soybeans to overcome the lodging barrier to yield, combined with modified crop management, has led to higher yield potential in soybean in high-yielding environments (Cooper, 1985). In high-yielding sugarcane, Muchow *et al.* (1994) identified an early yield plateau well before scheduled crop harvest even though climatic conditions and resource supply were favourable for continued crop growth. Stalk death and subsequent loss of biomass associated with lodging and smothering were significant contributors to the cessation in yield accumulation, with consequent poor efficiency of utilisation of radiation, water and nutrients due to poor late growth in these crops.

Resource Use Efficiency, Productivity and Sustainability

Recent developments

One of the main socio-economic constraints for agricultural development is labour availability and cost. This has resulted in the replacement of labour by the use of external inputs (e.g. herbicides) and energy (e.g. mechanisation) and led to what is generally called intensive agricultural systems. However, depending on the level of development, agriculture has to achieve a multiple set of objectives besides production of food and income for the farmer. Management of agricultural production systems has been changed to meet the demands from society with respect to environmental issues.

In intensive agricultural systems, water and nutrient management has changed in the past decade because of environmental problems. Examples include N in groundwater and the competition for water resources among sectors in society like industry, households and agriculture (WRR, 1995). Nitrogen is mobile in the soil system especially in dryland crops, and with the required high rates of N application for high yields, the possibility exists for losses (e.g. leaching and volatilisation) and negative environmental impact. In rice it appeared essential to match the seasonal pattern of N supply to the N demand of the crop at each stage of development to achieve full yield potential but also to minimise N losses to the environment (Cassman *et al.*, 1993; Berge *et al.*, 1994). Optimisation of resource use, therefore, often serves different objectives that seem to conflict. This will be discussed in more detail later.

Concepts in crop protection in intensive agricultural systems have changed from exclusion or destruction of pests to pest management. Serious problems with pesticides including development of pest resistance, environmental effects of residues, and cost, have resulted in new approaches to pest management based on improved knowledge of pest dynamics, their natural enemies, and of the interaction between pest and crop (Kropff *et al.*, 1995b). Better understanding of the system will help identify new control techniques by indicating intervention points and can be useful in decision making in pest management (Rabbinge *et al.*, 1989; Boote *et al.*, 1993; Pinnschmidt *et al.*, 1994).

Decision support systems range from simple decision rules to complex multiple-criteria optimisation software. In simplest form, a decision support tool could be a threshold pest infestation level calculated from empirical relations based on field data. Empirical models and thresholds have been developed for many pests (Zadoks, 1985). Successful decision support systems have been developed based on these relationships (e.g. EPIPRE; Rabbinge and Rijsdijk, 1984; Rijsdijk *et al.,* 1989), and have reduced sprayings on wheat in The Netherlands to 2.5 compared to 8.5 in the UK and 7 in Germany (Rabbinge, 1987).

The possibility of linking pest-crop models opens new options to improve decision making in pest management in intensive agricultural production systems which are designed to serve multiple objectives: the analysis of risk and uncertainty associated with alternative decisions and strategies for pest management with minimum environmental impact from pesticides.

Resource use efficiency

A major question related to agricultural intensification is the efficiency of the production systems, i.e. the external inputs needed per unit of product. Examples are labour productivity, energy efficiency, amount of fertiliser used per unit product, etc. These are important measures with respect to sustainability at a global scale. However, for environmental issues, other measures have to be used as well. For example, in a highly productive system with a very high efficiency of N use (N uptake/kg product), the amount of N lost to the environment per unit of area might be higher than in low input systems with lower productivity and efficiency. Single measures of the system sustainability therefore cannot be defined.

In general, the efficiency of resource use increases with the level of production. Therefore, it has been argued (De Wit, 1979) that all resources are used most efficiently when all factors are optimally available (Fig. 2.2). The three-quadrant approach is given for N fertiliser treatments at three levels of reclamation (poor drainage, proper drainage and optimum water supply). The initial slope of N uptake versus yield is the same for all situations, as N in the crop is limiting yield. However, at high N uptake rates, the yield is higher at better reclamation levels. At the same time, the recovery of N increases as a result of better water management. Thus, at the optimum fertiliser rate for farmers (dashed lines) the efficiency of N applied versus yield increases with the level of reclamation, and thus production level.

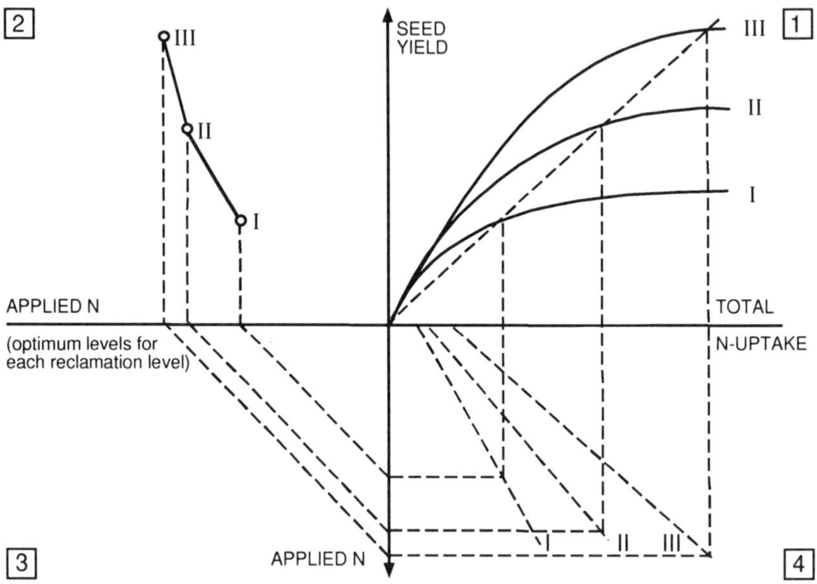

Fig. 2.2. Yield-uptake fertiliser rate diagram for nitrogen (N) at three levels of land reclamation (I. poor drainage; II. proper drainage; III. optimum water supply). Quadrant numbers (within squares) and the line in quadrant 2 represent the relationship between seed yield and fertiliser rate when the latter is adapted to the reclamation level. [After De Wit (1979)]

Especially with respect to environmental issues, N recovery should be optimised to minimise losses to the environment. In The Netherlands, N recovery in grassland crops increased from 30-50% in the early 1960s to 80-90% in the 1990s (Meer and Van Ulm-Van Lohuysen, 1986). On experimental farms, with research focused on losses, N losses have been reduced from 427 to c.150kg N h^{-1} (Lantinga and Rabbinge, 1996). In rice, a better split increased N recovery from 25% to 50% (Kropff *et al.*, 1994d).

An often used measure of the efficiency of agricultural systems is the ratio of energy outputs to energy inputs. Such analyses show that the output/input ratio increases with mechanisation and production level (De Wit *et al.*, 1987; Pimentel and Hall, 1984). However, the usefulness of such a comparison by adding all types of energy ranging from food for human labour to gasoline for tractors can be questioned. De Wit *et al.* (1987) indicated that energy is involved in several production factors that have different impacts at different production levels and that increasing returns on energy with progressive innovation in technology can be obtained. They argued that the different production factors can be most efficiently used at the potential production level. At lower production levels, where the production process is less understood, optimisation of inputs is much more complex. This is illustrated for sugarbeet in two production situations in Table 2.3. There are trade-offs, e.g. in a better production situation, the use of resources per unit of product is higher but the emission to the environment may be higher in terms of emission per ha. However, in many cases, a high resource use efficiency at high production levels may be combined with lower emissions to the environment as well. For example, in the sugarbeet grown on a sandy soil versus a clay soil, the resource use efficiency per unit of produce is lower, but the emission with drainage water to the groundwater is higher.

Table 2.3. Input, output and technical efficiency of sugarbeet grown in two production situations in The Netherlands. [Van Ittersum and Rabbinge, unpublished data]

	Central clay	Eastern sand
Input		
- N (kg ha^{-1})	140	160
- Pesticides (kg a.i. ha^{-1})	8	11.5
Output		
- yield (t ha^{-1})	68	51
Technical efficiency		
- yield per kg N	0.49	0.32
- yield per kg of pesticides	8.5	4.4

Opportunities to achieve multiple goals with respect to profitability and environmental issues have been explored in The Netherlands (WRR, 1992, 1995; Jansma *et al.*, 1994; Habekotté and Schans, 1996). For arable farming, profitability of different farming systems was determined at different maximum levels of N emission to the environment and pesticide use (Habekotté and Schans, 1996). In the model, a wide range of production technologies was generated with varying input-output relationships and different use of N fertilisers or manure and pesticides with the help of crop simulation models and expert knowledge. The model optimises income at different levels of constraints with respect to N emission and pesticide use by selecting the most profitable combination of crops in the rotation (Fig. 2.3). Rotation systems changed dramatically when the maximum N loss was set to 70kg ha^{-1}, as no manure was used in the systems. Because of the manure surplus in The Netherlands an objective was to maximise manure application. In that case, 15t ha^{-1} of manure was used at a cost of Dfl 1000 ha^{-1}. This clearly shows the strength of the approach to explore options and the exchange of different objectives; however, socio-economic studies are also needed to determine the possibilities to implement such systems.

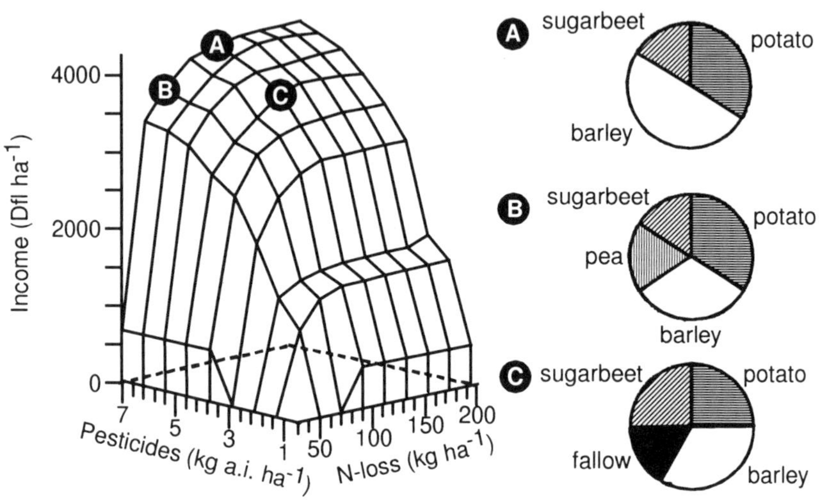

Fig. 2.3. Income (Dfl) as a function of pesticide use and N losses on arable farms in The Netherlands as determined by Interactive Linear Multiple Goal programming. The 3 pie charts indicate the optimised rotation schemes for situations A, B and C in the main graph. [After Habekotté and Schans, 1996]

Productivity and sustainability

Agricultural production systems should be productive, efficient and effective. From the ecological viewpoint, this implies that renewable resources are maintained, that non-renewable resources are used with foresight, and that the intrinsic value of the environment and its biodiversity is recognised and protected. From a socio-economic viewpoint, this requires that farmers make a decent living and that increasing and changing demands for agricultural products can be met at affordable prices (De Wit, 1990). These goals often conflict with productivity and sustainability demands resulting in continuous political debate. The results depend on ever-changing ideologies and political views, on socio-economic conditions, but also on technical possibilities. Political views are changing rapidly. For example, in 1990, the Dutch Ministries responsible for agriculture, environment and economics together with the agricultural sector developed a strategic plan to reduce the active ingredient volume of biocides by 40% in the year 1995 and by 60% in the year 2000. Objectives were also set to reduce the dependency on biocides and to reduce the emission of biocides to the environment. In volume reduction, the overall objective has been met to date with some exceptions for specific biocides. However, the other two objectives could not be met. More recently in 1995, the Dutch Minister of Agriculture set the 'ecologisation' of farming systems as a major objective. This included multi-functional land use systems in which nature conservation had a place. Such major changes in policy require technical means for implementation and effective and efficient agricultural research systems to adequately address changing requirements.

To enhance the sustainability of high input systems, the resource use efficiency has to be optimised. Crop/soil models can help match supply and demand of resources to obtain optimum yields from target environments, based on accurate predictions of the attainable yield. Much is known about potential yields. Knowledge and tools to optimise yields with minimum environmental impact and minimum use of natural resources are required as the socio-economic limits to intensive agricultural production systems become more and more important. As shown in the previous section, systems approaches are required to analyse the effects of multiple inputs with respect to achieving the necessary goals. Sustainability does not just relate to the environment, but also to the relationship between man and the environment (WRR, 1995). Attitudes towards sustainability differ because of different weights attached to facts, uncertainties, and environmental and social risks. Therefore, an analysis of the impact of different action perspectives, which at first sight can all be labelled as sustainable, can help to evaluate long term scenarios (WRR, 1995).

Conclusion

Agricultural productivity has increased tremendously, particularly since the World War II. The biophysical limits to yield are being reached in some intensive agricultural production systems. Although there are still many opportunities for yield increases world-wide, there seems to be a biophysically determined maximum with respect to agricultural production of which estimates have hardly changed since the 1960s (WRR, 1995). This potential yield level can be accurately quantified using models, and this knowledge can help to identify opportunities to increase actual and attainable yield levels by improved crop protection and management of inputs of water and/or nutrients, respectively. The challenge remains to achieve increased productivity and profitability but in a sustainable manner.

Sustainability problems occur through under-use of resources (soil degradation) or over-use of resources (environmental effects of nutrients and pesticides). Optimisation of resource use is one of the key elements for sustainable agricultural development. This is, counter-intuitively, often reached at higher production levels. Systems approaches to optimise farming systems with respect to multiple objectives, based on a comprehensive scientific analysis may help to identify options for sustainable agricultural development. Explorative land use studies are probably most efficient in the development of efficient and effective production systems at higher levels of integration.

References

Arthington, A.H., Marshall, J.C., Rayment, G.E., Hunter, H.M. and Bunn, S.E. (1997*)* Potential impact of sugarcane production on riparian and freshwater environments. In: Keating, B.A. and Wilson, J.R. (eds), *Intensive Sugarcane Production: Meeting the Challenges Beyond 2000*, CAB International, Wallingford, UK, pp. 403-422.

Berge, Ten, H.F.M., Wopereis, M.C.S. and Shin, J.C. (eds) (1994*) Nitrogen Economy of Irrigated Rice: Field and Simulation Studies.* SARP Research Proceedings, IRRI, Los Baños, Philippines, AB-DLO, Wageningen, The Netherlands, 306 pp.

Boote, K.J., Batchelor, W.D., Jones, J.W., Pinnschmidt, H. and Bourgeois, G. (1993) Pest damage relations at the field level. In: Penning De Vries, F., Teng, P. and Metselaar, K. (eds), *Systems Approaches for Agricultural Development.* Kluwer Academic Publishers, Dordrecht, The Netherlands, pp. 277-296.

Cassman, K.G., Kropff, M.J., Gaunt, J. and Peng, S. (1993) Nitrogen use efficiency of rice reconsidered: What are the key constraints? *Plant and Soil* 155/156, 359-362.

Cooper, R.L. (1985) Breeding semi-dwarf soybeans. *Plant Breeding Reviews* 3, 289-311.

Cramer, H.H. (1967) *Plant Protection and World Crop Production.* Bayer, Leverkusen.

Crossland, C.J., Done, T.J. and Brunskill, G.J. (1997) Potential impacts of sugarcane production on the marine environment. In: Keating, B.A. and Wilson, J.R. (eds), *Intensive Sugarcane Production: Meeting the Challenges Beyond 2000*, CAB International, Wallingford, UK, pp. 423-436.

Evans, L, T. (1993) *Crop Evolution, Adaptation and Yield.* Cambridge University Press, Cambridge, UK, 500 pp.

Garside, A.L., Smith, M.A., Chapman, L.S., Hurney, A.P. and Magarey, R.C. (1997) The yield plateau in the Australian sugar industry: 1970-1990. In: Keating, B.A. and Wilson, J.R. (eds), *Intensive Sugarcane Production: Meeting the Challenges Beyond 2000*, CAB International, Wallingford, UK, pp. 103-124.

Habekotté, B. and Schans, J. (1996) Explorative modeling for opportunities of integrated arable farming. Ab-DLO. Report no. 64. 68 pp. (in Dutch)

IRRI (1993) *Rice Research in a Time of Change.* The International Rice Research Institute, Los Banos, Philippines, 79 pp.

Jansma, J.E., Rossing, W.A.H., Ruijter, F.J. de and Schans, J. (1994) *The Bulb Rolling.* Agricultural University, Wageningen, The Netherlands, 60 pp. (in Dutch).

Kropff, M.J., Cassman, K.G., Van Laar, H.H. and Peng, S. (1993) Nitrogen and yield potential of irrigated rice. *Plant and Soil* 155/156, 391-394.

Kropff, M.J., Van Laar, H.H. and Matthews, R.B. (eds) (1994a) *ORYZA1: An Ecophysiological Model for Irrigated Rice Production*. SARP Research Proceedings, IRRI, Wageningen, The Netherlands, 110 pp.

Kropff, M.J., Cassman, K.G., Van Laar, H.H., Torres, R.O., Liboon, S.P., Williams, R.L. and Horie, T. (1994b) Evaluation of the model ORYZA1. In: Kropff, M.J., Van Laar, H.H. and Matthews, R.B. (eds), *ORYZA1: An Ecophysiological Model for Irrigated Rice Production*. SARP Research Proceedings, IRRI, Wageningen, The Netherlands, pp. 57-65.

Kropff, M.J., Cassman, K.G., Peng, S., Matthews, R.B. and Setter, T.L. (1994c) Quantitative understanding of yield potential. In: Cassman, K.G. (ed.), *Breaking the Yield Barrier: Proceedings of a Workshop on Rice Yield Potential in Favourable Environments*. IRRI, 29 November - 4 December 1993, Los Baños, Philippines, pp. 21-38.

Kropff, M.J., Cassman, K.G. and Van Laar, H.H. (1994d) Quantitative understanding of the irrigated rice ecosystems and yield potential. In: Virmani, S.S. (ed.), *Hybrid Rice Technology: New Developments and Future Prospects,* IRRI, Los Baños, Philippines, pp. 97-114.

Kropff, M.J., Williams, R.L., Horie, T., Angus, J.F., Singh, U., Centeno, H.G. and Cassman, K.G. (1995a) Predicting the yield potential of rice in different environments. In: Humphreys, E., Murray, E.A., Clampett, W.S. and Lewinn, L.G. (eds), *Temperate Rice - Achievements and Potential.* NSW Agriculture, Griffith, Australia, pp. 657-664.

Kropff, M.J., Teng, P.S. and Rabbing, R. (1995b) The challenge of linking pest and crop models. *Agricultural Systems* 49, 413-434.

Lantinga, E.A. and Rabbinge, R. (1996) The renaissance of mixed farming systems: a way towards sustainable agriculture. In: Van Ittersum, M.K., Venner, G.E.G.T., van der Geijn, S.C. and Jetten, T.H. (eds), *Abstracts of the Fourth Congress of the European Society for Agronomy*, 7-11 June, 1996, Velthoven, Wageningen, The Netherlands, pp. 428-429.

Meer, H.G. van der (1991) N balances. In: Meer, H.G. van der (ed.), *N-use and Losses in Dairy Farming Systems*. DLO, Wageningen, The Netherlands, pp. 5-12 (in Dutch).

Meer, H.G. van der and Van Ulm-Van Lohuysen, M.G. (1986) The relationships between inputs and outputs of nitrogen in intensive grassland systems. In: Meer, H.G. van der, Rijden, J.C. and Ennik, G.C. (eds), *Nitrogen Fluxes in Intensive Grassland Systems. Developments in Plant and Soil Sciences*. Nijhoff/Junk, The Netherlands, pp. 1-18.

Monteith, J.L. (1977) Climate and the efficiency of crop production in Britain. *Philosophical Transactions Royal Society London* B281, 277-94.

Muchow, R.C. (1989) Comparative productivity of maize, sorghum and pearl millet in a semi-arid tropical environment. II. Effect of water deficits. *Field Crops Research* 20, 207-219.

Muchow, R.C. (1994) Effect of nitrogen on yield determination in irrigated maize in tropical and subtropical environments. *Field Crops Research* 38, 1-13.

Muchow, R.C. and Carberry, P.S. (1989) Environmental control of phenology and leaf growth in a tropically adapted maize. *Field Crops Research* 20, 221-236.

Muchow, R.C., Hammer, G.L. and Carberry, P.S. (1991) Optimizing Crop and Cultivar Selection in Response to Climatic Risk. In: Muchow, R.C. and Bellamy, J.A. (eds), *Climatic Risk in Crop Production: Models and Management for the Semi-Arid Tropics and Subtropics*. CAB International, Wallingford, UK, pp. 235-262.

Muchow, R.C. and Kropff, M.J. (1997) Assessing the potential yield of tropical crops: role of field experimentation and simulation. In: Kropff, M.J., Teng, P.S., Aggarwal, P.K., Bouma, J., Bouwman, B.A.M., Jones, J.W. and van Laar, H.H. (eds) *Applications of Systems Approaches at the Field Level.* Kluwer Academic Publishers, Dordrecht, The Netherlands, pp. 101-112.

Muchow, R.C., Robertson, M.J. and Keating, B.A. (1997) Limits to the Australian Sugar Industry. In: Keating, B.A. and Wilson, J.R. (eds), *Intensive Sugarcane Production: Meeting the Challenges Beyond 2000.* CAB International, Wallingford, UK, pp. 37-54.

Muchow, R.C. and Sinclair, T.R. (1994) Nitrogen response of leaf photosynthesis and canopy radiation use efficiency in field-grown maize and sorghum. *Crop Science* 34, 721-727.

Muchow, R.C., Spillman, M.F., Wood, A.W. and Thomas, M.R. (1994) Radiation interception and biomass accumulation in a sugarcane crop grown under irrigated tropical conditions. *Australian Journal Agricultural Research* 45, 37-49.

Oerke, E.C., Dehne, H.W., Schönbeck, F. and Weber, A. (1994) *Crop Production and Crop Protection. Estimated Losses in Major Food and Cash Crops.* Elsevier, Amsterdam, The Netherlands, 808 pp.

Oldeman, L.R., Hakkeling, R.T.A. and Sombroek, W.G. (1991) *World Map of the Status of Human-Induced Soil Degradation: Global Assessment of Soil Degradation.* ISRIC, Wageningen, The Netherlands, 36 pp.

Pimentel, D. and Hall, C.W. (1984) *Food and Energy Resources.* Academic Press, New York, USA, 268 pp.

Pinnschmidt, H.O., Batchelor, W.D. and Teng, P.S. (1994a) Simulation of multiple species pest damage on rice. *Agricultural Systems* 48, 193-222.

Rabbinge, R. (1987) Implementation of integrated crop protection systems. In: Royle, D.J., Rabbinge, R. and Flückiger, C.R. (eds), *Pest and Disease Models in Forecasting, Crop Loss Appraisal and Decision Supporting Crop Protection Systems.* International Organisation of Biological Control Bulletin, Wageningen, The Netherlands, pp. 69-74

Rabbinge, R. (1993) The ecological background of food production. In: *Crop Protection and Sustainable Agriculture.* Ciba Foundation Symposium 177, John Wiley and Sons, New York, USA, pp. 2-23.

Rabbinge, R. and Rijsdijk, F.H. (1984) Epidemiological and crop physiological basis of EPIPRE. In: Gallagher, E.J. (ed.), *Cereal Production.* Butterworths, London, UK, pp. 227-235.

Rabbinge, R., Rossing, W.A.H. and Van der Werf, W. (1994) Systems approaches in pest management: The role of production ecology. In: Rajan, A. and Ibrahim, Y. (eds), *Proceedings of the Fourth International Conference on Plant Protection in the Tropics*, pp. 25-46.

Rabbinge, R., Ward, S.A., and Van Laar, H.H. (eds) (1989) *Simulation and Systems Management in Crop Protection.* Simulation Monographs, Pudoc, Wageningen, The Netherlands, 420 pp.

Rijsdijk, F.H., Zadoks, J.C. and Rabbinge, R. (1989) Decision making and data management. In: Rabbinge, R., Ward, S.A. and Van Laar, H.H. (eds), *Simulation and Systems Management in Crop Protection.* Simulation Monographs, Pudoc, Wageningen, The Netherlands, pp. 265-277.

Setter, T.L., Peng, S., Kirk, G.J.D., Virmani, S.S., Kropff, M.J. and Cassman, K.G. (1994) Physiological considerations and hybrid rice. In: Cassman, K.G. (ed.), *Breaking the*

Yield Barrier: Proceedings of a Workshop on Rice Yield Potential in Favourable Environments, IRRI, Los Baños, Philippines, pp. 39-62.

Williams, W.A., Loomis, R.S. and Lepley, C.R. (1965) Vegetative growth of corn as affected by population density. I. Productivity in relation to interception of solar radiation. *Crop Science* 5, 211-215.

Wit, C.T. De (1972) Food production: past, present and future. *Stikstof* 15, 65-80.

Wit, C.T. De (1979) The efficient use of labour, energy and land in agriculture. *Agricultural Systems* 12, 279-287

Wit, C.T. De (1990) Understanding and managing changes in agriculture. In: Jones, J.W.G. and Street, P.R. (eds), *Systems Theory Applied to Agriculture and the Food Chain*. Elsevier, London, UK, pp. 235-250.

Wit, C.T. De (1992) Resource use efficiency in agriculture. *Agricultural Systems* 40, 125-151.

Wit, C.T. De, Huisman, H. and Rabbinge, R. (1987) Agriculture and its environment: are there other ways? *Agricultural Systems* 23, 211-236.

Wonder, B. and Fisher, B.S. (1990) Agriculture in the Economy. In: Williams, D.B. (ed.), *Agriculture in the Australian Economy*. Sydney University Press, Sydney, Australia, pp. 50-67.

World Commission for Environment and Development (1987) *Our Common Future*. Oxford University Press, Oxford, UK.

World Resource Institute (1992) World Resources 1992-1993. World Resources Institute, 1992. Oxford University Press, Oxford, UK.

WRR (1992) *Ground for Choices*. Netherlands Scientific Council for Government Policy, The Hague, The Netherlands, 144 pp.

WRR (1995) *Sustained Risks: a Lasting Phenomenon*. Netherlands Scientific Council for Government Policy, The Hague, The Netherlands, 205 pp.

Zadoks, J.C. (1985) On the conceptual basis of crop loss assessment: The threshold theory. *Annual Review of Phytopathology* 23, 455-473.

Chapter three:

Limits to the Australian Sugar Industry: Climatic and Biological Factors

R.C. Muchow, M.J. Robertson and B.A. Keating

CSIRO Tropical Agriculture and CRC for Sustainable Sugar Production, 306 Carmody Road, St Lucia, Qld 4067, Australia

Abstract

Climate impacts directly on the amount of sugar produced, and indirectly on industry profitability through an effect on the scheduling of field operations, and on the supply of cane and the marketing of sugar. Application of knowledge on the biological limits to yield accumulation, particularly through the use of crop growth simulation, allows the climatic effects on productivity to be assessed. This paper assesses the climatic and biological limits to sugarcane production in the context of the Australian sugar industry.

The APSIM-Sugarcane simulation model, encapsulating knowledge of crop response to temperature and radiation, was coupled to historical climatic data to establish potential yields for different locations in the Australian sugar industry. In tropical regions, average sugar yield ranges from 22 to 31t ha^{-1} y^{-1}, with some lowering of potential (e.g. down to 18 to 24t ha^{-1} y^{-1}) in the cooler environments of southern Queensland and northern New South Wales. In tropical regions, low solar radiation associated with high rainfall results in potential sugar yields not being closely related to latitude. In contrast in the subtropics, yields decline with increasing latitude in response to decreasing temperature and radiation.

Climate is shown to be only one factor limiting the performance of the Australian sugar industry. At all locations, simulated potential yields are above actual yields except for a small proportion of growers' fields. Consequently, from a climate perspective, there is considerable scope to raise productivity per hectare. Scope also exists at the farm, mill district and industry scales for growers, millers and marketers all to benefit from forecasting and scheduling practices that better utilise the climatic resources available to the industry.

In: Keating, B.A. and Wilson, J.R. (eds) *Intensive Sugarcane Production: Meeting the Challenges Beyond 2000*. CAB International, Wallingford, UK, 1997.

Introduction

Sugarcane is a tropical plant that is grown under diverse climates throughout the world from sea level to 1500m at latitudes between 36.7°N and 31.0°S. Humbert (1968) describes the ideal climate as a long, warm growing season and a fairly dry, sunny, cool, but frost-free, ripening and harvest season, free from hurricanes and typhoons. In Australia, sugarcane is grown in climates ranging from the semi-arid tropics in north Western Australia (15° S) to the tropical and subtropical areas of the east coast ranging from 16 to 29° S. The availability of land in relation to water resources, transport infrastructure and milling capacity limit the potential for further expansion in Australia beyond current regions (Chapman and Milford, 1997). Application of knowledge on the biological limits to yield accumulation allows quantitative assessment of climatic effects on productivity.

The impact of climate on the sugarcane production system can be considered in terms of three aspects. Firstly, climate directly determines the processes of yield accumulation and the amount of sugar produced. Secondly, climate indirectly limits industry performance by affecting the scheduling of field operations, and the supply and marketing of sugar. Thirdly, climatic conditions control the development and spread of fungal diseases, insects, pests and weeds which can restrict crop growth. Assessing climatic limits is complex due to the interaction of variety, crop management and these elements. This paper focuses on the first two aspects, with the impacts of variety and pests and diseases being considered in other chapters of this volume.

The climate of Australia's sugar growing regions has been analysed in several publications (e.g. Kingston *et al.*, 1995) and is only briefly described here. Our primary focus is on describing the impacts on sugar yield. An important issue in assessing limits to productivity is temporal and spatial variability because climatic conditions vary from season to season, and within and across regions. In addition to direct effects on crop productivity, rainfall variability indirectly influences mechanical harvesting and the scheduling of cane supply during the harvest season. Rainfall also sets the potential for runoff and deep drainage with possible environmental impact associated with the movement of nutrients and pesticides. The sugar industry is in a part of Australia that has inherently high climatic variability, within a continent whose climate is one of the most variable in the world (Angus, 1991). Dealing with this climatic variability is important to profitable and sustainable sugar production because stability of income from year to year affects the riskiness of farming and milling operations (Wegener, 1997). This paper assesses the limits imposed by the climatic constraints and the opportunities for better managing such variability to improve the performance of the Australian sugar industry.

Climate of Australia's Sugar Growing Regions

The three main climatic elements affecting crop yield are radiation, temperature and rainfall; annual values for selected locations in Table 3.1 are based on historical records ranging from 80 to 114 years. The years shown represent the availability of complete daily rainfall records for given locations. The record length for temperature is much shorter, and many data were generated as outlined in Appendix 1. Few solar radiation data are available, and data in Table 3.1 were also generated as outlined in Appendix 1. Whilst we consider these data the best currently available, effort is needed to assess and improve the reliability of generated radiation data as longer records of observed data become available.

Annual mean daily temperature decreases with latitude, and whilst there is a similar trend for solar radiation, exceptions do occur in sites of high rainfall, e.g. Mareeba compared to Cairns, and Innisfail and Ingham compared to Ayr. A similar analysis for the South African sugar industry (Inman-Bamber, 1995) showed annual mean daily solar radiation decreasing from 20 to 16.7MJ m^{-2} between 25 to 31°S, but with extended periods of low cloud or mist reducing radiation in the south and at high altitudes. An important feature of Table 3.1 is the variability across the historical record. Whilst this is most pronounced for annual rainfall, solar radiation and temperature also vary from year to year.

The data in Table 3.1 are of restricted value in assessing climatic limits to productivity. The growth processes of sugarcane have differing sensitivities to radiation, temperature and water supply and since these climatic elements vary from day to day, their consequences on productivity are difficult to assess from annual mean data. The question of how representative the recording site is of the surrounding region is also an important issue. Rainfall is highly variable within a region and mill district. For southern regions, the occurrence of frost is variable and usually confined to low-landscape inland areas south of Bundaberg (Kingston *et al.*, 1995), and the frequency of frosts is difficult to assess from standard meteorological data. The occurrence of drought and waterlogging is dependent on soil type, the presence of water tables and rainfall pattern.

Events such as cyclones and high winds are not captured in climatic records, yet they can have a major impact on productivity. Given these complexities, it is appropriate to consider the hierarchy of production levels outlined by Kropff *et al.* (1997) in assessing climatic limits. It is feasible to assess the climatic limits to 'potential' production as determined by radiation and temperature, and the limits to 'attainable' yield as determined by yield potential, and water and nutrient supply. 'Actual' yield may be further discounted due to possible yield restrictions from pests, diseases, mechanical damage (e.g. stool damage associated with wet weather harvesting) and calamities. Additionally, it may not be economic to operate at maximum production levels, or environmental considerations may limit the management applications of water and nutrients. Social factors such as the availability of labour at key times and lifestyle decisions also determine actual yields.

Table 3.1. Daily mean solar radiation and temperature [(maximum+minimum)/2], annual mean rainfall and range (in parentheses) for selected locations in the Australian sugar industry. [Sources of data are given in Appendix 1]

Location (lat., long.)	Years	Radiation (MJ m^{-2} d^{-1})	Temperature (°C)	Rainfall (mm)
1. Kununurra 15.65°S, 128.72°E	1907-1988	22.6 (21.4-25.3)	28.0 (26.8-29.0)	765 (403-1448)
2. Cairns 16.88°S, 145.75°E	1888-1993	18.2 (16.8-20.0)	24.8 (23.9-25.5)	2098 (928-4031)
3. Mareeba 17.00°S, 145.42°E	1904-1993	19.5 (17.7-21.0)	24.9 (23.9-25.5)	915 (348-1864)
4. Babinda 17.35°S, 145.92°E	1912-1993	17.0 (15.6-19.0)	23.6 (22.5-24.4)	4219 (2243-6714)
5. Innisfail 17.52°S, 146.03°E	1888-1993	17.9 (16.4-20.0)	24.9 (23.9-25.5)	3580 (1775-5895)
6. Ingham 18.65°S, 146.17°E	1893-1993	18.8 (16.8-20.8)	24.1 (23.3-25.2)	2066 (815-4557)
7. Ayr 19.57°S, 147.40°E	1888-1993	19.9 (18.0-21.4)	24.2 (22.8-25.2)	1079 (261-2432)
8. Proserpine 20.50°S, 148.53°E	1889-1993	19.9 (18.4-21.6)	23.2 (22.2-24.0)	1803 (698-3499)
9. Mackay 21.17°S, 149.12°E	1891-1993	19.6 (16.2-21.4)	22.6 (21.8-23.3)	1684 (632-3455)
10. Bundaberg 24.85°S, 152.38°E	1888-1993	18.7 (16.8-20.5)	21.0 (20.3-21.9)	1128 (338-2360)
11. Maryborough 25.55°S, 152.68°E	1888-1993	18.8 (17.4-20.7)	21.0 (20.3-21.9)	1155 (325-2248)
12. Nambour 26.62°S, 152.97°E	1894-1993	18.4 (16.7-20.4)	20.4 (19.9-21.5)	1706 (379-3777)
13. Murwillumbah 28.35°S, 153.38°E	1891-1993	18.5 (16.4-20.4)	20.0 (19.2-20.7)	1687 (743-2868)
14. Grafton 29.70°S, 152.93°E	1880-1993	18.5 (16.6-19.9)	18.6 (17.5-19.6)	981 (372-1818)

Crop growth simulation models have a role in assessing climatic limits to yield and in assessing variability (e.g. Muchow and Bellamy, 1991). These can be used to establish a benchmark yield for a given season, location, soil type, management strategy and crop duration, or to generate probabilistic estimates of crop yield across the whole historical climatic record. The availability of high quality long term climatic data is central to this approach, and this limitation has been discussed above. Furthermore, crop models encapsulate our current understanding of the determinants of productivity and should be used with caution beyond the bounds of that understanding. Analysis of the determinants of sugarcane yield have been presented in several publications (e.g. Muchow *et al.*, 1994, 1995, 1996a, 1996b, Robertson *et al.*, 1996). Here, we discuss the application of that knowledge to assess the impact of climate on sugar yield.

Benchmarking Potential Yield

The effect of the interaction of radiation and temperature on potential yields at different locations can be examined using growth simulation models. The APSIM-Sugarcane model was configured as described in Appendix 2 and run for the historical climate record for the locations in Table 3.1. Predictions were made of potential sucrose yields after 12 months for crops ratooned on 1 October. For four selected locations, Fig. 3.1 shows sucrose yield for each year as a function of annual mean daily radiation receipt.

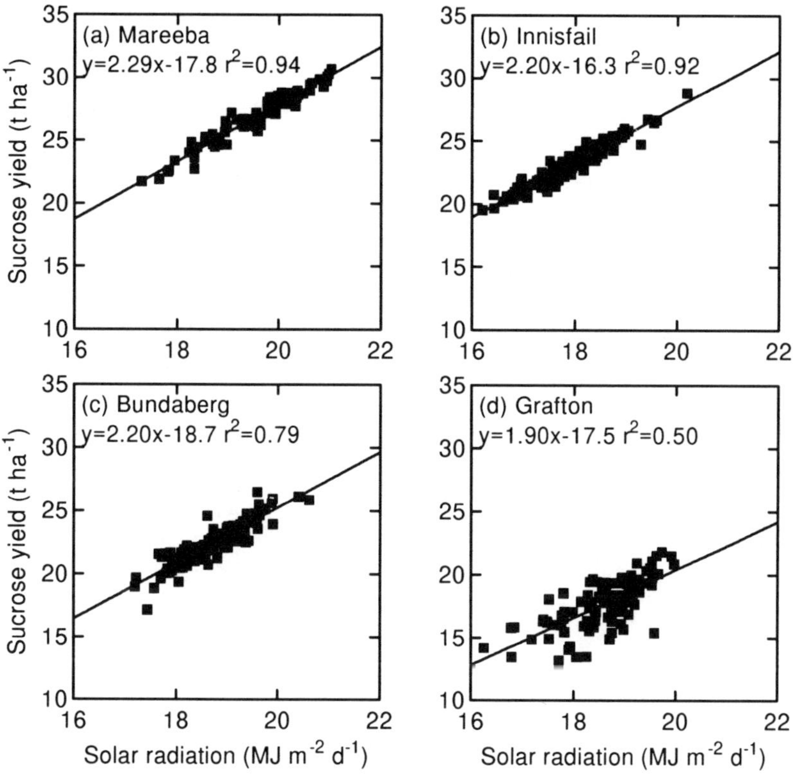

Fig. 3.1. Simulated annual potential sucrose yield as a function of annual mean daily radiation for (a) Mareeba, (b) Innisfail, (c) Bundaberg and (d) Grafton. [Data are for the historical record shown in Table 3.1, based on simulations outlined in Appendix 2]

Yield was linearly related to radiation with the proportion of the variation accounted for declining from northern to southern sites due to low temperature effects. However, even at the northern sites (Figs 3.1a,b), there was moderate variation in yield at a given level of radiation, limiting the predictive value of the relationships. Nevertheless, the data show that much of the year-year and location-location variation in northern regions is associated with annual radiation receipt, with higher yields predicted, for example, at Mareeba compared to Innisfail. For the Hawaiian sugar industry, Stanford (1963) similarly fitted a response surface for cane yield at 24 months dependent on average incident radiation for that period. Inman-Bamber (1995) accounted for 96% of the variation in annual potential sucrose yield in South Africa using a relationship based on latitude and altitude, with yields ranging from 22t ha^{-1} in the north to 12t ha^{-1} in the high altitude south. For different locations in the Australian sugar industry (Table 3.1), the simulated 12 month potential sucrose yields across years were meaned, and the variation across latitude is shown in Fig. 3.2.

The approach of Inman-Bamber (1995) is not appropriate to the northern regions of the Australian industry given the yield range for closely related latitudes. However, the trend for potential sucrose yield to decline with latitude in southern regions is similar to that observed by Inman-Bamber (1995) in South Africa. The data in Fig. 3.2 can be used to calculate a benchmark potential annual mean sucrose yield for different Australian regions, given the specifications of the simulations (Appendix 2), and noting that mean data do not account for the variability in potential yield from year to year as seen in Fig. 3.1.

The simulated yields for different locations accord with experimental observations, e.g. the annual mean sucrose yields predicted for Ingham and Ayr of 24.2 and 27.3t ha^{-1} (Fig. 3.2) are similar to experimental observations of 24.5 and 26.7t ha^{-1}, respectively (R.C. Muchow *et al.*, unpublished data).

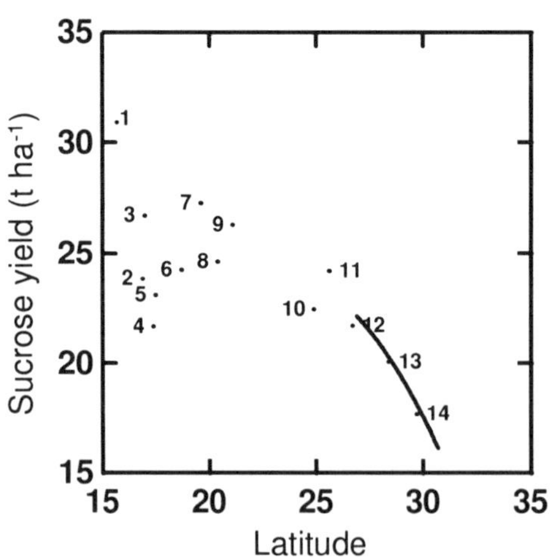

Fig. 3.2. Mean annual potential sucrose yield simulated for sites across the latitudinal range of the Australian sugar industry. [Site numbers are given in Table 3.1. Estimates are for 12 month old ratoon crops simulated using APSIM-Sugarcane as outlined in Appendix 2]

Thompson (1988) reports annual sucrose yield of 19.7t ha^{-1} at Pongola (27.4° S) and Inman-Bamber (1995) simulated 17.1t ha^{-1} at Mt Edgecombe (29.7° S) in South Africa. These are comparable with 20.1t ha^{-1} predicted at Murwillumbah (28.4° S) and 17.7t ha^{-1} predicted at Grafton (29.7° S) (Fig. 3.2).

Coping with Variability

Production variability

Production variability can be expressed at a range of temporal and spatial scales (Muchow *et al.*, 1997). Insights can be gained by examining the variability across individual fields. Geographic Information Systems and Productivity Databases are being used in the Australian sugar industry to examine performance (Lee-Lovick *et al.*, 1992; Wegener *et al.*, 1992). Examples of yield variation across fields of Q120, the major variety in Mossman (north of Cairns), Mareeba and Babinda (north of Innisfail), and of Q141, the major variety in Bingera (near Bundaberg), are given in Fig. 3.3. First ratoon crops (IR) were used to minimise crop age effects and older ratoon losses.

Figure 3.3 indicates a wide distribution in 'actual' yield in a given year, and differences in the yield distribution across years and locations. Considerable improvement in productivity would occur if it was possible to skew yields towards the upper end of the distribution. Climatic, soil and management factors contribute to the range in actual yields, and we expect that the observed highest yields are approaching the climatically determined attainable yield.

For the first ratoon of Q120 at Mossman and Babinda, the average yield was higher in 1992 than in 1993, and the whole distribution of yields was lower in 1993 (Fig. 3.3a,b). At Babinda, yields in 1995 (Fig. 3.3c) were lower than in 1992 and 1993 (Fig. 3.3b). Both are wet tropical sites, and 1992 was a drier year than 1993 and 1995 (annual rainfall at Babinda was 2465, 3461 and 4364mm in 1992, 1993 and 1995, respectively). Historical climatic data were available until 1993 (Table 3.1), and the simulated potential sucrose yield in 1992 and 1993 was 24.9 and 22.9t ha^{-1} at Cairns (near Mossman), and 22.9 and 21.8t ha^{-1} at Babinda. The simulated yields were obtained for mean annual daily radiation in 1992 and 1993 of 18.6 and 17.7MJ m^{-2} d^{-1} at Cairns, and 17.4 and 16.6MJ m^{-2} d^{-1} at Babinda. These simulated yields were close to the observed 'actual' yields at the upper end of the distribution (Fig. 3.3a,b), and the lower simulated yield in 1993 indicates that climatic differences contributed to some of the yield variation among years. Both observed and simulated data suggest that sugar yields > 20t ha^{-1} are achievable at these wet tropical sites.

At Mareeba on the Atherton Tablelands, there was a greater frequency of higher yields, and maximum yields in 1995 were very much higher, than was observed at Babinda, reflecting higher radiation receipt (Fig. 3.3c and Table 3.1). At Bingera near Bundaberg (Fig. 3.3d), a greater proportion of higher yields were obtained in 1993 than in 1992, and this difference was reflected in the simulated sucrose yields. Again, the simulated yields were close to the maximum observed yields and climatic differences contributed to some of the variation among years.

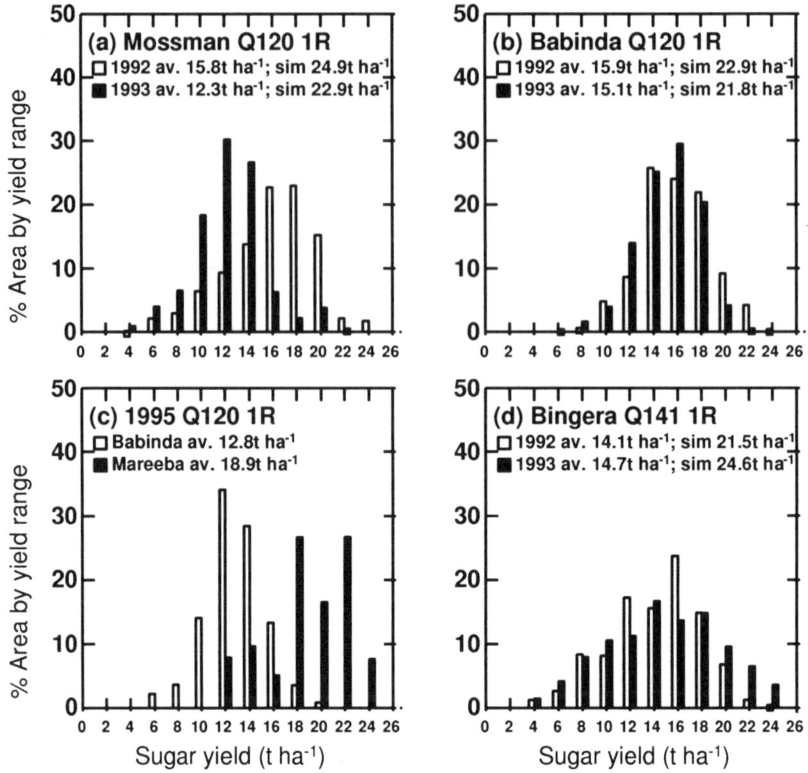

Fig. 3.3. Variation in sugar yield (t ha^{-1}) between individually harvested blocks (a) Q120 1R at Mossman (coastal areas) in 1992 and 1993, (b) Q120 1R at Babinda in 1992 and 1993, (c) Q120 1R at Mareeba and Babinda in 1995, and (d) Q141 1R at Bingera in 1992 and 1993. [Average actual and simulated potential sugar yields (t ha^{-1}) are also shown for each year. Blocks smaller than 0.5ha were excluded. Data sources: Mossman and Mareeba (A.V. Rudd, A.W. Ford); Babinda (G.F. Longden); and Bingera (P. Ledger)]

The challenge now is to assess how much of the demonstrated variability in production is associated with climatic variability from year to year across locations, and what scope there is for yield improvement given the climatic limit. To answer these questions, an approach using crop growth simulation and historical climatic data to assess potential and attainable yields is described below. Such analyses can be specified for a given crop age, but in assessing climatic limits, account needs to be taken of the variable crop age at harvest in commercial production. As well as a direct effect on yield, rainfall affects harvesting operations, runoff and deep drainage, and possible environmental impacts due to movement of nutrients and pesticides. Examples of these likely impacts are given below.

Impacts of climatic variability

In Australia, sugarcane harvest dates vary from June to December. It is not possible here to consider all permutations of the production system in assessing the impact of climatic variability. Rather, a single production system is examined for a first ratoon crop grown for 12 months under potential (maximum growth) conditions harvested on 1 October each year using the APSIM-Sugarcane model (Appendix 2). As an example of year to year variability, simulated yields for Ingham in north Queensland over 100 years are shown in Fig. 3.4a. Also, simulated data for each year of the historical climatic record for selected locations (Table 3.1) were sorted into ascending order, and the cumulative 0, 25, 50 (median), 75 and 100% probability values associated with different levels of yield were estimated (Fig. 3.4b).

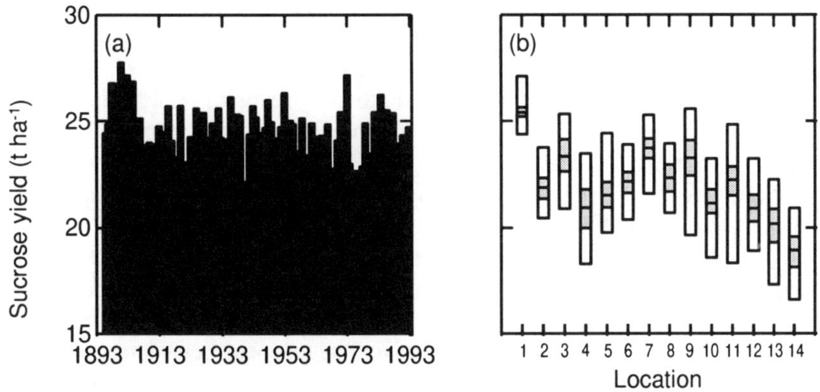

Fig. 3.4. Variation in potential sucrose yield (t ha^{-1}) simulated using historical climatic records for (a) year to year at Ingham and (b) all sites in Table 3.1. [The median value is shown as the centre horizontal bar, the hatched area is between the 25 and 75% cumulative probability value, and the open bars show the upper and lower yield limits. Twelve month IR crops simulated using APSIM-Sugarcane (Appendix 2)]

For Ingham, the potential sucrose yield varied between 20.7 and 27.8t h^{-1} from year to year (Fig. 3.4a), with a median value of 24.3t ha^{-1} (location 6, Fig. 3.4b). For sites in the tropical region, yields were predicted to be highest at Kununurra, followed by Ayr and Mackay, and lowest in the wetter Ingham, Cairns, Innisfail and Babinda environments. However, with variations from year to year, there was considerable overlap between locations. For subtropical sites, potential yields were more related to latitude (Fig. 3.2), and declined from Bundaberg to Grafton (Fig. 3.4b) in response to lower temperature and radiation (Table 3.1).

Rainfall causes additional yield variability particularly in rainfed supplementary-irrigated districts as demonstrated by Robertson *et al.* (1997) for a 12 month ratoon crop at Mackay. Under rainfed conditions, sucrose yield varied from 0-24.7t ha^{-1} with a mean of 14.9t ha^{-1}. With 4Ml ha^{-1} irrigation, variation in yield was reduced to 13.9-29.3t ha^{-1}, with a mean of 24.5t ha^{-1}. Rainfall also influences deep drainage and runoff, e.g. at Mackay, simulated seasonal deep drainage varied from 0-1444mm and 0-1529mm for the rainfed and supplementary-irrigated situations, while runoff varied from 1-1679mm and 3-1347mm, respectively. Consequences of water losses for leaching of chemicals, and ground- and surface-waters are considered by Keating *et al.* (1997).

Rainfall amount and distribution prior to and during the harvest season can also markedly influence CCS, and thus the price premium paid for cane (Russell *et al.*, 1991). Higher CCS tends to be associated with lower rainfall (Russell *et al.*, 1991). Rainfall variability also affects fibre levels and the availability of fuel for mills. Variation in CCS over the harvest season in different zones of a mill district can be related to rainfall, e.g. in the Herbert district (Fig. 3.5a), CCS is higher at Bambaroo than at Halifax early in the harvest season, but at Bambaroo it falls dramatically near the end of the season. Rainfall tends to be lower at Bambaroo earlier in the season, and higher towards the end of the season, than at Halifax (Fig. 3.6a). In another mill district (e.g. Mossman in north Queensland), CCS is higher throughout the season at Mareeba on the Atherton Tablelands compared to coastal Mossman (Fig. 3.5b), with rainfall much lower in the irrigated Mareeba area (Fig. 3.6b). An important feature of this variation (Fig. 3.5) is that it offers the opportunity to improve whole mill area profitability by scheduling harvest and cane supply from different areas to optimise whole season CCS, although other factors apart from CCS are important in scheduling cane supply, including infrastructure and transport systems, and economic and social considerations. Also, the issue of maximising sugar yield per hectare as well as CCS needs to be considered.

Rainfall can disrupt harvesting and mechanical harvesting under wet conditions can cause loss of cane stool, soil structural damage, compaction and rutting in paddocks. This reduces the yield and profitability of the following year's ratoon crop. Currently, the Australian sugar industry is expanding the sugarcane area in existing mill districts and new sugarcane growing regions (e.g. Atherton Tablelands and Ord Irrigation Area) (Chapman and Milford, 1997). Knowledge of rainfall probabilities in different mill areas, particularly beyond the current season length, can assist decision-making on optimising the schedule of harvest. Muchow and Wood (1996) analysed historical rainfall records to determine the rainfall risk for different times in the Burdekin, Herbert and Ord regions, and Hughes and Muchow (1996) presented data for the New South Wales sugar industry. Data are usually presented on a probabilistic basis (see Fig. 3.6), and the selection of probability level is dependent on the attitude to risk of the decision maker. For the sites demonstrated in Fig. 3.6, the rainfall risk towards the end of the harvest season is much higher than at the beginning. The Herbert example (Fig. 3.6a) also shows that within-region variation occurs, with an earlier start and finish favoured at Bambaroo compared to Halifax. Rainfall risk to harvest interruption is only one aspect of optimising harvest schedules.

Cane yield and CCS of current (e.g. Fig. 3.5) and subsequent ratoon crops associated with different times of harvest is another important component. Evidence suggests significant yield losses of subsequent ratoons following late harvest (Chapman and Leverington, 1976; Leverington *et al.*, 1978).

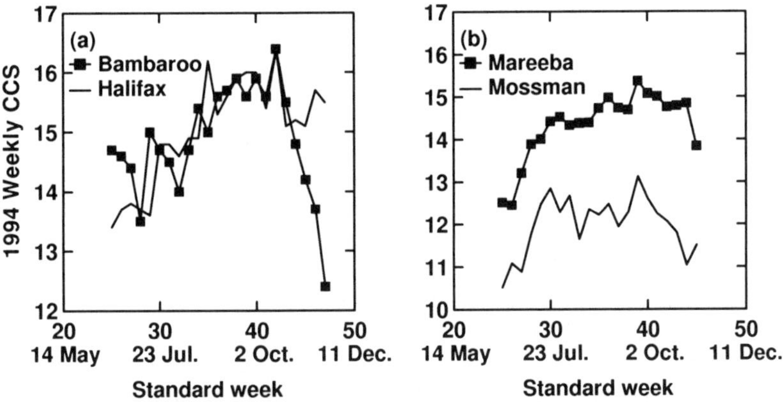

Fig. 3.5. Variation in weekly CCS during the harvest season for (a) Bambaroo and Halifax in the Herbert region in 1994, and (b) Mareeba (tablelands) and Mossman (coast) in the Mossman district in 1995. [Data sources: Bambaroo and Halifax (M.J. Sefton, A.W. Wood); and Mareeba and Mossman (A.V. Rudd, A.W. Ford)]

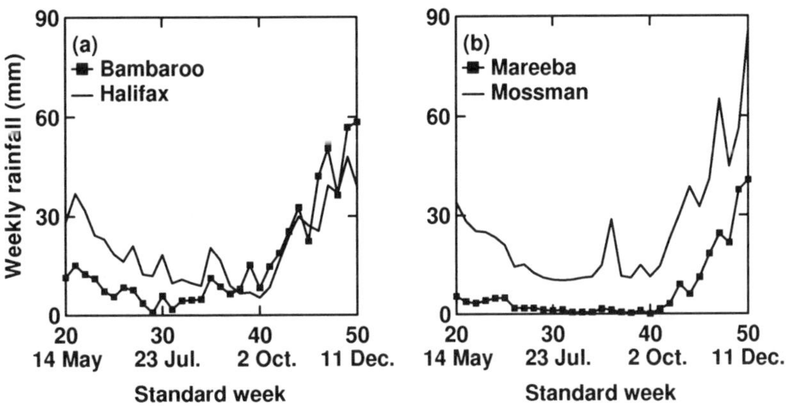

Fig. 3.6. Weekly rainfall expressed in terms of the cumulative probability of receiving less than the specified amount in 80% of years, for (a) Bambaroo and Halifax; and (b) Mareeba and Mossman. [Based on 72 to 93 years of records for the different locations]

The analysis in this section indicates that there is scope to use knowledge of rainfall variability, and of the climatic and biological limits to yield in different locations for different harvest dates and crop ages, to develop options for optimising crop schedules and further enhancing the performance of the Australian sugar industry.

Conclusions

Application of knowledge on the biological limits to yield accumulation in association with historical climatic data has shown that the annual production potential for sugarcane in Australia is high. In tropical regions, average sugar yield ranges from 22 to 31t ha^{-1} y^{-1}, with some lowering of potential down to 18 to 24t ha^{-1} y^{-1} in the cooler environments of southern Queensland and northern New South Wales. Low solar radiation, associated with high rainfall, in the wet tropics results in potential yield not being closely related to latitude, but potential yields are more closely related to latitude in subtropical environments. The potential productivity in all locations still far exceeds the production actually achieved. Clearly, the most significant limits to current productivity lie other than in the radiation and temperature environment; that is, in the availability of water for irrigation, suitable soils, efficient agronomic practices, influence of pests and pathogens, cost/price pressures, lifestyle and environmental considerations, harvest scheduling, milling and transport infrastructure, marketing issues and the effects of competing land uses.

Potential productivity will rarely equate to actual farm yields, although our confidence in simulated estimates is boosted by individual blocks of a few producers achieving production levels comparable with the estimates of potential. The main value of defining potential productivity lies in the opportunity for growers, productivity boards or the industry to 'benchmark' what is possible and investigate reasons for actual productivity falling short of the potential. The Australian industry will continue to have, at least in the short-term, opportunities to expand the area of cultivation. Forces external to the industry may ultimately limit this expansion onto new lands (see Johnson *et al.*, 1997) and future increases in production might be more dependent on raising productivity on existing lands. This paper has shown that from a climatic perspective, there is considerable scope to raise the productivity of existing germplasm at the block level. Scope also exists at the farm, mill district and industry scales for growers, millers and marketers to all benefit from forecasting and scheduling practices that better utilise the climatic resources available to the industry.

The availability of climatic data and/or robust approaches to the generation of such data over space and time has proved to be one of the most challenging aspects of this paper. The main problem is that we have insufficient baseline data for variables such as solar radiation to be fully confident in the application of generated data in circumstances of site to site and year to year comparisons where the climatic data will have a strong impact on the outcome of the analysis. Crop simulation outcomes as demonstrated here, together with models of other elements of the production system, databases, geographic information systems, and enhanced climate prediction will contribute to on-going efforts to raise industry productivity and profitability at block, farm, mill and industry scales. Further advances in our knowledge of the climatic and biological limits of the sugarcane production system will benefit the international competitiveness of the Australian sugar industry.

Acknowledgements

We thank Heidi Vogelsang (data analysis and presentation), Neil Huth (computer simulations) and the Agricultural Production Systems Research Unit of CSIRO/QDPI/QDNR (climatic data). Much of the research presented in this paper was conducted in partnership with the Sugar Research and Development Corporation and CSR Technical Field Department.

References

Angus, J.F. (1991) The evolution of methods for quantifying risk in water limited environments. In: Muchow, R.C. and Bellamy, J.A. (eds), *Climatic Risk in Crop Production: Models and Management for the Semi-Arid Tropics and Subtropics.* CAB International, Wallingford, UK, pp. 39-53.

Chapman, L.S. and Leverington, K.C. (1976) Optimising harvest schedules in the Mackay Area. *Proceedings of Queensland Society of Sugar Cane Technologists, 43rd Conference,* pp. 33-38.

Chapman, R.N. and Milford, B.J. (1997) Economic constraints and opportunities of the Australian sugar industry. In: Keating, B.A. and Wilson, J.R. (eds), *Intensive Sugarcane Production: Meeting the Challenges Beyond 2000.* CAB International, Wallingford, UK, pp. 55-70.

Hughes, R.M. and Muchow, R.C. (1996) Rainfall variability in the New South Wales sugar industry. In: Wilson, J.R., Hogarth, D.M., Campbell, J.A. and Garside, A.L. (eds), *Sugarcane: Research Towards Efficient and Sustainable Production.* CSIRO Division of Tropical Crops and Pastures, Brisbane, Australia, pp. 28-30.

Humbert, R.P. (1968) *The Growing of Sugar Cane.* Elsevier, Amsterdam, The Netherlands.

Inman-Bamber, N.G. (1995) Climate and water constraints to production in the South African sugar industry. *Proceedings of the South African Sugar Technologists' Association,* pp. 55-59.

Johnson, A.K.L., McDonald, G.T., Shrubsole, D.A. and Walker, D.H. (1997) Sharing the land: the sugar industry as part of the wider landscape. In: Keating, B.A. and Wilson, J.R. (eds), *Intensive Sugarcane Production: Meeting the Challenges Beyond 2000.* CAB International, Wallingford, UK, pp. 361-380.

Keating, B.A., Verburg, K., Huth, N.I. and Robertson, M.J. (1997) Nitrogen management in intensive agriculture: the case for sugarcane in Australia. In: Keating, B.A. and Wilson, J.R. (eds), *Intensive Sugarcane Production: Meeting the Challenges Beyond 2000.* CAB International, Wallingford, UK, pp. 221-242.

Kingston, G., Hughes, R.M. and Wood, A.W. (1995) Climate, water and nitrogen as constraints to production in the Australian sugar industry. In: Robertson, M.J. (ed.), *Research and Modelling Approaches to Assess Sugarcane Production Opportunities and Constraints*, CSIRO Division of Tropical Crops and Pastures, St Lucia, Australia, pp. 11-18.

Kropff, M.J., Muchow, R.C., Rabbinge, R. and Aggarwal, P.K. (1997) Limits to intensive agricultural systems and opportunities for sustainable agricultural development. In: Keating, B.A. and Wilson, J.R. (eds), *Intensive Sugarcane Production: Meeting the Challenges Beyond 2000.* CAB International, Wallingford, UK, pp. 17-35.

Lee-Lovick, G., Saunders, M., Willcox, T. and Bent, M. (1992) The establishment of a geographic information system in the Bundaberg district. *Proceedings of Australian Society of Sugar Cane Technologists,* pp. 107-115.

Leverington, K.C., Hogarth, D.M. and Ham, G.J. (1978) The influence of time of harvest on yields in the Burdekin District. *Proceedings of the Queensland Society of Sugar Cane Technologists, 45th Conference,* pp. 27-30.

Muchow, R.C. and Bellamy, J.A. (1991) (eds), *Climatic Risk in Crop Production: Models and Management for the Semi-Arid Tropics and Subtropics.* CAB International, Wallingford, UK.

Muchow, R.C., Robertson, M.J., Wood, A.W. and Keating, B.A. (1996a) Effect of nitrogen on the time-course of sucrose accumulation in sugarcane. *Field Crops Research* 47, 143-153.

Muchow, R.C., Robertson, M.J. and Wood, A.W. (1996b) Growth of sugarcane under high input conditions in tropical Australia. II. Sucrose accumulation and partitioning, and commercial yield. *Field Crops Research* 48, 27-36.

Muchow, R.C., Robertson, M.J., Wood, A.W. and Keating, B.A. (1997) Assessing limits to sugarcane yield. *Proceedings of the Australian Society of Sugar Cane Technologists* 19, (In press).

Muchow, R.C., Spillman, M.F., Wood, A.W. and Thomas, M.R., (1994) Radiation interception and biomass accumulation in a sugarcane crop grown under irrigated tropical conditions. *Australian Journal of Agricultural Research* 45, 37-49.

Muchow, R.C. and Wood, A.W. (1996) Rainfall risk and scheduling the harvest of sugarcane. In: Wilson, J.R., Hogarth, D.M., Campbell, J.A. and Garside, A.L. (eds), *Sugarcane: Research Towards Efficient and Sustainable Production.* CSIRO Division of Tropical Crops and Pastures, Brisbane, Australia, pp. 26-27.

Muchow, R.C., Wood, A.W. and Robertson, M.J. (1995) Does stalk death set the yield ceiling in high-yielding sugarcane crops? *Proceedings of the Australian Society of Sugar Cane Technologists* 17, 142-148.

Robertson, M.J., Inman-Bamber, N.G. and Muchow, R.C. (1997) Opportunities for improving the use of limited water by the sugarcane crop. In: Keating, B.A. and Wilson, J.R. (eds), *Intensive Sugarcane Production: Meeting the Challenges Beyond 2000.* CAB International, Wallingford, UK, pp. 287-304.

Robertson, M.J., Wood, A.W. and Muchow, R.C. (1996) Growth of sugarcane under high-input conditions in tropical Australia. I. Radiation use, biomass accumulation and partitioning. *Field Crops Research* 48, 11-25.

Russell, J.S., Wegener, M.K. and Valentine, T.R. (1991) Effect of weather variables on C.C.S. at Tully simulated by the AUSCANE model. *Proceedings of the Australian Society of Sugar Cane Technologists* 13, 157-163.

Stanford, G. (1963) Sugarcane quality and nitrogen fertilisation. *Hawaiian Planters' Record* 56, 289-333.

Thompson, G.D. (1988) Comparisons of the growth of plant and first ratoon crops at Pongola. *Proceedings of the South African Sugar Technologists Association* 62, 180-184.

Wegener, M.K. (1997) Opportunities to improve economic performance of sugarcane farms. In: Keating, B.A. and Wilson, J.R. (eds), *Intensive Sugarcane Production: Meeting the Challenges Beyond 2000.* CAB International, Wallingford, UK, pp. 343-360.

Wegener, M.K., Bruce, D. and Fuelling, T.G. (1992) Application of geographic information systems software for productivity evaluation. *Proceedings of the Australian Society Sugar Cane Technologists* 14, 99-106.

Appendix 1. Generation of Historical Climate Data

Historical daily records of rainfall, temperature and solar radiation are required to conduct the simulation analyses presented in this paper. Rainfall records are available for many locations throughout Australia, but temperature data are less frequently available and usually for shorter durations, while recorded solar radiation is scarce. Climate files were developed for the locations listed in Table 3.1 using the following methods.

Rainfall

Daily rainfall data were obtained from Australian RAINMAN, with data quality checked and missing data infilled using methods described by Clarkson and Owens (1991).

Temperature and Solar Radiation

The procedure for infilling and generating temperature data and its adequacy for the purpose of this study is outlined by Meinke *et al.* (1995). Briefly, electronic temperature records (daily maximum and minimum temperatures) are available from the Bureau of Meteorology from 1957 onwards. Prior to that, only hand-written records exist in the archives. At each location, the five years containing the least gaps in temperature data were used to derive the necessary monthly input coefficients for the weather data generator WGEN (Richardson and Wright, 1984). For locations without long term temperature records in electronic form, WGEN input parameters were derived from the nearest temperature record of comparable altitude.

The procedure for generating solar radiation was similar to that of Meinke *et al.* (1995) except that monthly mean solar radiation values were derived from the method of N. Flood (*pers. comm.* 1996) rather than ESOCLIM (Hutchinson, 1991) as in Meinke *et al.* (1995). The Flood method is based on a multiple regression of monthly average radiation for all sites with radiation data in Australia such that:

Daily radiation = Extraterrestrial radiation x transmission/100 Eq. (1)

where, Extraterrestrial radiation is a function of latitude and day of year,

Transmission (%) = 30.58 +1.12 * temperature difference + 18.96 * $e^{-rainfall}$ Eq. (2)

Temperature difference = daily maximum temperature - daily minimum temperature (°C) and rainfall is daily rainfall in millimetres. Monthly averages were calculated from the available long term temperature and rainfall data.

WGEN was then used to infill the missing data back to the beginning of the available rainfall records. Briefly WGEN takes long term monthly maximum and minimum temperature and solar radiation for a site and, together with values from other nearby weather stations, generates daily values. Because WGEN applies a probabilistic treatment to the generated temperature values, it can generate maximum and minimum temperature outside the historical range recorded for a site. This is undesirable when generated weather files are being used for a crop simulation model, such as sugarcane, as only a few extremes can initiate frosting which has major consequences for the crop simulation. To minimise this problem, all generated temperature data were bounded between the absolute highest and lowest temperature recorded for a site.

Another point to note is that care must be taken in direct comparisons between individual years in these generated weather files. While rainfall data are directly measured and provide a valid basis for site comparisons in individual years, and the algorithms used for temperature and solar radiation incorporate a rainfall factor, temperature and solar radiation are generated to reflect the long term probabilistic nature of the climate at a site and do not incorporate any space by time correlations that might exist.

Appendix 2: Simulation Using APSIM-Sugarcane

The APSIM-Sugarcane model, using a daily time step, is designed to simulate a uniform block of cane and predict on an area basis, sucrose yield, CCS, crop biomass, crop nitrogen uptake, and partitioning to leaf and stalk (Robertson *et al.*, 1995). The model is designed as a 'plug-in / pull-out' module of a systems model called APSIM (McCown *et al.*, 1996). The model is 'driven' by soil factors such as depth, water holding capacity, and nitrogen status; management factors such as planting/ratooning date, fertiliser and irrigation; climatic factors including rainfall, temperature and solar radiation; and genetic factors. Enhancement of modelling capability is on-going, and the predictive ability of the current model is shown in Fig. 3.7.

APSIM-Sugarcane was configured to simulate rainfed and fully irrigated production, and hence irrigation requirement for a standard first ratoon crop with genetic characteristics of variety Q117 for 14 locations in the Australian sugar industry (see Table 3.1). The manager module was configured to plant into a full profile of soil water on the first year of the climatic record, and then ratoon the crop, every year on 1st October. In this way, the model simulated carry-over effects of soil water storage from one season to the next. The crop was burnt before harvest, except at Cairns, Babinda, Innisfail and Ingham. Nitrogen was supplied at 250kg ha^{-1} as urea on the day of ratooning, buried at a depth of 100mm. The first year of output for the plant crop was omitted from the subsequent analysis of yields, etc. The simulations were conducted for the soil type described in Table 3.1 of Robertson *et al.* (1997), i.e. 162mm available water to 1.5m depth. Under full irrigation, water was applied when 30% of the available water in the root zone was depleted. This frequency of application is not realistic in commercial practice (except under trickle), but ensured that no water stress occurred and that simulated yields were those attainable under fully irrigated, potential production. No water was applied during 30 days preceding harvest, to simulate a drying-off period.

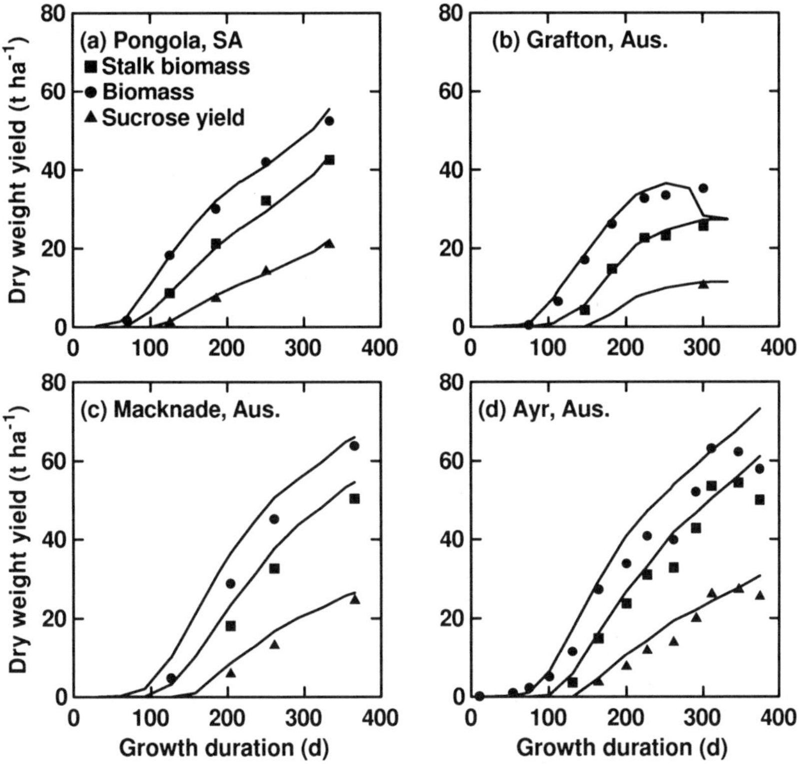

Fig. 3.7. Predictive ability of APSIM-Sugarcane in diverse production environments. Observed data are shown as symbols and daily simulated values are shown by continuous line. [Data from South Africa (a) Thompson (1988) for variety N14 and Australia (b) R.C. Muchow and R.M. Hughes for variety Q117; (c) Muchow *et al.* (1993) for Q117 at Ingham; (d) R.C. Muchow, M.J. Robertson and A.W. Wood for Q117 at Ayr]

References

Clarkson, N.M. and Owens, D.T. (1991) *Rainman - Rainfall Information for Better Management.* Queensland Department of Primary Industries, Brisbane, Australia.

Hutchinson, M.F. (1991) Climatic analyses in data sparse regions. In: Muchow, R.C. and Bellamy, J.A. (eds), *Climatic Risk in Crop Production - Models and Management for the Semi-Arid Tropics and Subtropics.* CAB International, Wallingford, UK, pp. 55-71.

McCown, R.L., Hammer, G.L., Hargreaves, J.N.G. Holzworth, D.P. and Freebairn, D.M. (1996) APSIM: A novel software system for model development, model testing and simulation in agricultural systems research. *Agricultural Systems* 50, 255-271.

Meinke, H., Carberry, P.S., McCaskill, M.R., Hills, M. and McLeod, I. (1995) Evaluation of radiation and temperature data generators in the Australian tropics and sub-tropics using crop simulation models. *Agricultural and Forest Meteorology* 72, 295-316.

Muchow, R.C., Robertson, M.J., Wood, A.W. and Spillman, M.F. (1993) Effect of soil fumigation on sugarcane productivity under high-yielding conditions in north Queensland. *Proceedings of the Australian Society of Sugar Cane Technologists* 15, 187-92.

Richardson, C.W. and Wright, D.A. (1984) *WGEN: A Model for Generating Daily Weather Variables*. US Department of Agriculture, Agricultural Research Service, ARS-8, USA.

Robertson, M.J., Inman-Bamber, N.G. and Muchow, R.C. (1997) Opportunities for improving the use of limited water by the sugarcane crop. In: Keating, B.A. and Wilson, J.R. (eds), *Intensive Sugarcane Production: Meeting the Challenges Beyond 2000*. CAB International, Wallingford, UK, pp. 287-304.

Robertson, M.J., Keating, B.A. and Muchow, R.C. (1995) APSIM-Sugar: its history, conceptual basis, uses and wider applications. In: Robertson, M.J. (ed.), *Research and Modelling Approaches to Assess Sugarcane Production Opportunities and Constraints*. CSIRO Division of Tropical Crops and Pastures, St Lucia, Australia, pp. 35-42.

Thompson, G.D. (1988) Comparisons of the growth of plant and first ratoon crops at Pongola. *Proceedings of the South African Sugar Technologists Association* 62, 180-184.

Chapter four:

Economic Constraints and Opportunities in the Australian Sugar Industry

R.N. Chapman[1] and B.J. Milford[2]

1. *Deputy General Manager - Industry Management*
2. *Manager - Economic & Technical Services, CANEGROWERS, 190-194 Edward Street, Brisbane, Qld 4000, Australia*

Abstract

This paper considers the major economic constraints and opportunities of the Australian sugar industry. It sets these in the context of the historical and cultural circumstances of the industry and likely future arrangements.

It considers the industry within the broad framework of the value chain. It first examines the international market place, looking at demand and supply issues and how these impact on the potential for Australia within the global market. It then reviews the Australian system of marketing, looking at its likely development based on economic and structural forces currently prevailing. The paper then moves onto raw sugar production in Australia, considering the constraints including natural resources (predominantly land and water), capital resources and human resources (including R&D) within the context of the existing and emerging structural arrangements.

Background

The sugar producing areas of Australia are scattered along the eastern coast from Mossman (16° S) in north Queensland to Yamba (29° S) in New South Wales, a distance of 2100 kilometres. In 1996, Australia will produce about 40Mt of cane and from this make over 5Mt of sugar. With a domestic market of around 920,000t, by far the major share of this production is therefore destined for export. Thus the focus of Australia's 6800 cane farmers is on world price. This

In: Keating, B.A. and Wilson, J.R. (eds) *Intensive Sugarcane Production: Meeting the Challenges Beyond 2000*. CAB International, Wallingford, UK, 1997.

injects very strongly the discipline of international competitiveness into the Australian industry.

Cane farming in Queensland is characteristically a family-owned, farmer-operated enterprise. About 95% of farms are owned by sole proprietors or family partnerships; less than 2.5% of cane land is operated by sugar milling companies. Individual farms produce up to 100,000t but the average grower produces 4,346t (5-year average) from 68ha of cane. Total area harvested is 460,000ha. Approximately 40% of the cane land in Queensland is irrigated.

The International Market Place

Sugar has been described as the most political commodity in the world and the establishment of sugar industries in most countries, developed and developing, was characterised by government intervention and a drive for internal self sufficiency. In most large sugar-producing countries (and many small ones), the sugar industry has been isolated from the impact of movements in the international sugar price because market price signals were not being passed to producers. Industries set up in this way tend to be indifferent to world market conditions. Production levels, if not constrained geographically, typically rise to the requirements of the national market and beyond. Because excess production is produced at marginal prices, it can be exported with relative indifference to market conditions. This has led to the world market being a residual one with an accelerated, cyclical pattern of wild swings in price; the sugar market is regarded as the most variable of any commonly traded international commodity.

Production

The inclusion of agriculture in multilateral trade talks conducted under General Agreement on Trade and Tariffs (GATT) is evidence that governments and regulators are becoming aware of the economic distortions caused by support for domestic agricultural industries. Unfortunately, there was little specific reference to sugar in the outcome of the Uruguay Round. Nor have the two most influential economies that have large, protected sugar markets, the European Union and the United States, made any substantial commitment to reform their sugar policies. However, it is now six years since there has been any change in the nominal level of support and in this time inflation (and, in the case of the USA, some tinkering in the latest Farm Bill) has had an effect on real levels of support in both cases. The practical effects of this are that European exports have not changed significantly in the last four years and the level of US imports has remained relatively constant.

At the same time the collapse of the centrally planned economies of eastern Europe has led to some relaxation of sugar policies in these countries. There has been a drastic reduction in production from eastern Europe, and from Cuba whose industry relied on Soviet support. However, a recovery is now apparent in areas such as Poland, the former East Germany, and in Cuba.

The collapse of industries in some countries which formerly were major exporters or self-sufficient (e.g. Philippines and Taiwan) and the recent extended drought and rising domestic consumption in many exporting countries have had two major effects. The first was a decrease in sugar traded at prices subject to special arrangements between countries which now only accounts for 10% of world trade compared to c. 30% in the early 1980s. The second was a concentration of exports arising from Brazil, Thailand, Australia and Europe in particular. Figure 4.1 shows the changing share of the world free market held by these major exporters. It can be seen that the European share of sugar exports has changed only marginally, leaving most of that remaining to be shared by Australia, Brazil and Thailand.

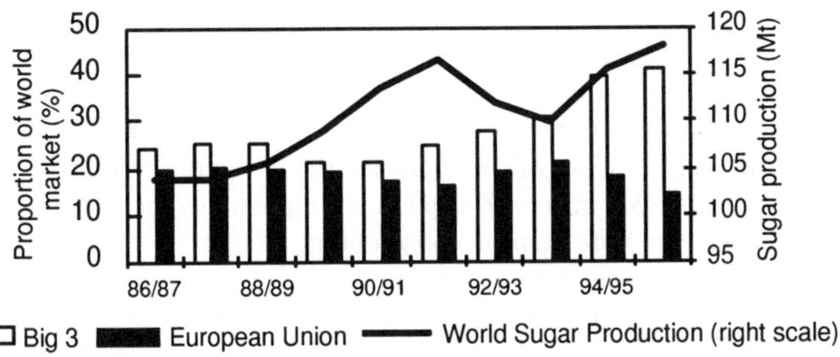

Fig. 4.1. Proportion of exports from the Big 3 (Australia, Brazil, Thailand) compared with European Union, and total world production. [USDA, 1995]

Consumption

Licht (1995) has predicted that, by the end of the century, world sugar consumption will be about 124.8Mt raw value. This compares with an estimate of consumption in 1995/96 of 116.6Mt. Figure 4.2 shows the growth of world consumption of sugar that, post-war up to the late 1980s, grew by 1.5% to 2% per annum.

Historically, consumption has not varied much with changes in price as approximately two thirds of sugar production was consumed by developed countries who tended to purchase their sugar needs irrespective of price. In the 1990s, however, consumption per capita in the developed world ceased growing (USDA, 1994) and alternative sweeteners took part of the market. Population growth in these countries also slowed and this has resulted in static or declining consumption in developed countries.

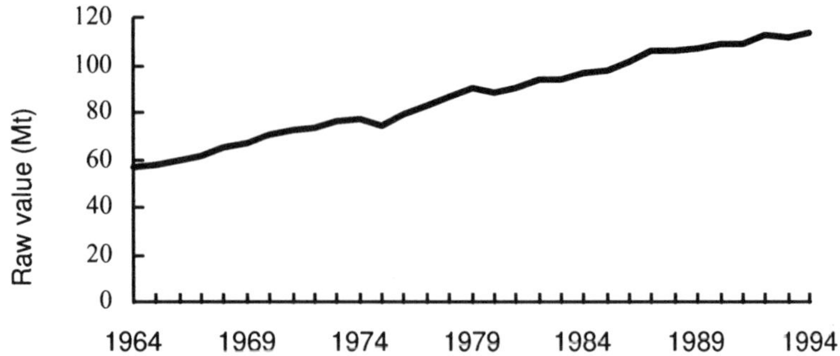

Fig. 4.2. Trends in world sugar consumption. [ISO 1994]

In general, developed countries have higher levels of consumption (Table 4.1), but the rate of growth is small or dropping. In contrast, consumption for developing countries has increased dramatically. In particular, Asia has seen significant growth with a regional increase in consumption 24% larger than total increase in world consumption over the five year period. This has been balanced by, *inter alia,* the former states of the USSR recording significant falls.

Much of the growth in world sugar consumption has occurred in South America and Asia. In Asia, annual consumption has grown from 11.1kg per person in 1985 to 12.8kg per person in 1994, with a compound growth rate of 1.6%. Although this increase in consumption may not seem large, it should be considered in the context that Asia represents 58% of the world's population but in 1995 consumed only 34% of the world's sugar. The impact of this growth compared to the rest of the world is shown in Fig. 4.3.

The net effect of these changes has been an increasing share of world sugar production consumed in countries such as China and India. These countries are much more inclined to decrease consumption when prices increase. As developing rather than developed countries now account for two thirds of sugar consumption, the high price peaks experienced up to 1950 have not occurred to the same degree over the past decade or so. This increase in price sensitivity and demand elasticity is best illustrated by the rapid decline in consumption in the former USSR states (Table 4.1) once the centralist planned economy failed.

The absence of price spikes has, it is believed, contributed to the non-recurrence of disastrously low price levels like those of the mid 1980s. In the past, high sugar prices often led governments to encourage investment in sugar industries which would otherwise be uneconomic. The difficulty experienced in winding back these investments when prices eventually fall has, in the past, contributed to long periods of low prices.

Table 4.1. Growth in total world sugar consumption (Mt) and consumption per person from 1990/91 to 1995/96. [USDA, 1994; Licht, 1995]

	Consumption total		Change per annum	Share of growth in world consumption	Per capita use	
	90/91	95/96			1990	1994
	(Mt)		(%)	(%)	(kg)	
North & Central America	15.85	16.69	5.3	18	37.9	37.7
South America	12.08	13.94	15.5	40	39.5	42.4
European Union	13.85	14.00	1.1	3	38.5	38.1
Other Western Europe	0.50	0.50	-0.8	0	37.5	36.3
Eastern Europe	17.76	13.32	-25.0	-95	47.2	36.3
Africa	8.99	9.36	4.2	8	14.6	13.5
Asia	34.16	39.96	17.0	124	12.2	12.8
Oceania	1.12	1.22	8.8	2	41.6	41.6
Total World	111.93	116.61	4.2		21	20.3

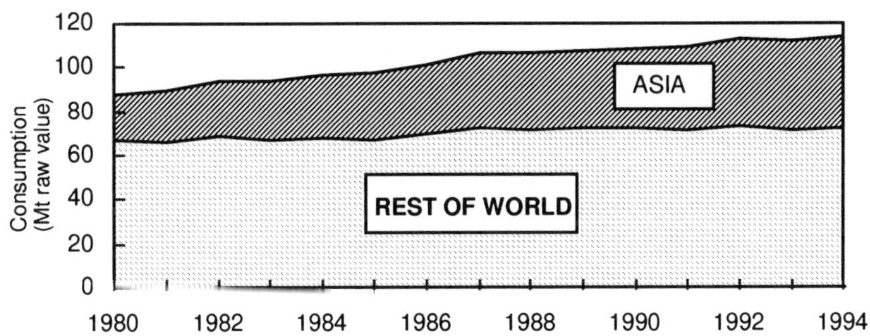

Fig. 4.3. Asia's growing share of world sugar consumption. [After Licht 1995]

In some countries, the extra demand that results from rising Gross Domestic Product may not necessarily be translated into growth in sugar demand. Alternative options such as High Fructose Corn Syrup (HFCS), or synthetic sweeteners such as saccharin may take up some of this slack. It appears that this is occurring in some Asian countries, notably China (high intensity sweeteners) and Korea (HFCS).

The substantial change in the behaviour of the international sugar market arising from the demographic changes discussed above means that growth in consumption for the rest of this decade is likely to continue to be close to 1% per annum or perhaps a little more. This has implications for Australia in that it will restrict the overall world market for our sugar. However, since most of the growth in consumption is expected to be in countries in Australia's immediate region, we can look forward to continuing demand for our sugar provided our marketing and service package is acceptable to customers. It is to Australia's marketing position that we now turn.

Australian Sugar Marketing

The Australian marketing system is characterised by single desk marketing operations, year-round reliability of supply, long term trading relationships with customers and good, reliable quality. This marketing package has enabled Australia to become the preferred supplier in virtually all of the countries to which it markets sugar. This means that we have the flexibility to determine to which destinations its sugar shall be sent. This is particularly valuable as production expands.

Australia has a sustainable competitive advantage over Brazil and Thailand in transport and storage of raw sugar because of the close proximity of its production areas to its ports. This is enhanced by the ability of the Australian industry to store, schedule, and ship bulk sugar in a way which provides for rapid turnaround of large ships. This provides scale and efficiency advantages which helps overcome Australia's comparative geographic isolation. Advantages that accrue from these arrangements include:

1. Ability to manage the level of supply to match demand in premium markets without extinguishing the premium by oversupply (see later section).
2. Benefits of scale in marketing and shipping charter.
3. Because of reliability of supply and the ability to deal directly with government authorities the industry can negotiate long term contracts which, even if they do not produce higher returns, provide a measure of risk management.
4. The ability to obtain a premium from our markets because our marketing package adds value for our customers.

It has been argued that if the Queensland raw sugar marketing system was further deregulated, the industry would benefit in the following ways:

1. Individual producers of raw sugar would be able to market their own sugar rather than through the common pool system.
2. Individual producers would better be able to manage their own level of risk.
3. Pooling of resources and costs dampens price signals which may therefore reduce opportunities to differentiate products and add value.

Despite these possible advantages, the authors believe that a single desk selling arrangement is preferable because of the competitive advantages it provides in the poorly functioning international sugar market.

The major economic factors imposed by the international market on the Australian sugar industry are:

1. The price of sugar.
2. Australia's ability to obtain a premium on the market.
3. The competitiveness of Australia compared to other major producers.
4. Australia's ability to expand production to obtain economies of scale.

The first three of these is examined in this section, the last in a later section.

Price of sugar

Even though Australia is one of the world's largest exporters of sugar, it still has little control over world prices. World price is dependent on the level of available stocks and the market's expectations of future production and consumption levels. As discussed above, the market is volatile.

In most countries, the domestic market absorbs most of the production at a significant premium when compared to the world market price. Australia's domestic market is the smallest of any of the major exporters, comprising only around 17% of current production with minimal tariff protection.

Since 1990, Australia has expanded sugar production at approximately 5% per annum or 276,000t of sugar per annum. However, expansion will only occur if price expectations are reasonable. Given such expectations, studies by the CANEGROWERS organisation have predicted that Australia will produce 6.5Mt of sugar by 2003. Will the world market be able to sustain such expansion?

There has also been vigorous expansion by Australia's competitors between 1990 and 1995. Although there were reductions in production in some countries (notably Cuba, China and the former Soviet Union), expansion proceeded rapidly in countries such as Thailand, Brazil and India. Using unsmoothed data (which does not take account of the effect of drought in Australia and Thailand during the period), the annual rates of growth in these four major producing countries are shown in Table 4.2.

During this period, growth in world consumption averaged 1.3Mt per annum. Clearly, the rate of growth in these four countries alone was far in excess of the rate of growth of world consumption. If these countries continue to expand at the current rate, a surplus would build up very quickly and prices would fall. This 'doomsday scenario' can be modified if we look at reasons for expansion in these countries and prospects for the current rate of expansion continuing.

Table 4.2. Annual rates of increase in sugar production ('000t) in selected countries, 1990-1995.

Australia	276
Brazil	1007
India	277
Thailand	289
TOTAL	1928

The Indian sugar industry is largely insulated from the world market and has responded to price signals within that country. There are recent indications that these price signals have started pushing in the opposite direction, leading to a reduction in cane supply. Indian production is generally price responsive; there are literally millions of growers, most of whom replant for every crop. Most of the world's sugar stocks are currently held in India and it is considered unlikely that Indian expansion will continue.

Good prices in international markets have set the pace for expansion in Thailand but the rate of growth there is likely to slacken. As production moves to the north east of the country cane yields may fall and costs for the transport of sugar to ports will increase. Thailand is also faced with a rapid change to its cane harvesting system and transport infrastructure; away from manual systems to mechanical systems. This will impact on growth because of the increased losses of cane in the field, and the requirement for capital and technically proficient human resources.

The expansion of sugar production in Brazil has accompanied the rationalisation of the ethanol market and privatisation of marketing. Brazil continues to have a strong commitment to produce ethanol and the rate at which producers can switch to sugar will be determined by the interplay between the world sugar price, the ethanol and oil markets and Brazil's foreign exchange demands.

It would therefore be reasonable to expect the rate of growth in all of these countries to moderate over the next few years. The extent of this moderation will determine the course of the world price in the medium term. This will in turn feed back into the sentiments of Australia's producers and will be an important issue in determining how quickly Australia expands its production.

Ability to maintain a competitive advantage through marketing

There are a number of elements in Australia's sugar marketing strategy which allow us to achieve a competitive advantage. These include the added value of the Australian marketing package to customers and the economies of scale that we obtain in bulk handling and shipping. Perhaps the most important advantage that Australia has over its competitors is the 'Far East Premium' that it receives on sugar sales (Boston Consulting Group, 1996). This 'Far East Premium' arises because of shipping costs for sugar, from suppliers such as Brazil, to the developing markets of East Asia. The actual level of the premium varies but in recent years has been in the range of US1-4c lb^{-1}. This premium does not always reflect a statistical deficit of sugar in the region. If Australia were to sell all of its sugar onto East Asian markets in an attempt to capture the premium for as much of its crop as possible, the premium would fall, or even become negative, as markets became saturated. Because Australia is a preferred supplier in its markets, it can therefore choose to sell some of its sugar out of the Far East region to customers in other parts of the world. Tonnages shipped to these markets can be adjusted to ensure that a deficit remains in the Far East market and premiums on sugar sold in that market are maximised. This premium can only be achieved whilst single desk seller arrangements remain in place.

Australia's position in world sugar trade

The emerging market concentration in export sugar, coupled with the changes to key economic drivers on the world market, creates economic constraints on the Australian industry. We must be able to produce sugar as cheaply as our main competitors, Brazil and Thailand. A significant percentage of Australian sugarcane growers may go out of production if the price they receive is lower than their average variable costs per tonne of cane in the short run and lower than the minimum point on their average total cost curve in the long run and they perceive that this situation will not change in the near future.

In some international benchmarking studies, Australia had similar cost of production levels for sugar to Brazil and Thailand. This provides an indication of relative competitiveness based on cost of production and indicates that we should be able to weather the storms, if we keep downward pressure on prices and continue to implement innovations which enhance our competitiveness.

Industry personnel travelling among producing countries indicate that most major producer countries would have difficulty in maintaining sustainable production at an income level below US10c lb^{-1} (given current exchange rates). Any prolonged period of price below this level would create a reduction in production growth, not only for Australia but also its major competitors.

Australian Production System

The major economic constraints faced by the Australian sugar industry relate to the use and development of resources. Major resources include: land; water; capital; human resources; and research and development.

The annual rate of industry growth of c. 5% in the past four to five years exceeds Australia's historical rate of expansion of 1-2% and is providing considerable challenges to key stakeholders in ensuring that adequate land, water, and capital resources are available. To an extent, the current rate of expansion is being driven by the perception of producers that sugar production is profitable, based on the past three years of higher market prices.

History would suggest that the area of land planted to cane tends not to fall as prices decline. Area harvested is relatively elastic in an upward direction and a response to prices can be readily demonstrated. However, when prices fall, cane area falls only slowly, if at all. This is partly due to the long term nature of the crop but also to the cyclical nature which the sugar market has shown over time. A producer facing a low price would have a reasonable expectation that higher prices will be achieved during the life of the ratoons for the crop that he or she is contemplating planting. Farmers facing a low price with a crop in the ground would regard the planting cost as sunk costs. They would therefore continue to produce cane from ratoon crops so long as the price is above the variable costs of ratooning and harvesting.

Downward inelasticity is also a function of the availability of alternative uses for the land. In areas where alternative crops (such as bananas in the wet tropics and tomatoes in the southern areas) are available, there is more movement out of cane in periods of low price. However, the demand for these alternative products is fairly inelastic and the opportunity to move out of cane to take advantage of these markets is only available to a relatively small number of growers.

CANEGROWERS organisation recently conducted a survey of Queensland cane growers seeking to clarify, amongst other things, growers' attitudes to expansion. The survey showed that 51% of growers were likely to increase production. Of this group 58% believed this could be achieved by improving productivity per hectare, 44% by bringing new land into production and 33% by buying a new farm (some growers chose more than one option) (AGB McNair, 1996). These results indicate that, under present circumstances, growers are prepared to continue to increase production and invest the necessary resources.

Land

At present in Queensland sugarcane must be grown on land that has been 'assigned' to a mill under the *Sugar Industry Act*. There is therefore a system for allocating of land to canegrowing. This system allows for environmental checks to occur on the land and allows for a district to ensure that the capacity of the mill is sufficient to meet the expected production from new lands. The assignment system is one of the aspects of the industry being studied in the current review. The two alternatives for expansion of cane production are that suitable new land is located within 100km of an existing mill or that new mills are viable.

Potential production from existing mill areas

Sugarcane has a value of between \$20 and \$40 t^{-1}, depending on sugar price and sugar content (CCS); the 10 year average price is around \$28. Typical costs for hauling sugarcane a distance of 100km are about \$9 to \$11 t^{-1}. CANEGROWERS' modelling shows that the approximate margin over processing cost for new cane is around \$10 t^{-1}; it is therefore uneconomic to haul cane further than around 100km.

As discussed previously, Australian sugar production is currently increasing at an annual rate of about 5%. The major areas of expansion are located in the Central region around Mackay and Proserpine, in the Burdekin and Herbert River valleys, around Tully, and on the Atherton Tableland. These areas represent about 50% of the industry. Therefore the actual annual rate of expansion in these areas exceeds 10%. More land is available in each of these areas although supply of land is not unlimited in the medium to long term.

Increased production outside existing mill areas

Work carried out by CANEGROWERS indicates some potential to develop new mills. Real internal rate of return for new milling capital is highly sensitive to throughput, the price of cane, and the speed with which maximum design capacity can be reached. Based on sugar price of US10c lb^{-1} and an exchange rate of US75c/A\$1, real internal rates of return between 8% and 13% are likely. Given that this 'hurdle' rate is likely to be below the return achieved by alternative investments with similar risk profiles, at first glance it would appear unlikely that new mills would be built. However, there may be other considerations in the equation.

Under existing cane payment arrangements growers' internal rates of return would be in the vicinity of 4% to 10% depending on the cost of land and its development, the haulage cost of cane to the mill delivery point, and the cost of development of water infrastructure. Compared with other agricultural system, this is a reasonable rate of return. Thus CANEGROWERS organisation may choose to invest in a mill if it allows opening up of new land to cane. Alternatively, there may be synergistic or strategic reasons for a proprietary miller to build a mill. Some new mills may be built in the next 10 years, but this is likely to occur only at the margin of existing sugar producing areas where infrastructure costs can be kept to a minimum.

The most likely new development is on the Atherton Tableland which could be producing up to 150,000t of sugar within the next ten years. Potential sites for further mills in the Burdekin irrigation area and for the area south of Tully also exist. There are other sites in the Northern Territory adjacent to the Ord River Irrigation Area and in the Fitzroy River basin in Queensland, with suitable land resources but water and other infrastructure would need to be developed.

Water resources

In virtually all areas with major expansion potential (with the exception of Tully and the Herbert River), expansion depends on the availability of water at prices which ensure continued international competitiveness of sugar production. It is no accident that the most productive areas tend to be the areas with access to adequate irrigation supplies.

The capacity of the Atherton Tableland scheme is such that an increase in production of 110,000 to 120,000t of sugar by the year 2003 is possible. In the Burdekin district, the existing capacity of the Burdekin Dam will allow a further 3.3Mt of cane, which would yield an extra 0.5Mt of sugar. For the Proserpine district, it is expected that potential expansion due to the Peter Faust Dam will yield an extra 700,000t of cane which will provide an increase in production of 110,000t of sugar. For the Bundaberg district, and much of the Mackay district, increased production depends substantially on the availability of increased water supplies. Because large extra water storage schemes are becoming more difficult to justify economically and environmentally, these areas are less likely to be able to expand.

In future, the development of new water schemes will probably be based on a significant contribution of the capital cost by users. For established farms in established mill areas, the economic upper limit for the cost of water including the capitalised value of capital contribution is approximately \$60M l^{-1} for supplementary irrigation. This is based on the outcome of negotiations between Queensland Water Resources and canegrower users of water in the Proserpine and Mackay districts. The likely upper limit of water cost in regions such as the Burdekin, which relies on full irrigation, would be substantially less. For new farms and new mills, the capacity to pay for water would be substantially below \$60M l^{-1}. For instance, based on CANEGROWERS' calculations for the Mareeba irrigation area, the capacity to pay would be approximately \$40M l^{-1}. We conclude that any development of new water storage is likely to be marginal at a sugar price of US10c lb^{-1}.

Milling costs - matching capacity

In an unregulated free market, matching supply of cane with milling capacity would occur through growers expanding until the returns from cane equated with the marginal cost of production. This occurs because as the season length increases, the average sugar content across the season decreases; if growers are paid on the basis of the sugar content of their cane, not only will the return from the marginal tonnes of cane fall as season length is increased but the average price growers are paid for all cane supplied during the season also declines. There are a number of inherent structural aspects in the sugar industry which frustrate any attempt to set up a 'free market' in cane. One of these is the monopsonistic nature of the sugar miller operating as a single buyer in an area. Another is the growers' ability to bargain collectively. These counter-balancing monopolies therefore provide an environment where the parties agree to work together to determine optimum season length.

Increasing throughput is of value to a miller; the CANEGROWERS organisation estimates that additional tonnage throughput returns a margin over processing costs of \$10 t^{-1} to an existing miller based on price of sugar of \$320 t^{-1} (10-year average). Mills therefore have an incentive to seek higher tonnage. If the constraint on this extra tonnage is season length, mills would be obliged to increase their capital investment to allow for more tonnage to be processed within the existing season. The capital required to increase a mill's throughput varies from case to case and depends on the location of 'bottlenecks' in the milling process. Inevitably, there comes a time in an expansion program when a single large expensive piece of equipment is required (e.g. a new large boiler, estimated to cost around \$15 million). Such an investment will only be made when the projected rate of return is above a company's 'hurdle' rate. The projected rate of return will, in turn, be dependent on projected sugar prices. There are indications that this capital restraint on milling capacity may, within the next few years, become a major constraint on additional area expansion. There is a need for the industry to work together to devise ways in which the issue of the optimum season length can be determined and achieved.

Human Resources and Research & Development

Human resources can be a major determinant of competitiveness. The Australian sugar industry has a strategic and sustainable advantage in this area in the ability of its personnel and, in particular, its farmers. Australian farmers have a depth of experience and understanding in the use and management of complex cultural, mechanical, chemical and irrigation systems. They are also getting younger. A recent survey (AGB McNair, 1996) showed that the average age of cane growers is 50 years: this has fallen by one year over the past three years. The expertise of this former resource makes it difficult at present for some competitors to simulate Australia's competitive advantage by purchasing more sophisticated agricultural technology.

The Australian sugar industry is now investing in excess of \$38M in research and development per year. This is close to 2% of the gross value of product and compares more than favourably with its competitors. The industry has largely relied on its ability to implement innovation to sustain its competitive position against low labour costs in other countries or highly subsidised producers.

Figure 4.4 shows that Australia's productivity per hectare is almost twice that of its major competitors in the raw sugar market, Thailand, Brazil, India and the EU. Although Australia continues to retain a significant advantage in sugar productivity, there is evidence that Brazil and the European Union are slowly closing the gap. This is particularly apparent if comparisons are made with particular industry segments, such as the central southern area of Brazil or the French beet industry.

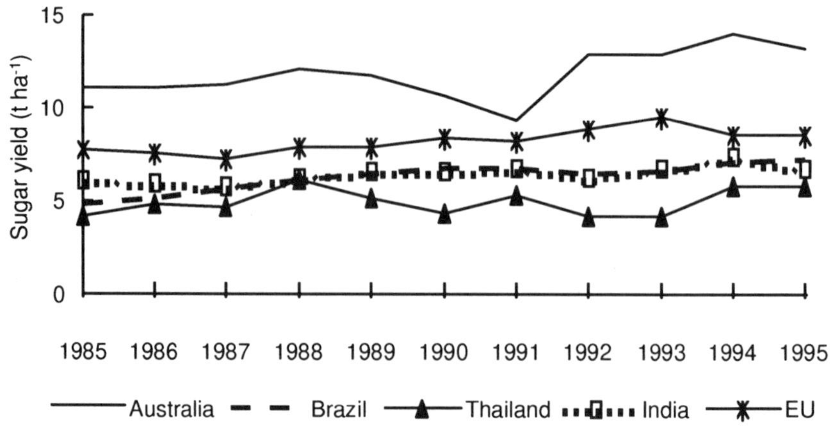

Fig. 4.4. Sugar yields (t ha^{-1}) in selected countries. [USDA, 1995]

Significant examples of the implementation of innovation which has provided competitive advantages for the Australian sugar industry include the development of mechanical harvesting and bulk terminal systems. More recently, the reduction of losses in harvesting has led to an estimated annual saving of $38M (Lindale and Ridge, 1996) with the potential for significant further savings.

If Australia is to remain competitive, it must continue to invest and implement targeted R&D where this is judged to achieve the best return on investment. This is likely to include non-traditional areas such as a systems approach to the analysis of the structural arrangements covering growers and millers or research into understanding the competitive strengths and weaknesses of the Australian industry. Competitive strengths are best judged within the context of the emerging marketplace signals. The administration of R&D policy and implementation of outcomes from research must continue to provide both comparative and competitive advantages if Australia is to retain its position as one of the world's largest and most efficient sugar industries.

Conclusions

The fundamentals which impact on the world market of sugar have changed, with supply and demand being more elastic than has previously been the case. This means that very high and low prices are unlikely to occur to the degree experienced in the past with long run prices more likely to reflect more closely the marginal cost of production for efficient producers.

Another major change is the increasing concentration of sellers of sugar on the international market with Brazil, Europe, Thailand and Australia taking 55% of the market between them. This will increase pressure on less efficient producers who in turn will be faced with declining production in the absence of price support mechanisms. This is increasingly likely with an effective cap on price support mechanisms and a downwards trend of support levels as indicated by the reducing number of special bi-lateral agreements incorporating non world market prices. An example of this was the collapse of the arrangement between the former USSR states and Cuba.

Australia also faces increasing risk because of its low level of support from domestic sales on an increasingly competitive international market. The Queensland sugar industry's structural marketing arrangements have provided a competitive strength through the operations of a single desk seller coupled with an efficient transport and storage system.

The current Australian system provides a sharing of decision-making between growers and millers which has facilitated a 30% increase in production since 1989. The comparatively rapid growth has highlighted a number of emerging constraints. Those identified include: the availability of suitable land within 100km of existing mills, water resources, capital and human resources. How the industry manages these considerations will have a significant impact on its future competitiveness.

Mill owners are finding it increasingly difficult to justify the increasing capital commitment needed to match crushing capacity with the rapid expansion of caneland in the short term whilst growers are struggling to obtain expensive and increasingly scarce water and land resources and to develop transport infrastructure. New mills are proving difficult to establish because of comparatively low internal rates of return caused by the prolonged start up phase and capital intensive nature of this investment.

To achieve continued expansion and a share of the increasing world market Australia must continue to achieve high levels of innovation in milling, sugarcane production and transport and storage. This places a particular challenge on industry research and development providers as innovation is undoubtedly one of the most likely significant sources for Australia's continued competitiveness.

References

AGB McNair (1996) *A Monitor of Queensland Growers' Perceptions of CANEGROWERS.* AGM McNair, Brisbane, Australia.

Boston Consulting Group (1996) *Report to the Sugar Industry Review Working Party: Analysis of Issues and Identification of Possible Options.* Main Report, BCG, Sydney, Australia, pp. 82-86

ISO (1994) *Sugar Year Book*, International Sugar Organisation, London, UK.

Licht, F.O. (1995) Sugar consumption at the end of this century, *F.O. Licht's International Sugar & Sweetener Report,* Vol. 127, p. 495.

Lindale, A.I. and Ridge, D.R. (1996) A successful campaign to minimise harvesting losses within the Queensland sugar industry. *Proceedings of the Australian Society of Sugar Cane Technologists* 18, 1-5.

USDA (1995) *World Sugar & Sweetener Situation & Outlook Yearbook.* United States Department of Agriculture, Washington DC, USA.

Chapter five:

Social Factors Shaping the Australian Sugar Industry

G. Lawrence[1] and I. Gray[2]

1. *Central Queensland University, Rockhampton, Qld 4700, Australia*
2. *Charles Sturt University, Wagga Wagga, NSW 2678, Australia*

Abstract

Structural change is affecting the sugar industry in ways which are altering on-farm production processes, the health of producers, and the links between producers and the wider rural community. This paper reports the findings of social research conducted among Central Queensland cane growers in the early 1990s. To cope with change, producers adopted a number of strategies among which were: postponing capital purchases, working off the farm, reducing labour and other costs, buying or selling land, and intensifying production. While a combination of these measures helped to ensure that many would remain in cane farming, such measures were also responsible for high stress levels among producers and for the undermining of traditional values. Conditions in the industry were thought also to be limiting a move by producers toward more sustainable options. The paper points to the need for further sociological research to be undertaken in an effort to clarify these and other issues.

Background: Agricultural Restructuring in Australia

The restructuring of Australian agriculture is bringing profound social changes. This process is creating such strain as to move many industries - including canegrowing - towards the limits of what has been termed 'family-farm' agriculture. The changes are considered by sociologists to be of two kinds: structural and cultural. The former, those arising from economic change, are leading to the polarisation of farming into capital-intensive, agribusiness-linked

In: Keating, B.A. and Wilson, J.R. (eds), *Intensive Sugarcane Production: Meeting the Challenges Beyond 2000*. CAB International, Wallingford, UK, 1997.

enterprises which derive an increasing share of sector income, and those enterprises which are less able to compete in a high input-output regime (Bonanno *et al.*, 1994). The cultural changes are those which contribute to detraditionalisation of farming: the replacement of traditional practices, values and beliefs, from which rural communities derived much of their characteristic social stability, with a more instrumental culture - one oriented to survival of the farm purely as a business (Gray and Phillips, 1996). These structural and cultural processes are linked. The capacity to survive during economic hardship has been underpinned by a culture of continuity and perseverance shared and fostered among individuals, families and communities. Economic downturn brings threats to the social sustainability as well as the longer term economic viability of the sugar industry at each of those three levels.

Another factor to be considered is the extent to which moves toward more sustainable production in farming systems complements or undermines family-based farming systems. There is no doubt that the evolution and incorporation of such things as Total Catchment Management, Ecologically Sustainable Development, Landcare and other natural resource management programs represents the setting of new agendas for farmers. Farmers are now but one stakeholder group among competing groups in relation to resource use. More importantly, despite evidence that many producers hold firmly to an ethos of 'stewardship' of the land (see Vanclay and Lawrence, 1995) they have been implicated in past rounds of environmental degradation. Farmers are being forced to address issues of on-farm and downstream pollution/destruction. However, the chemical and other strategies they have employed to maintain their viability cannot easily be replaced without undermining their already fragile economic position. Moves toward a more sustainable form of production will require new sorts of inputs - such as genetically engineered and more environmentally benign plants or animals - and new forms of agricultural equipment, but these can only be purchased at a cost. In the context of the economic problems, the ability of producers to move away from the current productivist strategies to more sustainable options has been questioned (Lawrence, 1995).

The lower incomes which have tended to accompany structural change in farming not only place a limit on the ability of producers to move quickly to more environmentally sound options, but also mean that less money is available for family-farm based social and community activities. So-called 'belt-tightening' brings greater stress on family and community relationships. Those facing particular difficulty may be trapped by continuing high input costs and low product prices while the value of their land falls. Perceptions of hopelessness give rise to further stress. Avenues of escape are few. For many, the only options are departure from agriculture or subsidising life on the land via increased levels of off-farm income. For some there may be avenues for on-farm diversification, but all these options rely on farm attributes, the possession of skills and the availability of market opportunities for viability. These actions are all part of what is euphemistically termed 'structural adjustment' (Lawrence, 1994). What appears - in the macro-statistical analyses of economists - as a slow process of change is, for many people, an abrupt shock which brings stress and trauma. When mental and physical energies are focused almost exclusively on how to feed and clothe a family and how to remain on the land, the pursuit of 'sustainability' - the very approach which is being endorsed by government and industry officials alike - takes something of a back seat (Share *et al.*, 1991; Lawrence, 1995). Such changes threaten the sugar industry.

Case Study: Central Queensland Sugar Producers

As part of a sociological study of four industries in selected areas of three states, research into the processes of change in the sugar industry has been carried out at the Centre for Rural Social Research at Charles Sturt University (Gray *et al.*, 1993). The project involved interviews with 245 farm people on 106 farms in the dairy industry in Northern Victoria (60 respondents), mixed farming and horticulture in the New South Wales Riverina (61 respondents in each) and canegrowing in Central Queensland (63 respondents). The survey sought information on experience of change and plans for the future. It was carried out in 1992, during a period of financial difficulty for many farmers brought about by, among other things, a high interest rate regime. Drought was adding an additional dimension to hardship in the sugar industry of Central Queensland. The sample was constructed to cover the breadth of experience among farms at the time by including those which were in a sound financial condition for comparison with those which were struggling. The circumstances prevailing at the time allowed investigation of the conditions in which social sustainability was under greatest threat.

A wide range of information was obtained to develop a comprehensive view of the forces moving canegrowing towards its social limits. Interviews covered farm outputs and size of operation, perceived recent climatic conditions and output levels, farm financial position, farm management and personal coping strategies, plans for the future of the farm and levels of personal stress.

Sugar was produced on all 24 farms in Central Queensland on which interviews were conducted. Other outputs were few: five had cattle, one produced mangoes and another had a quarry. The other areas were similarly narrow in their range of products, except the mixed farming areas of the Riverina where some 15 outputs were mentioned. The mean area of Central Queensland farms was 220 hectares. Respondents were asked how favourable the weather had been for their farm production during the prior 12 months. Replies were placed on a five-point scale from very favourable to very unfavourable. Central Queensland had clearly suffered more from the weather (specifically drought) than the rest of the sample (Table 5.1).

Table 5.1. Reported weather conditions over 12 months prior to survey.

	Central Qld canegrowers	Total sample
Very favourable	0% of farms	22% of farms
Favourable	4	33
About average	4	16
Unfavourable	17	11
Very unfavourable	75	18

Interviewers asked respondents how their production levels compared with average years on a scale from well below average to well above average. Reported levels of farm output were notably lower in the Central Queensland canegrower sample than in the other areas studied, in parallel with the responses regarding weather conditions (Table 5.2).

Table 5.2. Reported farm output levels over previous 12 months.

	Central Qld canegrowers	Total sample
Well below average	67% of farms	17% of farms
Below average	17	24
About average	4	29
Above average	4	27
Well above average	8	3

Farm financial position is indicated by equity ratios (Table 5.3) and farm cash incomes (Table 5.4). Despite attempts in the survey to obtain representation of the range of experience, including the financial condition of farms, the sample drawn in Central Queensland appears to have been suffering more heavily than the other areas.

Table 5.3. Levels of equity held in farms.

Equity	Central Qld canegrowers	Total sample
91-100%	21% of farms	34% of farms
81-90	25	20
71-80	13	16
61-70	13	13
60% or less	28	17

Table 5.4. Farm cash incomes.

$	Central Qld canegrowers	Total sample
Less than -40,000	37% of farms	16% of farms
-39,999 to -20,000	13	9
-19,999 to 0	13	18
1 to 19,999	21	19
20,000 to 39,999	8	17
40,000 to 69,999	0	16
70,000 or higher	8	5

Coping strategies

A range of strategies was adopted to cope with financial difficulties. These are classified into the categories used by Stayner and Gow (1992) in Table 5.5, in terms of the number of farms on which they were reported. All respondents reported they had reduced their costs: 'belt tightening'. A large majority had also postponed capital purchases. The seeking of work off the farm was reported on 70% of all farms and on 79% of the cane farms sampled. The only substantial difference between canegrowers and the sample overall was the incidence of diversification into different farm outputs. The overall result, however, was affected substantially by the relatively high degree of diversification in the mixed-farming area of the Riverina. The other areas covered by the study tended to be dominated by single or related industries.

Table 5.5. Farm management coping strategies.

Strategy	Central Qld canegrowers	Total sample
Postpone capital purchases	75% of farms	84% of farms
Debt restructuring	42	50
Cost reduction	100	100
Output expansion	4	13
Production intensification	25	10
Change enterprise mix	4	3
Diversification	8	35
Intergenerational transfer	4	1
Buy or sell land	79	61
Retraining	17	8
Off-farm employment	79	70
Off-farm investment	4	4
Share capital expenditure	0	1
Obtain government help	4	2

Respondents were asked to recall difficult events and discuss their personal reaction to them. Almost two-thirds of all the respondents were able to recollect at least one especially worrisome event. The most frequently mentioned were news of low prices and low product yields, being mentioned as often as each other among the canegrower sample. Across the whole sample, low prices was the most frequently mentioned problem, totalling 56% of all responses.

Questions were then asked about reactions to such news. Again it was possible for respondents to mention up to three. About one-third of respondents from the whole sample (about one-quarter of the canegrowers) said that they did nothing (Gray *et al.,* 1993). Of the other responses offered, more than half constituted direct resolution-action oriented approaches, such as cutting expenses, improving management and seeking employment, but the remainder included talking it over with family and friends (which was not, however, mentioned at all in the canegrower sample) and in a few cases, attending meetings and working towards a community response by political activity. Emotion-focused responses constituted 8% of the total. These included hoping for better times, trying to forget the bad news and, simply, 'worrying'. There were no significant differences in responses between the canegrower sample and the others, except the point mentioned above, that talking with family and friends was not mentioned as a personal coping strategy.

Plans for the future of the farm

The interviews sought information about plans for the farm including how long respondents planned to stay on it. Where there was more than one respondent on a farm, respondents were asked jointly. Responses were recorded on a five-point scale from 'indefinitely' to 'only until we can sell'. A rather higher proportion of canegrowers than that of the total sample reported that they planned to stay only until they could sell (Table 5.6). This may reflect a sense of despair, but it could also be a result of 'pull factors' such as the price of land in some instances.

Table 5.6. Planned length of stay on farm.

	Central Qld canegrowers	Total sample
Indefinitely	46% of farms	43% of farms
Until retirement	33	31
Until children have grown up	4	12
For a short time while we look elsewhere	0	7
Only until we can sell	17	7

Respondents were equally divided between those who expected their child or children to remain in farming and those who were at least doubtful. Of 58 respondents who had children, 50% said that they expected one or more children to continue farming, but 32% said that did not expect a child to succeed them and 18% were undecided.

Personal stress

The survey measured levels of personal stress by applying a 25 item scale which asked for the frequency, on a five-point scale from 'never' to 'very often', of events and feelings which are related to stress. The scale was adapted from those used by Weigel and Weigel (1987) and Walker and Walker (1987) with some farm-specific items added. The scale divides into two sub-scales, one measuring unforseen events and inconvenient problems (hassles) which could occur at any time on any farm (11 items yielding an alpha coefficient of 0.81) and the other sub-scale measuring symptoms of stress which may be related to financial difficulty (14 items yielding an alpha coefficient of 0.87). This scale approaches stress as impairment to the social life of the farm and family. It does not measure clinical symptoms, such as headaches and sleeplessness, but rather follows the definition of Hedlund and Berkowitz (1979, p. 235) in which 'stress responses are manifested in terms of interpersonal relationships among family members, inability to meet family goals, or impaired ability to meet family goals or impaired ability to perform necessary tasks'. The canegrower family members, while on average scoring a little higher on the stress symptoms scale, do not differ substantially from the sample average on either scale (Table 5.7).

Table 5.7. Mean scores on personal 'hassles' and stress symptoms, based on a rating scale of 1 (never) to 5 (very often).

	Central Qld canegrowers	Total sample
'Hassles'	2.00	2.03
Stress symptoms	2.31	2.26

This does not mean, however, that stress is insignificant. There were many instances of people in the sample of canegrower families indicating a high degree of stress. Items from the stress symptoms scale illustrate the reported extent of this problem: 18% said that in the last few months they had often had difficulty balancing work and family responsibilities; 40% said that they often worried about keeping the farm in the family; 22% had often had some conflict with their spouse over spending priorities; 49% had often felt that they had too much to do in too little time; 52% had often worried about owing money; 57% had often worried about the continued financial viability of the farm; 8% had often found relations with their spouse to have worsened; and 3% reported that they had often felt that they wanted to end their life. Some felt that they were losing control over their lives: 10% agreed with the statement 'I have little control over the things that happen to me'; 8% agreed with the statement 'all in all I am inclined to think that I am a failure', indicating a loss of self-esteem.

The canegrower sample is too small on its own to derive explanatory and predictive factors for stress. Analysis of the whole sample has shown that financial condition, strength of perception of what is at stake at a time of crisis, the combination of on-farm and off-farm obligations and the general frustrations of farm life are found to predict stress among both men and women (see Gray and Lawrence, 1996). Gender role conflict arising from increasing and conflicting demands and their degree of attachment to farm life are stronger predictors for women than for men. Farm women suffer self-questioning about their multiple roles as mothers, farmers and often also as employees off the farm (Gray *et al.*, 1993). They may find greater personal conflict, as they see their homes and farms as good places for children to grow up, but dangerous places for children without adequate supervision. Relative youth is a strong predictor of stress for men. This may be a result of pressure on farmers at the stage of their life cycle in which they are attempting to expand their farm to provide for a growing family. Farm men still confront expectations of themselves as farmers whose hard work can support the farm family unit.

The taking of work off the farm appears as a significant factor bringing strains to family relationships which could push them to breaking point. It can precipitate many personal conflicts for an adult within the structure of the farm family. It is ironic that attempts to develop and preserve the family farm by seeking income elsewhere may be creating stressful situations which threaten the family relations upon which the farming system is based.

The social limits to family farm-based canegrowing will be approached when the practices of succession and inheritance cease. While it is difficult to imagine a situation in which families cease to operate sugar farms, there are forces at work at both family and community levels which are creating a downward spiral. The necessity for some families to obtain off-farm income may be precipitating marital disruption which results in farm break-up and the end of family succession. Succession and inheritance have also been found to be related to the strength of local community ties which are themselves dependent on succession and inheritance (Gray and Phillips, 1996). Those families active in community networks are more likely to have children take over the farm, but as that practice declines, so do the community ties which are dependent on it.

Constraints upon Change in Australian Agriculture

The structural conditions relating to the state of the agricultural economy are very important in constraining on-farm change in agriculture. Yet - overlooked by those economists and bureaucrats who believe farmers should somehow simply 'obey' price signals and, where they are unable to become more productive or efficient, exit from agriculture - are the cultural or social aspects such as those discussed in the case study above, which must be viewed as being of equal significance.

Farmers live and work within a complex social world, one in which their particular values, attitudes, preferences and concerns are made sense by reference to what is considered important, desirable and acceptable within a community. Patterns of social interaction are based on social structures and power relations which can, with detailed research, be understood by social scientists (Gray, 1992). Rather than speaking, for example, about the constraints financial capital has on farmer options and actions, behaviour might be best understood as relating to cultural capital or symbolic capital. A 'good' farmer - in cultural capital terms - may not be financially well endowed, but may have the respect of other farmers within the district. The farmer and her/his predecessors may have successfully run a property for several generations, come through a period of economic downturn without visible signs of 'stress', or may have local knowledge which ensures that production decisions are made in an environmentally sensitive manner (Phillips and Gray, 1995). Once others recognise and give social prestige to a 'good' farmer, the cultural capital which that person possesses will be an important determinant in what is considered acceptable farm practice in the district.

In much previous rural sociological and agricultural extension-based research, it was assumed that all that was necessary in understanding farming social systems and driving farmers towards more desirable production-based 'ends' was to concentrate upon those producers identified as the innovators. In the adoption/diffusion literature, while some farmers would be innovators, many would not be prepared immediately to adopt new technology; still others would show extreme caution or reluctance (the so-called laggards). The role of the extension agent would be to work with the innovators to help 'lead' others towards new and desirable approaches to farming. Innovation would 'trickle down' over time. Extension agents would seek to identify 'barriers to adoption' so that the trickle down process might be enhanced (Buttel *et al.*, 1990).

The adoption/diffusion model has been subjected to a sustained critique by sociologists (Buttel *et al.*, 1990; Vanclay and Lawrence, 1995). One inadequacy relates to the faulty notion that new innovations represent 'improvements' to existing systems but the question which is begged is: 'Based on whose judgement?' Other problems with the model are that the impacts of technological change have been poorly considered; local knowledge is often subordinated and trivialised in the face of 'technology'; and there is an implied rationality in adoption - that it is unambiguously in the interests of the farmer and society to use the latest technologies and management regimes. In terms of the new agenda of sustainable agriculture there is another, more devastating criticism. The adoption/diffusion model so beloved of traditional extension does not appear to have applicability to the uptake of conservation innovations. That is, a model based on promoting technical solutions to on-farm problems does not appear capable of promoting - in a more holistic manner - agricultural sustainability. It has been shown that different motives and different circumstances drive farmers to adopt conservation technologies as distinct from production technologies (Pampel and van Es, 1977; and see Buttel *et al.*, 1990). The adoption/diffusion model has little efficacy in the development of more sustainable options for agriculture. Indeed, to continue to apply it is, ultimately, to promote the very strategies which have led to the environmental degradation which agricultural extension would seek to ameliorate! For this reason alone, it is necessary to abandon the older 'technology transfer' models and move toward more recent participatory approaches (Frank and Chamala, 1992). To ensure that they serve the interests of farmers, participatory extension methods must be based on an understanding of the social structures, characteristics and processes which underpin farming culture.

Social Factors Limiting Change in Farming Systems

Sociologists have sought to determine what demographic and other social/cultural factors characterise rural production. Education is one such 'variable'. Participation rates of Australian farmers in post-secondary education appear to be much lower than those of producers in comparable overseas nations such as New Zealand and Britain (Woodford and Collins, 1993). In an increasingly complex world, the ability of those with minimal education to apply some of the recent information technologies - the very things which might offer insights into new approaches and management regimes - would appear to be limited. This is particularly so in relation to environmental innovations which require a sophisticated, holistic understanding of a great many relationships (agronomic, economic, on-farm, downstream, etc.). Yet, difficulties understanding these relationships will limit the extent to which producers accept (and adopt) such innovations (Vanclay and Lawrence, 1995).

Women appear to be of great importance to the future of Australian agriculture. Their number in agriculture doubled in the past 15 years (*Australian*, 1995) and their on-farm and off-farm roles have increased both their economic status and, it would seem, their influence in farm decision-making. As farms become increasingly dependent upon off-farm earnings, and much of the off-farm earnings are obtained by women, so it would seem that women have the potential further to increase their influence in farm management decisions. Given that women tend to have higher levels of formal education than men (Gooday, 1995) and to hold less conservative views (*Australian*, 1995), new opportunities and options might arise from their further integration into farming, farm organisations and agro-politics (Alston, 1995). To the extent that women are not as involved as they might be in cane production and cane organisations, their potential to alter the trajectory in that industry must be viewed, consequently, as being limited.

Age is also of importance. The research of Gooday (1995) has suggested that average farm age in broadacre agriculture is increasing (from 50 in 1978-79 to 53 in 1993-94) and it might be hypothesised that this is occurring in the canegrowing industry. Many older producers have fixed ideas about what constitutes appropriate farming methods. They may dismiss new trends and tendencies (such as the drive toward sustainability) as being simply a fashionable, city-based phenomenon which is not deserving of their support. In any case, many possess what has been called 'agrarianism' (Halpin and Martin, 1996) or 'countrymindedness' (Share, 1995). These ideologies highlight the virtues of rural living and position the farmer as the real 'environmentalists', those who care most about the land. In construing environmentalists and 'greenies' to be little more than urban-based radicals, the action of producers in joining Landcare may be a defence of their position as rural decision-makers, rather than a genuine attempt to seek measures to overcome environmental degradation (Morrisey and Lawrence, 1995). Some anecdotal information indicates that some older producers - although controlling vast tracts of the Australian landscape - are in something of a 'holding pattern'. They do not see that they will capture the benefits of any new investments in the farm, are somewhat suspicious about government-supported environmental programs, have children who may not be interested in taking over the farm, and yet feel that they do not want to retire. There is evidence that farmers who have successors willing to take over the farm are more likely to adopt conservation practices (Gray and Phillips, 1996).

There is another possible interpretation in regard to supposed inaction relating to environmental damage. Farmers may not at all be convinced that it is their own farming practices which may be causing off-farm environmental degradation. They tend to espouse views that they are nurturers of the soil and that, if any 'problems' have occurred, these are due to the vagaries of nature - not to the characteristics of traditional farming practices. Furthermore, there may be very little 'proof' that it is their own land-use techniques which have contributed to silting in the local river system or their own chemical applications which have caused algal blooms kilometres away from the property. They are told, on the one hand, that chemicals are essential for increased productivity and, on the other, that they must seek to reduce their chemical applications. Whose 'knowledge' will they trust? And, where agricultural extension is becoming increasingly the province of commercial firms (P. Wylie, unpublished data, 1992), they may not have any 'alternative' position to consider. Finally, as Vanclay (1992) has pointed out, what constitutes environmental degradation (presented in the media as barren salt-scalded paddocks, dried up dams, and the bones of dead cattle) may not be what constitutes their own 'reality'. It is also important to recognise that even in situations where farmers do understand and acknowledge that there are problems occurring on their own land they might consider that the risk of further degradation fails to outweigh the risk of what might be expensive remedial action.

'Farming Styles' Research: A New Approach in Understanding Social Factors in Agriculture?

Sociological work has identified what are described as 'farming styles' (van der Ploeg, 1993). It is argued that because of the heterogeneity of rural production it is impossible to generalise about 'the farmer' and his or her motivation (de Bruin and Roex, 1994). While coming close to the 'labelling' criticised in the adoption/diffusion approach, and being based on the yet-to-be demonstrated notion that attitudes directly inform practices, the use of styles makes it possible to suggest the ways in which lifestyle choices influence on-farm decision-making to produce certain observable farming outcomes. Thus, although some farming style taxonomies tend to highlight orientation to the market as the most significant distinguishing factor among farmers, it is possible to incorporate cultural orientation.

Recent sociological research among dairy farmers in The Netherlands, for example, has revealed that there are four different sorts of farmers in the industry (de Bruin and Roex, 1994). The types identified vary in terms of the degree to which they took a business-like approach, were reliant on farm income and their approach to technological developments. The authors' basic conclusion is that while the business-like farms are considered to be the 'vanguards' for a more intensive and scientific agriculture, the strategies they adopt do not necessarily assist in the maintenance of the natural environment. In New Zealand/Aoteoroa, Fairweather and Keating (1994) identified three management styles among farmers in the Canterbury region of the South Island. The researchers established that there was significant variation in the degree to which producers were dedicated to farming, their attitude to diversification and their interest shown in resolving environmental problems. The work of Mesiti and Vanclay (1996) brings this type of analysis closer to home. Focus groups and unstructured interviews among Sunraysia producers identified six types of farming styles. These were similar to those discussed above with the addition of large corporate farmers and those who sought a rural lifestyle but showed little interest in farming. This approach too offers opportunities to consider the environmental implications of the particular farming styles which were identified.

While the farming styles approach offers only a static model, it may be possible to consider change in farm practices which will arise in association with broader patterns of change in farm populations and the values which they hold. This would, however, require a much deeper understanding of the cultural and structural foundations of farming styles. While being aware that certain biases are present in any 'labelling' of farming styles and that the pictures which emerge may be a very simple and static representation of a complex reality, sociologists would argue that this approach does, at one and the same time, appreciate the heterogeneous nature of farming, the values which underpin particular approaches to farming, and the ways cultural factors inform specific behaviours.

Future Sociological Research Needs for the Sugar Industry

Considerable advantage might accrue from understanding the micro-sociological characteristics of sugar cane production in the 1990s. Such research would be based on an appreciation of the 'cultural capital' sought to be gained by farmers in their day-to-day interactions with peers, as well as upon the 'styles' of farming and their environmental implications, the motives and attitudes of producers and the links between producers and other stakeholders in natural resource management. It would also build upon local surveys - e.g. the Central Queensland case study - which explore the coping strategies of farmers under stress. Little is understood about social aspects of cane production because few rural sociologists are involved in farm-level social research in Australia; the reasons for this is discussed by Lawrence (1995). We need to move beyond conventional technology transfer models which fail to appreciate that it may be appropriate and rational for farmers to reject the very technologies and practices which are being 'sold' to them (Frank and Chamala, 1992; Vanclay and Lawrence, 1995).

The Rural Social and Economic Research Centre of Central Queensland University has begun studies of the sugar industry in Australia, currently in relation to structural change in the sugar industry in Bundaberg and community change in the Pioneer Valley, Mackay. Three other projects are envisaged. One, a social 'audit' of the sugar industry, is a necessary first step to understand the social characteristics (including the attitudes and opinions) of those in the industry. Another, the examination of producer/stakeholder attitudes to the extension of the cane harvest season, will provide the industry with important information about grower needs and preferences in relation to anticipated change. A third will identify attitudinal and behavioural profiles of producers relating to on-farm water and nutrient management. This information will help to identify the issues of concern *from the growers' point of view*, and will therefore allow those concerns and interests to be built into future research programs. The survey will allow differences between districts to be identified. Analysis will produce a sophisticated knowledge of the influence of age and education on farming practice, gender differences in farm decision-making and the influence of such things as size of property and location on attitudes to sustainable practices. The social audit will allow identification of certain industry problems and concerns (e.g. inappropriate attitudes to chemicals, lack of knowledge about sustainable options). Follow up work which targets particular groups and/or locations might then be taken up, in subsequent activities, in an effort to move the industry toward a more sustainable future. While the social audit will reveal views on such things as extended season length it will be necessary, given the likely community impacts of changes within the industry, to undertake wider community consultation. Again, sociologists should take a leading role in such activity.

Conclusion

Agricultural restructuring is occurring in all Australian rural industries. The sugar industry has not been exempted. Sociological research has indicated the extent to which external (structural) pressures for 'adjustment' have affected those within the industry. The Central Queensland study revealed that cane producers were coping with change in a number of ways - by reducing costs, postponing capital purchases, and by taking off-farm work. For many, they were remaining in the industry only until the opportunity came for them to sell. Others suffered stress which has significant impacts. Many were disillusioned with what was occurring in sugar growing, and evidence of a 'downward spiral' emerged among both farming families and the communities which were dependent upon cane production. In this context, it was argued that the structural conditions affecting sugar producers might militate against attempts to move the industry toward more sustainable options.

Rather than looking at so-called 'barriers to adoption' *within* production (as if there were some unambiguous future to which all producers should inevitably head) it was suggested that cognisance be taken of the reasons why it might not be 'rational' for producers to accept new ideas and to introduce new practices. Among the considerations were the age and education of producers, the complexity of some of the new - sustainable - options, the source of information, their own perceptions of the state of their properties, the degree to which they believed that change would be personally beneficial, and the socio-dynamics of farmer-to-farmer interaction. In relation to the latter, it was suggested that more research was necessary to understand how 'culture' related to 'structure' in the sugar industry. An understanding of patterns of social interaction is required, highlighting not only community power relations, but also the extent to which some farmers are imbued with 'cultural capital'. The various farming 'styles' and their relation with environmentally sound practice might provide a useful start to a more thoroughgoing sociological analysis of contemporary sugar production.

Acknowledgment

Support from the Rural Industries Research and Development Corporation in the funding of the study on coping strategies, some results from which were detailed in the first section of this paper, is gratefully acknowledged.

References

Alston, M. (1995) *Women on the Land: the Hidden Heart of Rural Australia*, University of NSW Press, Kensington, Australia, 160pp.

Australian, 13 December 1995: 7.

Bonanno, A., Busch, L., Friedland, W., Gouveia, L. and Mingione, E. (eds.) (1994) *From Columbus to ConAgra: The Globalization of Agriculture and Food.* University Press of Kansas, Lawrence, Kansas, USA, 294pp.

de Bruin, R. and Roex, J. (1994) Farmer initiatives as countervailing power: a new approach to sustainable agricultural development. In: Symes, D. and Jansen, A. (eds), *Agricultural Restructuring and Rural Change in Europe.* Agricultural University, Wageningen, The Netherlands, pp. 195-208.

Buttel, F., Larson, O. and Gillespie, G. (1990) *The Sociology of Agriculture.* Greenwood, Westport, 263pp.

Fairweather, J. and Keating, N. (1994) Goals and management styles of New Zealand farmers. *Agricultural Systems* 44, 181-200.

Frank, B. and Chamala, S. (1992) Effectiveness of extension strategies. In: Lawrence, G., Vanclay, F. and Furze, B. (eds), *Agriculture, Environment and Society.* Macmillan, Melbourne, Australia, pp. 122-140.

Gooday, J. (1995) Women in agriculture: broadacre and dairy farms 1993-94. In: ABARE, *Farm Surveys Report.* ABARE, Canberra, Australia, pp.70-75.

Gray, I. (1992) Power relations in rural communities: implications for environmental management. In: Lawrence, G., Vanclay, F. and Furze, B. (eds), *Agriculture, Environment and Society: Contemporary Issues for Australia.* Macmillan, Melbourne, Australia, pp. 141- 156.

Gray, I. and Lawrence, G. (1996) Predictors of stress among Australian farmers. *Australian Journal of Social Issues* 31, 173-189.

Gray, I., Lawrence, G. and Dunn, A. (1993) *Coping with Change: Australian Farmers in the 1990s.* Centre for Rural Social Research, Charles Sturt University, Wagga Wagga, Australia, 144pp.

Gray, I. and Phillips, E. (1996) Sustainability and the restructuring of tradition: analysis of farm practice and social change in three rural localities. In: Lawrence, G., Lyons, K. and Momtaz, S. (eds), *Social Change in Rural Australia: Perspectives from the Social Sciences.* Rural Social and Economic Research Centre, Central Queensland University, Rockhampton, Australia, pp. 276-289.

Halpin, D. and Martin, P. (1996) Agrarianism and farmer representation: ideology in rural Australia. In: Lawrence, G., Lyons, K. and Momtaz, S. (eds), *Social Change in Rural Australia*, Rural Social and Economic Research Centre, Central Queensland University, Rockhampton, Australia, pp. 9-24.

Hedlund, D. and Berkowitz, A. (1979) The incidence of social psychological stress in farm families. *International Journal of the Sociology of the Family* 9, 233-243.

Lawrence, G. (1994) Rural adjustment revisited: in defence of a sociological approach. *Rural Society* 4, 11-16.

Lawrence, G. (1995) *Futures for Rural Australia: From Agricultural Productivism to Community Sustainability.* Rural Social and Economic Research Centre, Central Queensland University, Rockhampton, Australia, 33pp.

Mesiti, L. and Vanclay, F. (1996) Farming styles among grape growers of the Sunraysia district. In: Lawrence, G., Lyons, K. and Momtaz, S. (eds), *Social Change in Rural Australia*, Rural Social and Economic Research Centre, Central Queensland University, Rockhampton, Australia, pp. 55-63.

Morrisey, P. and Lawrence, G. (1995) A critical assessment of landcare in a region of central Queensland. In: Vanclay, F. (ed.), *With a Rural Focus*, Centre for Rural Social Research, Charles Sturt University, Wagga Wagga, Australia, pp. 63-75.

Pampel, F. and van Es, J. (1977) Environmental quality and issues of adoption research. *Rural Sociology* 42, 57-71.

Phillips, E. and Gray, I. (1995) Farming practice as temporally and spatially situated intersections of biography, culture and social structure. *Australian Geographer* 26, 127-132.

Ploeg, J. van der (1993) Rural sociology and the new agrarian question: a perspective from the Netherlands. *Sociologia Ruralis* 33, 240-260.

Share, P. (ed.) (1995) *Communication and Culture in Rural Areas*, Centre for Rural Social Research, Charles Sturt University, Wagga Wagga, Australia, 303pp.

Share, P., Campbell, H. and Lawrence, G. (1991) The vertical and horizontal restructuring of rural regions: Australia and New Zealand. In: Alston, M. (ed.), *Family Farming: Australia and New Zealand*. Centre for Rural Social Research, Charles Sturt University, Wagga Wagga, Australia, pp. 1-23.

Stayner, R and Gow, J. (1992) *A Review of the Process of Farm Adjustment in Australia.* Rural Development Centre, University of New England, Armidale, Australia, 59pp.

Vanclay, F. (1992) The social context of farmers' adoption of environmentally sound farming practices. In: Lawrence, G., Vanclay, F. and Furze, B. (eds), *Agriculture, Environment and Society: Contemporary Issues for Australia.* Macmillan, Melbourne, Australia, pp. 94-121.

Vanclay, F. and Lawrence, G. (1995) *The Environmental Imperative: Eco-social Concerns for Australian Agriculture.* Central Queensland University Press, Rockhampton, Australia, 204pp.

Walker, L. and Walker, J. (1987) Stressors and symptoms predictive of distress in farmers. *Family Relations* 36, 374-378.

Weigel, R. and Weigel, D. (1987) Identifying stressors and coping strategies in two-generation farm families. *Family Relations* 36, 379-384.

Woodford, K. and Collins, R. (1993) Farm management education for the future. In: Boylan, C. and Alston, M. (eds), *Rural Education Issues: An Australian Perspective.* Keypapers No. 3, Centre for Rural Social Research, CSU, Wagga Wagga, Australia, pp. 91-110.

Chapter six:

Yield 'Plateaus' in Grain Crops: The Topography of Yield Increase

T.R. Sinclair

USDA-ARS, Agronomy Department, University of Florida, P.O. Box 110840, Gainesville, FL 32611-0840, USA

Abstract

An analogy to the topography of a mountain range is used to explore the various levels of crop yield. The peaks in crop yield are set by biophysical factors. That is, crop yields are ultimately constrained by the amount of light energy that is available for use by plants to assimilate carbon. The total amount of intercepted light energy depends on the solar radiation climate, the development and persistence of crop leaf area, and the length of the growing season. Although theoretical peak yields have been achieved experimentally, most crop yields are at lower levels constrained by the availability of resources. The resource plateaus for yield are set in many cases by the availability of either water or minerals, especially nitrogen. In industrialised countries, it appears that indeed yield plateaus have developed in the past ten years that are closely associated with fertiliser application plateaus. Nitrogen, in particular, has a dramatic influence on crop yields and nitrogen application rates parallel a levelling in crop yields. While it is possible for growers to raise the fertiliser plateau, socioeconomic factors seem to have become important in establishing this plateau. Consequently, socioeconomic plateaus, which are based on economic decisions to maximise profits rather than maximise yields, have become a new, important reality in levelling of crop yields. Also, societal pressures resulting from environmental and food safety concerns encourage crop production at less than maximum levels. Peak, or maximum, crop yields do not appear to be a realistic future objective for many cropping systems in view of these emerging socioeconomic plateaus.

In: Keating, B.A. and Wilson, J.R. (eds), *Intensive Sugarcane Production: Meeting the Challenges Beyond 2000*. CAB International, Wallingford, UK, 1997.

Introduction

Grain crop yields during the twentieth century have steadily risen to achieve remarkable levels. The increases in yields have been especially great since World War II, and these increases have been generally associated with the introduction of chemical pesticides, widespread use of genetically improved cultivars, and availability of inexpensive chemical fertilisers. A critical question in planning the well-being of humankind's future is whether the dramatic increases in crop yields will continue into the twenty-first century.

The intent of this paper is to examine the 'topography' of grain crop yields. The yields of grain crops will be examined in a rough analogy to the topography of a mountain range. At what levels are the 'peaks' and 'plateaus' that might be used to describe grain yields of various crops? What determines various peaks and plateaus in crop yields?

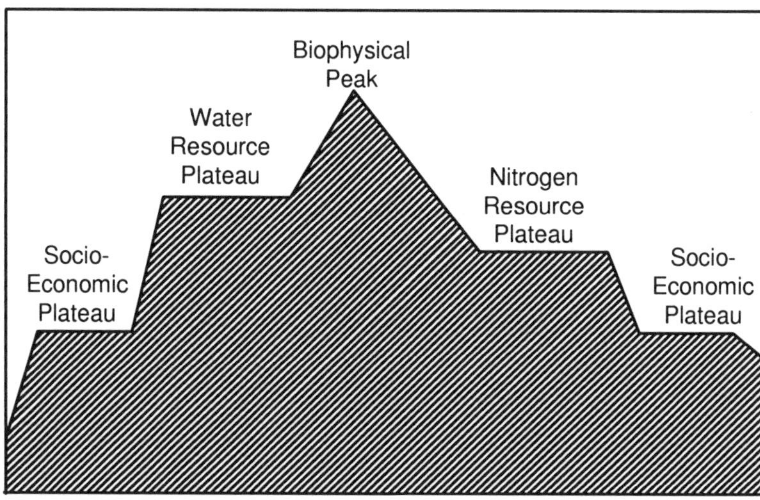

Fig. 6.1. Schematic drawing of the 'topography' mountain range analogy: peak, resource plateau, and socioeconomic plateau.

The peaks or maxima in crop yield can be identified by examining the biophysical factors required to produce grains (Fig. 6.1). The ultimate limitation is, of course, the amount of solar energy to fix carbon dioxide. Below the peak yields are plateaus associated with the availability of resources for crop production. The most important among these resources are water and minerals. At even lower yield levels, there appear to be important emerging plateaus related to socioeconomic factors that influence management decisions by growers. Each of these features of the topography of crop yield are discussed in this paper.

The Biophysical 'Peak' for Crop Yields

Considerable knowledge has been gained about the physiology of yield formation in crops, and it is now possible to quantitatively estimate limits to crop yield based on physiological and environmental constraints (Sinclair, 1993, 1994; Muchow *et al.*, 1997). An absolute peak in crop yields is imposed by the amount of light energy available to plants. This light energy, which is commonly referred to as solar radiation, is used to provide the energy required in the growth and maintenance of plant tissue. For annual crops, most of the captured solar radiation is used for new growth by assimilating carbon dioxide into plant mass.

Crop yield (Y, g m^{-2}) can be explicitly described by the following equation.

$$Y = HI \sum_{O}^{D} (I \times RUE) \qquad \text{Eq. (1)}$$

where: HI = harvest index (g g^{-1}),
D = duration of growing season (d),
I = daily intercepted solar radiation (MJ m^{-2} d^{-1}),
RUE = radiation use efficiency (g MJ^{-1}).

Maximising each term in Equation (1) allows peak crop yields to be calculated. Physical and physiological knowledge has progressed so that considerable understanding exists concerning the factors that influence each term in Equation (1). Starting from the right with RUE, experimental and theoretical evidence have converged to describe the amount of solar radiation required in the accumulation of crop mass. Experimentally, RUE can be measured for a given time period as the accumulated crop mass (g m^{-2}) divided by the amount of solar radiation intercepted by the crop (MJ m^{-2}). The value of RUE for non-stressed C_3 wheat and rice is about 1.4g MJ^{-1}, and for non stressed maize which has the C_4 system for initial carbon dioxide assimilation (as does sugarcane), RUE is about 1.75g MJ^{-1} (Kiniry *et al.*, 1989). These experimental values are consistent with RUE derived from theoretical considerations of photosynthetic capabilities and biochemical constituents of crops (Sinclair and Horie, 1989). Little opportunity appears to exist to increase RUE above those values currently achieved experimentally.

As expressed in Equation (1), the seasonal accumulation of crop mass results from the summation of the daily product of RUE and the amount of solar radiation that has been intercepted by the crop (I). Consequently, the daily fraction of the incident solar radiation that is intercepted by the crop through the season is a key variant. The fraction of solar radiation intercepted by a crop is determined by several factors related to crop leaf area. For annual plants commonly used in grain production, the leaf area development rate in the initial stage of crop development determines the interception of solar radiation in the early stage of crop growth.

Narrow row-spacing and increased plant density have been nearly fully exploited as methods to maximise early season leaf area in major grain crops. Freezing and low temperatures, however, can be especially important in inhibiting leaf development of annual crops early in the season.

The amount of solar radiation intercepted in the remainder of the growing season depends on leaf persistence. The persistence of leaf area is usually associated with the duration of the cropping season. Full season crops have been developed for temperate climates to maximise the duration of light interception. Cool and freezing temperatures late in the season, however, can abruptly terminate the physiological activity of leaves. Long growing seasons increase the opportunities for solar radiation interception and increase the total accumulated crop mass.

Finally, grain yield is obtained by multiplying the seasonal accumulated crop mass by the harvest index, the ratio of grain mass to total plant mass. Most crops that have been subjected to intensive breeding have been selected for increased harvest index resulting in shorter plants with high grain yields. Harvest indices of 0.20 and 0.25 that existed 50 years ago have been increased to 0.45 or greater. Austin (1994) concluded that the maximum harvest index of cereals was likely to be 0.62, which again defines a limit in setting the peak for crop yield.

The above approach has been used to calculate peak yields of various crops in different environments. For example, daily temperature and solar radiation data have been used to calculate the maximum yields anticipated in specific locations for soybean (Spaeth *et al.*, 1987), wheat (Amir and Sinclair, 1991), and maize (Muchow *et al.*, 1990). Not surprisingly, all three analyses concluded that peak yields were highly sensitive to climatic conditions. The highest yields were achieved as a result of cool temperatures, which extend the length of the growing season, and high solar radiation levels. No single yield estimate could be used to describe a global maximum yield based on the physiology of the crop, but rather, peak yield had to be assessed for each crop and each weather regime.

In maize and wheat, we have specifically searched for the 'Mt Everest' of yields. The highest recorded maize yield that could be substantiated was 17.3t ha^{-1}, which was achieved in 1982 at Grand Junction, CO, USA (Muchow *et al.*, 1990). In wheat, the highest yield was 15.2t ha^{-1} at Xiangride, in northwest China in 1978 (Sinclair and Bai, 1996). Both locations are at high elevations resulting in cool temperatures and the crop growing seasons were especially long. Also, these locations were nearly cloud-free and irrigation was essential for crop growth. Solar radiation levels were very high and the solar radiation was intercepted by the crops over long periods.

Importantly, in each situation the theoretical maximum yields, based on intercepted solar radiation and a high HI, resulted in yield estimates that matched the observed yields. That is, these theoretical, biophysical peaks in crop yields were achieved under optimum field conditions more than 10 years ago! Amir and Sinclair (1994) concluded, in fact, that the physiological components to achieve these peak yields may have existed in ancient wheat plants used for cropping more than 2000 years ago. Why then are not all crops now yielding at peak levels?

The 'Resource Plateaus' for Crop Yield

Not unlike the 'tree line' in the mountain range analogy, there are resource limitations under many circumstances that do not allow crop production at the peak level. Two resources that are especially critical under almost all cropping situations are water and mineral nutrients, especially nitrogen.

Water

Water deficits are one of the most important recurring abiotic stresses in crop production. The importance of water in plant growth is a result of the inescapable laws governing gas diffusion in and out of leaves. When stomata are open to allow carbon dioxide diffusion from the atmosphere into leaves for photosynthetic assimilation (*i.e.* nonstressed levels of RUE), transpiration also occurs as water vapour diffuses from the leaf interior to the atmosphere. The factors influencing the ratio of carbon assimilation and transpiration for crop canopies are well-defined theoretically (Sinclair *et al.*, 1984).

The close linkage between carbon assimilation and transpiration means that the limit for accumulated crop mass and yields are tied to the amount of water available for transpiration. That is, accumulated crop mass is constrained by the amount of water available to support carbon assimilation. In situations where water becomes limiting and stomata are closed, then accumulation of crop mass is necessarily decreased. A yield 'plateau' exists in this case because the amount of available water constrains yield. Under conditions where water deficits are severe, the drought can threaten crop survival and the yield plateau can approach zero. Irrigation is a solution to water deficit conditions that has been employed since nearly the beginning of crop cultivation. With irrigation, the yield plateau attributed to a limiting water resource is eliminated.

Nitrogen

A critical factor that is quantitatively linked to the remarkable history of crop yield increase is the increase in N supplied to crops. McCloud (1997) examined the historical development of various agricultural practices and crop yield. Each improvement in grain crop production technology, from the use of 'slash and burn' management through the 'Green Revolution' of the 1960s and 70s, was directly linked to increasing N availability to crops. Borlaug (1972) wrote that '*chemical fertiliser is the fuel that has powered [the Green Revolution's] forward thrust.*'

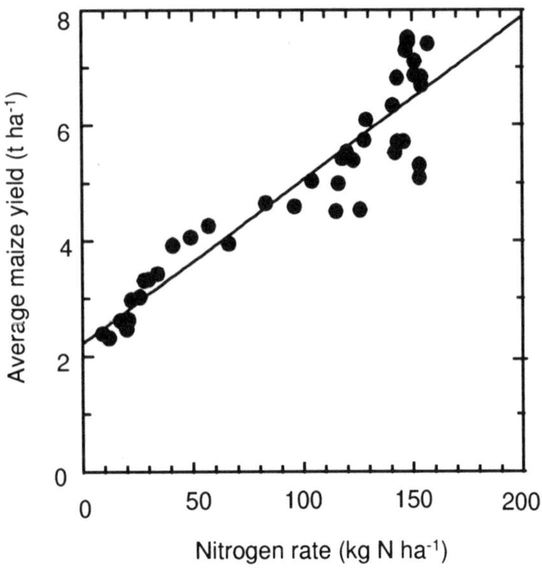

Fig. 6.2. Plot of annual average US maize yields versus the annual average amount of nitrogen fertiliser applied from 1945 to 1993.

The importance of the association between N supply and crop yield is illustrated by examining annual USA maize yields in relation to the average amounts of N fertiliser applied (Fig. 6.2). There was a tremendous yield increase from about 2t ha^{-1} to nearly 8t ha^{-1}. Similarly, the N applied to maize increased from nearly zero to almost 160kg N ha^{-1}. This graph shows a nearly linear relationship (r^2 = 0.86) between maize yields and applied N.

The critical importance of N fertility in increasing grain yields arises since each of the variables in Equation (1) is sensitive to N supply. Both theoretically and experimentally, RUE has been shown to depend on leaf N levels. The photosynthetic rates of individual leaves are closely correlated with the leaf N content expressed per unit area. Sinclair and Horie (1989) calculated curvilinear responses of RUE to leaf N content (Fig. 6.3).

The response of RUE to leaf N content differs among crop species (Fig. 6.3). First, the intersect on the leaf N axis varies among species with the lowest value for maize and then rice, and the highest for soybean. Second, the rate of increase in RUE with higher leaf N also varies among species. Maize has the greatest increase in RUE per increment of leaf N, and soybean the least. Such a response is consistent with the marked responsiveness of maize crops to the application of N fertiliser. Finally, the curvilinear response of RUE to leaf N approaches a maximum or saturated RUE that varies among species. The saturated RUE values are consistent with differences among species in RUE discussed previously for unstressed crops.

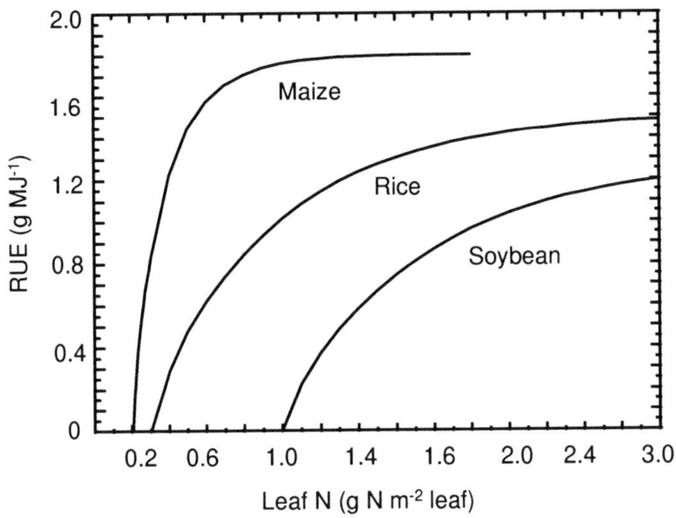

Fig. 6.3. Theoretical relationship between radiation use efficiency (RUE) and leaf nitrogen content per unit leaf area for maize, rice, and soybean. [Sinclair and Horie, 1989]

Experimental evidence is now accumulating to confirm the close linkage between RUE and leaf N content, although discrepancies with the exact relationship predicted by Sinclair and Horie (1989) have been reported. Curvilinear relationships between RUE and leaf N content have been observed in peanut (Wright *et al.*, 1993), soybean (Sinclair and Shiraiwa, 1993), maize and sorghum (Muchow and Sinclair, 1994), and sunflower (Hall *et al.*, 1995).

Leaf area development is also directly dependent on N availability. Inadequate N results in smaller leaves because N, as an essential constituent of proteins and nucleic acids, is required at a minimum concentration in the development of new cells and leaf area (Sinclair, 1990). The decrease in leaf area with decreased N availability has been shown in several crops including maize (Muchow, 1988) and wheat (Pearman *et al.*, 1977).

Finally, the harvest of grain is itself a harvest of N. In the grain of the three major grain crops, the N composition is roughly 13mg N g^{-1} in rice, 16mg N g^{-1} in maize, and 22mg N g^{-1} in wheat. A maize harvest of 8t ha^{-1}, for example, represents a substantial N harvest of about 130kg N ha^{-1}. Assuming about 40% of applied N fertiliser is lost and unrecoverable by the crop, an application of over 210kg N ha^{-1} is needed simply to meet the current cropping requirements.

Not surprisingly, many experiments have shown a close relationship between increasing amounts of applied N and increasing grain yields. Fig. 6.4 presents a typical response of yield to applied N. These data were for maize crops grown in Katherine and Lawes, Australia (Muchow and Sinclair, 1995). Importantly, for many growers there is a large increase in yield at the initial, low levels of N application. At these low levels, there is substantial opportunity to greatly increase crop yields by removing the N resource plateau with the application of fertiliser N. As the levels of N application increase, however, the yield increases are diminished. Ultimately, the yield response to applied N is saturated, which in Fig. 6.4 occurred above about 240kg ha^{-1} nitrogen application.

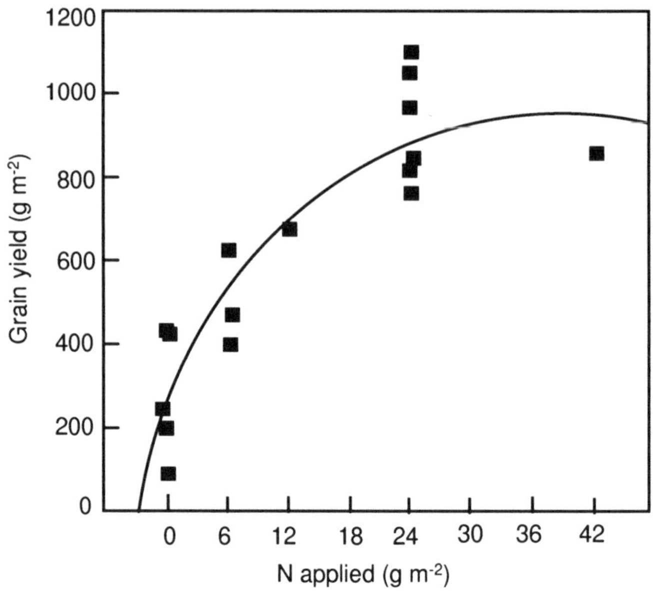

Fig. 6.4. Response of maize yield to N fertiliser rate. [Muchow and Sinclair, 1995]

An essential ingredient for the increase in grain yield in response to increasing N fertilisation has been the development of plants that are able to respond to increased availability of N. The Green Revolution, for example, was intimately associated with the selection of N-responsive plants (Chandler, 1969). Larger reproductive structures were selected to accommodate more grain. The stems of modern plants are shorter and stiffer than the stems of plants grown 50 years ago, so that the new cultivars can physically support the heavier grain load. These changes in plant morphology are often summarised as increases in harvest index resulting from plant selection. Importantly, the N harvest index (ratio of amount of N in the grain to that in the whole plant) has also increased dramatically in modern cultivars with values now over 0.9.

Interestingly, an important perspective on the role of plant breeding in increasing crop yields can be gained by studying yield increases in a crop that is not dependent on N fertilisation. Soybean, which obtains its N through symbiotic fixation of atmospheric N, has had a steady increase in yield similar to grain crops of 14 to 19kg ha^{-1} y^{-1} with the progression of variety releases (Boerma, 1979;

Specht and Williams, 1984). Salado-Navarro *et al.* (1993) found the same trend in increasing yields for soybean varieties released between 1945 and 1983 when the varieties were grown in the intended production area (southeastern USA). When the varieties were grown in a virgin area with a similar climate (Argentina) where there were no established populations of soybean pathogens, yields of all varieties were greater than those obtained in the USA. Most importantly, there was no trend in yield increase with year of variety release. The implication of this study was that soybean breeders had effectively selected against pathogens endemic in the southeastern USA to achieve a steady increase in soybean yield, but that the inherent physiological capability of the species (*i.e.* its peak yield) had not been altered. Similarly, in grain crops it is expected that along with the selection of plants that are responsive to increased N fertility, a considerable portion of the plant breeding contribution to increasing crop yields has been the selection of plants to overcome pathogens that can damage crops.

Growers and Resource Plateaus

Since the availability of resources can establish plateaus in crop yields, do these plateaus exist in the yields achieved by growers? An approach to answering this question is to examine yields achieved by growers over a number of years. Park and Sinclair (1993) used this approach in studying the mean yields achieved for maize in Champaign County, Illinois, from 1965 to 1988. Champaign County is in the heart of the US Corn Belt and has deep, rich soils.

The observed yields in Champaign County (Fig. 6.5) showed considerable variation over years. Most of the variation was associated with variability in seasonal precipitation. That is, the water resource restricted yields to a lower level during drier years. Water availability appeared to be an important constraint on yield in a number of years during this 24-year period. Examining only the wetter years, maize yields from Champaign County (Fig. 6.5) increased steadily over this 24-year period, although the rate of increase appears to have slowed in more recent years. This yield increase was likely associated with a steadily increased availability of a constraining resource. As indicated in Figs 6.2 and 6.4, increasing N fertilisation, along with new cultivars able to respond to greater N fertility, is associated with maize yield increases.

To gain some perspective on the changes in maize yields in Champaign County, an attempt was made to compare 'growers' yields to 'peak' yields expected based on the weather in each year. Park and Sinclair (1993) compared in each year observed yields against what could be expected based on peak yield calculations adjusted by the water resource. The model of Muchow and Sinclair (1991) was used to calculate maximum yield using an approach very similar to that represented by Equation (1). The model also incorporated the 'resource plateau' of water by incorporating sensitivities in crop development and growth to soil water deficits.

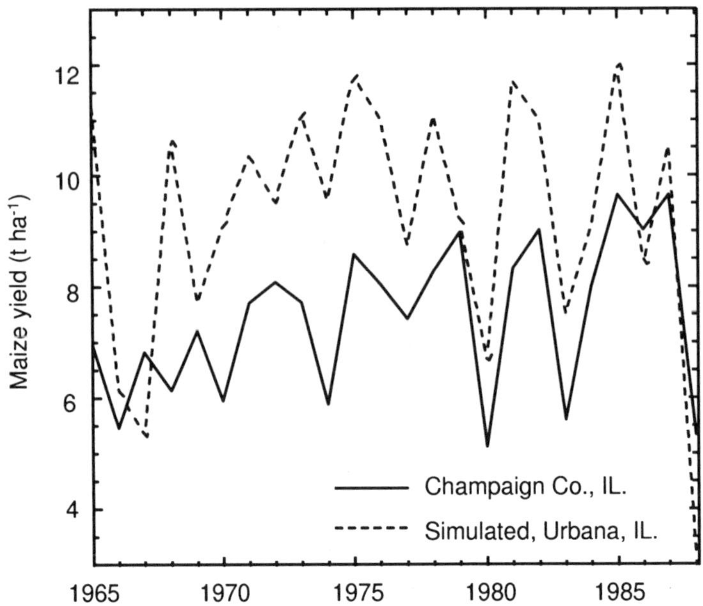

Fig. 6.5. Average maize yields obtained and simulated in Champaign County, Illinois, USA over the 1965-1988 period.

Much of the observed variation in yield among years was reproduced by the model (Fig. 6.5), indicating the importance of water deficits in yield variation. In contrast to observed increases in 'growers' yields over the wetter years, the model predicted little trend in yield over this 24-year period. This, of course, was not unexpected from the model because no resource limitation other than water was introduced into the calculations.

The difference between simulated and actual yields is a critical comparison in these results from Champaign County. This difference between the two data sets diminished over the 24 years (Fig. 6.5) as the actual yields increased. Interestingly, in the last few years it appeared that the difference between the simulated yields and growers' yields may have stabilised. That is, the actual yields were still less than simulated yields but progress in narrowing the difference between the two data sets had slowed, if not stopped.

The apparent stabilisation in the yield difference between calculated yields and 'growers' yields in the later years of the comparison in Fig. 6.5 is important. Clearly, in the early years growers had been gradually increasing a resource plateau, presumably one associated with N. In the last few years, it seems that the growers were no longer successful in increasing yields and may well have accepted yields less than those allowed by the water resource plateau. The possibility is explored in the next section that growers were no longer increasing N applications and were accepting yields on an N resource plateau.

Widespread Evidence for Nitrogen-Resource Plateaus

National, annual yields since 1950 for several grain crops indicate the possibility of the development of yield plateaus. Three cases from industrialised countries with intensive agricultural practices are examined: rice in Japan, wheat in the United Kingdom, and maize in the United States.

The annual rice yields for Japan peaked with a maximum yield of 4.7t ha^{-1} in 1984 (Fig. 6.6). There has been no further trend towards increasing yield in the last ten years. Wheat yields obtained by growers in the United Kingdom showed much the same results as the rice in Japan (Fig. 6.6). There was a steady rise in yields until a peak of 7.5t ha^{-1} was reached in 1984. Since 1984, wheat yields in the United Kingdom have not again reached 7.5t ha^{-1}.

Finally, average maize yields in the United States also indicate a plateau (Fig. 6.6). Maize yields reached 7.5t ha^{-1} in 1986 and 1987 and have remained essentially at that level or lower since. The levelling in maize yields in recent years for the entire US is not unlike that shown for maize yields for Champaign County (Fig. 6.5).

The data from these three examples all strongly indicate that national yield averages have stabilised in roughly the past ten years. There is no evidence of a dramatic change in the global environment that would have stabilised crop yields. If anything, increases in atmospheric carbon dioxide would have been expected to stimulate yields. Also, government programs in these three countries to constrain production would favour concentrating the crops on the better land. These trends would seemingly favour increased yields per unit land area. Counter to these favourable factors for increased yields, the evidence indicates a plateauing in yields.

The plateauing of yields in these three important cases seems to indicate the development of a resource plateau. There is substantial evidence that this resource plateau is closely linked to a stabilisation in the levels of N applications by growers. Figure 6.7 shows the remarkable increase in N fertiliser application per unit land area on maize in the United States since 1950. The amount of N fertiliser applied peaked at less than 160kg ha^{-1} in 1985 and has decreased slightly since that time. A similar pattern exists in the UK for N application on wheat (Austin, 1994) where N fertilisation peaked in 1985 at about 190kg ha^{-1}.

Also included in Fig. 6.7 are the rather surprising data that world-wide, fertilisation has levelled. World-wide, the application of all minerals on fertilised lands peaked in 1989 at 211kg ha^{-1}. This overall change in increasing fertiliser application seems to have resulted for two major reasons. One was the stabilisation of fertiliser application levels in the most industrialised countries as discussed above. A second reason was that economic factors in several countries that formerly relied heavily on fertilisation have changed, resulting in decreased fertiliser use. This is especially true in countries of the former Soviet Union, where subsidies for fertiliser have been eliminated. Economic and policy changes have also decreased fertiliser use in China and India (Brown and Kane, 1994).

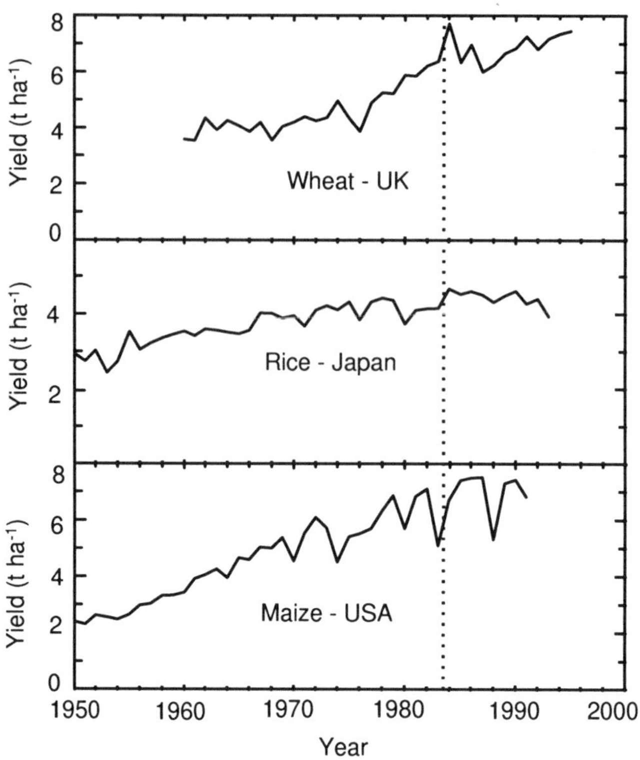

Fig. 6.6. National average yields for rice in Japan, wheat in UK and maize in USA over the 1950 to 1993 period.

The close association between the levelling of both crop yields (Fig. 6.6) and N fertiliser use (Fig. 6.7) provides evidence of a yield plateau associated with an N limitation for some grain crops in industrialised countries. The high sensitivity of physiological processes in crops to N levels, as discussed previously, provides the fundamental explanation for the N influence on grain yields.

The N-resource yield plateaus are below the biophysical yield peaks that are ultimately possible in crop growth. Interestingly, the current N-resource yield plateaus seem to be something entirely new when considering national crop yields. That is, the effort by growers to achieve ever increasing crop yields appears now to be supplanted by a willingness to accept lower yields on a 'socioeconomic' plateau.

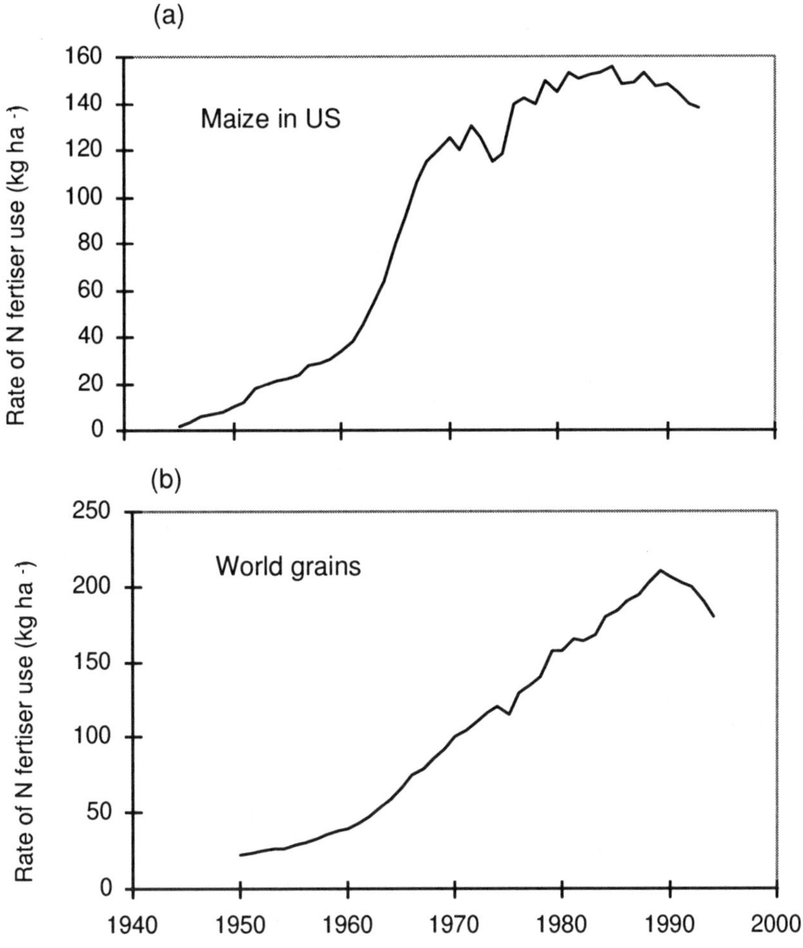

Fig. 6.7. Trends in fertiliser use. (a) Nitrogen applied to maize in the US (b) All fertilisers applied to grains world-wide.

The Socioeconomic Yield Plateaus

Certainly, fertiliser applications at levels above what growers are applying result in further yield increases. The example of the response of maize to applied N (Fig. 6.4) illustrates a benefit under these conditions of added N fertiliser up to 240kg ha^{-1}. These results show, however, that the response curve asymptotically approaches a maximum yield, which presumably is the yield peak for this environment. This response curve is a classic diminishing return in yield for additional N application. Seemingly, growers have decided not to pursue the peak yield but have settled for yields that are obtained at somewhat lower levels of N application.

The decision by growers to accept yields at a level below that possible with further application of resources is understandable. Growers are driven by the need to maximise profits. Until recent times, maximum profits were invariably associated with maximising yields. This no longer seems to be the case. The diminishing-return response curve for N application in Fig. 6.4 is an appropriate example. The various costs of increasing fertiliser application on these maize crops from 150 to 200kg ha^{-1} may not be economically justified when the expected yield increase is only 0.7t ha^{-1}. Therefore, maximising crop yield is no longer the appropriate paradigm, but rather a maximising of profits appears to have led to plateaus in crop yields. In many countries economic incentives and policies to encourage fertiliser applications have changed and further discouraged large N applications by growers.

Societal pressures against high levels of fertiliser applications also contribute to these new socioeconomic plateaus of crop yields. The general public has become aware of the environmental consequences resulting from losses of applied fertilisers from crop lands. Nitrate and phosphate leaving crop lands enter streams, wetlands, and lakes. These minerals can stimulate undesired plant growth and change the ecology of natural habitats. Legislation in industrialised countries is in place or is being contemplated to restrict fertilisation to regulate this environmental conflict. Also, in many countries there is considerable concern about the nitrate levels in groundwater under agricultural lands. High nitrate levels in drinking water have been indicated as a health risk, particularly for children. In the US, legislation has been enacted to set allowable limits of nitrate in the groundwater under crop lands.

Consequently, socioeconomic plateaus seem to be becoming major, new factors in defining limits on crop yields. These economic and societal pressures appear to be directly linked to a levelling in fertiliser applications, that in turn have been associated with plateauing crop yields in industrialised countries. Other socioeconomic factors can also result in major constraints on crop production. Competition for water with urban and industrial needs may well result in restrictions on the water resource for crop production. Mandated tillage practices and restricted pesticide may also play a role in setting the socioeconomic plateau for crop yields. While technology exists to eliminate most of the resource plateaus and allow peak yields under experimental conditions, the new reality in crop production may well be one of yield plateaus defined by socioeconomic constraints.

References

Amir, J. and Sinclair, T.R. (1991) A model of the temperature and solar-radiation effects on spring wheat growth and yield. *Field Crops Research* 28, 47-58.

Amir, J. and Sinclair, T.R. (1994) Cereal grain yield: Biblical aspirations and modern experience in the Middle East. *Agronomy Journal* 86, 362-364.

Austin, R.B. (1994) Plant breeding opportunities. In: Boote, K.J. *et al.* (eds), *Physiology and Determination of Crop Yield.* American Society of Agronomy, Madison, WI, USA, pp. 567-586.

Boerma, H.R. (1979) Comparison of past and recently developed soybean cultivars in Maturity Groups VI, VII, and VIII. *Crop Science* 19, 611-613.

Borlaug, N.E. (1972) *The Green Revolution, Peace and Humanity.* CIMMYT Reprint and Translation Series No. 3, Londres, Mexico.

Brown, L.R. and Kane, H. (1994) *Full House.* W.W. Norton & Co., NY, USA.

Chandles, R.F., Jr. (1969) Plant morphology and stand geometry in relation to nitrogen. In: Eastin, J.D., Haskins, F.A., Sullivan, C.Y. and van Bavel, C.H.M. (eds), *Physiological Aspects of Crop Yield.* American Society of Agronomy, Madison, WI, USA, pp. 265-285.

Hall, A.J., Connor, D.J. and Sadras, V.O. (1995) Radiation-use efficiency of sunflower crops: effects of specific leaf nitrogen and ontogeny. *Field Crops Research* 41, 65-77.

Kiniry, J.R., Jones, C.A., O'Toole, J.C., Blanchet, R., Cabelguenne, M. and Spanel, D.A. (1989) Radiation-use efficiency in biomass accumulation prior to grain-filling for five grain-crop species. *Field Crops Research* 20, 51-64.

McCloud, D.E. (1997) Development of agricultural ecosystems. In: Sinclair, T.R. and Gardner, F.P. (eds), *Environmental Issues in Plant Production.* Marcel Dekker, NY, USA (In Press).

Muchow, R.C. (1988) Effect of nitrogen supply on the comparative productivity of maize and sorghum in a semi-arid tropical environment. I. Leaf growth and leaf nitrogen. *Field Crops Research* 18, 1-16.

Muchow, R.C., Robertson, M.J. and Keating, B.A. (1997) Limits to the Australian sugar industry: climatic and biological factors. In: Keating, B.A. and Wilson, J.R. (eds), *Intensive Sugarcane Production: Meeting the Challenges Beyond 2000.* CAB International, Wallingford, UK, pp. 37-54.

Muchow, R.C. and Sinclair, T.R. (1991) Water deficit effects on maize yields modeled under current and 'greenhouse' climates. *Agronomy Journal* 83, 1052-1059.

Muchow, R.C. and Sinclair, T.R. (1994) Nitrogen response of leaf photosynthesis and canopy radiation use efficiency in field-grown maize and sorghum. *Crop Science* 34, 721-727.

Muchow, R.C. and Sinclair, T.R. (1995) Effect of nitrogen supply on maize yield: II. Field and model analysis. *Agronomy Journal* 87, 642-648.

Muchow, R.C., Sinclair, T.R. and Bennett, J.M. (1990) Temperature and solar radiation effects on potential maize yield across locations. *Agronomy Journal* 82, 338-343.

Park, W.I. and Sinclair, T.R. (1993) Consequences of climate and crop yield limits on the distribution of corn yields. *Review of Agricultural Economics* 15, 483-493.

Pearman, I., Thomas, S.M. and Thorne, G.N. (1977) Effects of nitrogen fertilizer on growth and yield of spring wheat. *Annals of Botany* 41, 93-108.

Salado-Navarro, L.R., Sinclair, T.R. and Hinson, K. (1993) Changes in yield and seed growth traits in soybean cultivars released in the southern USA from 1945 to 1983. *Crop Science* 33, 1204-1209.

Sinclair, T.R. (1990) Nitrogen influence on the physiology of crop yield. In: Rabbinge, R. *et al.* (eds), *Theoretical Production Ecology: Reflections and Prospects.* Pudoc, Wageningen, The Netherlands, pp. 41-55.

Sinclair, T.R. (1993) Crop yield potential and fairy tales. In: Buxton, D.R. *et al.* (eds), *International Crop Science I.* Crop Science Society of America, Madison, WI, USA, pp. 707-711.

Sinclair, T.R. (1994) Limits to crop yield? In: Boote, K.J. *et al.* (eds), *Physiology and Determination of Crop Yield.* American Society of Agronomy, Madison, WI, USA, pp. 509-532.

Sinclair, T.R. and Bai, Q. (1996) Analysis of high wheat yields in northwest China. *Agricultural Systems* (In Press).

Sinclair, T.R. and Horie, T. (1989) Leaf nitrogen, photosynthesis, and crop radiation use efficiency: A review. *Crop Science* 29, 90-98.

Sinclair, T.R. and Shiraiwa, T. (1993) Soybean radiation-use efficiency as influenced by nonuniform specific leaf nitrogen distribution and diffuse radiation. *Crop Science* 33, 808-812.

Sinclair, T.R., Tanner, C.B. and Bennett, J.M. (1984) Water-use efficiency in crop production. *Bioscience* 34, 36-40.

Spaeth, S.C., Sinclair, T.R., Ohnuma, T. and Konno, S. (1987) Temperature, radiation, and duration dependence of high soybean yields: Measurement and simulation. *Field Crops Research* 16, 297-307.

Specht, J.E. and Williams, J.H. (1984) Contributions of genetic technology to soybean productivity. Retrospect and prospect. In: Fehr,. W.R. (ed.), *Genetic Contributions to Yield Gains of Five Major Crop Plants.* Crop Science Society of America Special Publication 7, Madison, WI, USA, pp. 49-74.

Wright, G.C., Bell, M.J. and Hammer, G.L. (1993) Leaf nitrogen content and minimum temperature interactions affect radiation-use efficiency in peanut. *Crop Science* 33, 476-481.

Chapter seven:

The Yield Plateau in the Australian Sugar Industry: 1970-1990

A.L. Garside[1], M.A. Smith[2], L.S. Chapman[1,3], A.P. Hurney[1,4] and R.C. Magarey[1,4]

1. *Bureau of Sugar Experiment Stations, Davies Laboratory, PO Aitkenvale, Townsville, Qld 4814, Australia*
2. *Bundaberg Sugar Ltd, Bundaberg, Qld 4670, Australia*
3. *BSES, Mackay, Qld 4740, Australia*
4. *BSES, Tully, Qld 4854, Australia*

Abstract

The Australian sugar industry was on a productivity (yield of sugar per harvested hectare) plateau between 1970 and 1990. In this paper we attempt to undertake an objective assessment of reasons for this plateau. Both changes to the farming system and the phenomenon of yield decline - *the loss of productive capacity of sugarcane growing soils under long term monoculture* - are believed to have contributed to the productivity plateau, with the former being the major contributor. However, it is impossible from available data to apportion the relative importance of each. The period was characterised by the most substantial changes to the farming system in the history of the sugar industry. Changes such as the removal of assignment restrictions, introduction of chopper harvesters, improved varieties, green cane trash blanket, herbicide usage, major diseases such as Fiji, rust and *Pachymetra*, increased use of nitrogen fertiliser, and calcium, magnesium and zinc nutrition can all be shown to have contributed to the productivity plateau. The significance of these changes is discussed. Further, although productivity may have plateaued during the period, profitability probably increased as many of the changes to the farming system allowed for reductions in labour requirements and production costs. Finally, it is recommended that the sugar industry needs to institute a programme to collect, collate, and maintain data that will permit valid analyses of, and the identification of reasons for, long term industry trends.

In: Keating, B.A. and Wilson, J.R. (eds) *Intensive Sugarcane Production, Meeting the Challenges Beyond 2000*. CAB International, Wallingford, UK, 1997.

Introduction

Sugar industry statistics show that productivity (yield of sugar per harvested hectare) plateaued for the 20 year period between 1970 and 1990 (Fig. 7.1) (SRDC, 1995). This caused major concern within the industry as it was a divergence from the norm; productivity having consistently increased over the previous 70 years, with the exception of the period of the Second World War (1941-1947) when extenuating circumstances applied (Hogarth, 1972). Many possible reasons have been put forward for the plateau with some of the more popular being: industry expansion onto more marginal soils, introduction of chopper harvesters, a decrease in the ratio of plant to ratoon crops, increased labour costs, increased age structure of the cane-growing community, and the phenomenon known as yield decline (Anon., 1991). However, there has been no attempt to apportion the relative importance of the various possible causes. In this paper, these and other factors are examined in an attempt to quantify their contribution to the yield plateau.

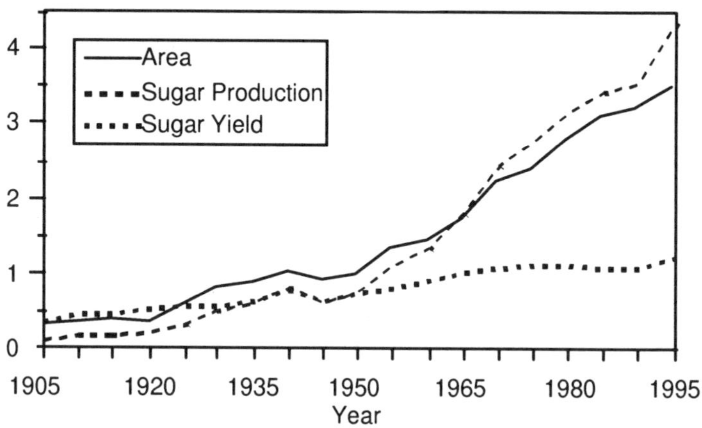

Fig. 7.1. Area (00,000's ha), sugar production (000,000's t) and sugar yield (10's t ha^{-1}) for the Australian sugar industry 1905-1995. [Data are averages for consecutive five year periods. (SRDC, 1995)]

In approaching this review it became apparent that three broad aspects influence the productivity plateau: (i) the bio-physical environment, how it has been amended by and for the sugarcane farming system, and the implications these amendments have had for productivity; (ii) industry development with associated regulatory and infrastructural changes that have influenced the farming system and productivity; and (iii) the effect of socio-economic factors on the input/output mix. Each of these broad areas can be identified at different levels of importance within the various factors that are considered to have influenced productivity.

Finally, the current period of 1990 to 1995 has once again seen a trend towards increasing productivity (SRDC, 1995); possible reasons for this are discussed.

Important Features of the Sugarcane Production System

As described by Sinclair (1997), yield plateaus are becoming common in agricultural crops and can normally be related to bio-physical factors and their influence on the yield of the reproductive portion. This is not, however, the case with sugarcane. Sugarcane varies from most other agricultural crops in that it is a C_4 grass whose economic product, sucrose, is contained in the vegetative structure. Productivity is dependent on the mix of sugar content and cane yield.

Sugarcane also varies from most other Australian agricultural crops in that it is grown in an intensive monoculture. The monoculture system developed for a number of reasons including the need to be close to sugar mills due to high transport costs, the assignment system, and the lack of alternative enterprises as profitable as sugarcane (Courtenay, 1978). In modern times the system revolves around a plant and four to five ratoon crops before the root stock is ploughed out. In most situations, a maximum of a six month fallow (either bare, as weeds, a sown legume, or a mixture of all three) is incorporated before the cycle is repeated.

The monoculture system has been intensified with industry development. Inorganic fertilisers, insecticides, and herbicides have been used extensively as the economics of sugarcane production have justified their costs. In particular, the heavy use of some of the major elements, such as phosphorus, has produced very high soil concentrations resulting in an unbalanced nutritional composition in many sugarcane soils. Further, nitrogen rates well in excess of that deemed necessary for maximum yield are often used. Other changes to the farming system include trash retention. Until the last decade the industry was largely based on burnt cane harvesting so the only organic matter returned to the system was in decomposing roots and stools. Green cane harvesting and trash retention has been a major innovation in recent years and the benefits of such a system in terms of organic matter improvement, soil health and erosion control are now starting to emerge (Wood, 1985, 1991; Prove *et al.*, 1986; Spain and Hodgen, 1994). In addition, heavy machinery is used to harvest and transport the cane, often under wet conditions, resulting in substantial soil compaction (Braunack and Hurney, 1996).

Yield Plateaus in the Australian Sugar Industry

In an analysis based on mill statistics, Smith (1991) demonstrated the existence of the 20 year productivity plateau between 1970 and 1990. During this period sugarcane yield increased at the rate of only 0.01t ha^{-1} y^{-1} while sugar content (CCS) actually decreased by 1 unit. These results contrast with the increasing productivity demonstrated by Hogarth (1972) for the 20 year periods 1899-1919, 1920-1940, and 1947-1970 when respective yield increases were 0.6, 0.7, and 0.9t cane ha^{-1} y^{-1} and 0.09, 0.15, and 0.15t sugar ha^{-1} y^{-1}. The Second World War years of 1941 to 1947 were deleted because they represented periods of reduced inputs, particularly in terms of labour.

However, the 1970 to 1990 productivity plateau is not the first major plateau recorded by the Australian sugar industry, as very poor and declining productivity late last century was a major reason for the study - *Investigation into the Condition of the Sugar Industry in Queensland* (Maxwell, 1900) - which concluded that poor and declining yields were due to soil-related factors, particularly severe nutrient deficiencies associated with lack of fertilisation, and pest and diseases associated with too many ratoon crops. Bell (1935, 1938) reported on a productivity plateau for the Bundaberg-Gin Gin area which was apparent for 40 years from 1898 to 1937 when sugarcane yields stabilised at around 37t ha^{-1}. Bell noted: *'In short the native fertility is being rapidly lost by growing a crop which is a gross feeder and requires constant cultivation which brings about fertility depletion and soil erosion; the soil is becoming dead'.* Basically, both Bell and Maxwell attributed the problem to bio-physical factors associated with the monoculture.

The most recent yield plateau may also be attributable to bio-physical factors. However, it is possible to identify a whole range of technological changes that have occurred during the period 1970-1990 which should have resulted in productivity improvements. For example, breeding programmes have been estimated to provide productivity improvements of 1% per year (Berding and Skinner, 1987; Roach and Daniels, 1987; Bull *et al.*, 1993; Chapman, 1996). Even allowing for farm yields to be only 76% of experimental yields (Davidson, 1962) the productivity increase from varieties alone should have been 15% over the twenty year period. Further, the percentage of the crop irrigated increased by 6%, mean nitrogen fertiliser rates by 40% and other fertilisers by various amounts (Smith, 1991). This suggests that the causes of the most recent yield plateau may be more diverse than the bio-physical factors identified by Maxwell (1900) and Bell (1935, 1938), although root diseases were common to all periods (Bell, 1938; Magarey, 1996).

History of Sugarcane Production Systems in Australia

Throughout its history the sugar industry has undergone some major changes which have impacted on productivity. Significant changes to the farming system during the 20th century include the introduction of:

1. Inorganic fertilisers during the 1920s and their subsequent widespread adoption.
2. Interspecific (*Saccharum officinarum* × *S. spontaneum)* hybrids as the prime source of varietal material.
3. Burnt cane harvesting during the late 1930s.
4. BHC (benzene hexachloride) for the control of cane grubs in the mid-1940s, and the subsequent use of a range of organo-chlorine pesticides for the next 40 years.
5. Mechanical harvesting on a whole stick basis during the 1950s and early 1960s.

However, changes to the farming system during the period from the mid-1960s to the early 1990s far outweighed those of earlier periods. In particular:

1. Regulations which indirectly forced farmers to fallow 25% of their assigned area each year were relaxed to 15% forced fallowing in 1964 and abandoned in 1975 (Wegener, 1985).
2. Chopper harvesters have dominated the industry from the early 1970s.
3. Varieties with greater ratooning ability have emerged .
4. Green cane harvesting has accounted for an increasingly larger proportion of the crop.
5. Herbicide usage has increased substantially with major reductions in cultivation, particularly with ratoon crops.
6. Major diseases such as Fiji disease (southern region), and rust (*Puccinia melanocephala*) and pachymetra root rot (*Pachymetra chaunorhiza)* emerged during the 1970s.
7. Use of nitrogen fertiliser was substantially increased.
8. Calcium, magnesium and zinc deficiencies emerged as problems, particularly in north Queensland.
9. Organo-chlorine insecticides were banned.

Influence of Changes in the Farming System on Productivity

Removal of assignment restrictions

The removal of assignment restrictions encouraged more intensive cultivation of sugarcane on assigned land with plough-out/re-plant being practised by many producers. In 1969 only 69% of the assigned area was harvested, yet this increased to more than 80% by 1974. The area has varied between 80 and 88% since (Smith, 1991). Yields per harvested hectare have been reduced in plough-out/re-plant as opposed to fallow plant (Roach *et al.*, 1981; Ridge, 1983; Patch *et al.*, 1986; Wegener and Pollock, 1986). For example, the reduction in sugar yield per harvested hectare has been 10.7, 4.1, 5.4 and 4.5% for plant, first, second and third ratoon respectively in the Tully mill area (Anon., 1993). In the Burdekin region Wegener and Pollock (1986) showed an overall average increase in cane yield of 35t ha^{-1} or 24% for fallow-plant over plough-out/re-plant.

Although both cane and sugar yield per harvested hectare are reduced in plough-out/re-plant, productivity per assigned hectare has improved. In fact, Ridge (1983) showed that production per assigned area was maximised by planting all crops as re-plant. Consequently, the increase in plough-out/re-plant during the period 1970-1990 has contributed to the yield plateau, but has also contributed to improved farm profitability (Smith, 1991).

Further, the accepted adverse effects on productivity of plough-out/re-plant warrant questioning as no direct comparative studies have been conducted. In nearly every case comparative studies have been confounded by time of planting, fallow plantings being earlier and so older at harvest (Wegener and Pollock, 1986). However, observations suggest that there is a definite direct effect (A.P.Hurney, unpublished data).

Mechanical harvesting

The popularity of chopper harvesters grew rapidly during the 1960s accounting for 10% of the crop in 1962, 50% in 1968 (King, 1969) and virtually the whole crop by 1972 (D.R. Ridge, *pers. comm.*). With the advent of chopper harvesters substantial yield losses and/or increases in extraneous matter appear to have occurred, although it is only recently that these have been quantified (Linedale *et al.*, 1993; Linedale and Ridge, 1996). Numerous estimates of cane and sugar loss have been made for chopper harvesters and these have varied between burnt and green cane. Hurney *et al.* (1984) recorded cleaning losses of 5-8t ha^{-1} in burnt cane while green cane losses of between 4.6 and 18% were recorded from a range of experiments (Ridge and Dick, 1988; Shaw and Brotherton, 1992). Sturgess *et al.* (1972) recorded substantial and consistent CCS decline of around 0.5 units on the wet tropical coast between cane harvested in billet form and that harvested as whole stalk.

The recent studies by Linedale and Ridge (1996) have quantified the magnitude of harvesting losses with chopper harvesters. Surveys confirmed average losses of 7.4 and 3.4t ha^{-1} in green and burnt cane, respectively. As the result of a major extension campaign between 1992 to 1995, surveys estimate that cane losses have been reduced by more than one-third of their previous level (Linedale and Ridge, 1996).

On the basis of these data there would appear to be little doubt that the productivity plateau of 1970-1990 was partly associated with harvesting losses.

Ratooning

A feature of the farming system between 1970 to 1990 has been an increase in the number of ratoons grown and the subsequent relative decrease in plant crop area (Chapman, 1988; Smith, 1991). Industry wide, the ratoon : plant crop ratio almost doubled from 2.2 in 1966 to 4.3 in 1986 (Chapman, 1988). Smith (1991) attributed the slower increase in productivity compared with previous periods (Hogarth, 1972) to the fact that more of the area was in relatively low yielding ratoons. Although able to identify plant and ratoon cane yield increases of 0.08 and 0.30t ha^{-1} y^{-1} between 1970 and 1990, Smith (1991) suggested that the proportions of each in the ground at any one time resulted in industry wide cane yield increases of only 0.01t ha^{-1} y^{-1} because of the lower yield of ratoon compared with plant crops.

However, as was the case with plough-out/re-plant, profitability can be enhanced by longer ratooning cycles provided yields can be maintained at an acceptable level. Planting is a considerably more expensive operation than ratooning (Hardman *et al.*, 1985). Cessation of ratooning and re-planting decisions are ideally made when the cumulative lost production, that is, returns from past ratoons and forward estimates for the next ratoon crop, exceeds the cost of re-planting (Simms, 1982).

In recent studies it has been shown that yields from the sixth and seventh ratoons can be as high as those from plant and first ratoon crops depending on the varieties used. More recent variety releases (Q124 in 1984 and Q138 in 1990) are showing a much better ratooning capability due to reduced stool losses (Chapman *et al.*, 1992; Chapman and Wilson, 1996). The impact of such varieties would not have been realised for much of the 1970-1990 period as by 1990 they represented only 9.5% (Q124) and 2.7% (Q138) of the crop (BSES, 1991). Interestingly, by 1994, Q124 represented 25% and Q138, 6% of the crop.

Green cane trash blanket (GCTB)

Green cane harvesting was practised in the Australian sugar industry until the late 1930s when burning was introduced to combat the spread of Weil's disease amongst cane cutters. The re-birth of green cane harvesting occurred during the mid-1970s with the advent of suitable harvesters combined with deficiencies in the burnt system such as the inability to burn during wet weather and the loss of burnt cane following heavy overnight rain (Wood, 1991). Expansion of the area harvested green continues annually to the present time. GCTB was promoted because of a whole range of perceived benefits. Smith *et al.* (1984) listed these as reduced soil erosion, no burnt cane penalty, no deterioration as occurs after burning, cane fire problems avoided, wet weather harvesting facilitated which led to a decrease in season length, environmental criticism of cane fires avoided, increased CCS, increased yields, reduced growing costs and reduced soil degradation. Smith (1993) subsequently questioned the effect of GCTB on CCS, yield, growing costs and soil degradation and in general concluded that there was probably a small yield and CCS reduction with GCTB but these were compensated for by growing cost savings. However, confounding factors precluded any firm conclusions, in particular a general feeling that smaller crops were often cut green. The conclusion from these studies were that GCTB lowers the cost of production, increases the gross margin of ratoon relative to plant cane, and increases the gross margin and profits from a ratoon. Consequently, a grower using GCTB with lower yields is likely to obtain similar returns to a conventional farmer at higher yields. In effect the influence of GCTB may be reduced productivity but improved profitability through lower farm costs (Wood, 1991).

In reviewing a number of unreplicated test strips, Smith (1993) showed that there appeared to be an overall positive yield response to GCTB, particularly under dry conditions, but the data were inconclusive. However, greater harvesting losses appear to have reduced the attractiveness of GCTB when compared to burnt cane harvesting (Fuelling *et al.*, 1978; Ridge and Dick, 1987, 1988; Stewart and McComiskie, 1988; Linedale and Ridge, 1996).

GCTB has affected several soil properties. Firstly, there appears to have been a small increase in organic carbon levels in the surface soil (Wood, 1986; Stewart, 1987; Wood, 1991). Secondly, allelopathic affects have been associated with the large quantities of trash returned to the soil surface. Hurney and Ridge (1992) identified a number of effects from a series of glasshouse experiments including: (i) negative effects on germination which appeared variety specific, were enhanced by poor internal drainage, and disappeared with time after application of the blanket; (ii) adverse effects on root growth; and (iii) the identification of several phenolic acids in extracts from leaf and trash which are known to have allelopathic effects on other plant species. Wood (1991) also showed the negative effects of GCTB combined with poor internal drainage. Thirdly, physical degradation appears to have been initially enhanced as the trash blanket has permitted harvesting under wetter conditions resulting in more soil compaction and stool damage. Further, the reduction in cultivation for weed control has reduced any alleviation of compaction through soil disturbance (Wood, 1986). However, these effects may be short term as experience in other agricultural systems with zero tillage and stubble retention suggests that often a period of 10 years is required before improvements in soil physical properties are recorded. Finally, Wood (1986, 1991) reported increases in soil biota, less diurnal fluctuation in soil temperature, and better soil water conservation under GCTB compared with conventional cultivation.

Overall, there appears to have been little contribution from GCTB towards the productivity plateau outside the previously mentioned harvesting losses which have been substantial. However, changes measured under GCTB such as improved organic matter, improved soil biotic activity, and enhanced soil water conservation are generally associated with soil and productivity improvement but often take considerable time to be realised. Of immediate significance is that GCTB, in conjunction with strategic herbicide use, has resulted in substantial labour savings in terms of cultivation and burning which has enhanced profitability.

Herbicides

During the period under review there was a substantial change in the method of weed control on sugar farms with chemicals largely replacing cultivation. For example, in the central district (Mackay, Queensland) 70% of farmers had adopted chemical weed control by 1982 (Hyde and Teske, 1987) whereas very little chemical weed control had been used in 1969 (Chapman, 1969). The BSES Soil Monitoring Data Base (SMDB) (Chapman *et al.*, 1981) clearly shows that major reductions in mechanical weed control occurred in the late 1970s (L.S. Chapman, unpublished data).

Two important points emerge from this change to chemical weed control. Firstly, there are substantial labour savings in terms of cultivation costs and presumably long term improvements in soil structure. Secondly, at least some of the chemicals used have phytotoxic effects on the cane plant. Such effects, caused by, for example, paraquat, ametryn, diuron and velpar, have been generally regarded as transient with the plant seemingly recovering. However, no studies have been carried out to objectively assess the effects and some limited test strips suggest that yield losses may be occurring through phytotoxicity (A.J. Linedale, *pers. comm.*).

Diseases

Fiji disease, a virus, became a major problem in the southern (Bundaberg, Queensland) growing area during the early 1970s in the previously very productive variety NCO310 (Toohey and Neilsen, 1972; Egan and Fraser, 1977; Egan, 1982). By the mid-1970s substantial yield losses of up to 30% were being measured (Chardon and Hogarth, 1977; Turner and Churchward, 1977). In the late 1970s sugarcane rust *(Puccinia melanocephala)* emerged initially as a problem in north Queensland but rapidly spread to all mill areas in Queensland and New South Wales by mid-1979 (Ryan and Ledger, 1980). Several major commercial cultivars were susceptible and yield losses up to 50% were recorded (Taylor *et al.*, 1986). Pachymetra (*Pachymetra chaunorhiza*) root rot was isolated as part of a complex associated with northern poor root syndrome, the cause of substantial yield losses in north Queensland during the 1970s (Egan *et al.*, 1984). It has subsequently been identified in most sugarcane growing areas and has been a major factor in the yield decline phenomenon. Yield losses of up to 40% have been demonstrated in susceptible varieties (Magarey, 1994). Although resistant varieties were developed in the case of all three diseases, substantial yield losses were sustained for a number of years.

Nitrogen fertiliser

Fertiliser rates increased during the period covered by this review. In particular there were substantial increases in the amount of nitrogen applied with often adverse results in terms of both cane and sugar yields. Ridge (1984) reported sugar yield increases up to 150kg ha^{-1} nitrogen but no increases when rates moved above that level. Industry-wide nitrogen rates below 150kg ha^{-1} were used up until 1969, were increased to just over 150kg ha^{-1} around 1975, and then substantially increased to around 210kg ha^{-1} in 1981 in response to improved sugar prices. They remained at this level until the mid 1980s, after which they appear to have been reduced during the second half of the 1980s, presumably in response to low sugar prices (L.S. Chapman, unpublished data). However, recent data from farmer surveys (R. Beattie, *pers. comm.*) and the SMDB (L.S. Chapman, unpublished data) suggest that rates of up to 300kg ha^{-1} are being used by some producers.

The industry accepts the importance of nitrogen fertiliser and tends to use excess as a risk avoidance mechanism (Keating *et al.*, 1994, 1997; Johnson, 1995). Further, the advent of GCTB has complicated the application of nitrogen fertiliser with losses through surface application a factor (Prammanee *et al.*, 1989; Denmead *et al.*, 1990; Wood *et al.*, 1990) that has encouraged growers to increase rates. However, in doing so the risk of cane and sugar yield losses is increased as the likelihood of lodging increases. Hurney (1984) found in experiments at Tully that nitrogen application rates in excess of 130kg ha^{-1} on ratoon crops could reduce sugar yields due to a decline in CCS of lodged crops. More recently, Muchow *et al.* (1995) have associated lodging with stalk death and yield loss. Moller (1974) also linked reduction in CCS with lodging of larger, irrigated plant cane crops at Bundaberg when compared with more erect ratoons. More recent observations in north Queensland where sugarcane and banana rotations are becoming common indicates severely lodged crops with low CCS in the first year cane following bananas (A.J. Benson, *pers. comm.*). This has been attributed to residual supplies of nitrogen from the banana crop, along with a healthier sugarcane root system, better able to exploit available nutrients. Conversely, more recent data suggest that if the crop does not lodge or is harvested relatively early, high rates of nitrogen *per se* do not adversely affect sugar yield (Muchow and Robertson, 1994).

Excess nitrogen fertilisation may well be one of the factors that contributed to the yield plateau as the major contributor to the yield plateau has been reduced CCS (Smith, 1991).

Calcium, magnesium and zinc

Calcium, magnesium (Ridge *et al.*, 1980; A.P. Hurney, unpublished data) and zinc (Reghenzani, 1990, 1993) have all emerged as deficiencies influencing cane yield, particularly in north Queensland. In the case of calcium and magnesium, deficiencies appear to have been the result of many years of cropping without nutrient addition reinforced by the release of varieties more susceptible to deficiency and an increase in the use of high analysis fertilisers. Furthermore, although both calcium and magnesium are now widely applied, there was initially slow adoption of ameliorative practices. Zinc deficiencies tended to become more important with the adoption of high analysis fertiliser mixes and the demise of superphosphate as a phosphorus source, because zinc was often a contaminant of superphosphate.

Varieties

Varieties changed substantially between 1970 and 1990, with more than 50 new varieties released by plant breeders. On the basis of productivity increases of 1% per year estimated by sugarcane plant breeders (Berding and Skinner, 1987; Roach and Daniels, 1987; Bull *et al.*, 1993; Chapman, 1996) a productivity improvement of 20% (or 15% commercially - Davidson, 1962) should have been realised between 1970 and 1990). That clearly has not been the case. Measured yield improvements in variety evaluation and other research plots have not been transferred to commercial cane fields (Berding and Skinner, 1987; Chapman, 1996). Berding and Skinner (1987) noted that '*new clones were released in the sugar industry in the hope of achieving real increases in productivity but the net effect has been little more than maintaining productivity in a declining agro-environment*'. It appears then that the farming system may have imposed a limitation on potential productivity increases through genetic improvement. In particular, over-adoption of better varieties has in some cases led to their earlier demise and replacement by potentially less productive varieties with specific characteristics, *viz.* NCO310 and Fiji disease in southern Queensland, Q90 and rust in north Queensland.

Lodging, once a factor selected against, has become a less important selection criterion since the advent of chopper harvesters. However, under the wet, humid conditions in north Queensland there is increasing evidence that lodging can lead to both loss of cane yield and reduction in sugar content (Hurney, 1984; Muchow *et al.*, 1995; Garside and Nable, 1996).

It is difficult to gauge the impact of varieties *per se* on the productivity plateau as genetic improvement in a stable environment has not been able to be measured. Diseases such as Fiji disease and rust have created important diversions in the realisation of genetic gain and have resulted in potentially less productive but more resistant varieties being grown. However, it is generally agreed that the most recent yield plateau may well have been a decline but for the influence of improved varieties.

Climate

There has been nothing unique about the seasonal conditions impinging on productivity over the period 1970-1990, although the 1970s were a period of relatively high rainfall. Seasonal variation is a feature of Australian agriculture and the sugar industry is affected in the same way as other crop industries, except that parts of the industry regularly suffer from excess rainfall and waterlogging. Smith (1991), using regression analyses, showed that much of the annual variation in sugar yield between 1970 and 1990 could be accounted for by annual rainfall variability. However, he could not identify any major climatic variability between the 20 year period he considered and the earlier periods considered by Hogarth (1972), suggesting that long term rainfall patterns were not having an influence on productivity trends. There has been general speculation that the combination of chopper harvesters and wet harvesting conditions during the early 1970s may have adversely affected yield.

Influence of Yield Decline on Productivity

In sugarcane, yield decline is defined as the loss of productive capacity of sugarcane-growing soils under long term monoculture. It has been regarded as largely associated with soil microbiological factors (Magarey, 1996). Yield decline is well known in agricultural crops and has been recognised in such diverse species as potatoes (Vos and Van Loon, 1989), asparagus (Blok and Bollen, 1995), corn (Johnson *et al.*, 1992), tobacco (Hendrix *et al.*, 1992), and cherries (Pitcher *et al.*, 1966). There is little doubt that yield decline occurs in sugarcane as is evidenced by better growth on new versus old land (Anon., 1935; Garside and Nable, 1996), and following soil fumigation (Bell, 1935; Egan *et al.*, 1984; Muchow *et al.*, 1994; Magarey and Croft, 1995) and crop rotation with other species (Bieske, 1965; Chinloy and Hogg, 1968).

Yield decline appears to have been with the Australian sugar industry since the very start. Studies by Maxwell (1900) and Bell (1935, 1938) demonstrated major losses in productive capacity under the monoculture system. It was also the subject of considerable study in international sugar research and development programmes during the period 1940-1970 when the focus was on variety yield decline, an apparent decline in productivity of cane varieties (Arceneaux and Hebert, 1943; King and Steindl, 1953; Coleman, 1974). Much of the decline recorded then was related to ratoon stunting disease (King and Steindl, 1953) and there was no evidence of genetic shift within varieties (Mangelsdorf, 1959; Moore *et al.*, 1993).

The recent history of yield decline in sugarcane in Australia dates back to 1967 when poor root syndrome was recognised as a problem in sugarcane on Queensland's wet tropical coast (Egan *et al.*, 1984). Subsequent studies into a range of possible causes focused on pathogenic fungi (Croft and Magarey, 1984).

This resulted in the isolation of root pathogen *Pachymetra chaunorhiza* as one of the causes with 40% yield increases in resistant varieties (Magarey, 1994). However, even greater increases (>100%) were obtained when soil from the same site was fumigated with methyl bromide, suggesting that factors other than Pachymetra were involved (Croft *et al.*, 1984). Further, when Pachymetra-susceptible and resistant sugarcane varieties were grown on fumigated and unfumigated soil, the resistant variety outyielded the susceptible variety but still showed a further 36% yield increase after soil fumigation (A.P. Hurney, unpublished data). More recently, it has been established that substantial sugarcane yield responses to fumigation can be measured in all sugar growing areas of Queensland, whether Pachymetra is present or not (Magarey and Croft, 1995). It has also been shown that the yield increase in response to fumigation of sugarcane land is largely specific to sugarcane; other monocotyledon species (maize, sorghum) are only slightly responsive while dicotyledons are unresponsive (Garside *et al.*, 1995). This suggests that yield increases in response to fumigation are not simply an artifact of the fumigation process but an effect of improving the soil environment for sugarcane growth. However, there are negative aspects of the use of fumigation as beneficial soil biota are also removed. This has been recently demonstrated with respect to vesicular arbuscular mycorrhiza (R.C. Magarey, unpublished data).

Since the isolation of Pachymetra, there has been limited success in isolating other pathogens responsible for sugarcane yield decline (Magarey *et al.*, 1995). Consequently, the approach which assumes that root pathogens are the primary cause of yield decline has been questioned. Indeed, it has been suggested that a build-up of root pathogens may simply be the ultimate expression of other factors being out of balance in the farming system (Garside, 1995).

In an attempt to identify the causes of yield decline, a Joint Venture group involving the Sugar Research and Development Corporation, the Bureau of Sugar Experiment Stations, the CSIRO Division of Soils and Queensland Department of Primary Industries was established in 1993 to study soil and plant properties in paired old and new sugarcane sites; following the breaking of the monoculture with different species for different periods of time, and after new land is first planted and continues to grow sugarcane. Soil chemical, physical and biological properties and their relationship with crop growth and yield are being measured in each of these three systems.

This work is incomplete but results from studies into paired old and new land sites suggest that crop growth, as measured by stalk development, is reduced on old land. Whether this is reflected in ultimate yield appears to depend on a number of factors associated with the growing conditions (Garside and Nable, 1996). However, some general trends emerged in soil properties that indicated old land was more degraded in terms of acidification, soil structure and organic components. In general, old land was shown to be more acidic, have lower CEC, more exchangeable aluminium and manganese, less copper and zinc (Bramley *et al.*, 1996), less microbial biomass (J.A. Holt, unpublished data), greater soil strength, lower infiltration and water holding capacity (Ford and Bristow, 1995a,b), and more root pathogens (R.C. Magarey, unpublished data). More specific projects, including studies of sugarcane root systems, are being developed based on these findings.

Discussion

We have attempted in this paper to highlight the factors that may have contributed to the productivity plateau that occurred between 1970 and 1990. We believe they fall into two broad categories, *viz.* changes to the farming system and the issue of yield decline. There is little doubt that both have played a part in the productivity plateau but it is impossible with the data available and the confounding of that data by numerous factors to actually apportion causes. We believe there is an urgent need for the development of an industry wide system that will allow the consistent collection, collation and maintenance of data to permit valid analyses of industry trends and allow the reasons for such trends to be identified. A group dedicated to such activity appears to be an urgent requirement for the sugar industry. As part of such an approach a basic resource for the recording of block data, the Soil Monitoring Data Base sites, was established by BSES in all mill areas in 1975. This resource should be maintained and upgraded to provide more accurate and up to date information.

The period 1970-1990 has probably seen the most substantial and wide reaching changes to the farming system in the history of the industry. Changes such as the lifting of assignment restrictions, plough-out/re-plant, more ratoons, chopper harvesting, green cane trash blanketing, chemical weed control, and increased nitrogen rates can all be shown to have had some adverse effects on productivity. However, they may not have all had adverse effects on profitability. Plough-out/re-plant increased productivity per assigned area, while more ratoons, green cane trash blanket, and chemical weed control all reduced costs, largely through labour savings, an important consideration when costs are increasing more rapidly than sugar price. This of course begs the question - should productivity continue to be measured in terms of sugar yield per harvested hectare or should a combination of both productivity and profitability per harvested hectare be adopted as a more meaningful measure?

That changes to the farming system have been an integral part of the recent productivity plateau is evidenced by the fact that in the most recent 5 year period (1990-1995) the industry appears to have moved off the plateau (SRDC, 1995). Again it is difficult to apportion the reasons for this improvement in productivity, but commencing in the late 1980s a number of important changes in the farming system, that have probably contributed, can be identified. Firstly, recent expansions have largely been onto better quality soil and/or irrigated situations, *viz.* the Burdekin River Irrigation Area, the Murray area of north Queensland and the Atherton Tablelands. Secondly, the harvesting losses have been addressed with an estimated 2% increase in harvesting efficiency since 1990 (Linedale and Ridge, 1996). Thirdly, the good ratooning variety Q124 has rapidly expanded as the major variety across the industry. Fourthly, the percentage of the crop receiving at least some irrigation has increased substantially. Fifthly, below average rainfall throughout the first half of the decade has substantially reduced the yield losses associated with waterlogging in north Queensland and increased yield potential in irrigated areas due to higher radiation. Finally, now that considerable time has elapsed since the establishment of GCTB in many areas, the benefits of such a system may be starting to be realised.

The significance of yield decline in the current productivity plateau cannot be apportioned. There is little doubt that it has had an impact, particularly during the period when poor root syndrome was causing major problems in north Queensland during the 1970s. However, it is questionable as to whether it has simply re-emerged as a problem in the 1960s after causing such major problems in some areas in the 1930s (Bell, 1935, 1938). It must be considered that yield decline may well be a latent problem that has been with the industry throughout and that the productivity increases of the past may have been achieved in spite of yield decline. This is not to say that the problem should not be researched and overcome. The fumigation responses clearly indicate there are potentially major productivity increases that need to be exploited.

The causes of yield decline need to be understood and overcome. Although there is little doubt that root pathogens are a very important component, approaches that suggest the only solution to yield decline lies in the identification of, and breeding resistance to, specific soil pathogens need to be reassessed as an appropriate strategy. A more contemporary approach, being taken by the Yield Decline Joint Venture, is to understand how to manage the farming system in such a way that soil pathogens do not become a dominant factor. The end result may well be that it is not possible to manage soil pathogens through the farming system without suffering productivity losses. However, at this stage there is insufficient data to make objective judgements.

The lack of commercial response to variety improvement should be of major concern to the industry. Until the emergence of improved ratooning ability in the more recent varieties such as Q124, significant productivity improvement through varietal development has been difficult to quantify. Berding and Skinner (1987) questioned whether variety improvement was providing real advancement or simply providing a mechanism by which a declining agro-environment was being sustained. This question can still not be answered. If the suggestion of Berding and Skinner (1987) is correct, it must call into question the long term sustainability of the current production system. How long can a continually degrading system be maintained through genetic manipulation? The industry needs to know the answers to such questions so it can take appropriate action. Conversely, changes to the farming system may well allow genetic potential to be better exploited.

Two other factors associated with current varieties are worthy of mention. Firstly, the rapid expansion in the area of Q124. The industry should view this with concern as previous instances of single varieties controlling large portions of the industry have resulted in serious consequences, *viz.* NCO310 with Fiji disease and Q90 with rust. Secondly, the propensity of current varieties to lodge. Biologically, it is hard to imagine that yield losses are not incurred in lodged crops. Certainly, there appears to be adequate anecdotal evidence to indicate that lodged crops in wet humid environments suffer from stalk decay, yield and sugar losses. However, in the more arid environments, it is generally accepted that lodging is of little consequence in terms of productivity loss. Unfortunately, there are no objective data upon which to base such a belief. The significance of lodging requires quantification.

One final issue that has received virtually no mention in this review but forms an integral part of the farming system is the environment. Environmental issues have been addressed by other speakers at this symposium but they need to be considered in any discussion of changes to the farming system. It may well emerge that environmental considerations will place limitations on production practices in the sugar industry. A better understanding of the farming system may permit more environmentally acceptable farming methods to be developed.

Conclusions

It had been our intention in this review to identify the cause/s of the 1970 to 1990 productivity plateau and to rank the factors in order of importance. Clearly, we have achieved neither. However, we have provided sufficient evidence to suggest that the productivity plateau is not simply due to the declining productive capacity of sugarcane growing land under long term monoculture (the current definition of yield decline), although that is an important component. We have put forward many possibilities, all of which have probably contributed, and in doing so demonstrated the complexities of the system. There is little doubt that by its very nature, in having both plant and ratoon crops, sugar production is a most complex agricultural system. If limitations are to be identified and strategies put in place to overcome them there is a real need for growers, millers, extension providers and researchers to fully appreciate the complexities and have the resources and capacity to accurately attribute causal factors. It will only be with such understanding that more objective decisions can be made on industry priorities for research, development, and extension.

Acknowledgments

We thank the numerous colleagues who allowed us access to unpublished data.

References

Anon. (1935) *Cane Growers Quarterly Bulletin*, Brisbane, Australia p. 85.

Anon. (1991) *Yield Decline and the Productivity Plateau in the Queensland Sugar Industry.* Bureau of Sugar Experiment Stations, Occasional Publication, 91-0296, Brisbane, Australia.

Anon. (1993) *CAPA Productivity Review 1993 Crushing Season.* Tully Sugar Ltd. Publication A.C.N. 011 030 256, Tully, Australia, p.15.

Arceneaux, G. and Hebert, L.P. (1943) A statistical analysis of varietal yields of sugarcane obtained over a period of years. *Journal American Society of Agronomy* 35,148-160.

Bell, A.F.(1935) Sick Soils. *Proceedings of the Queensland Society Sugarcane Technologists* 6, 9-18.

Bell, A.F. (1938) Crop rotation with special reference to the principles of green manuring. *Cane Growers Quarterly* 1, 138-148.

Berding, N. and Skinner, J.C. (1987) Traditional breeding methods. *Proceedings Copersucar International Sugarcane Breeding Workshop,* Piracicaba, Brazil, pp. 269-320.

Bieske, G.C. (1965) Soil organic matter. In: King, N.J., Mungomery, R.W. and Hughes, C.G. (eds), *Manual of Cane Growing.* Angus and Robertson, Sydney, Australia, pp. 110-132.

Blok, W.J. and Bollen, G.J. (1995) Fungi on roots and stem bases of asparagus in the Netherlands: Species and pathogenicity. *European Journal of Plant Pathology* 1 (1), 15-24.

Bramley, R.G.V., Ellis, N., Nable, R.O. and Garside, A.L. (1996) Changes in soil chemical properties under long term sugarcane monoculture and their possible role in sugarcane yield decline. *Australian Journal of Soil Research,* 34, 967-984.

Braunack, M.V. and Hurney, A.P. (1996) Reducing soil physical constraints under sugarcane. In: Wilson, J.R., Hogarth, D.M., Campbell, J. and Garside, A.L. (eds), *Sugarcane: Research Towards Efficient and Sustainable Production.* CSIRO, Division of Tropical Crops and Pasture, Brisbane, Australia, pp. 217-219.

BSES (1991) Bureau of Sugar Experiment Stations, Annual Report 1991, Brisbane, Australia.

Bull, J., Mungomery, V.E. and Hogarth, D.M. (1993) Realisation of genetic gain for yield in commercial production. In: Imrie, B.C. and Hacker, J.B. (eds), *Focussed Plant Improvement: Towards Responsible and Sustainable Agriculture. Proceedings 10th Australian Plant Breeders Conference,* Gold Coast, Australia, 18-23 April, 1993, pp. 92-103.

Chapman, L.S. (1969) An economic evaluation of pre-emergent weed control at Mackay. *Proceedings of the Queensland Society of Sugarcane Technologists, 1969 Conference,* pp. 131-141.

Chapman, L.S. (1988) Constraints to production in ratoon crops. *Proceedings of the Australian Society of Sugarcane Technologists* 9, 189- 192.

Chapman, L.S. (1996) Increase in sugar yield from plant breeding from 1946 to 1964. In: Wilson, J.R., Hogarth, D.M., Campbell, J. and Garside, A.L. (eds), *Sugarcane: Research Towards Efficient and Sustainable Production.* CSIRO, Division of Tropical Crops and Pastures, Brisbane, Australia, pp. 37-38.

Chapman, L.S., Ferraris, R. and Ludlow, M.M. (1992) Ratooning ability of cane varieties: Variation in yield and yield components. *Proceedings of the Australian Society of Sugarcane Technologists* 14, 130-138.

Chapman, L.S., Haysom, M.B.C. and Chardon, C.W. (1981) Checking the fertility of Queensland sugarcane land. *Proceedings of the Australian Society of Sugarcane Technologists* 3, 325-332.

Chapman, L.S. and Wilson, J.R. (1996) Economics of ratoon cycle length in sugarcane. In: Wilson, J.R., Hogarth, D.M., Campbell, J. and Garside, A.L. (eds), *Sugarcane: Research Towards Efficient and Sustainable Production.* Publication CSIRO, Division of Tropical Crops and Pastures, Brisbane, Australia, pp. 169-171.

Chardon, C.W. and Hogarth, D.M. (1977) A technique to assess yield losses attributable to Fiji disease. *Proceedings of the Queensland Society of Sugarcane Technologists, 1977 Conference,* pp. 67-69.

Chinloy, T. and Hogg, B.M. (1968) Rotation of pangola grass and sugarcane on a Jamaican sugar estate. *Proceedings International Society of Sugarcane Technologists* 13, 636-642.

Coleman, R.E. (1974) Ten years of yield decline research. *Proceedings International Society of Sugarcane Technologists, XV Cong.,* Durban, South Africa, pp. 885- 891.

Courtenay, P.P. (1978) Agriculture in north Queensland. *Australian Geographical Studies* 16, 29-42.

Croft, B.J. and Magarey, R.C. (1984) Pathogenic fungi associated with northern poor root syndrome of sugarcane. *Proceedings of the Australian Society of Sugarcane Technologists* 6, 55-61.

Croft, B.J., Reghenzani, J.R. and Hurney, A.P. (1984) Northern poor root syndrome of sugarcane - studies on soil transmission and the effects of various fungicidal, nutritional, and agronomic treatments. *Proceedings of the Australian Society of Sugarcane Technologists* 6, 69-77.

Davidson, B. (1962) Crop yields in experiments and on farms. *Nature* 194, 458-459.

Denmead, O.T., Freney, J.R., Jackson, A.V., Smith, J.W.B., Saffigna, P.G., Wood, A.W. and Chapman, L.S. (1990) Volatilisation of ammonia from urea and ammonium sulphate applied to sugarcane trash in north Queensland. *Proceedings of the Australian Society of Sugarcane Technologists* 12, 72-78.

Egan, B.T. (1982) The theoretical basis for the development of the Bundaberg Fiji disease epidemic. *Proceedings of the Australian Society of Sugarcane Technologists* 4, 97-101

Egan, B.T. and Fraser, T.K. (1977) The development of the Fiji disease epidemic at Bundaberg. *Proceedings of the Queensland Society of Sugarcane Technologists, 1977 Conference,* pp. 43-48.

Egan, B.T., Hurney, A.P., Ryan, C.C. and Matthews, A.A. (1984) A review of the northern poor root syndrome of sugarcane in north Queensland. *Proceedings of the Australian Society of Sugarcane Technologists* 6, 1-10.

Ford, E.J. and Bristow, K.L. (1995a) Soil physical properties of several sugarcane producing soils in north Queensland. I. Soil strength, soil moisture, infiltration, saturated hydraulic conductivity, and bulk density. CSIRO Australia Division of Soils, Technical Report, 6/1995.

Ford, E.J. and Bristow, K.L. (1995b) Soil physical properties of several sugarcane producing soils in north Queensland. II. Soil moisture retention, bulk density, and particle size distribution. CSIRO Australia Division of Soils, Technical Report, 7/1995.

Fuelling, T.G., Henkel, C.R., Leverington, K.C. and Wegener, M.K. (1978) Sugarcane harvester performance. *Proceedings of the Queensland Society of Sugarcane Technologists, 1978 Conference,* pp. 209-216.

Garside, A.L. (1995) *Yield Decline Joint Venture. BSES Farmer Information Meetings 1995*, BSES Occasional Publication, Brisbane, Australia.

Garside, A.L., Magarey, R.C. and Nable, R.O. (1995) Growth of different plant species in fumigated/sterilised and untreated sugarcane soil with varying cropping histories. *Proceedings of the Australian Society of Sugarcane Technologists* 17, 123-127.

Garside, A.L. and Nable, R.O. (1996) Sugarcane growth and yield comparisons in paired old and new land sites. In: Wilson, J.R., Hogarth, D.M., Campbell, J. and Garside, A.L. (eds), *Sugarcane: Research Towards Efficient and Sustainable Production.* CSIRO, Division of Tropical Crops and Pastures, Brisbane, Australia, pp. 248-250.

Hardman, J.R., Tilley, L.G.W. and Glanville, T. (1985) Agronomic and economic aspects of various farming systems for sugarcane in the Bundaberg district. *Proceedings of the Australian Society of Sugarcane Technologists* 7, 147-153.

Hendrix, J.W., Jones, K.J. and Nesmith, W.C. (1992) Control of pathogenic mycorrhizal fungi in maintenance of soil productivity by crop rotation. *Journal of Production Agriculture* 5, 383-386.

Hogarth, D.M. (1972) Production trends in Queensland from 1899 to 1970. *Canegrowers Quarterly Bulletin,* Vol. XXXV, No. 4, pp. 113-116.

Hurney, A.P. (1984) Excess nitrogen is detrimental to crop yields. *BSES Bulletin* 8, pp. 18-19.

Hurney, A.P. and Ridge, D.R. (1992) Phytotoxic substances associated with the decomposition of sugarcane trash residues. BSES Internal Report, Brisbane, Australia.

Hurney, A.P., Ridge, D.R. and Dick, R.G. (1984) Evaluation of the efficiency of cane harvesters in removing extraneous matter and in limiting cane loss during the cleaning process. *Proceedings of the Australian Society of Sugarcane Technologists* 6, 11-19.

Hyde, R. and Teske, L. (1987) Surface management systems in the central district canelands. *Proceedings of the Australian Society of Sugarcane Technologists* 9, 45-48.

Johnson, A.K.L. (1995) Risk perceptions and nutrient management responses in the Australian sugar industry: preliminary results from the Herbert River District. *Proceedings of the Australian Society of Sugarcane Technologists* 17, 172-178

Johnson, N.C., Copeland, P.J., Crookston, R.K. and Pfleger, F.L. (1992) Mycorrhizae: possible explanation for yield decline with continuous corn and soybean. *Agronomy Journal* 84, 387-390.

Keating, B.A., Vallis, I., Parton, W.J., Catchpoole, V.R., Muchow, R.C. and Robertson, M.J. (1994) Modelling and its application to nitrogen management and research for sugarcane. *Proceedings of the Australian Society of Sugarcane Technologists* 16, 131-142.

Keating, B.A., Verburg, K., Huth, N.I. and Robertson, M.J. (1997) Nitrogen management in intensive agriculture: the case for sugarcane in Australia. In: Keating, B.A. and Wilson, J.R. (eds), *Intensive Sugarcane Production: Meeting the Challenges Beyond 2000.* CAB International, Wallingford, UK, pp. 221-242.

King, N.J. (1969) Historical review of sugarcane mechanisation in Australia. *Proceedings South African Sugar Technologists Association,* pp. 93-98.

King, N.J. and Steindl, D.R.L. (1953) The relationship between variety yield deterioration and ratoon stunting disease. *Proceedings International Society of Sugarcane Technologists, VIII Congress,* British West Indies, pp. 851-860.

Linedale, A.I. and Ridge, D.R. (1996) A successful campaign to minimise harvesting losses within the Queensland sugar industry. *Proceedings of the Australian Society of Sugarcane Technologists* 18, 1-5.

Linedale, A.I., Ridge, D.R. and Chapman, F.C. (1993) A coordinated extension approach for maximising returns from mechanical harvesting. *Proceedings of the Australian Society of Sugarcane Technologists* 15, 45-51.

Magarey, R.C. (1994) The effect of Pachymetra root rot on sugarcane yield. *Plant Disease* 78, 475-477.

Magarey, R.C. (1996) Microbial aspects of sugarcane yield decline. *Australian Journal of Agricultural Research* 47, 307-322.

Magarey, R.C. and Croft, B.J. (1995) A review of root disease research in Australia. In: Cock, J.H. and Brekebbaum, T. (eds), *Proceedings International Society of Sugarcane Technologists, XXII Congress,* Colombia, Vol. 2, pp. 505-513.

Magarey, R.C., Yip, H.Y., Bull, J.I. and Johnson, E.J. (1995) Recent studies into the soil biology of yield decline. *Proceedings Australian Society of Sugarcane Technologists* 7, 128-133.

Manglesdorf, A.J. (1959) Genetic aspects of yield decline. *Proceedings International Society of Sugarcane Technologists, X Congress,* pp. 72-76.

Maxwell, M. (1900) *Report upon an Investigation into the Condition of the Sugar Industry in Queensland.* Department of Agriculture, Brisbane, Australia.

Moller, R.B. (1974) Some factors influencing CCS of irrigated cane in southern Queensland. *Proceedings of the Queensland Society of Sugarcane Technologists, 1974 Conference,* pp. 37-41.

Moore, P.H., Heinz, D.J. and Osgood, R.V. (1993) The yield decline conundrum: variety yield decline, fact or fiction? *Hawaiian Sugar Planters Association, Experiment Station Journal,* Paper No. 763.

Muchow, R.C. and Robertson, M.J. (1994) Relating crop nitrogen uptake to sugarcane yield. *Proceedings of the Australian Society of Sugarcane Technologists* 16, 122-130.

Muchow, R.C., Robertson, M.J., Wood, A.W. and Spillman, M.F. (1994) Effect of soil fumigation on sugarcane productivity under high yielding conditions in north Queensland. *Proceedings of the Australian Society of Sugarcane Technologists* 16, 187-192.

Muchow, R.C., Wood, A.W. and Robertson, M.J. (1995) Does stalk death set the yield ceiling in high yielding sugarcane crops? *Proceedings of the Australian Society of Sugarcane Technologists* 17, 142-148.

Patch, G., Fuelling, T.G. and Williams, N.D. (1986) Farm productivity - results from Tully CAPA. *Proceedings of the Australian Society of Sugarcane Technologists* 8, 115-121.

Pitcher, R.S., Way, D.W. and Savory, B.M. (1966) Specific replant diseases of apple and cherry and their control by soil fumigation. *Journal of Horticultural Science* 41, 379-396.

Prammanee, P., Saffigna, P.G., Wood, A.W. and Freney, J.R. (1989) Loss of nitrogen from urea and ammonium sulphate applied to sugarcane crop residues. *Proceedings of the Australian Society of Sugarcane Technologists* 11, 76-84.

Prove, B.G., Truong, P.N. and Evans, D.S. (1986) Strategies for controlling caneland erosion in the wet tropical coast of Queensland. *Proceedings of the Australian Society of Sugarcane Technologists* 8, 77-84.

Reghenzani, J.R. (1990) *Effect of Zinc Deficiency as a Factor Limiting Sugarcane Growth.* BSES Publication, Final Report, SRDC Project BS14S, Brisbane, Australia.

Reghenzani, J.R. (1993) *Extent of Zinc Deficiency in Cane Growing Soils of North Queensland.* BSES Publication, Final Report SRDC Project BS67S, Brisbane, Australia

Ridge, D.R. (1983) Comparison of different cropping cycles at Tully and Bundaberg. *BSES Bulletin* No. 4, pp. 14.

Ridge, D.R. (1984) *Review of Research into Sugarcane Nutrition.* BSES Project Report, No. 102.30.000, Brisbane, Australia.

Ridge, D.R. and Dick, R.G. (1987) A new method for testing cane harvester performance. *Proceedings of the Australian Society of Sugarcane Technologists* 9, 87-92.

Ridge, D.R. and Dick, R.G. (1988) Current research on green cane harvesting and dirt rejection by harvesters. *Proceedings of the Australian Society of Sugarcane Technologists* 10, 19-25.

Ridge, D.R., Hurney, A.P. and Haysom, M.B.C. (1980) Some aspects of calcium and magnesium nutrition in north Queensland. *Proceedings of the Australian Society of Sugarcane Technologists* 2, 55-61.

Roach, B.T. and Daniels, J. (1987) A review of the origin and improvement of sugarcane. *Proceedings Copersucar International Sugarcane Breeding Workshop,* Piracicaba, Brazil, pp. 15-21.

Roach, B.T., Quinan, D.J. and Parsons, D.H. (1981) Some aspects of land utilisation or rotation in relation to farm production and productivity. *Proceedings of the Australian Society of Sugarcane Technologists* 3, 319-24.

Ryan, C.C. and Ledger, P.E. (1980) Studies and observations of rust disease in Queensland. *Proceedings of the Australian Society of Sugarcane Technologists* 2, 71-73.

Shaw, G.R. and Brotherton, G.A. (1992) Green cane harvesting - a dilemma. *Proceedings of the Australian Society of Sugarcane Technologists* 14, 1-7.

Simms, E. (1982) When to replant: a re-examination. *Sugarcane* 12, 1- 8.

Sinclair, T.S. (1997) Yield 'plateaus' in grain crops: the topography of yield increase. In: Keating, B.A. and Wilson, J.R. (eds) *Intensive Sugarcane Production, Meeting the Challenges Beyond 2000*, CAB International, Wallingford, UK, pp. 87-102

Smith, M.A. (1991) *Recent Production Trends in the Queensland Sugar Industry.* BSES Occasional Publication, Brisbane, Australia, pp. 65.

Smith, M.A. (1993) Is green cane trash blanketing as productive as burnt cane alternatives? BSES Technical Report, TE93003, Brisbane, Australia.

Smith, N.J., McGuire, P.J., Mackson, J. and Hickling, R.C. (1984) Green cane harvesting - a review with particular reference to the Mulgrave mill area. *Proceedings of the Australian Society of Sugarcane Technologists* 6, 21-27.

Spain, A.V. and Hodgen, M.J. (1994) Changes in the composition of sugarcane harvest residues during decomposition as a surface mulch. *Biological Fertility of Soils* 17, 225-231.

SRDC (1995) *Research and Development Plan 1995-2000.* Sugar Research and Development Corporation, Brisbane, Australia.

Stewart, R.L. (1987) Agronomic aspects of green cane farming. *Proceedings Green Cane Symposium,* CSR Technical Field Department., Brisbane, Australia, pp. 5-10.

Stewart, R.L. and McComiskie, G. (1988) A comparison of green and burnt cane harvesting in the Burdekin. *Proceedings of the Australian Society of Sugarcane Technologists* 10, 27-31.

Sturgess, O.W., Hogarth, D.M., Hurney, A.P. and Leverington, K.C. (1972) Factors affecting sugar content of sugar crops in the wet tropics of Queensland - statistical analyses of CCS data at Babinda and Tully. BSES Research Report, December 1972, Brisbane, Australia.

Taylor, P.W.J., Croft, B.J. and Ryan, C.C. (1986) Studies into the effect of sugarcane rust *(Puccinia melanocephala)* on yield. *Proceedings International Society of Sugarcane Technologists, XIX Congress,* Jakarta, Indonesia, pp. 411-419.

Toohey, C.L. and Neilsen, P.J. (1972) Fiji disease at Bundaberg. *Proceedings of the Queensland Society of Sugarcane Technologists, 1972 Conference,* pp. 191-196.

Turner, G.N. and Churchward, E.H. (1977) Assessing the direct losses from Fiji disease. *Proceedings of the Queensland Society of Sugarcane Technologists, 1977 Conference,* pp. 71-73.

Vos, J. and van Loon, C.D. (1989) Effect of cropping frequency on potato production. In: Vos, J., van Loon, C.D. and Bollen, G.J. (eds), *Effects of Crop Rotation on Potato Production in the Temperate Zones.* Kluwer Academic Publishers, Dirdrecht, The Netherlands, 310 pp.

Wegener, M.K. (1985) The contribution of science to Australian tropical agriculture. IV. The sugar industry. *Journal Australian Institute of Agricultural Science* 51, 29- 41.

Wegener, M.K. and Pollock, J.S. (1986) Yield of ploughout-replant crops in the Burdekin district. *Proceedings of the Australian Society of Sugarcane Technologists* 8, 109-114.

Wood, A.W. (1985) Soil degradation and management under intensive sugarcane cultivation in north Queensland. *Soil Use and Management* 1, 120-124.

Wood, A.W. (1986) Green cane trash management in the Herbert Valley. *Proceedings of the Australian Society of Sugarcane Technologists* 8, 85-94.

Wood, A.W. (1991) Management of crop residues following green harvesting of sugarcane in north Queensland. *Soil and Tillage Research* 20, 69-85.

Wood, A.W., Saffigna, P.G., Prammanee, P. and Freney, J.R. (1990) Loss of nitrogen from urea applied to sugarcane trash in north Queensland. *Proceedings International Society of Sugarcane Technologists, XX Congress,* pp. 481-488

Chapter eight:

Transgenic Sugarcane: Opportunities and Limitations

R.G. Birch
Department of Botany, The University of Queensland, Brisbane, Qld 4072, Australia

Abstract

An efficient system is now in place for improvement of diverse sugarcane cultivars by genetic transformation: the insertion of useful new genes into single cells followed by the regeneration of genetically modified (transgenic) plants. The method has already been used to introduce genes for resistance to several major diseases, insect pests and a herbicide. Field testing has commenced to establish the stability of introduced gene expression, and the extent of random genetic variation associated with the genetic transformation process in sugarcane. Research is underway to identify and introduce other genes for increased environmental stress resistance, agronomic efficiency and yield of sucrose or other valuable products. Experience in other crops has shown that genetically improved varieties which provide genuine environmental and consumer benefits are welcomed by producers and consumers. The regulatory processes to ensure product safety are likely to become more efficient as the technology becomes more widely understood. A substantial research investment is required into useful new genes and sequences controlling expression of introduced genes in sugarcane. These new gene technologies will reshape the sugar industry and determine the international competitive efficiency of producers. They are also eligible for patent protection, allowing control of the technology for at least 20 years by the inventors. Industries must therefore consider whether to participate as creators of key technologies which they will need to access for continued profitability.

In: Keating, B.A. and Wilson, J.R. (eds), *Intensive Sugarcane Production: Meeting the Challenges Beyond 2000*. CAB International, Wallingford, UK, 1997.

Genetic Transformation for Sugarcane Improvement

The first field tests of sugarcane varieties improved by genetic transformation commenced in 1996, and the first products from other crops improved in this way are already in the marketplace. This technology will revolutionise the sugarcane industry, and it is vital for those who will participate beyond the year 2000 to be aware of the technological and commercial challenges and opportunities from transgenic sugarcane.

What is genetic transformation?

Plant genetic transformation in a broad sense can include any process, other than pollination, which results in the formation of stable new combinations of heritable material in living plants. Most recent progress has been made in techniques to introduce specific isolated genes into an existing plant variety, and this is the context in which the term will be used throughout this paper. Synonymous terms include plant genetic modification, genetic manipulation, genetic engineering, molecular improvement or molecular plant breeding.

Why genetic transformation is useful in sugarcane

Modern commercial sugarcane varieties are obtained through a multi-stage selection scheme over a period of approximately 10 years to identify a few elite clones in very large populations of seedlings. The costs of selection approach $1M per cultivar from advanced breeding programs. Some elite clones have to be completely abandoned because of a single fault such as disease susceptibility. Traditional breeding approaches to correct such faults in an existing cultivar are impractical in sugarcane, because of the genetic complexity of cultivars and the long periods needed for each round of selection for some agronomically important traits such as ratooning ability. Genetic transformation can correct single faults in elite cultivars, by insertion of a specific gene. It can help us to understand the role of existing sugarcane genes in complex traits such as yield or sugar content, and it can introduce valuable 'novel' genes for new properties in sugarcane.

It is now clear that genes from virtually any source can be modified for expression in plants. This makes genetic transformation a powerful accessory for sugarcane improvement, allowing:

1. Transfer into sugarcane of novel useful genes (e.g. genes for insect or pest resistance, or genes for value-added sugars).
2. Inactivation of known harmful sugarcane genes, even if linked to useful genes (e.g. inactivation of genes for lignification resulting in excessive fibre content).
3. Molecular switches for genes required only under special circumstances (e.g. ability to produce higher yielding non-flowering commercial versions of varieties while maintaining versions with reliable flower induction as elite parents).

4. Direct testing of hypotheses about genetic factors limiting yield (e.g. photosynthesis *vs* sugar transport *vs* storage capacity *vs* remobilisation as factors limiting sugar accumulation).
5. Specific alteration of existing genes to modify expression patterns or products (e.g. altered enzyme properties for enhanced sucrose synthesis, or altered expression patterns for reduced remobilisation out of storage tissues).

Transgenic Sugarcane Will Reshape the Industry

The first generation of transgenic sugarcane varieties will improve agronomic efficiency through higher, more consistent yields and reduced need for emergency measures such as pesticide application. However, there is also a real prospect of increasing the value of the crop by engineering sugarcane to produce useful products in addition to sucrose (Fig. 8.1). Indeed this latter prospect has the potential to dramatically reshape, expand and diversify the industry as we harness the extraordinary capacity of this crop for both high biomass and high value chemical production.

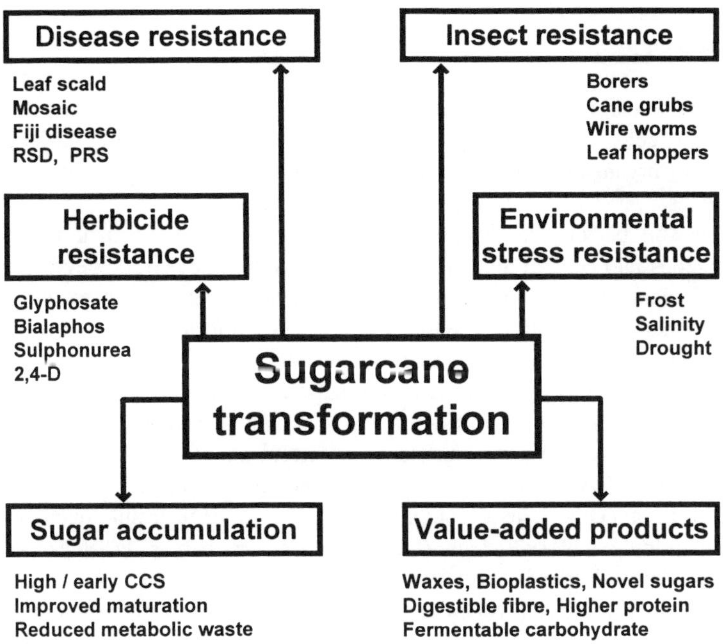

Fig. 8.1. Applications of sugarcane transformation for sugarcane improvement. [Modified from Birch, 1993]

In the past, the international leaders in sugar production have relied on a competitive technological edge in superior varieties from breeding programs, skilled extension workers and farmers implementing best practices in agronomy and crop protection, high efficiency from on-farm mechanisation, and advanced milling and refining technology. These technologies will be essential in the future, but for the most part they are impossible to protect, so the technological gap will narrow through wider adoption of these practices until the winning combinations of the past several decades become the baseline for competitive sugar production.

The technologies supporting production of transgenic sugarcane varieties can be protected internationally under patent laws. These gene technologies are a quantum advance for plant improvement. They will determine competitive efficiency in the decades from the year 2000 in the same way that the quantum advance to mechanical harvesting determined competitive efficiency from the 1960s. The industries or companies that control and apply these technologies will have the competitive edge in sucrose production, and the lead in development of new markets for value-added products from sugarcane.

Technology for Sugarcane Transformation

Gene transfer

All genetic transformation techniques work by inserting genetic material (DNA) into single cells, followed by the regeneration of genetically modified (or **transgenic**) organisms from individual cells which have stably incorporated the added genes.

Some soil bacteria in the genus *Agrobacterium* have the natural ability to transfer genes into cells of host plant species. This has allowed the development of a very efficient process using modified *Agrobacterium* strains to transfer useful genes into many dicot plants (Grant *et al.*, 1991). Unfortunately, monocot plants including sugarcane are not natural hosts for *Agrobacterium*. Attempts to apply *Agrobacterium*-mediated transformation to sugarcane have not yet been successful (Birch and Maretzki, 1993), but recent success with rice may indicate conditions which can be adapted for use with sugarcane (Hiei *et al.*, 1994).

When *Agrobacterium* is not effective, the alternative is to transfer DNA directly into plant cells. This is complicated by the presence around plant cells of a wall containing cellulose, which is a very effective barrier to entry by DNA molecules. One solution is to use enzymes to strip off the cell walls, and transfer DNA into the stripped cells (or **protoplasts**) after treatment with chemicals such as polyethylene glycol (PEG) or with high voltage electrical pulses (**electroporation**) to make temporary pores in the cell membrane. This has allowed the production of transformed sugarcane cells, but it has proved practically impossible to regenerate the protoplasts back to intact plants (Chen *et al.*, 1987; Rathus and Birch, 1992; Taylor *et al.*, 1992a). The breakthrough in production of transgenic sugarcane plants came with the development of techniques to shoot DNA into intact plant cells on high velocity microprojectiles (Franks and Birch, 1991a; Bower and Birch, 1992).

To genetically transform sugarcane, DNA containing the purified genes of interest is precipitated onto the surface of tungsten or gold microprojectiles, about 1μm in diameter. The DNA-coated particles are accelerated to velocities around 1000km h^{-1}, which gives them sufficient momentum to penetrate the cell walls. Embryogenic sugarcane callus is bombarded in a vacuum chamber, which helps to maintain particle velocity. Many cells are penetrated by DNA-coated particles during each bombardment, and survive to express the introduced genes. The development and practical application of particle bombardment for plant transformation has been reviewed elsewhere (Birch and Franks, 1991; Franks and Birch, 1991b; Birch and Bower, 1994). Fig. 8.2a shows a schematic drawing of a typical **particle bombardment** device.

Recently, evidence has been presented for transfer of foreign genes into intact and regenerable sugarcane cells by electroporation (Arencibia *et al.*, 1995). This will be of substantial interest if unequivocal molecular tests confirm the efficient production of stably transformed plants, as previously shown for particle bombardment.

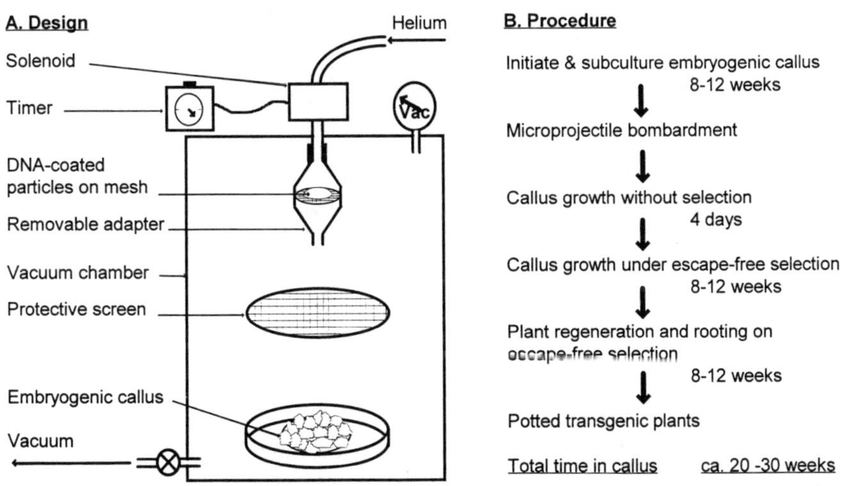

Fig. 8.2. a) Schematic design of apparatus, and b) Procedure for selection of transgenic sugarcane plants following particle bombardment. [Modified from Birch, 1993]

Regeneration and selection of transgenic plants

Most gene transfer methods, including particle bombardment, can only transfer genes into a small proportion of treated cells. Only a small proportion of these cells will typically survive, integrate the foreign DNA, and be capable of regeneration into a plant. Careful background work to establish conditions which optimise these frequencies underlies most successful transformation methods. Fortunately, research into sugarcane tissue culture has established an excellent embryogenic callus culture and plant regeneration system (Ahloowalia and Maretzki, 1983; Chen *et al.*, 1988; Fitch and Moore, 1990; Taylor *et al.*, 1992b). **Embryogenic callus** is a proliferating mass of cells, produced in the laboratory from young sugarcane leaves. Individual cells in this callus can divide and organise embryo-like structures which can further develop into sugarcane plants identical to the variety used to establish the callus (Fig. 8.3). Embryogenic callus culture is simple and reproducible to apply. Varieties differ substantially in efficiency of callus initiation, proliferation, maintenance and regeneration, but embryogenic callus suitable for transformation has been obtained for most tested sugarcane cultivars by experimentation with key parameters such as phytohormone levels.

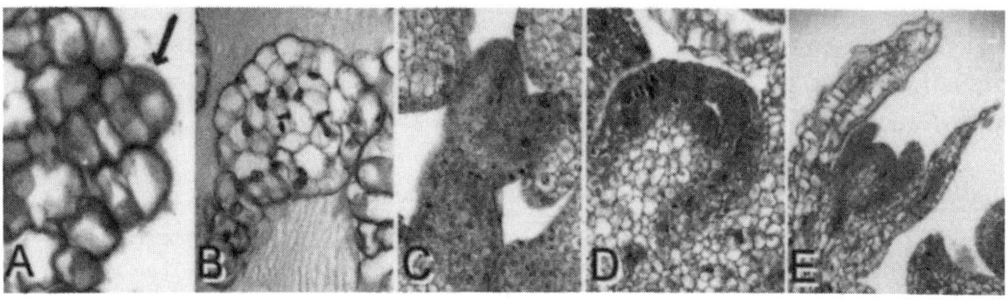

Fig. 8.3. Stages in regeneration of a sugarcane plant from a single cell in embryogenic callus. A) Four cell tier (arrow) after two cell divisions. B) Globular embryo structure. C) Torpedo embryo stage. D) Shoot apex with coleoptile sheath. E) Advanced shoot with leaves. [Micrographs by Isnaini Nurwahyuni]

In sugarcane, particle bombardment conditions have been established which result in expression of introduced genes by thousands of cells per 'shot' into highly regenerable embryogenic callus tissue (Fig. 8.4A). The bombarded callus is allowed to proliferate so that embryos form from individual cells which have stably incorporated the introduced genes (Fig. 8.4B), allowing regeneration of plants consisting entirely of transformed cells (Fig. 8.4C). The anthocyanin pigment **'reporter' genes** allowing visual selection of transformants (Fig. 8.4) are not useful for practical transformation in sugarcane, because the pigmented plants are weakened. With the help of a low light camera, visual selection of light emitting transformants is possible following introduction of a luciferase reporter gene. However, this approach is inefficient because of the need to maintain and screen many untransformed calli and plants in order to detect the low proportion of transformants. It is more efficient to use a **'selectable marker' gene** which renders transformed cells resistant to an antibiotic which can be applied to kill untransformed cells soon after bombardment. For example, the *aphA* gene allows selection of transformants based on geneticin antibiotic resistance (see Bower *et al.*, 1996 for details of these genes). A typical selection regimen is shown in Fig. 8.2b.

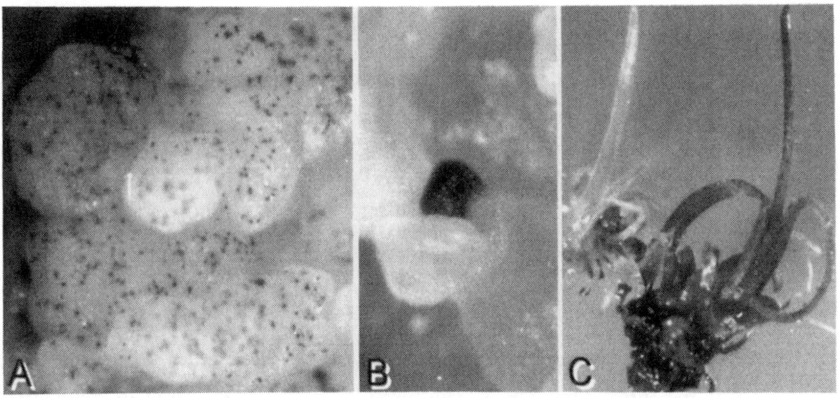

Fig. 8.4. Anthocyanin pigment production from expression of a maize reporter gene in transformed sugarcane. A) Transient expression in bombarded callus. B) Stable transformation in an embryo. C) Uniformly transformed plant.

The sugarcane transformation system described above has been used to produce transgenic sugarcane plants expressing several reporter and marker genes. Stable integration of these genes into sugarcane chromosomes has been demonstrated using unequivocal molecular tests. Transgenic plants have been produced from diverse sugarcane cultivars in Queensland (including NCo310, Pindar, Q63, Q87, Q96, Q117, Q124, Q138, Q142, Q155, 78N-146), and the technique appears to be readily applicable to any cultivar. The current transformation frequency is around ten transgenic plants per bombardment, which is sufficient for introduction of known useful genes for sugarcane improvement (Birch *et al.*, 1995; Bower *et al.*, 1996). The method with modifications has also been successfully applied in Hawaii (Sun *et al.*, 1993), Texas (Gallo-Meagher and Irvine, 1995) and South Africa (F.C. Botha, *pers. comm.*)

Characterisation of Transgenic Sugarcane Lines

Cotransformation and coexpression of unselected genes

With the exception of herbicide resistance, it is generally not possible to select directly for transformation with genes of agronomic interest. However, such genes can be mixed with a selectable marker gene in the DNA solution precipitated onto the particles for bombardment. A proportion of transformed plants selected for resistance to the antibiotic will also have integrated the desired agronomic gene (**cotransformation**). If necessary this can be determined by molecular tests, to reduce the number of transformants grown to the stage where they can be tested for expression of the desired agronomic gene (**coexpression**). For practical purposes, it is desirable to have a high frequency of cotransformation and coexpression. Fortunately, this appears to be the case for the optimised particle bombardment system. When *luc* and *aphA* on separate plasmids were coprecipitated onto microprojectiles before bombardment, 70 to 80% of lines selected for geneticin resistance also showed luciferase activity. The number of copies of the introduced genes varied between lines, with a mean of approximately 5 copies (Bower *et al.*, 1996).

Stability of transgene expression

An unexpected finding from the first studies of transgenic sugarcane plants was that expression of the introduced genes was turned off after some time in many lines (**transgene silencing**) (Birch *et al.*, 1995). Gene expression is controlled by sequences (**promoter regions**) generally located immediately 'upstream' of the gene, which determine the strength, developmental timing and tissue specificity of expression of the adjacent gene. Sometimes this control is influenced by more distant sequences (**enhancers or silencers**), or by broader influences such as condensation of the chromosome surrounding the site of transgene insertion. Control can also operate later in the process of gene expression, for example on stability of the protein product from the gene. The promoters used in early sugarcane transformation studies were shown to be strongly expressed in bombarded cells, and thought to be expressed more or less continuously in most cell types (strong **constitutive promoters**). Silencing of introduced genes has been observed in a proportion of transformants in other plant species, but in sugarcane silencing occurred at around the time of plant regeneration in virtually every line with some foreign promoters. As a result there was no correspondence between promoter strength as measured soon after bombardment of genes into sugarcane cells (**transient expression**), and the presence or level of expression in transgenic plants (**stable expression**) (Birch *et al.*, 1995).

This is of substantial practical concern, because predictable and stable expression of introduced genes is an absolute prerequisite for use of genetic engineering for sugarcane varietal improvement. One explanation for silencing is that foreign or artificial promoters may lack some undefined features necessary for stable gene expression in sugarcane. For many genes, we will also need tissue-specific or inducible rather than strong constitutive expression. For these reasons it makes sense to isolate desired promoter types from sugarcane. Unfortunately, recovering promoters from sugarcane is complicated by the highly polyploid genome. Several promoter-like regions recovered and tested to date have also been inactive or silenced when coupled to a reporter gene and reintroduced into sugarcane (Birch *et al.*, 1995).

Of promoters tested to date, the artificial Emu promoter has driven stable expression of the selectable *aphA* gene in many transformed lines, and the maize ubiquitin 1 (Ubi) promoter has driven stable expression of several unselected reporter genes in a high proportion of transformants. In both cases, these genes have been transmitted normally to progeny through setts and expression has been maintained in the plant lines over several years of testing. However, expression levels in transgenic sugarcane are low relative to reported levels from strong constitutive promoters in transgenic dicotyledonous plants. As discussed below, research into promoters for sugarcane is a priority for practical application of transformation technology.

Genetic integrity and field performance of transgenic cultivars

The greatest immediate benefits from sugarcane genetic engineering are anticipated from introduction of one or a few genes into elite commercial or near-commercial varieties, to correct faults or add desired traits. Sugarcane is a clonally propagated crop with a long and expensive selection process to reveal elite lines from conventional breeding with highly heterozygous parent lines. Back-crossing to move an introduced gene into commercial varieties is less practical than in genetically simpler crops. In sugarcane, we need to be able to introduce genes into elite cultivars while avoiding any undesired changes to cultivar performance as a result of the gene transfer process. Under some conditions, plants regenerated from tissue culture can show genetic changes (**somaclonal variation**), and in highly selected varieties such random changes are likely to be harmful to commercial performance.

Plants regenerated from tissue culture may also show unusual growth due to lingering effects of altered physiological responses or patterns of gene expression in the tissue culture environment, without any permanent genetic changes. Provided plant lines resume normal growth patterns within a few vegetative generations in the glasshouse or field, these temporary changes (**epigenetic effects**) are not a problem. Major genetic rearrangements can be detected by fast molecular tests (e.g. RAPD analysis), but single base mutations could also affect varietal performance detectable only by field testing. One way to reduce the frequency of somaclonal variation is to minimise the time in tissue culture, as in the procedure shown in Fig. 8.2b.

Under glasshouse conditions, up to 5% of transformed sugarcane lines show changes in growth habit in primary regenerants, but in most cases these appear to be epigenetic effects which are not evident in vegetative progeny. RAPD analysis reveals drastic genetic changes in protoplast-derived sugarcane callus lines, but no detectable changes in a range of plant lines transformed by particle bombardment (Taylor *et al.*, 1995). The first results from replicated yield trials, including some 70 transformed lines of five sugarcane cultivars at two locations, should be available during 1996. These results will be very important to establish whether somaclonal variation occurs at a troublesome frequency following sugarcane transformation. Some of these lines will also be used to test the pattern of inheritance of the introduced genes through seed, and whether there is any transmission from sugarcane to other grasses through pollen. These results will indicate the potential for use of introduced genes in conventional breeding, and for undesired movement of introduced genes from sugarcane into potential weeds.

Genetic Transformation and Conventional Breeding

It is important to note that genetic transformation complements, but does not replace conventional breeding for cultivar improvement. A complex multicellular organism such as sugarcane is the product of controlled expression of hundreds of thousands of genes. At present, genetic transformation technology allows us to manipulate only one or a few genes at a time. Single genes can have major effects, and part of the power of genetic engineering is its ability to make very precise changes. Single novel genes will revolutionise the sugarcane industry. However, our knowledge of the genetic and physiological basis for many agronomic traits is rudimentary. For maximum benefit from genetic transformation, research to improve this understanding, and thereby to reveal the genes which should be targeted for modification, will be essential.

For some agronomic traits, molecular tools other than transformation will be most important for variety improvement. For example, international efforts are under way to find molecular markers linked to agronomically important genes in sugarcane, and perhaps to provide a map of the locations of these genes on the sugarcane chromosomes (D'Hont *et al.*, 1995; Moore, 1995). In the medium term this should help the selection process in conventional sugarcane breeding, and in the long term it may allow specific useful genes to be identified, isolated and modified for cultivar improvement. Where a large number of genes each contributes a little to a characteristic, improvement by conventional breeding may be by far the most efficient approach.

Regulatory Environment

Internationally, over 1000 field trials of transgenic plants were conducted by 1993, involving genes intended to improve insect pest resistance, viral and fungal pathogen resistance, weed control, hybrid production, product flavour or other post-harvest qualities, in several crop species (Fig. 8.5). Testing approximately doubled during 1994, involving many additional species, including sugarcane.

There have been no reports of any harmful environmental effects or other unforseen behaviour of transgenic plants in the field. Regulatory processes are being streamlined, with the focus on products rather than processes of plant genetic modification (Huttner *et al.*, 1992). In most countries, planned field releases are first scrutinised and approved by regulatory authorities established by the federal government, to ensure that products are safe to the environment and consumers. Depending on the efficiency of the regulatory authorities, this can slow some phases of research, but it is an important mechanism to reassure the general public of the safety of a new technology which is not well understood by most citizens. Like any promising new crop variety, genetically modified lines will need thorough field testing before commercial cultivation, to assure growers that they will perform as expected, with stable expression of the modified trait and no changes in other desired characteristics of the parent material.

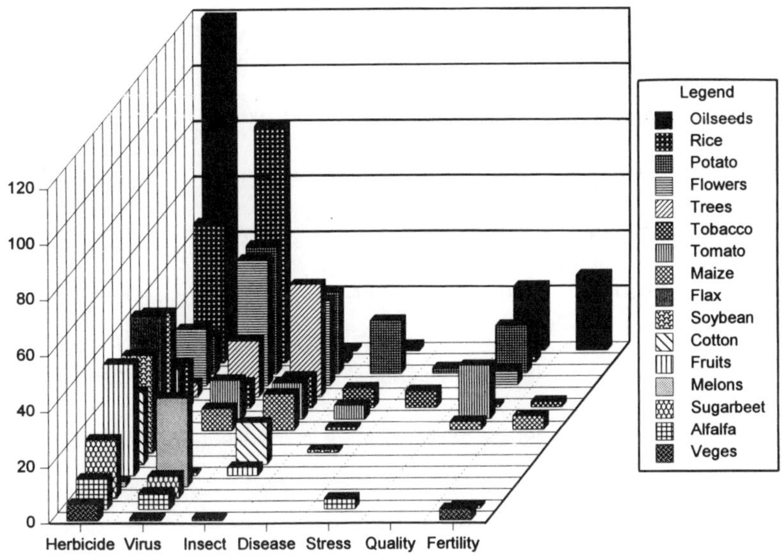

Fig. 8.5. Field trials of transgenic plants until 1993. [Data updated from Gunary, 1992]

As with conventional breeding, it is highly undesirable to move to excessive genetic uniformity in current varieties in any crop. Even a single gene in all varieties can create problems if it leads to unforseen disease susceptibility. For example, the USA maize crop was devastated in 1970 because of disease susceptibility accompanying a gene which was made ubiquitous by conventional breeding to simplify hybrid seed production (Zadoks and Schein, 1979). This lesson should not be forgotten as valuable genes are introduced into cultivars by transformation. In several countries, release of transgenic insect-resistant plant varieties has been linked to mandatory programs of insect monitoring and industry responses to avoid premature loss of useful insect control genes due to a build-up of resistant insect populations.

Consumer response to transgenic plant products has now been tested with the commercial release of improved varieties in a range of crops (Table 8. 1). In each case, consumers have responded positively to quality or price advantages, and demand has outstripped supply in the first season.

Table 8. 1. Commercial releases of transgenic plant varieties.

Name	Crop	Trait	Company	Release status
Flavr Savr	Tomato	Quality (vine ripened flavour, shelf life)	Calgene / Monsanto	Released 1994
Endless Summer	Tomato	Quality (vine ripened flavour, shelf life)	DNA Plant Technology	Blocked by patent claims
Laurical	Canola	Oil characteristics	Calgene	Released 1994
BXN	Cotton	Herbicide resistance	Calgene	Approved 1994
Freedom II	Squash	Virus resistance	Asgrow	Released 1995
Roundup Ready	Soybean	Herbicide resistance	Monsanto	Released 1995
Bollgard	Cotton	Insect resistance	Monsanto	Approved 1995
NewLeaf	Potato	Insect resistance	Monsanto	Approved 1995
Maximiser	Maize	Insect resistance	Ciba Seeds	Approved 1996

Intellectual Property

As technical limitations are overcome, it is possible that commercial limitations will become more serious barriers to exploitation of genetic transformation. New technologies developed in this area are effectively inventions, and are therefore eligible for patent protection which may require intending users to negotiate licence agreements. For example, patents have already been issued on use of many isolated genes and promoters, and on techniques for gene manipulation.

In contrast to some other major crop species (Wrage, 1992a), no patents have been issued on sugarcane transformation *per se*, and this technology is believed to be in the public domain. However, some of the underlying technology may be protected in some countries. This can be difficult to establish without periodic patent searching, or even legal challenges (Hecter, 1995). For example, the process of particle bombardment for gene transfer into plant cells has been patented in the USA (Sanford *et al.*, 1990) but not in Australia. Apparatus for particle bombardment involving a macroprojectile and stop plate has been patented in Australia (Sanford *et al.*, 1992), but not apparatus involving particle acceleration in a gas pulse, unless it is directed through a fine pore or capillary to precisely target a small group of cells such as a meristem (Sautter *et al.*, 1993).

Patent coverage for specific genes and promoters can be similarly restricted. For example, the maize ubiquitin promoter is the subject of granted patents or applications in Europe, Japan and the USA (Quail *et al.*, 1994), but apparently not in major sugarcane producing countries including Australia.

Under these circumstances, a transgenic sugarcane variety produced and used commercially without infringing any patent rights in one country could infringe certain patent claims if used (even for research or other non-commercial purposes) in another country with a different array of granted patents. To complicate matters further, patent applications are not available publicly in some countries until the patent is granted, which can be years after the application is filed.

Of course, patents are intended to encourage and reward useful invention and technical innovation, and the new technology enters the public domain after a period of 16 to 20 years. In the interim, however, commercial restrictions can appear quite ruthless as patent holders adopt commercialisation strategies to capture the value of protected intellectual property. There is a trend to cross-licensing agreements between organisations with claims to mutually useful technologies (Wrage, 1992b). Intending users who have not invested in research generating new technologies to trade may have no alternative to payment of royalties if patents are enforced, assuming they are able to afford royalties acceptable to the patent holders.

Research and Development Needs

Sugar industries internationally have been slow to grasp and respond to the combined implications of the power of transformation technology for varietal improvement and the possibility of patent protection of genes, promoters, techniques and improved varieties. The rewards will flow, not to those who have developed the underlying technologies which are now in the public domain, but to those who invest in and control useful new genes, promoters, techniques and varieties. There are signs that this is being recognised as industries with effective conventional breeding programs move for the first time to secure plant variety rights for new cultivars, which are the likely targets for further molecular improvement. However, it remains to be seen whether sugar industries have the vision and agility to respond individually or collectively to this challenge and opportunity, or whether the opportunity will be taken instead by the large biotechnology companies. It must be understood that if the latter occurs, there will be a substantial redirection of profits away from the traditional industry towards the brokers of the new technology. There are no prizes for second in a race to develop and patent key technologies.

The availability of an efficient genetic transformation system makes research to develop useful genes for transfer into sugarcane a priority. The methods used to locate, isolate and purify useful genes for transfer are highly varied, and beyond the scope of this paper. However, it is not a trivial matter to isolate useful genes, and a substantial investment in research will be required to develop each new gene for transfer. Several genes for resistance to insect pests, diseases or herbicides have already been isolated and shown to be effective in field trials with transgenic plants of other crop species. Some of these genes could readily be adapted for testing in sugarcane, but there will generally need to be some commercial negotiation with the organisations which made the investment in the research to isolate these useful genes. Projects aimed at the isolation and testing of genes for resistance to major viral and bacterial diseases of sugarcane are well advanced, and genes likely to influence sugar accumulation are also being explored. Research to isolate key promoter types needed for expression of the genes likely to be used commercially in transgenic sugarcane must also be an industry priority. It is now technically feasible to isolate genes influencing almost every characteristic listed in Fig. 8.1. Sugar industries are well advised to consider how they will be affected by the availability of such genes, and whether they wish to invest now to participate as creators of the key technologies, or gamble on being able to pay later as consumers of key technologies controlled for profit by others.

Acknowledgments

Research into sugarcane transformation in the author's laboratory has been supported by grants from SRDC, ARC and UQ, with major inputs from PhD students Tricia Franks, Robert Bower and Giovanni Cordeiro, and postdoctoral scientists Adrian Elliott, Bernard Potier and Lianhui Zhang. Field studies of transgenic sugarcane are in collaboration with Drs Mac Hogarth, Nils Berding and Mike Cox of BSES.

References

Ahloowalia, B.S. and Maretzki, A. (1983) Plant regeneration via somatic embryogenesis in sugarcane. *Plant Cell Reports* 2, 21-25.

Arencibia, A., Molina, P.R., de la Riva, G. and Selman-Housein, G. (1995) Production of transgenic sugarcane (*Saccharum officinarum* L.) plants by intact cell electroporation. *Plant Cell Reports* 14, 305-309.

Birch, R.G. (1993) Genetic transformation for varietal improvement - current status and implications for sugarcane. In: Egan, B.T. (ed.), *Proceedings of the Australian Society of Sugarcane Technologists 15th Conference, Cairns, April 1993*. Watson Ferguson Co, Brisbane, Australia, pp. 1-6.

Birch, R.G. and Bower, R. (1994) Principles of gene transfer using particle bombardment. In: Yang, N.-S. and Christou, P. (eds), *Particle Bombardment Technology for Gene Transfer*. Oxford University Press, New York, USA, pp. 3-37.

Birch, R.G., Bower, R., Elliott, A.R., Potier, B.A.M., Franks, T. and Cordeiro, G. (1995) Expression of foreign genes in sugarcane. In: Autrey, J.C. (ed.), *Proceedings of the International Society of Sugar Cane Technologists XXII Congress*. Cartagena, Columbia, September 1995, Vol 2, pp. 368-373.

Birch, R.G. and Franks, T. (1991) Development and optimization of microprojectile systems for plant genetic transformation. *Australian Journal of Plant Physiology* 18, 453-469.

Birch, R.G. and Maretzki, A. (1993) Transformation of sugarcane. In: Bajaj, Y.P.S. (ed.), *Plant Protoplasts and Genetic Engineering IV. Biotechnology in Agriculture and Forestry*, Vol. 23. Springer Verlag, Heidelberg, Germany, pp. 348-360.

Bower, R. and Birch, R.G. (1992) Transgenic sugarcane plants via microprojectile bombardment. *The Plant Journal* 2, 409-416.

Bower, R., Elliott, A.R., Potier, B.A.M. and Birch, R.G. (1996) High-efficiency, microprojectile-mediated cotransformation of sugarcane. *Molecular Breeding*, 2, 239-249.

Chen, W.H., Davey, M.R., Power, J.B. and Cocking, E.C. (1988) Control and maintenance of plant regeneration in sugarcane callus cultures. *Journal of Experimental Botany* 39, 251-261.

Chen, W.H., Gartland, K.M.A., Davey, M.R., Sotak, R., Gartland, J.S., Mulligan, B.J., Power, J.B. and Cocking, E.C. (1987) Transformation of sugarcane protoplasts by direct uptake of a selectable chimaeric gene. *Plant Cell Reports* 6, 297-301.

D'Hont, A., Grivet, L., Lu, Y.H., Roques, D., Feldmann, P., Rao, P.S., Daugrois, J.H., Dufour, P., Berding, N., Walker, D.I.T., Hamon, P. and Glaszmann, J.C. (1995) The genome of modern sugarcane varieties. In: Autrey, J.C. (ed.), *Proceedings of the International Society of Sugar Cane Technologists XXII Congress*. Cartagena, Columbia, September 1995, Vol 2, pp. 363-367.

Fitch, M.M.M. and Moore, P.H. (1990) Comparison of 2,4D and picloram for selection of long-term totipotent green callus cultures of sugarcane. *Plant Cell Tissue and Organ Culture* 20, 157-163.

Franks, T. and Birch, R.G. (1991a) Gene transfer into intact sugarcane cells using microprojectile bombardment. *Australian Journal of Plant Physiology* 18, 471-480.

Franks, T. and Birch, R.G. (1991b) Microprojectile techniques for direct gene transfer into intact plant cells. In: Murray, D. (ed.), *Advanced Methods in Plant Breeding and Biotechnology*. CAB International, Wallingford, UK, pp. 103-127.

Gallo-Meagher, M. and Irvine, J.E. (1995) Selecting sugarcane plants transformed with the *bar* gene. In: Autrey, J.C. (ed.), *Proceedings of the International Society of Sugar Cane Technologists XXII Congress*. Cartagena, Columbia, September 1995, Vol 2, pp. 445-449.

Grant, J.E., Dommisse, E.M., Christey, M.C. and Conner, A.J. (1991) Gene transfer to plants using *Agrobacterium*. In: Murray, D.R. (ed.), *Advanced Methods in Plant Breeding and Biotechnology*. CAB, Wallingford, UK, pp. 50-73.

Gunary, D., (1992) *Group of National Experts on Safety in Biotechnology: Analysis of field release experiments*. OECD Science Directorate, Committee for scientific and technological policy, Paris, France.

Hecter, A. (1995) Genetically engineered plant patents are under attack. *Genetic Technology News* January 1995, 5-6.

Hiei, Y., Ohta, S., Komari, T. and Kumashiro, T. (1994) Efficient transformation of rice (*Oryza sativa* L.) mediated by *Agrobacterium* and sequence analysis of the boundaries of the T-DNA. *The Plant Journal* 6, 271-282.

Huttner, S.L., Arntzen, C., Beachy, R., Breuning, G., Nester, E., Qualset, C. and Vidaver, A. (1992) Revising oversight of genetically modified plants. *Bio/Technology* 10, 967-971.

Moore, P.H. (1995) Progress in sugarcane molecular biology. In: Autrey, J.C. (ed.), *Proceedings of the International Society of Sugar Cane Technologists XXII Congress.* Cartagena, Columbia, September 1995, Vol 2, pp. 353-362.

Quail, P.H., Christiansen, A.H., Hershey, H.P., Sharrock, R.A. and Sullivan, T.D. (1994) *Plant Ubiquitin Promoter System.* Europe Patent EP 342,926 3 NOV 1994. Assignee Lubrizol Genetics Inc., Priority appl. USA 194,824 17 MAY 1988.

Rathus, C. and Birch, R.G. (1992) Stable transformation of callus from electroporated sugarcane protoplasts. *Plant Science* 82, 81-89.

Sanford, J.C., Wolf, E.D. and Allen, N.K. (1990) *Method for Transporting Substances into Living Cells and Tissues and Apparatus Therefor.* USA Patent 4,954,050 31 JUL 1990. Assignee Cornell Research Foundation, Inc., Priority appl. USA 670,771 13 NOV 1984.

Sanford, J.C., Wolf, E.D. and Allen, N.K. (1992) *Biolistic Apparatus for Delivering Substances into Cells and Tissues in a Non-lethal Manner.* Australia Patent AU 621,561 19 JUL 1992. Assignee Biolistics, Inc., Priority appl. USA 161807 29 FEB 1988.

Sautter, C., Waldner, H. and Potrykus, I. (1993) *Process and Apparatus for the Genetic Transformation of Cells.* Australia Patent AU 636,422 29 APR 1993. Assignee Ciba-Geigy AG, Priority appl. Switzerland 4562/89 19 DEC 1989.

Sun, S.S.M., Maretzki, A., Nagai, C., Houtchens, K. and Bidney, D. (1993) Transformation of *Saccharum spontaneum* by particle bombardment. *Sugar Cane* 5, 1-7.

Taylor, P.W.J., Geijskes, J.R., Ko, H.-L., Fraser, T.A., Henry, R.J. and Birch, R.G. (1995) Sensitivity of random amplified polymorphic DNA analysis to detect genetic change in sugarcane during tissue culture. *Theoretical and Applied Genetics* 90, 1169-1173.

Taylor, P.W.J., Ko, H.-L. and Adkins, S.W. (1992a) Factors affecting protoplast isolation and the regeneration of shoot-like structures from protoplast-derived callus of sugarcane (*accharum* spp. hybrids). *Australian Journal of Botany* 40, 863-876.

Taylor, P.W.J., Ko, H.-L., Adkins, S.W., Rathus, C. and Birch, R.G. (1992b) Establishment of embryogenic callus and high protoplast yielding suspension cultures of sugarcane (*Saccharum* spp. hybrids). *Plant Cell Tissue and Organ Culture* 28, 69-78.

Wrage, K. (1992a) Agracetus claims patent on 'all' genetically engineered cotton. *AgBiotechnology News* DEC 92, 1-4.

Wrage, K. (1992b) The licensing game: without patents, you don't get to play. *AgBiotechnology News* DEC 92, 1-5.

Zadoks, J.C. and Schein, R.D. (1979) Southern corn leaf blight. In: *Epidemiology and Plant Disease Management.* Oxford University Press, New York, USA, pp. 331-335.

Chapter nine:

Potential for Overcoming Physio-Biochemical Limits to Sucrose Accumulation

P.H. Moore[1], F.C. Botha[2], R.T. Furbank[3] and C.P.L. Grof[4]

1. *USDA-ARS, 99-193 Aiea Heights Drive, Aiea, HI 96701, USA*
2. *South African Sugar Association Experiment Station, Private Bag X02, Mount Edgecombe 4300, Natal, South Africa*
3. *CSIRO, Division of Plant Industry, GPO Box 1600, Canberra, ACT 2601, Australia*
4. *CSIRO, Tropical Agriculture, 306 Carmody Road, St Lucia, Qld 4067, Australia*

Abstract

Sugarcane yields are among the highest for any crop. Nevertheless, record yields are only approximately 65% of the theoretical maximum. There is a significant potential for increasing sucrose accumulation in sugarcane if the physiological and biochemical limits can be identified and modified. Limits undoubtedly exist in the major physiological processes of photosynthesis, sucrose transport, and metabolism. Enzymes involved in these steps have been identified, characterised, and evaluated for their roles as key regulatory steps. Genes encoding these enzymes are being cloned and used to transform plants to increase or decrease enzyme activity and sucrose accumulation. Results obtained thus far are encouraging but suggest that control in major metabolic pathways is widely shared and is thus more difficult to modify than had been appreciated.

In: Keating, B.A. and Wilson, J.R. (eds) *Intensive Sugarcane Production: Meeting the Challenges Beyond 2000*. CAB International, Wallingford, UK, 1997.

Introduction

Crop yields are manifestations of complex interactions among the physical and biological environments affecting expression of the estimated more than 100,000 plant genes. When viewed from this perspective it is obvious that there is a multitude of factors that can limit sucrose yields. The purpose of this review is to evaluate the potential for biotechnology to overcome physiological and biochemical constraints in order to achieve greater production of sucrose.

Thermodynamic Yield Limits

The ultimate limit to crop yields is set by principles of energy transformation described by the laws of thermodynamics. Thus, the uppermost theoretical gross yield limits are set as the product of the amount of solar radiation available during the growing season, the plant efficiency in absorption of solar radiation, and the efficiency of the crop to fix CO_2. The net yield will be reduced by the amount of fixed CO_2 lost to respiration, pests, and diseases.

The amount of solar radiation available is easily measured but the other parameters of CO_2 assimilation are difficult to quantify on the time scale of a growing season. Loomis and Williams (1963) used instantaneous photosynthetic rates and several assumptions about photosynthetic efficiency to calculate a potential net production of 77g m^{-2} d^{-1} (770kg ha^{-1} d^{-1} = 281t ha^{-1} y^{-1}). The amount of biomass accumulated per unit of intercepted solar radiation, i.e. the radiation use efficiency (RUE), calculates to 2.3g MJ^{-1} (Sinclair and Horie, 1989).

Several conversion values in the Loomis and Williams model for estimating maximum yield have been criticised as being unrealistically high. Nevertheless, the RUE calculated from their model is fairly well supported by field experiments. The RUE of Australian sugarcane is 1.75g MJ^{-1} (Muchow *et al.*, 1994). Since the RUE of sugarcane was based on only the total above ground biomass and the stubble plus root mass of a one year crop can exceed 15% of the total biomass (HSPA, 1948), the RUE for the entire plant should exceed 2.0g MJ^{-1}.

Record Yields

There are various kinds of 'record' sucrose yields. Records of small hand-harvested experimental plots processed in a laboratory are understandably higher than those of large fields harvested by machine and processed in a sugarcane mill. Record field yields reported in Hawaii are 48.5t sucrose ha^{-1} in a crop of 28.2 months of age and 47.4t sucrose ha^{-1} in a crop of 24.1 months (Leffingwell,

1985). These field record yields are 20.6 and 23.6t sucrose ha^{-1} y^{-1}. A 25 month old experimental crop in Hawaii yielded 549.3t cane ha^{-1} and 73.3t sucrose ha^{-1} or 35.2t sucrose ha^{-1} y^{-1}. A 12 month old experimental crop in Australia yielded 211.8t cane ha^{-1} and 32.8t sucrose ha^{-1} y^{-1} (Ham, 1970). Neither dry weight nor fibre values were reported in either experimental test. However, if one assumes that sucrose represents 0.41 of the total dry matter above ground (Muchow *et al.*, 1996), then the Hawaii dry weight (dw) production would be 85.6t ha^{-1} y^{-1} and the Australia dw production would be 80t ha^{-1} y^{-1}.

Sugarcane's estimated dw record yield of 85.6t ha^{-1} y^{-1} is 39% of the theoretical maximum calculated by Loomis and Williams (1963) or 41% of the maximum expected from sugarcane's RUE. On these theoretical bases, we suggest that 60% higher yields might be achieved by removing yield-limiting constraints.

Physiological Processes

Under conditions favouring sucrose accumulation, the storage parenchyma tissue of the sugarcane stem can store sucrose exceeding 62% dw or 25% fresh weight (fw) (Bull and Glasziou, 1963). However, wild relatives of sugarcane store less than 2% fw as sucrose. These striking differences in storage sink accumulation of sucrose are not offset by differences in biomass accumulation, nor are they paralleled by differences in leaf photosynthetic activity. Photosynthetic rates of *Saccharum spontaneum* are reported as nearly twice those of *S. officinarum* and 30% greater than those of hybrid cultivars (Irvine, 1975). On this basis, differences in sucrose storage appear to be regulated within the stem or within the translocation system regulating the partitioning of photoassimilates among alternative sinks. On the other hand, during periods of rapid growth the high sucrose-storing types of sugarcane exhibit lower levels of sucrose than during periods of slow growth; this indicates that sometimes source activity also limits assimilate accumulation. There are thus multiple pathways in which there might be physio-biochemical reactions rate-limiting to sucrose accumulation:

1. Leaf reactions including photosynthetic rate, sucrose metabolism, and carbon partitioning into different pools.
2. Phloem reactions including loading in the leaf, translocation to and unloading in the stalk.
3. Stalk reactions including membrane transport, sucrose metabolism, carbon partitioning into different pools, and remobilisation of stored sucrose.
4. Varietal genetic and developmental constraints including duration and timing of maturation.

Transgenic Plants for Metabolic Design

The organisational complexity at the whole plant level and the metabolic compartmentation within cells and tissues require the use of a combination of molecular tools to successfully manipulate the metabolic processes involved in sucrose accumulation. Directed changes by genetic transformation require a transformation system, a suitable gene, suitable promoter sequences for cell-and tissue-specific expression, and effective targeting signals to direct the protein product to its final destination within the cell. The design to metabolically alter sugarcane to increase sucrose accumulation can involve either inhibition or overexpression of the native plant genes or expression of foreign genes.

The most effective method for transformation of sugarcane involves particle bombardment of embryogenic calli (Birch and Maretzki, 1993). To date, relatively little effort has been directed towards developing transgenic sugarcane plants altered metabolically for increased sucrose accumulation. Consequently, the following discussion is based primarily on studies on regulation of metabolism in plants other than sugarcane (Stitt and Sonnewald, 1995).

To efficiently manipulate a metabolic process the gene must be expressed in a predictable and suitable manner. This may require promoters for organ- and/or cell-specific expression, expression under the appropriate environmental conditions and at the correct developmental stage. Organ-specific promoters have been reported for storage sink tissues such as seeds, potato tubers, and fruits (Edwards and Coruzzi, 1990). Among the most extensively studied are the leaf-specific genes which encode the small subunit of Rubisco (Fluhr *et al.*, 1986). Sugarcane small subunit Rubisco (*rbcs*) genes have been isolated, sequenced and the promoter elements isolated and fused with a reporter gene to determine tissue specificity of expression (Tang *et al.*, 1996). The promoter elicited expression specifically in leaves and preferentially in bundle sheath cells (BSC). A similar expression pattern has been observed in transgenic sugarcane using the rice *rbcs* promoter (Grof *et al.*, 1996b).

Transgenes may need to be expressed as a fusion protein containing a suitable sequence to target heterologous proteins to the appropriate cellular site (Nakamura and Matsuoka, 1993). Since sucrose is unloaded in the apoplasm, passes through the cytoplasm, and is stored in the vacuole, it may be desirable to direct the expression of specific genes to specific sites using targeting constructs.

Metabolism can be altered by decreasing or increasing enzyme activity through genetic transformation. Antisense is used to alter the expression of endogenous genes by transforming plants to produce transcripts in its reverse, or antisense, orientation to reduce the amount of enzyme. Endogenous genes can also be down-regulated by expression of homologous sense-transcripts which is referred to as co-suppression (Napoli *et al.*, 1990). Alternatively the gene may be up-regulated or overexpressed by transformation either with additional copies of endogenous genes driven by stronger promoters or by heterologous genes for ectopic expression so that more enzyme is produced. However, overexpression of genes in plants has sometimes failed due to co-suppression (Napoli *et al.*, 1990; van der Krol *et al.*, 1990) or inactivation by endogenous regulatory systems (Sonnewald *et al.*, 1994). To overcome the problems of co-suppression and inactivation, transformations can be made with prokaryotic or 'lower eukaryotic' genes. Expression of novel genes not found in higher plants has the advantage that regulatory elements are likely to be absent so that the transgene enzyme will not be down-regulated by endogenous effector molecules.

Manipulating Plant Metabolic Pathways

Photosynthesis

The key feature of C_4 photosynthesis is the compartmentation of activities into two chloroplast types and two specialised types of cells, the inner ring of the BSC and the outer ring of the mesophyll cells (MC). The key regulatory enzymes unique to the C_4 pathway, PPdK, PEPC, and NADP-dependent malate dehydrogenase (NADP-MDH) are localised in MC while Rubisco and enzymes of the photosynthetic carbon reduction cycle are localised in the BSC (Fig. 9.1). Activities of these enzymes *in vivo* are increased markedly by illumination (Hatch, 1988) suggesting that each of these enzymes is potentially rate limiting. A direct demonstration of which enzymes are important in controlling photosynthetic flux in C_4 plants has been addressed using recombinant DNA technology.

The NADP-ME-type C_4 dicot *Flaveria bidentis* was transformed with antisense constructs targeted to the key photosynthetic enzymes PPdK, NADP-MDH and Rubisco (Furbank and Taylor, 1995; Furbank, 1996; Furbank *et al.*, 1996). Surprisingly, levels of PPdK and NADP-MDH were reduced considerably with no discernible effect on photosynthetic rate whereas Rubisco activity was closely linked with photosynthetic performance.

In antisense FBPase potato plants (Kossmann *et al.*, 1995) and in antisense *rbc*S tobacco plants (Stitt and Schulze, 1994) maximal photosynthetic rate was unrelated to biomass accumulation. In contrast, Masle *et al.* (1993) observed a penalty to biomass accumulation in all *rbc*S antisense tobacco plants which displayed reduced Rubisco activity and reduced relative growth rate early in development.

Sucrose phosphate synthase (SPS) is potentially rate-limiting for sucrose formation in the leaf (Stitt and Quick, 1989; Krause and Stitt, 1992). Attempts have been made to overcome this potentially rate-limiting step by over-expressing SPS. A number of transgenic tomato lines, in which the maize SPS gene is driven by the Rubisco small subunit promoter (*rbcs*) from tobacco, were produced and evaluated (Worrell *et al.*, 1991). Transformed tomato plants expressing maize SPS exhibited leaf SPS activity up to six times greater than that of control plants (Worrell *et al.*, 1991). Increased rates of photosynthesis and sucrose synthesis were observed (Galtier *et al.*, 1995; Micallef *et al.*, 1995), however, a strict relationship between photosynthesis and yield did not exist. At lower levels of CO_2 the transformed plants exhibited more fruit, earlier maturing fruit and a higher total dw of fruit than did control plants. However, at higher CO_2, fruit yield was unaffected (Micallef *et al.*, 1995). Galtier *et al.* (1993) report a positive correlation between SPS activity and the sucrose to starch ratio, and an increased ratio of shoot to root dw in the transformed plants. These phenotypes could prove beneficial in sugarcane by partitioning a greater portion of the photosynthetically fixed carbon into stem and sucrose. This possibility is being explored by over-expressing SPS in sugarcane (Grof *et al.*, 1996a).

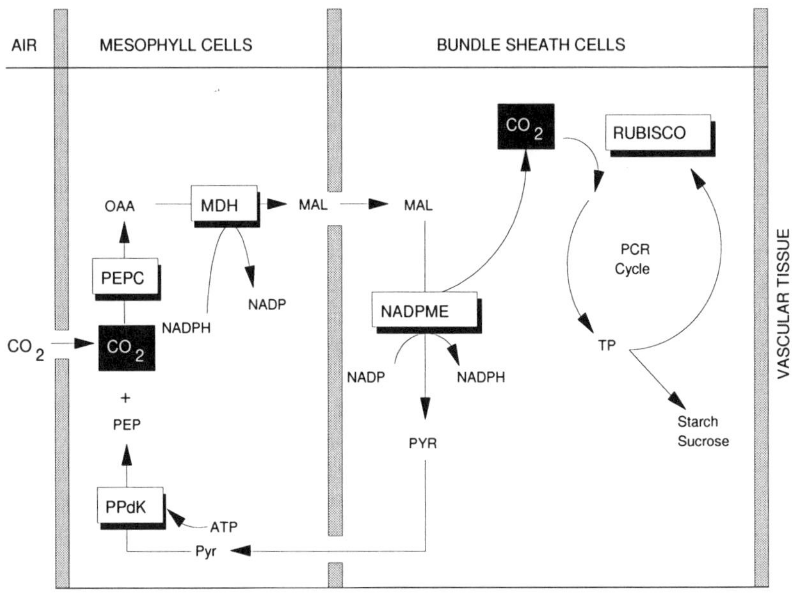

Fig. 9.1. Simplified scheme of the sugarcane C_4 photosynthetic pathway.

[The Sugarcane C_4 pathway is NADP-malic enzyme type which transports malate from mesophyll to bundle sheath chloroplasts. Compartmentation of enzymes between the two cell types is shown. CO_2, as bicarbonate, is fixed by the enzyme PEPC to form the C_4 acid, oxaloacetate (OAA), which is reduced by NADPH from the light reactions to form malate (MAL), the C_4 acid that is transported to the bundle sheath cells. Malate is decarboxylated in bundle sheath cells, where CO_2 released is fixed by the PCR cycle in much the same way as in C_3 plants. The 3-carbon compound pyruvate (PYR) diffuses back to the mesophyll where it is phosphorylated by ATP to regenerate the carbon acceptor phosphoenolpyruvate (PEP). Phosphoenolpyruvate carboxylase (PEPC) catalyses PEP + HCO^{3-} $\rightarrow$ oxaloacetate; malate dehydrogenase (MDH) catalyses OAA + NADP $\rightarrow$ malate + $NADP^+$; NADP-malic enzyme (NADPME) catalyses malate + $NADP^+$ $\rightarrow$ pyruvate + NADPH; pyruvate orthophosphate dikinase (PPdK), catalyses pyruvate + ATP + Pi $\rightarrow$ PEP + AMP + PPi. (Modified from Furbank and Taylor, 1995.)]

Sucrose transport pathway and membrane barriers

The pathway of sucrose accumulation in sugarcane involves the translocation of sucrose from the leaves through the phloem sieve elements to the stem internodes. Experiments with labelled sugars showed that sucrose is unloaded from the phloem into the apoplast where it is cleaved by acid invertase to produce hexoses (Sacher *et al.*, 1963; Hawker and Hatch, 1965). Alternatively, radial transport of sucrose may be through the symplast continuum of the parenchyma tissues. The high concentrations of sucrose and hexoses in the apoplast are possibly the result of leakage from the storage parenchyma cells (Glasziou and Gayler, 1972b; Moore, 1995) and may function to maintain low turgor in storage cells to facilitate sucrose translocation from the phloem by mass transfer (Patrick and Offler, 1996). The hexoses which occur in the apoplast can be taken up into the cells by hexose/symporter carrier sites in the plasmalemma. This idea became a salient feature of a sucrose accumulation model by Glasziou and Gayler (1972a). However, this model has recently been questioned on the basis of anatomical (Jacobsen *et al.*, 1992) and physiological (Welbaum *et al.*, 1992) data.

Sucrose transporter cDNAs have been isolated from spinach (Riesmeier *et al.*, 1992), potato (Riesmeier *et al.*, 1993), and *Arabidopsis* (Sauer and Stolz, 1994). Hexose transporter cDNAs have been isolated from *Arabidopsis* (Sauer *et al.*, 1990), tobacco (Sauer and Stadler, 1993) and sugarcane (Bugos and Thom, 1993). Most of these clones are source-specific and thought to be involved in loading of the phloem in leaf tissue. However, some are specifically expressed in sink tissues (Sauer and Stadler, 1993). As the technology for transformation of sugarcane becomes more efficient we can expect to see sugarcane transformed with genes regulating membrane transport of sugars in attempts to increase the level of sucrose accumulation.

Sucrose accumulation

When sucrose arrives at a recipient cell it forms the carbon skeleton for essentially every metabolic process that the cell will conduct. Depending on the physiological demands of the plant, the arriving sucrose may be catabolised through the glycolytic pathway to be utilised for growth or structural materials, or it may be resynthesised via the gluconeogenic pathway and placed into storage (Sung *et al.*, 1988; Fig. 9.2). During both glycolysis and gluconeogenesis, alternative enzymes are present at various steps to carry out parallel pathways forming a network of reactions for the interconversion of sucrose and pyruvate. Plants use these cytoplasmic alternatives as they develop and respond to environment changes. Thus, plant metabolism can be quite plastic in that there may be different routes to the same end.

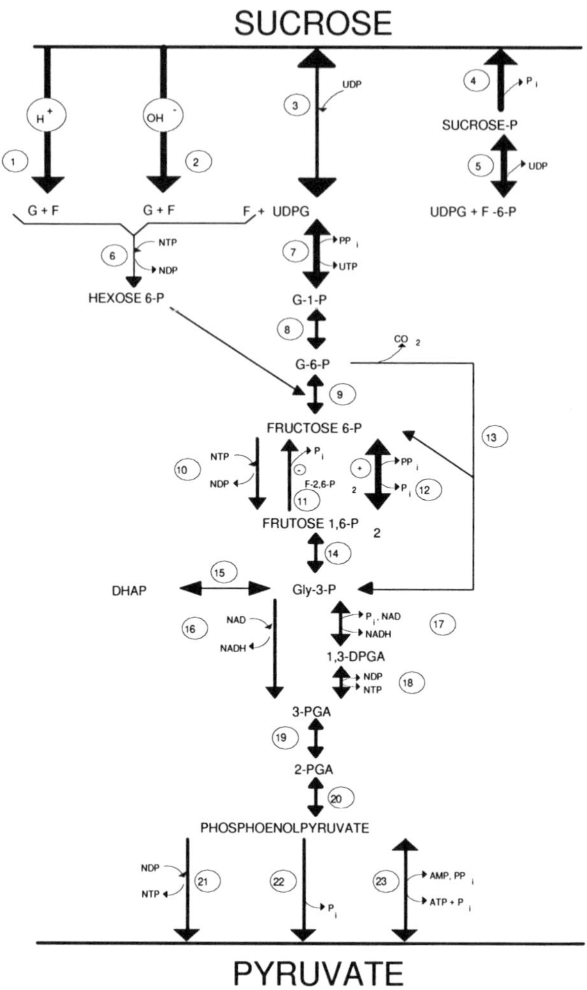

Fig. 9.2. Alternative steps in the network of reactions for the glycolytic and gluconeogenic interconversions of sucrose and pyruvate in plant cells.

[Enzymes which catalyse the numbered steps are: ***1)*** *soluble acid invertase, AI;* ***2)*** *alkaline or neutral invertase, NI;* ***3)*** *sucrose synthase, SS;* ***4)*** *sucrose 6-P phosphatase;* ***5)*** *sucrose phosphate synthase, SPS;* ***6)*** *glucokinase and fructokinase, G, F;* ***7)*** *UDPG pyrophosphorylase;* ***8)*** *phosphoglucomutase;* ***9)*** *phosphohexoseisomerase;* ***10)*** *nucleotide triphosphate-dependent phosphofructokinase, PFK;* ***11)*** *F-1,6-P_2, 1-phosphatase;* ***12)*** *PPi-dependent phosphofructokinase, PPi:PFP;* ***13)*** *the pentose phosphate shunt;* ***14)*** *aldolase;* ***15)*** *triose phosphate isomerase;* ***16)*** *non-phosphorylated NAD^+-dependent Gly-3-P dehydrogenase;* ***17)*** *Gly-3-P dehydrogenase;* ***18)*** *3-PGA kinase;* ***19)*** *3-PGA mutase;* ***20)*** *enolase;* ***21)*** *pyruvate kinase;* ***22)*** *PEP phosphatase;* ***23)*** *pyruvate Pi-dikinase. (Modified from Sung et al., 1988)]*

Sucrose hydrolysis in the storage parenchyma is believed to be accomplished by one of the invertases. Acid invertase (AI) is high in the apoplast (Sacher *et al.*, 1963) and vacuole of young actively growing internodes but its activity is virtually absent from mature internodes of high sucrose storing clones (Hatch and Glasziou, 1963). Cell-wall bound AI is active at low levels in both young and older internodes (Hawker and Hatch, 1965). Neutral invertase (NI) is active at low levels in very young tissue and in greater levels in older tissues (Sacher *et al.*, 1963). The potential roles of invertases in regulation of sucrose accumulation were proposed by correlating enzyme activity with the rate of sucrose accumulation (Hatch and Glasziou, 1963). It was suggested that the vacuolar form of AI is involved with regulation of turgor and the internal sugar pools; the cell wall form of AI was suggested as the controller of dry matter import accompanying cell extension growth; and NI was proposed to be involved in controlling sugar flux in mature storage tissue.

The strong relationship between invertase activity, growth, and sucrose storage indicates a good potential for regulation of the expression of the invertase genes to increase the sink strength of storage organs. This potential was demonstrated in potato (Frommer and Sonnewald, 1995) and tobacco (Sonnewald *et al.*, 1991). Potato plants transformed with a yeast invertase gene expressed in the apoplast showed improved tuber growth and increased accumulation of reducing sugars. For sugarcane, the AI genes are being isolated, cloned, and characterised (Albert *et al.*, 1996; Peters *et al.* 1996; Zhu *et al.*, 1996). Experiments are under way on sugarcane to modify the level of expression of AI both in the apoplastic space and in the vacuole to discover if there are limits to productivity imposed at this level (Albert *et al.*, 1996).

Sucrose synthase (SS) activity is low in storage parenchyma cells but relatively high in the vascular strands from both immature (Hatch *et al.*, 1963) and mature storage tissue (Hawker and Hatch, 1965). Hawker and Hatch (1965) recognised that SS might be involved in sucrose breakdown but did not speculate about the possible role of SS in sucrose accumulation. Buczynski *et al.* (1993) and Lingle and Smith (1991) suggested that SS activity is related to sink strength based on tissue demand for UDPG. In the developing roots of sugarbeet plants, SS was reported to contribute to increasing the sink capacity while sucrose phosphate synthase (SPS) provided the steady increase in sucrose content (Fieuw and Willenbrink, 1987). A similar temporal difference in functioning of the two sucrose synthesis pathways, first SS and then SPS, might exist in sugarcane but this has not been demonstrated.

In most plants investigated, SS appears to be encoded by at least two genes which are differentially expressed among different genotypes, organs, tissues, or stages of development. In sugarcane, total SS activity was lowest in the mature leaf and generally increased in the stalk from a low value at the apex to a maximum in internode 14, the oldest internode measured (Buczynski *et al.*, 1993). Western blots indicated that all of the SS activity in the older internodes was due to the sugarcane isozyme homologous to the maize SS1. The sugarcane isozyme homologous to maize SS2 appeared to be down-regulated in older internodes. The mechanisms of regulation of transcription and translation of the SS genes are not yet known but they may be activated through carbohydrate or sugar status of the plant (Koch, 1996).

Each of the sucrose cleavage steps leads to the production of hexose 6-P. From here alternative enzyme steps again exist in the cytoplasm. Glucose 6-P can be oxidised via the pentose phosphate shunt which may feed back in at fructose 6-phosphate (F-6-P) and at glyceraldehyde 3-phosphate (Gly-3-P). However, it is more probable that F-6-P is converted to F-1,6-P_2 by phosphofructokinase (PFK) or pyrophosphate:fructose-6-phosphate-1-phosphotransferase (PPi:PFP). With the existence of several alternate/parallel steps for the interconversion of F-6P and F-1,6-P2, it is somewhat surprising that the enzyme PFP has been shown to be correlated with sucrose and fibre accumulation in sugarcane (Whittaker and Botha, 1995). The relationship however is not simple. Developmentally, down a stalk, PFP activity increased as sucrose concentration and fibre percentage increased. When the maximum PFP activity, and maximum sucrose and maximum fibre were compared among a collection of genotypes, there was a negative correlation between PFP activity and sucrose and a positive correlation between PFP activity and fibre (Whittaker and Botha, 1995). The implication is that PFP activity might be used to identify potential sucrose or fibre storage ability of potential cultivars.

Prospects

During the past 100 years the average sucrose yield of sugarcane has been substantially increased through plant breeding and optimisation of growing conditions. The improvement in sucrose yield was achieved because carbon assimilation by the leaves (total biomass) and sucrose accumulation in the stem (sucrose percent of juice) have been improved. During the 5 year period from 1910 through 1914 the average yield of sugarcane crops in Australia was 38.8t cane and 4.47t sugar ha^{-1}. This is in contrast to the average of the recent 5 year period of 1990 to 1995 when the yields were 82.6t cane and 11.96t sugar ha^{-1}. The increased yields in Australia were the result of both more than doubling biomass and increasing the fraction of yield in sucrose. The fraction of cane yield that was in sucrose increased from 11.5% in the early 1900s to 14.5% in recent times. Thus, biomass increased by 113% while the sucrose fraction increased by 26.1%.

The increase in total productivity of Australian sugarcane during the past century of breeding was due to higher CO_2 assimilation which was due to either increasing the light-intercepting area of the crop canopy and/or increasing the efficiency with which the leaves use the absorbed light to fix CO_2. This aspect of increasing biomass yields can be visualised as removing barriers to productivity from the source. On the other hand, the increase in the fraction of biomass that was stored as sucrose was at least partially due to a differential in the partitioning of the photosynthetically fixed CO_2 between structural elements of the plant and the sucrose placed into storage. This is visualised as the removal of productivity barriers from the sink. Thus, both source- and sink-limiting situations apparently existed in sugarcane production and have been reduced in the traditional breeding and selection programs.

Traditional breeding and selection programs are slow and costly and might be augmented by application of molecular techniques. By analysing the processes involved in the photosynthetic production, translocation, and metabolic interconversions of sucrose we have identified a number of sites that might be limiting sucrose accumulation. With the existence of several alternate/parallel routes through the glycolysis/gluconeogenic pathway it is apparent that there are difficulties in identifying a key regulatory enzyme. The site for control through a complex pathway might be recognised by quantifying enzyme activity and metabolite flux to calculate a flux control coefficient for each of the enzymes (Kacser and Porteous, 1987; Fell, 1992). Problems with this concept are not only the difficulties in accurately measuring enzyme activity and flux, but that these values depend on the environmental conditions to which the plant was exposed during development. For example, the flux control coefficient of Rubisco for photosynthesis in tobacco ranged from 0.1 to 0.2 under high nitrogen and moderate light and increased to 0.7 to 0.8 when the CO_2 was decreased suddenly or irradiance was increased suddenly (Stitt *et al.*, 1991). Therefore, an enzyme's contribution to control is not fixed.

Success in manipulating plant metabolism will be determined by several factors, including the extent that phenotype is affected by metabolic change, plasticity of plant metabolism, and the extent to which control is shared among the component reactions of individual pathways (ap Rees and Hill, 1994). The existence of alternative pathways means that any attempt to manipulate how much of the incoming sucrose is stored as sucrose, as compared with that being metabolised, may fail if only one route is manipulated and the enzyme controls a number of phenotypic traits. On the other hand, a switch to an alternate metabolic route may have a profound effect on a final phenotype by controlling processes we do not presently recognise as important. The only way to discover these processes is to alter the metabolic activity to see what modifications result. Therefore, although we have the technology for inserting genes to modify a pathway, there may be the surprise of either very small or unexpected change.

In spite of this somewhat pessimistic view for overcoming barriers to productivity by modifying metabolic enzymes we remain optimistic. Our ability to transform plants with foreign genes will likely allow us to direct some end-products of plant metabolism to novel products. For example, starch-storing plants have been modified by adding genes for the production of fructans, polyhydroxybutyrate, and trehalose, all products of higher value than the starch which they partially replace. The potential exists for similar modifications of the sucrose stored in sugarcane. Thus we may proceed with optimism, provided it is tempered with the realisation that the plasticity of plant metabolism and the sharing of control will sometimes defeat us.

References

Albert, H., Zhu, Y.J., Carr, J. and Moore, P. (1996) Structure and expression of soluble acid invertase genes in the stem of high- and low-sucrose accumulating *Saccharum* species and hybrids. In: Wilson J.R., Hogarth, D.M., Campbell, J.A. and Garside, A.L. (eds), *Sugarcane: Research Towards Efficient and Sustainable Production,* CSIRO Division of Tropical Crops and Pastures, Brisbane, Australia, pp. 102-104.

ap Rees, T. and Hill, S.A. (1994) Metabolic control analysis of plant metabolism. *Plant Cell and Environment* 17, 587-599.

Birch, R.G. and Maretzki, A. (1993) Transformation of sugarcane. In: Bajaj YPS (ed.), *Biotechnology in Agriculture and Forestry, 23: Plant Protoplasts and Genetic Engineering IV.* Springer-Verlag, Berlin, Germany, pp. 348-360.

Buczynski, S.R., Thom, M., Chourey, P. and Maretzki, A. (1993) Tissue distribution and characterization of *sucrose* synthase isozymes in sugarcane. *Journal of Plant Physiology* 142, 641-646.

Bugos, R.C. and Thom, M. (1993) Glucose transporter cDNAs from sugarcane. *Plant Physiology* 103, 1469-1470.

Bull, T.A. and Glasziou, K.T. (1963) The evolutionary significance of sugar accumulation in *Saccharum. Australian Journal of Biological Science* 16, 737-742.

Edwards, J.W. and Coruzzi, G.M. (1990) Cell-specific gene expression in plants. *Annual Review of Genetics* 24, 275-303.

Fell, D.A. (1992) Metabolic control analysis: a survey of its theoretical and experimental development. *Biochemical Journal* 286, 313-330.

Fieuw, S. and Willenbrink, J. (1987) Sucrose synthase and sucrose phosphate synthase in sugar beet plants (*Beta vulgaris* L. ssp. *altissima*). *Journal of Plant Physiology* 131, 153-162.

Fluhr, R., Kuhlemeier, C., Nagy, F. and Chua, N.H. (1986) Organ-specific and light-induced expression of plant genes. *Science* 232, 1106-1112.

Frommer, W.B. and Sonnewald, U. (1995) Molecular analysis of carbon partitioning in solanaceous species. *Journal of Experimental Botany* 46, 587-607.

Furbank, R.T. (1996) The C_4 pathway. In: Rhagavendra, A.S. (ed.), *Photosynthesis, a Comprehensive Treatise.* Cambridge University Press, Cambridge, UK (in press).

Furbank, R.T., Chitty, J.A., von Caemmerer, S. and Jenkins, C.L.D. (1996) Antisense RNA inhibition of rbcS gene expression reduces Rubisco level and photosynthesis in the C_4 plant *Flaveria bidentis. Plant Physiology* 111, 725-734.

Furbank, R.T. and Taylor, W.C. (1995) Regulation of photosynthesis in C_3 and C_4 plants: A molecular approach. *Plant Cell* 7, 797-807.

Galtier, N., Foyer, C.H., Huber, J.L.A., Voelker, T.A. and Huber, S.C. (1993) Effects of elevated sucrose phosphate synthase activity on photosynthesis assimilate partitioning and growth in tomato (*Lycopersicon esculentum* var. UC82B). *Plant Physiology* 101, 535-543.

Galtier, N., Foyer, C.H., Murchie, E., Alred, R., Quick, W.P., Voelker, T.A., Thepenier, C., Lasceve, G. and Betsche, T. (1995) Effects of light and atmospheric carbon dioxide enrichment on photosynthesis and carbon partitioning in the leaves of tomato (*Lycopersicon esculentum* L) plants over-expressing sucrose phosphate synthase. *Journal of Experimental Botany* 46, 1335-1344.

Glasziou, K.T. and Galyer, K.R. (1972a) Storage of sugars in stalks of sugar cane. *Botanical Review* 49, 471-490.

Glasziou, K.T. and Gayler, K.R. (1972b) Sugar accumulation in sugarcane. Role of cell walls in sucrose transport. *Plant Physiology* 49, 912-913.

Grof, C.P.L., Glossop, D., Quick, W.P., Sonnewald, U. and Campbell, J.A. (1996a) Molecular manipulation of sucrose phosphate synthase in sugarcane. In: Wilson, J.R., Hogarth, D.M., Campbell, J.A. and Garside, A.L. (eds), *Sugarcane: Research Towards Efficient and Sustainable Production.* CSIRO Division of Tropical Crops and Pastures, Brisbane, Australia, pp. 124-126.

Grof, C.P.L., Elliott, A., Glossop, D., Bertram, J.R., Campbell, J.A. and Birch, R.G. (1996b) Expression of β-glucuronidase in the C_4 monocot sugarcane directed by the rice Rubisco small subunit promoter. In: Wilson, J.R., Hogarth, D.M., Campbell, J.A. and Garside, A.L. (eds), *Sugarcane: Research Towards Efficient and Sustainable Production.* CSIRO Division of Tropical Crops and Pastures, Brisbane, Australia, pp. 120-123.

Ham, G.J. (1970) Water requirements of sugar cane. In: *Water Problems in Tropical Queensland.* Water Research Foundation of Australia Report No. 32, pp. 4.1-4.7.

Hatch, M.D. (1988) C4 photosynthesis: A unique blend of modified biochemistry, anatomy and ultrastructure. *Biochemica et Biophysica Acta* 895, 81-106.

Hatch, M.D. and Glasziou, K.T. (1963) Sugar accumulation cycle in sugar cane. II. Relationship of invertase activity to sugar content and growth rate in storage tissue of plants grown in controlled environments. *Plant Physiology* 38, 344-348.

Hatch, M.D., Sacher, J.A. and Glasziou, K.T. (1963) Sugar accumulation cycle in sugar cane. I. Studies on enzymes of the cycle. *Plant Physiology* 38, 338-343.

Hawker, J.S. and Hatch, M.D. (1965) Mechanism of sugar storage by mature stem tissue of sugar cane. *Physiologia Plantarum* 18, 444-453.

HSPA (1948) Hawaiian Sugar Planters' Association *Report of Committee in Charge of the Experiment Station*, p. 21.

Irvine, J.E. (1975) Relations of photosynthetic rates and leaf and canopy characters to sugarcane yield. *Crop Science* 15, 671-676.

Jacobsen, K.R., Fisher, D.G., Maretzki, A. and Moore, P.H. (1992) Developmental changes in the anatomy of the sugarcane stem in relation to phloem unloading and sucrose storage. *Botanica Acta* 105, 70-80.

Kacser, H. and Porteous, J. (1987) Control of metabolism, what do we have to measure. *Trends in Biochemistry Science* 12, 5-14.

Koch, K.E. (1996) Carbohydrate-modulated gene expression in plants. *Annual Review of Plant Physiology and Plant Molecular Biology* 47, 509-540.

Kossmann, J., Sonnewald, U. and Willmitzer, L. (1995) Reduction of the chloroplast fructose-1,6-bisphosphatase in transgenic potato plants impairs photosynthesis and plant growth. *Plant Journal* 6, 637-650.

Krause, K. and Stitt, M. (1992) Sucrose-6-phosphate levels in spinach leaves and their effects on sucrose phosphate synthase. *Phytochemistry* 31, 1143-1146.

Leffingwell, R.J. (1985) Hawaiian Report. *Sugar y Azucar* 80, 7-9.

Lingle, S.E. and Smith, R.C. (1991) Sucrose metabolism related to growth and ripening in sugarcane internodes. *Crop Science* 31, 172-177.

Loomis, R.S. and Williams, W.A. (1963) Maximum crop productivity: An estimate. *Crop Science* 3, 67-72.

Masle, J., Hudson, G. and Badger, M.R. (1993) Effects of ambient CO_2 concentration on growth and nitrogen use in tobacco (*Nicotiana tabacum*) plants transformed with an antisense gene to the small subunit of ribulose-1,5-bisphosphate carboxylase/oxygenase. *Plant Physiology* 103, 1075-1088.

Micallef, B.J., Haskins, K.A., Vanderveer, P.J., Roh, K.S., Shewmaker, C.K. and Sharkey, T.D. (1995) Altered photosynthesis, flowering, and fruiting in transgenic tomato plants that have increased capacity for sucrose synthesis. *Planta* 196, 327-334.

Moore, P.H. (1995) Temporal and spatial regulation of sucrose accumulation in the sugarcane stem. *Australian Journal of Plant Physiology* 22, 661-679.

Muchow, R.C., Spillman, M.F., Wood, A.W. and Thomas, M.R. (1994) Radiation interception and biomass accumulation in a sugarcane crop grown under irrigated tropical conditions. *Australian Journal of Agricultural Research* 45, 37- 49.

Muchow, R.C., Robertson, M.J., Wood, A.W. and Keating, B.A. (1996) Effect of nitrogen on the time-course of sucrose accumulation in sugarcane. *Field Crops Research* 47, 143-153.

Nakamura, K. and Matsuoka, K. (1993) Protein targeting to the vacuole in plant cells. *Plant Physiology* 101, 1-5.

Napoli, C., Lemieux, C. and Jorgensen, R. (1990) Introduction of a chimeric chalcone synthase gene into petunia results in reversible co-suppression of homologous genes *in trans*. *Plant Cell* 2, 279-289.

Patrick, J.W. and Offler, C.E. (1996) Post-sieve element transport of photoassimilates in sink regions. *Journal of Experimental Botany* 47, 1165-1177.

Peters, K.F., Grof, C.P.L., Botella, J. and Albert, H. (1996) Isolation and genetic manipulation of invertase genes in sugarcane. In: Wilson, J.R., Hogarth, D.M., Campbell, J.A. and Garside, A.L. (eds), *Sugarcane: Research Towards Efficient and Sustainable Production.* CSIRO Division of Tropical Crops and Pastures, Brisbane, Australia, pp. 127-129.

Riesmeier, J.W., Hirner, B. and Frommer, W.B. (1993) Potato sucrose transporter expression in minor veins indicates a role in phloem loading. *Plant Cell* 5, 1591-1598.

Riesmeier, J.W., Willmitzer, L. and Frommer, W.B. (1992) Isolation and characterization of a sucrose carrier cDNA from spinach by functional expression in yeast. *EMBO Journal* 11, 4705-4713.

Sacher, J.A., Hatch, M.D. and Glasziou, K.T. (1963) Sugar accumulation cycle in sugar cane. III. Physical and metabolic aspects of cycle in immature storage tissues. *Plant Physiology* 38, 348-354.

Sauer, N., Friedlander, K. and Graml-Wicke, U. (1990) Primary structure, genomic organization and heterologous expression of a glucose transporter from *Arabidopsis thaliana*. *EMBO Journal* 9, 3045-3050.

Sauer, N. and Stadler, R. (1993) A sink-specific H^+/monosaccharide co-transporter from *Nicotiana tabacum*: cloning and heterologous expression in baker's yeast. *Plant Journal* 4, 601-610.

Sauer, N. and Stolz, R. (1994) SUC1 and SUC2: two sucrose transporters from *Arabidopsis thaliana*; expression and characterization in baker's yeast and identification of the histidine-tagged protein. *Plant Journal* 6, 67-77.

Sinclair, T.R. and Horie, T. (1989) Leaf nitrogen, photosynthesis, and crop radiation use efficiency: A review. *Crop Science* 29, 90-98.

Sonnewald, U., Brauer, M., von Schaewen, A., Stitt, M. and Willmitzer, L. (1991) Transgenic tobacco plants expressing yeast-derived invertase in either the cytosol, vacuole or apoplast: a powerful tool for studying sucrose metabolism and sink/source interactions. *The Plant Journal* 1, 95-106.

Sonnewald, U., Lerchl. J., Zrenner, R. and Frommer, W. (1994) Manipulation of sink-source relations in transgenic plants. *Plant Cell and Environment* 17, 649-658.

Stitt, M. and Quick, W.P. (1989) Photosynthetic carbon partitioning: its regulation and possibilities for manipulation. *Physiologia Plantarum* 77, 633-641.

Stitt, M., Quick, W.P., Schurr, U., Schulze, E.-D., Rodermel, S.R. and Bogorad, L. (1991) Decreased Rubisco in tobacco transformed with 'antisense' *rbs*S. II. Flux control coefficients for photosynthesis in varying light, CO_2 and air humidity. *Planta* 183, 555-565.

Stitt, M., and Schulze, E.-D. (1994) Does Rubisco control the rate of photosynthesis and plant growth? An exercise in molecular eco-physiology. *Plant Cell and Environment* 17, 465-488.

Stitt, M. and Sonnewald, U. (1995) Regulation of metabolism in transgenic plants. *Annual Review of Plant Physiology and Plant Molecular Biology* 46, 341-368.

Sung, S.J.S., Xu, D.P., Galloway, C.M. and Black, C.C. (1988) A reassessment of glycolysis and gluconeogenesis in higher plants. *Physiologia Plantarum* 72, 650-654.

Tang, W.D., Sun, S.S., Nagai, C. and Moore, P.H. (1996) Regulation of expression of β-glucuronidase in transgenic sugarcane by promoters of rubisco small subunit genes. In: Wilson, J.R., Hogarth, D.M., Campbell, J.A. and Garside, A.L. (eds), *Sugarcane: Research Towards Efficient and Sustainable Production.* CSIRO Division of Tropical Crops and Pastures, Brisbane, Australia, pp. 117-119.

Van der Krol, A.R., Mur, L.A., Beld, M., Mol, J.N.M. and Stultje, A.R. (1990) Flavonoid genes in petunia: Addition of a limited number of gene copies may lead to a suppression of gene expression. *Plant Cell* 2, 291-299.

Welbaum, G.E., Meinzer, F.C., Grayson, R.L. and Thornham, K.T. (1992) Evidence for and consequences of a barrier to solute diffusion between the apoplast and vascular bundles in sugarcane stalk tissue. *Australian Journal of Plant Physiology* 19, 611-623.

Whittaker, A. and Botha, F.C. (1995) Expression of PFP to high fibre and low sucrose levels. *Proceedings of the International Society of Sugar Cane Technology* 22, 286-287.

Worrell, A.C., Bruneau, J.M., Sommerfelt, K., Boersig, M. and Voelker, T.A. (1991) Expression of a maize sucrose-phosphate synthase in tomato alters leaf partitioning. *The Plant Cell* 3, 1121-1130.

Zhu, J., Komor, E. and Moore, P.H. (1996) Relationship of sucrose metabolism enzymes with sucrose storage in sugarcane. In: Wilson, J.R., Hogarth, D.M., Campbell, J.A. and Garside, A.L. (eds), *Sugarcane: Research Towards Efficient and Sustainable Production.* CSIRO Division of Tropical Crops and Pastures, Brisbane, Australia, pp. 89-91.

Chapter ten:

Nutrient Uptake and Metabolism: Prospects for Molecular Modification to Improve Efficiency

F.W. Smith

CSIRO Tropical Agriculture, 306 Carmody Road, St Lucia, Qld 4067, Australia

Abstract

Sustained high yields of sugarcane necessitates application of mineral nutrients. The costs of these inputs and community concerns regarding the use of chemical fertilisers in such environmentally sensitive areas as many of the sugar growing regions, require that fertilisers be used in the most efficient manner. New technologies are emerging which offer opportunities for improving the efficiency of uptake and utilisation of nutrients by crop, pasture, horticultural and forest species and for improving the quality of plant products.

The rate of uptake of many plant nutrients is limited by feedback control mechanisms. Recently, structural genes involved in the uptake of phosphate, potassium, nitrogen and sulfate by plants have been isolated and characterised. These clones provide the tools needed to unravel the mechanisms involved in the regulation of nutrient transport processes. When coupled to appropriate promoters which enable their expression in plants to be manipulated, they may provide a means for improving the efficiency of nutrient uptake. Other genes involved in the internal transport of nutrients in plants have also been identified. Molecular manipulation of these genes provides opportunities for improving the internal utilisation of nutrients and for changing the composition and quality of plant products.

Molecular approaches are also being applied to identify genes associated with efficient use of micronutrients by plants. The use of such genes as specific molecular markers in plant breeding programs is being explored. If successful, this approach provides an opportunity to rapidly screen breeding material for such traits and determine their heritability.

In: Keating, B.A. and Wilson, J.R. (eds) *Intensive Sugarcane Production: Meeting the Challenges Beyond 2000*. CAB International, Wallingford, UK, 1997.

Ways in which molecular approaches may contribute to manipulating the root environment to improve the acquisition of nutrients and protect plants against adverse soil factors such as salinity are also discussed.

Introduction

It is quite clear that sustained production of high yielding crops such as sugarcane requires inputs of nutrients even on fertile soils. A great deal of work has been devoted to defining the nutrient requirements of this crop and formulating fertiliser recommendations that meet these requirements. Annual applications of the order of 120 to 240kg N ha^{-1}, 30 to 50kg P ha^{-1}, 40 to 100kg K ha^{-1} and 15 to 25kg S ha^{-1} are common. These are substantial inputs of chemical fertiliser and lead one to ponder about the efficiency with which species grown in highly developed agricultural systems acquire and utilise mineral nutrients. There are indications that during the selection processes that have yielded modern varieties of annual crop species, some of the more conservative nutritional characteristics of their wild progenitors may have been lost. Inefficient use of fertiliser nutrients, particularly when applied in large amounts, is both costly to farmers and environmentally damaging.

The Australian sugar industry, largely located adjacent to rivers close to the coast, occupies environmentally sensitive areas. There are community concerns regarding inefficient use of mineral nutrients, particularly phosphorus, nitrogen and the heavy metal micronutrients (see Arthington *et al.*, 1997; Crossland *et al.*, 1997). Experience overseas with other highly developed intensive cropping systems indicates that community concerns will increase. These pressures have influenced plant nutritionists, internationally, to move beyond establishing fertiliser recommendations and diagnostic systems for monitoring the nutrient status of crops. In a new era, many plant nutritionists are concerning themselves more about the efficiency of nutrient uptake and use by plants. Questions are being asked about the regulation of the processes involved in transporting nutrients into plant cells, their assimilation into metabolically usable forms, their partitioning within the plant, their storage, their remobilisation from older tissues to actively growing tissues, and the overall control that growth and development have on these processes.

Although the traditional skills of plant physiologists have contributed a great deal to analysing these processes and quantifying their outcomes, they have proven inadequate for unravelling the basic mechanisms. Happily, physiologists and biochemists are now embracing powerful new approaches and tools that have arisen from advances in molecular biology over the past two decades and spectacular results are emerging. These approaches are beginning to reveal, at the molecular level, how plants acquire and use their nutrients. Importantly, these recombinant DNA approaches also provide an opportunity to intervene in the transport and metabolic processes to at least test outcomes of manipulations that may appear desirable. The necessary knowledge base and appropriate tools are being assembled for use in research programs that will determine whether it is going to be feasible to improve the efficient use of nutrients and the quality of plant products by manipulating plant processes. Areas being investigated include manipulating the transport of nutrients, both into the plant and within the plant, molecular probes for quickly identifying more nutrient efficient genotypes, modifying root architecture and rhizosphere organism associations to improve nutrient acquisition, and the excretion of agents from plant roots to modify the availability of soil nutrients.

Nutrient Transport

All cells are surrounded by membranes composed of a bilayer of lipid material. This presents a barrier to the movement of solutes and particularly to charged compounds such as nutrient ions. Proteins embedded in the cell membrane are believed to selectively mediate the passage of solutes across this otherwise impermeable barrier. The amount of solute being transported is therefore controlled by the number of transporter proteins that are operational and the rate at which the solute passes through them. Kinetic and electrophysiological studies have identified a number of classes of transporter proteins which operate in plant membranes. These include primary ion ATPases, H^+-cotransporters, exchangers, channels and ATP-binding-cassette transporters. Of these, H^+-cotransporters are particularly important in the nutrient uptake process in plants. In classic kinetic studies conducted in Emmanuel Epstein's laboratory over 40 years ago, it was demonstrated that these cotransporters were of two types, viz. high affinity transporters capable of capturing the low concentrations of nutrients that occur in soil solutions, and low affinity types that transport nutrients from high external concentrations (Epstein and Hagen, 1952).

The keys to modifying nutrient transport are to identify and isolate the transporter molecules and to understand how they are controlled. This has proven to be a very difficult task but breakthroughs have been made in the past three years as a result of application of molecular techniques. Genes that encode specific transporters for the nutrients potassium (Anderson *et al.*, 1992; Sentenac *et al.*, 1992; Schachtman and Schroeder, 1994), ammonium (Ninnemann *et al.*, 1994), sulfate (Smith *et al.*, 1995), phosphate (Smith *et al.*, 1997) and nitrate (Tsay *et al.*, 1993) have been identified, and functional and regulatory aspects of these clones are now being characterised. Sulfate transporters, isolated in my laboratory, serve as a model to illustrate their structure, expression and how they might be used.

The structure of transporters

Genes encoding three sulfate transporters have been isolated from the tropical forage legume *Stylosanthes hamata* (Smith *et al.*, 1995) and one from barley. These encode polypeptides that are approximately 640 amino acids in length. There is considerable identity among all four amino acid sequences. Analysis of the primary amino acid sequence of these polypeptides reveals that they contain 12 hydrophobic regions, each approximately 20 amino acids in length, that are α-helical in nature (Hawkesford and Smith, 1997). These regions are presumed to be embedded in the membrane and form 12 transmembrane domains in the polypeptides. The ammonium transporter (Ninnemann *et al.*, 1994), the phosphate transporters (Smith *et al.*, 1997), and the high affinity potassium transporter isolated from wheat roots (Schachtman and Schroeder, 1994), also contain 12 membrane spanning domains, a common feature of many transporter proteins isolated from other organisms (Griffith *et al.*, 1992). The plant phosphate transporters APT1 and APT2 (Smith *et al.*, 1997) differ from the sulfate transporters in that, in the phosphate transporters, the 12 membrane spanning domains are arranged in two groups of six separated by a long hydrophilic loop - a feature that places them in the superfacilitator family of transporters (Marger and Saier, 1993). This family also contains various sugar and organic acid transporters. By virtue of the membrane spanning domains, the transporter proteins are threaded back and forth through the membrane and form a shell of protein α-helices through which the specific ions traverse the otherwise impermeable membrane. Domains that determine the specificity of a transporter for a particular ion are still to be determined. Specific site-directed mutagenesis studies and, ultimately, crystallography data will elucidate these important structure-function relationships.

Regulation

Physiological data have shown that sulfate and phosphate transport are repressed when the external supplies of sulfate and phosphate, respectively, are adequate. The capacity to take up these ions is substantially increased when the external supplies are reduced. Thus, the potential sulfate uptake rates of siratro (*Neonotonia atropurpureum*) roots increased 6-fold 24 hours after being deprived of an external sulfur supply and 9-fold after sulfate was withheld for 48 hours (Clarkson *et al.*, 1983). The potential sulfate uptake rate rapidly declined to levels approaching that of plants that had remained adequately supplied with sulfur within eight hours of restoring the sulfate supply to deprived plants. Similar results have been reported with barley where it was demonstrated that the increase in sulfate uptake rates following sulfur deprivation was dependent upon protein synthesis (Clarkson *et al.*, 1992).

Phosphate uptake rates respond similarly to phosphate deprivation and resupply (Clarkson and Scattergood, 1982; Lefebvre and Glass, 1982; Cogliatti and Clarkson, 1983), but the magnitude of the response is less than that observed for sulfate uptake. Further, the reduced capacity for phosphate uptake following resupply of phosphate took two to three days to return to that of plants adequately supplied with phosphate throughout. Nitrate responds somewhat differently to nitrate supply where the presence of nitrate is required for induction of transporters involved in nitrate uptake (Jackson *et al.*, 1986).

Other inducible and repressible transport systems have been reviewed by Clarkson and Luttge (1991). These systems enable plants to match their rates of nutrient uptake to their current growth rates. This regulation may arise from two mechanisms. Allosteric regulation, as occurs with many enzyme catalysed reactions, involves attenuation of the activity of the protein by the binding of substrates or products. Such regulation occurs in very short time frames and could 'fine tune' the rate of transport of nutrients through transporters. There is evidence that potassium transport may be regulated in this way (Glass, 1983). Alternatively, the expression of the gene encoding the transporter may be regulated. This would result in the number of transporters being controlled and would occur in a time frame of hours or days depending upon the turnover rate of the transporter and its RNA message.

The expression of transporter genes is controlled by a promoter sequence upstream of the coding region of the gene. DNA-binding proteins and other transcription factors interact with sequences within the promoter to either facilitate or block transcription. Promoters may contain a number of binding sites enabling them to respond to different stimuli. Further, the same stimulus may interact with the promoters of a number of genes, thus permitting the coordinated regulation of several genes. In this way similar stimuli are able to coordinately regulate the expression of genes that transport and assimilate nutrients. The complexity of these regulatory mechanisms is illustrated in Fig. 10.1 by the control of sulfur transport into the fungus *Neurospora crassa* (Marzluf, 1993). Similar control circuits have not yet been elucidated for plant transport and assimilatory systems, but we should anticipate that they will be no less complicated.

Expression

Studies with the cloned sulfate and phosphate transporters help interpret the physiological observations on regulation of nutrient supply. Expression studies, in which the amount of messenger RNA transcribed by a gene is visualised, indicate that the genes encoding the SHST1 and SHST2 sulfate transporters are only expressed in plant roots (Fig. 10.2). Further, expression of these genes is considerably enhanced when the plants are deprived of sulfur (Fig. 10.2) and rapidly repressed when sulfate is resupplied. This pattern mirrors measured sulfate uptake rates (Clarkson *et al.*, 1983; Hawkesford and Smith, 1997) indicating that uptake of these nutrients is primarily controlled by transcription of genes encoding these transporters. Similarly with the genes encoding the phosphate transporters APT1 and APT2. These are also expressed in roots and their expression is enhanced when the plants are deprived of phosphate (Fig. 10.3). Another important point is illustrated by the data in Fig. 10.3 - regulation of the expression of the phosphate transporter genes responded only to the phosphorus status of the plants, neither nitrogen nor potassium deficiency having any effect.

These expression data suggest a strategy for enhancing nutrient transport into plant roots and improving nutrient efficiency. If the high level of expression observed in nutrient-deprived plants could be maintained when there was adequate nutrient available, it may be possible to accumulate the nutrient in the plant early in its life cycle. Molecular techniques can be used to link the transporters to promoters that are insensitive to the nutrient status of the plant, thus breaking the feedback control that normally occurs in plants adequately supplied with nutrients. Such studies on the overexpression of the sulfate and phosphate transporters are currently underway and we eagerly await their outcome. However, this strategy does pose some concerns which are discussed in the later section on 'Prospects'.

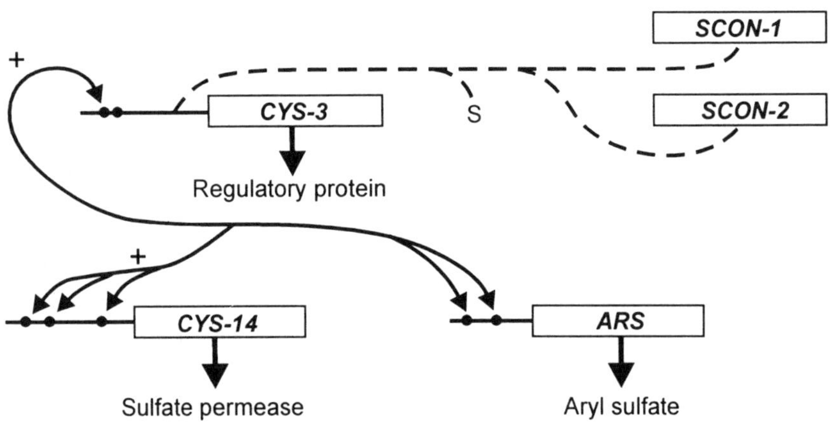

Fig. 10.1. The sulfur regulatory circuit *in Neurospora crassa.* [From Marzluf ,1993]

[Control of a sulfate transporter (coded by cys-14) and a catabolic enzyme, aryl sulfatase (coded by ars), is by three regulatory genes, scon1, scon2 and cys-3. The cys-3 gene is turned off when the cellular sulfur supply is limiting. The product of the cys-3 gene acts in a positive manner, binding to three sites in the promoter region of cys-14 and to sites in the ars region, turning on transcription of these genes. The cys-3 product also binds to a single duplex upstream of cys-3 itself, thereby enhancing its own transcription through autogenous regulation. The products of scon1 and scon2 act in a negative manner on cys-3 turning the expression of cys-3 off when the cellular sulfur supply is adequate. Thus the regulatory gene cys-3 is itself highly regulated, negatively by the products of scon1 and scon2, and positively by autogenous regulation by its own gene product.]

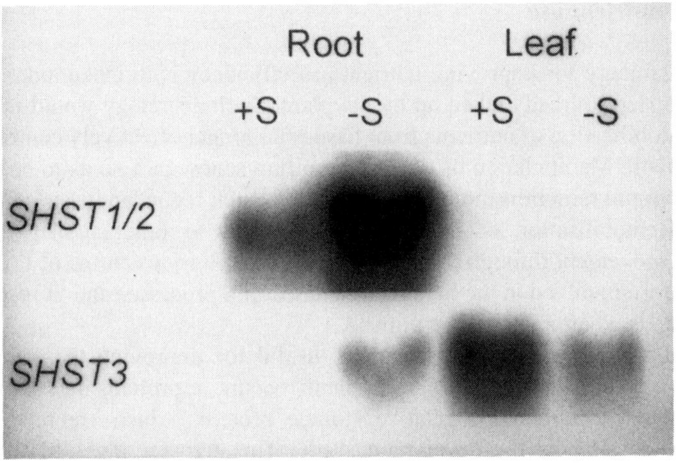

Fig. 10.2. Northern blot analysis of the expression of the sulfate transporter genes *SHST1, SHST2 and SHST3* in *Stylosanthes* roots and leaves of +S and -S grown plants.

[10μg of total RNA from plants grown continuously with 250μM sulfate (+S), deprived of sulfate for 72 hours (-S) was loaded onto each lane, separated and blotted onto Amersham Hybond. The blot was hybridized with random primed ^{32}P probe corresponding to SHST1 (indistinguishable from SHST2) or SHST3 and washed at high stringency.]

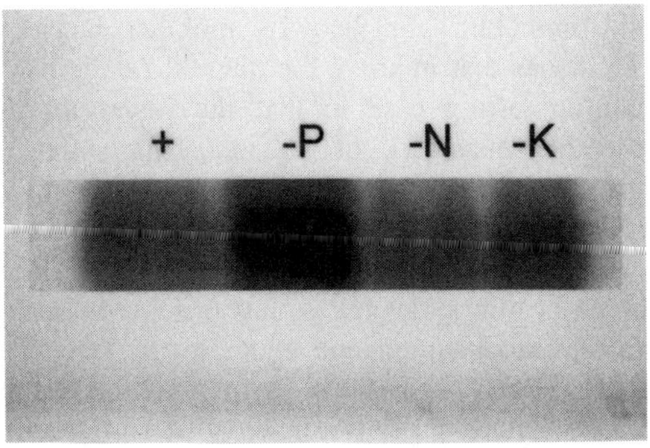

Fig. 10.3. Northern blot analysis of the expression of the phosphate transporter gene *APT1* in *Arabidopsis* roots under different conditions of nutrient supply.

[5μg of total RNA was loaded onto each lane, separated and blotted onto Amersham Hybond. The blot was hybridized with random primed ^{32}P corresponding to APT1 cDNA. Total RNA was purified from the roots of Arabidopsis plants that were grown in solution culture for 30 days with all nutrients supplied (+), without P for the last 72 h of growth (-P), without N for the last 72 h of growth (-N) or without K for the last 72 h of growth (-K). Other blots (not shown) indicated that APT1 and APT2 genes are not expressed in leaves, stems, flowers or siliques, irrespective of the P status of the plants.]

Internal nutrient use

Another strategy for improving nutrient use efficiency is to make more effective use of nutrients already taken up by the plant. Such a strategy would require the rapid remobilisation of nutrients from tissues no longer effectively contributing to plant growth. Manipulation of genes controlling senescence so as to enhance the early and rapid remobilisation of nutrients from such redundant tissues may prove useful. Remobilisation of nutrients and transfer to other tissues invariably involves movement through xylem and/or phloem. Various forms of transporters are therefore involved in the loading and unloading processes and in reabsorption into the cells where they will be utilised.

Such a strategy may be particularly useful for improving the efficiency of nitrogen use. Most of the nitrogen in plants occurs in proteins and amino acids. Many plants accumulate vegetative storage proteins which are released after flowering to support the developing seeds. This transfer involves amino acid transporters, a number of which have been cloned (Frommer *et al.*, 1994) and their regulation is being studied.

Many inorganic ions accumulate in the vacuoles of plant cells. It is felt that, in most cases, these accumulated ions form a buffer which can be called upon to meet the plant's demands when external nutrient supplies become limiting. Storage in and removal from vacuoles requires passage through the vacuolar membrane, the tonoplast, and the involvement of some type of transporter. To date, there have been no reports of the isolation of genes encoding tonoplast transporters so it is not yet possible to bring molecular techniques to bear on the manipulation of these transporters.

The so-called 'immobile' or 'slowly mobile' nutrients may benefit particularly from strategies that improve the internal remobilisation of nutrients. Sulfate presents an interesting case in that the lower leaves of plants can accumulate considerable quantities of sulfate. This sulfate is only slowly retranslocated to growing tissues, even when the plants are suffering from severe sulfur deficiency (Clarkson *et al.*, 1983). We observe the results of this in the symptoms displayed by sulfur deficient plants - yellowing young tissues at the top of the plant and green healthy older leaves towards the bottom. Compartmental analysis has shown that this phenomenon is due to the slow rate of turnover of sulfate in leaf vacuoles (Bell *et al.*, 1990). Remobilisation of this pool of sulfate is an obvious means of improving the efficiency of sulfate utilisation and may lead to substantial improvements in the sulfur content of grains. It is interesting to note that the gene encoding the SHST3 sulfate transporter isolated from *Stylosanthes* (Smith *et al.*, 1995) is expressed in leaf tissues. Unlike the root transporters SHST1 and SHST2 which are high affinity transporters (K_m=10μM) operating at low external sulfate concentrations, SHST3 is a lower affinity transporter (K_m=100μM) operating at high external sulfate concentrations. One would expect this characteristic of a transporter involved in the movement of nutrients between plant tissues or compartments within the plant. Expression of

SHST3 is repressed in leaves when the plant is deprived of sulfate (Fig. 10.2). There is a need to determine the sites of expression of the gene encoding SHST3 and clarify the role of this polypeptide in the internal transport of sulfate in plants as it may prove to be a useful model for other internal plant transporters.

Prospects

The isolation of genes encoding nutrient transporters in plants is quite recent. Attempts are being made to alter their expression in plants, but the results of such studies are not yet available. Most of the current studies are concerned with manipulation of nutrient uptake and, at this stage, are primarily aimed at gaining the knowledge base that will underpin the successful application of these new technologies. Nutrient transport is a fundamental plant process that is closely coupled to plant growth and development and the effects of intervening in the feedback mechanisms that control the process are not known at this time. Since physiologists have not had the capability to undertake such experiments before, it is difficult to predict their outcome.

There are reasons to believe that, at least for some nutrients, simply increasing their uptake alone may do little to improve nutrient efficiency. A common response to excessive uptake of some ions is an increase in efflux (Cogliatti and Santa Maria, 1990) which may result in little net gain. Further, some elements are known to be toxic when accumulated to high levels. Excessive accumulation of inorganic phosphate, for instance, can be toxic (Loneragan *et al.*, 1979) and has been shown to occur when plants previously deprived of phosphorus are resupplied with phosphorus (Cogliatti and Clarkson, 1983). Simply overexpressing a transporter responsible for loading inorganic phosphate into cells may therefore have disastrous consequences. Study of such problems is being hampered by the rather 'blunt instruments' that are presently being used. Rather than use the strong constitutive promoters that are widely available for transgenic plant work, there is an urgent need to identify and use inducible promoters that specifically drive expression in the tissues involved in the targeted nutrient transport process.

Linkages between the uptake of nutrients and their assimilation into usable metabolites suggests that there may be limited benefit from dramatically increasing the rate of nutrient uptake if the assimilatory pathways cannot handle the additional nutrient. The extent to which assimilatory pathways can be driven by increased nutrient supply rather than by demand, as they most often are, remains to be seen. That will be one of the most important results arising from the early transgenic plant studies with nutrient transporters. However, coordinated regulation of uptake and assimilation may offer an opportunity to up-regulate both processes simultaneously by manipulating the expression of genes that encode transcription factors that control both processes. Suggested linkages between regulation of nitrate uptake and nitrate reduction (Redinbaugh and Campbell, 1991) could provide a suitable model system, and progress is being made on the molecular genetics of nitrate uptake and assimilation (Crawford and Arst, 1993). The feasibility of such an approach can be gleaned from the well characterised regulatory circuits that operate in fungi (Fu *et al.*, 1990; Marzluf, 1993). However, such circuits have not been defined for plants and, without them, it is impossible to devise sensible strategies for intervening in transport-assimilatory pathways in ways that may improve nutrient efficiency.

Clarkson and Hawkesford (1993) have pointed out that storage proteins, synthesised early during plant growth, may provide a more useful reservoir of nitrogen and sulfur than accumulation of the inorganic ions. In the case of nitrogen, most plants would be unable to store sufficient inorganic nitrogen in their vacuoles to support more than a few days growth. However, large amounts of nitrogen and sulfur can be safely stored in proteins and released later in the growing season to support plant growth when external nutrient supplies become limiting. Synthesis of such proteins could also act as a sink to drive the early uptake and assimilation of nitrogen and sulfate through increasing the early demand for these nutrients.

Down-regulation of transporters

There are situations where restricting the amount of a substance entering into a plant may be desirable. Plant growth on saline soils or on sites polluted with heavy metals are examples. Antisense and 'gene shear' ribozyme technology permits the expression of target genes to be down-regulated. Genes encoding proteins involved in the entry of sodium and chloride into plant cells are being sought as tools to study mechanisms of salt tolerance. Work is also being done on the potassium/sodium specificity of a high affinity potassium transporter that has been isolated from wheat (Rubio *et al.*, 1995). Down-regulation of genes encoding transporters responsible for nitrogen uptake into legumes might increase the efficiency of nitrogen fixation by removing the feedback inhibition that nitrogen uptake from the soil has on the fixation process. In sugarcane, the quality of raw sugar might also benefit from down-regulating the uptake of nitrogen from the soil in the immediate pre-harvest period.

Probes to Identify Nutrient Efficient Genotypes

Another approach to improving nutrient efficiency is to improve the capacity of plants to take up nutrients under deficiency conditions. Problems in arriving at a consensus as to what 'nutrient efficiency' means and how to quantitatively evaluate it, particularly in relation to macronutrients, has made this difficult. A long running program to select for improved adaptation of white clover to low phosphorus soils (Caradus, 1994) exemplifies these problems. Further, there have been few incentives to make genotypes more nutrient efficient in the past, particularly for high value crops like sugarcane that are difficult and expensive to progress through selection and breeding programs. Undoubtedly, identification of genetic markers linked to aspects of nutrient efficiency would provide plant breeders with very valuable tools.

There have been some novel developments recently in the micronutrient area that may be of interest to those involved in improvement of difficult commercial crops. Arising from studies on the mechanism responsible for manganese efficiency in cereals (Huang *et al.*, 1994a,b, 1996), a protein that is only expressed in manganese-efficient genotypes has been identified. The cDNA encoding this protein has been isolated and is being evaluated for use as a simple probe for screening breeding material for the manganese efficiency trait and determining its heritability (R. D. Graham, *pers. comm.*). Preliminary work is also underway (Rengel and Graham, 1996) to identify probes that are specific for zinc efficient genotypes. For those nutrients for which there is considerable genetic variation in their efficiency of uptake or utilisation, this approach offers good prospects for success. It is a far more robust technology than molecular intervention and simpler to apply through traditional selection and breeding programs. It seems feasible that subtractive cDNA libraries made from efficient and inefficient genotypes could be used to identify similar nutrient specific proteins for use in other species.

Nutrient Acquisition

For many nutrients, and notably phosphorus, the rate of supply of nutrient to the root surface can be a major limitation to plant growth. In such situations the characteristics of the soil are a dominant factor. However, plants have developed responses which can ameliorate the supply of nutrients from the soil. Modification of pH in the rhizosphere adjacent to roots, excretion of organic compounds which specifically influence the availability of some nutrients and provide substrates for rhizosphere microorganisms, root growth and architecture, and associations with mycorrhizal fungi are all at least partially controlled by the plant.

Root excretions

Work in India on pigeonpea (*Cajanus cajan*), a crop that is very efficient in acquiring phosphate, has indicated that piscidic acid, excreted from the roots, may facilitate the acquisition of phosphate from normally insoluble iron phosphates (Ae *et al.*, 1990). Attempts are being made to verify the mechanisms associated with the action of piscidic acid and to use molecular techniques to elucidate its biosynthetic pathway and excretory process as a forerunner to transferring this capability to other plant species. But other more common organic acids such as citrate, malate and fumarate are also excreted by plant roots and are also capable of releasing phosphate from insoluble sources. The well known capacity of white lupins (*Lupinus albus*) to grow on low phosphorus soils is primarily due to mobilisation of insoluble soil phosphates by the large amounts of citrate excreted by the proteoid root structures of this species (Gardner *et al.*, 1983; Marschner, 1995). Excretion of citrate and malate is believed to be responsible for the efficient acquisition of phosphorus by chickpea (*Cicer arietinum*). There are good prospects for using molecular techniques to increase the excretion of these organic acids by plant roots. Research on this aspect is bolstered by the recent report that malic acid excretion may help protect plant roots from the damaging effects of aluminium toxicity in acid soils (Delhaize *et al.*, 1993). Variations in the capacity of roots to excrete phytosidcrophores, organic ligands important in iron nutrition (Romheld and Marschner, 1986a,b; Romheld, 1987), also presents prospects for improving the efficiency of iron uptake by plants through genetic means.

In a different strategy, more efficient use of the organic phosphorus pool in soils is envisaged. Large pools of organic phosphorus have built up in soils as a result of a history of phosphate fertilisation. Availability of this organic pool to plants is mediated through the activity of soil microoganisms. Inositol hexaphosphates (phytates) comprise a substantial part of the soil organic phosphorus pool. Recent work in Canberra (Richardson *et al.*, 1995) has characterised several isolates of soil bacteria that exhibit phytase activity and can liberate phosphate from inositol hexaphosphates. Moreover, in laboratory studies, plants inoculated with one of these bacterial isolates were able to derive their phosphorus supply from phytate. This opens up prospects for using molecular techniques to associate high, stable levels of phytase activity with plant roots, thus permitting them to draw more efficiently upon this large pool of accumulated organic phosphorus in the soil.

Root architecture

Root structure is a major determinant of the efficiency of nutrient acquisition. There are considerable species and genotypic differences in the root characteristics that influence the volume of soil that it is explored for nutrients. Factors such as root distribution in relation to the location of nutrients and water, root length, root diameter, the number of active root tips, the presence and characteristics of root hairs, rooting density and the effectiveness of any mycorrhizal associations have important consequences on nutrient acquisition. Research into root developmental biology is progressing with model plants like *Arabidopsis* (Benfey and Schiefelbein, 1994) and genes that control the characteristics of roots are being isolated. However, it is likely to be a long time before the basic knowledge arising from such studies can be routinely applied to tailor the roots of common agricultural species to improve their efficiency of nutrient acquisition.

Mycorrhizal associations

Vesicular arbuscular mycorrhizal fungi form symbiotic associations with the roots of most crop plants. The long, fine hyphae of these fungi are able to reach out into regions of the soil that have not been depleted of nutrients, effectively extending the root system considerably. These fungi play a crucial role in gathering nutrients such as phosphorus and zinc, whose uptake is often limited by their slow rate of diffusion through the soil to the root surface. For these nutrients, the effectiveness of the plant-mycorrhizal symbiosis and the efficiency with which the fungi can take up the nutrients and transfer them to the host plant often determines how efficiently they are used by a crop.

Work has started on the molecular mechanisms underlying this symbiosis. A high affinity phosphate transporter has recently been isolated from one of the common mycorrhizal fungi, *Glomus versiforme* (Harrison and van Buuren, 1995). This transporter appears to be expressed predominately in the external hyphae so is likely to be the primary transporter involved in phosphate uptake from the soil. Its amino acid sequence and primary structure are very similar to the APT1 and APT2 phosphate transporters isolated from plant roots themselves (Smith *et al.*, 1997), indicating that there has been a high degree of conservation in genes involved in these basic plant processes during evolution. Further research is needed to characterise the mechanisms involved in transfer of phosphorus from the mycorrhizae into root cells in order to determine whether molecular intervention might improve the efficiency of that process.

Conclusions

An impressive start has been made towards gaining an understanding of the molecular events underlying the uptake and utilisation of nutrients. But it is just that - a start. Genes encoding transporters, the key molecular species in these processes, are being isolated, their regulation characterised and the first tentative steps towards manipulating their expression in transgenic plant systems are being taken. These studies, and the more sophisticated ones that will undoubtedly follow, will provide the knowledge base that is essential for identifying reliable applications of these technologies. A large research effort is required and must be made if growing public demands for greater care in the use of chemical fertilisers in agricultural systems are to be met. High yielding perennial crop species such as sugarcane have evolved under conditions of much lower nutrient supply than they receive in the highly developed, intensive production systems of northern Australia and other parts of the world. They are likely to have retained many of the conservative nutritional characteristics of their wild-type progenitors; characteristics programmed to ensure their survival as perennials at the expense of efficient nutrient use for high production. There may not be major natural variation within these species that could be exploited to improve their nutrient efficiency. If this is the case, prospects for improving nutrient uptake and metabolism will rely upon molecular techniques to intervene in the control of these processes. We have a long way to go, but the opportunities have never been greater as a result of the new tools at our disposal. The road ahead promises to be very exciting!

References

Ae, N., Arihara, J., Okada, K., Yoshihara, T. and Johansen, C. (1990) Phosphorus uptake by pigeonpea and its role in cropping systems of the Indian subcontinent. *Science* 248, 477-480.

Anderson, J.J., Huprikar, S.S., Kochian, L.V., Lucas, W.J. and Gaber, R.F. (1992) Functional expression of a probable *Arabidopsis thaliana* potassium channel in *Saccharomyces cereviseae. Proceedings National Academy Science USA* 89, 3736-3740.

Arthington, A.H., Marshall, J.C. Rayment, G.E., Hunter, H.E. and Bunn, S.E. (1997) Potential impact of sugarcane production on the riparian and freshwater environment. In: Keating, B.A. and Wilson, J.R. (eds), *Intensive Sugarcane Production: Meeting the Challenges Beyond 2000.* CAB International, Wallingford, UK, pp. 403-422.

Bell, C.I., Cram, W.J. and Clarkson, D.T. (1990) Turnover of sulfate in leaf vacuoles limits retranslocation under sulfur stress. In: H. Rennenberg (ed.), *Sulfur Nutrition and Sulfur Assimilation in Higher Plants.* SPB Academic Press, The Hague, The Netherlands, pp. 163-165.

Benfey, P.N. and Schiefelbein, J.W. (1994) Getting to the root of plant development: the genetics of *Arabidopsis* root formation. *Trends in Genetics* 10, 84-88.

Caradus, J.R. (1994) Selection for improved adaptation of white clover to low phosphorus and acid soils. *Euphytica* 77, 243-250.

Clarkson, D.T., and Hawkesford, M.J. (1993) Molecular approaches to plant nutrition. *Plant and Soil* 155/156, 21-31.

Clarkson, D.T., Hawkesford, M.J., Davidian, J-C. and Grignon, C. (1992) Contrasting responses of sulphate and phosphate transport in barley (*Hordeum vulgare* L.) roots to protein-modifying reagents and inhibition of protein synthesis. *Planta* 187, 306-314.

Clarkson, D.T. and Lüttge, U. (1991) Mineral nutrition: Inducible and repressible nutrient transport systems. *Progress in Botany* 52, 61-83.

Clarkson, D.T. and Scattergood, C.B. (1982) Growth and phosphate transport in barley and tomato plants during development of and recovery from phosphate stress. *Journal Experimental Botany* 33, 865-875.

Clarkson, D.T., Smith, F.W. and Vanden Berg, P.J. (1983) Regulation of sulphate transport in a tropical legume, *Macroptilium atropurpureum*, cv. Siratro. *Journal Experimental Botany* 34, 1463-1483.

Cogliatti, D.H. and Clarkson, D.T. (1983) Physiological changes in phosphate uptake by potato plants during development of and recovery from phosphate deficiency. *Physiologia Plantarum* 58, 287-294.

Cogliatti, D.H. and Santa Maria, G.E. (1990) Influx and efflux of P in roots of wheat plants in non-limiting P concentrations. *Journal Experimental Botany* 41, 601-607.

Crawford, N.M. and Arst, H.N. Jr. (1993) The molecular genetics of nitrate assimilation in fungi and plants. *Annual Review of Genetics.* 27, 116-146.

Crossland, C.J., Done, T.J. and Brunskill, G.T. (1997) Potential impact of sugarcane production on the marine environment. In: Keating, B.A. and Wilson, J.R. (eds), *Intensive Sugarcane Production: Meeting the Challenges Beyond 2000.* CAB International, Wallingford, UK, pp. 423-436.

Delhaize, E., Ryan, P.R. and Randall, P.J. (1993) Aluminium tolerance in wheat (*Triticum aestivum* L.). II. Aluminium-stimulated excretion of malic acid from root apices. *Plant Physiology* 103, 695-702.

Epstein, E. and Hagen, C.E. (1952) A kinetic study of the absorption of alkali cations by barley roots. *Plant Physiology* 27, 457-474.

Frommer, W.B., Kwart, M., Hirner, B., Fischer, W.F., Hummel, S. and Ninnemann, O. (1994) Transporters for nitrogenous compounds in plants. *Plant Molecular Biology* 26, 1651-1670.

Fu, Y.-H., Lee, H.-J., Young, J.L., Jarai, G. and Marzluf, G.A. (1990) Nitrogen and sulfur regulatory circuits of Neurospora. In: *Developmental Biology.* Wiley-Liss, NY, USA, pp. 319-335.

Gardner, W.K., Barber, D.A. and Parbery, D.G. (1983) The acquisition of phosphorus by *Lupinus albus* L. III. The probable mechanism by which phosphorus movement in the root/soil interface is enhanced. *Plant and Soil* 70, 107-114.

Glass, A.D.M. (1983) Regulation of ion transport. *Annual Review Plant Physiology* 34, 311-326.

Griffith, J.K., Baker, M.E., Rouch, D.A., Page, M.G.P., Skurray, R.A., Paulsen, I.T., Chater, K.F., Baldwin, S.A. and Henderson, P.J.F. (1992) Membrane transport proteins: implications of sequence comparisons. *Current Opinion in Cell Biology* 4, 684-695.

Harrison, M.J. and van Buuren, M.L. (1995) A phosphate transporter from the mycorrhizal fungus *Glomus versiforme*. *Nature* 378, 626-629.

Hawkesford, M.J. and Smith, F.W. (1997) The molecular biology of higher plant sulphate transporters. In: J. Cram (ed.), *Sulfur Nutrition and Assimilation in Higher Plants.* 3rd International Workshop on Sulphur Metabolism, Newcastle-upon-Tyne, UK (in press).

Huang, C., Webb, M.J. and Graham, R.D. (1994a) Mn efficiency is expressed in barley growing in a soil system but not in solution culture. *Journal Plant Nutrition* 17, 83-95.

Huang, C., Webb, M.J. and Graham, R.D. (1994b) Effect of pH on manganese absorption among barley genotypes in a chelate-buffered nutrient solution. *Plant and Soil* 155/156, 437-440.

Huang, C., Webb, M.J. and Graham, R.D. (1996) Pot size affects expression of Mn efficiency in barley. *Plant and Soil* 178 (in press).

Jackson, W.A., Pan, W.L., Moll, R.H. and Kamprath, E.J. (1986) Uptake, translocation and reduction of nitrate. In: C.A. Neyra (ed.), *Biochemical Basis of Plant Breeding, Vol 2.* CRC Press, Boca Raton, USA, pp. 73-108.

Lefebvre, D.D. and Glass, A.D.M. (1982) Regulation of phosphate influx in barley roots; effect of phosphate deprivation and reduction of influx with provision of orthophosphate. *Physiologia Plantarum* 54, 199-206.

Loneragan, J.F., Grove, T.S., Robson, A.D. and Snowball, K. (1979) Phosphorus toxicity as a factor in zinc-phosphorus interactions in plants. *Soil Science Society of America Journal* 43, 966-972.

Marger, N.D. and Saier, M.H. (1993) A major superfamily of transmembrane facilitators that catalyse uniport, symport and antiport. *Trends in Biochemical Science* 18, 13-20.

Marschner, H. (1995) *Mineral Nutrition of Higher Plants, Second Edition.* Academic Press, London, UK.

Marzluf, G.A. (1993) Regulation of sulfur and nitrogen metabolism in filamentous fungi. *Annual Review Microbiology* 47, 31-55.

Ninnemann, O.W., Jauniauex, J.C. and Frommer, W.B. (1994) Identification of a high affinity ammonium transporter from plants. *EMBO Journal* 13, 3464-3471.

Redinbaugh, M.G. and Campbell, W.H. (1991) Higher plant responses to environmental nitrate. *Physiologia Plantarum* 82, 640-650.

Rengel, Z. and Graham, R.D. (1996) Uptake of zinc from chelate-buffered nutrient solutions by wheat genotypes differing in zinc efficiency. *Journal Experimental Botany* 47, 217-226.

Richardson, A.E., Saunders, C.L., Hadobas, P.A and Simpson, R.J. (1995) Microbial contribution to the availability of soil phosphorus. *Proc. First National Soil Phosphate Conference*, Wagga Wagga NSW, Australia, p.17.

Romheld, V. (1987) Different strategies for iron acquisition in higher plants. *Physiologia Plantarum* 70, 231-234.

Romheld, V. and Marschner, H. (1986a) Mobilization of iron in the rhizosphere of different plant species. *Advances in Plant Nutrition* 2, 155-204.

Romheld, V. and Marschner, H. (1986b) Evidence for a specific uptake system for iron phytosiderophores in roots of grasses. *Plant Physiology* 80, 175-180.

Rubio, F., Gassman, W. and Schroeder, J.I. (1995) Sodium-driven potassium uptake by the plant potassium transporter HKT1 and mutants conferring salt tolerance. *Science* 270, 1660-1663.

Schachtman, D.P. and Schroeder, J.I. (1994) Structure and transport mechanism of a high-affinity potassium uptake transporter from higher plants. *Nature* 370, 655-658.

Sentenac, H., Bonneaud, N., Minet, M., Lacroute, F., Salmon, J-M., Gaymard, F. and Grignon, C. (1992) Cloning and expression in yeast of a plant potassium ion transport system. *Science* 256, 663-665.

Smith, F.W., Ealing, P.E., Dong, B. and Delhaize, E. (1997) The cloning of two *Arabidopsis* genes belonging to a phosphate transporter family. *Plant Journal* (in press).

Smith, F.W., Ealing, P.M., Hawkesford, M.J. and Clarkson, D.T. (1995) Plant members of a family of sulfate transporters reveal functional subtypes. *Proceedings National Academy Science USA* 92, 9373-9377.

Tsay, Y.-F., Schroeder, J.I., Feldmann, K.A. and Crawford, N.M. (1993) The herbicide sensitivity gene CHL1 of *Arabidopsis* encodes a nitrate-inducible nitrate transporter. *Cell* 72, 705-712.

Chapter eleven:

Novel Approaches for Managing Pests and Diseases in Sugarcane

P.G. Allsopp[1] and J.M. Manners[2]

1. *Bureau of Sugar Experiment Stations, PO Box 651, Bundaberg, Qld 4670, Australia*
2. *CRC for Tropical Plant Pathology, The University of Queensland, Qld 4072, Australia*

Abstract

Pests and diseases are important constraints on the production of sugarcane. In Australia, pest control depends heavily on pesticides, but cultural practices do contribute. Disease control relies on cultural practices, chemical control and resistant cultivars. Novel approaches to minimising impacts of pests and diseases are considered. The major pests of Australian sugarcane are canegrubs and rodents, with weevil borer, soldier flies, wireworms and earthpearls less important. Management is moving from relying on synthetic pesticides to Integrated Pest Management (IPM) systems integrating cultural, biological and insecticidal controls. These aim to improve the efficiency of current controls, devise substitute controls and change the sugarcane system to minimise the impact of pests.

The ability to transform sugarcane plants has opened new opportunities for control of many diseases. In Australia, major targets for control through genetic-engineering strategies are ratoon stunting disease, pachymetra root rot and Fiji disease. Strategies for control of viral diseases using novel resistance genes encoding viral coat proteins, replicases and movement proteins are well documented in many other plants and as discussed here are now being applied to sugarcane.

In: Keating, B.A. and Wilson, J.R. (eds) *Intensive Sugarcane Production: Meeting the Challenges Beyond 2000*. CAB International, Wallingford, UK, 1997.

Introduction

Pests and diseases cause considerable losses in sugarcane production. Pest control depends heavily on use of synthetic pesticides, but cultural practices contribute to management systems. Disease control relies on cultural practices, chemical control and use of resistant cultivars. This review considers some novel approaches to minimising the impact of these organisms in general and on sugarcane in particular. Two different approaches, integrated pest management (IPM) and genetically engineered disease resistance are described for pests and diseases, respectively. The major pests of Australian sugarcane, canegrubs, rodents, sugarcane weevil borer, soldier flies, wireworms and pink earthpearls, are considered in descending order of importance. Viral, bacterial, fungal and oomycete pathogens and some novel options for controlling these diverse organisms will be described. Ultimately, integrated management systems for both pest and disease control will be used in conjunction with novel and improved genetic resistances to create overall packages of crop protection.

New Approaches to Pest Management

The last decade has seen a major change to the philosophy of how damage caused by pests to sugarcane in Australia can be minimised. This has come with the realisation that single-strategy controls, especially those based on curative application of pesticides, are unlikely to be viable in the long term. Withdrawal of organochlorines and thallium sulfate, a restricted range of alternative pesticides, potential resistance and environmental pressures have contributed to this realisation. The Australian sugar industry is now demanding alternatives to pesticides, alternative pesticides and responsible use of registered pesticides. These have spurred an increasing awareness of and a need for IPM strategies.

Integrated pest management uses all suitable techniques and methods in as compatible a manner as possible to maintain pest populations below those causing economic injury. It is a change from the philosophy of pest control by eradication to one of pest management, and to management of entire pest populations, rather than just localised ones. Emphasis is placed on use of a combination of methods aimed at providing cheap, but long term, reliable control with minimal harmful side effects. IPM first emphasises the efficient use of insecticides. Then follows a search for more benign substitutes for insecticides, such as sex attractants, traps and biological controls. Although useful as less intrusive curative methods, these still avoid confronting the causes of the problem. Most keep farmers dependent on experts and suppliers of products, and perpetuate an image of pests as enemies to be eliminated. The third phase in IPM recognises pests not as enemies, but as indicators of inappropriate design and management of systems. Systems should be redesigned to integrate a range of cultural and bio-ecological controls, so that potential pests remain at subeconomic levels.

Although the move towards IPM has not occurred to this extent with all sugarcane pests and may not be possible with some, there have been some significant changes. These and work underway are outlined below and illustrate different approaches for different pests. What is also important is a significant change in farmers' attitudes; the 'magic-wand' approach of applying a pesticide once a problem has developed is increasingly being questioned.

Canegrubs

Damage and pest control strategies

Canegrubs (Coleoptera: Scarabaeidae) have been pests of Australian sugarcane almost since the crop was first grown. Their ability to damage underground parts of plants, coupled with a large number of species, diversity in life history, distribution and behaviour, make canegrubs the most important pest complex (Robertson *et al.*, 1995a). For 40 years from the late 1940s canegrowers relied on benezene hexachloride (BHC) to control canegrubs. Recently, organophosphates, now the only insecticides used, have failed to control some species in some areas.

Novel strategies

Given the importance of canegrubs, IPM systems must be developed through a combination of improving the cost-efficiency of present controls, substitution of more benign controls for insecticides, and system redesign to reduce the impact of the pests (Robertson *et al.*, 1995a). All three approaches feature in the following list of priority areas, and all are dependent on a better knowledge of canegrub biology and ecology. Potential alternative strategies, such as the use of pheromones (Allsopp, 1993) or the use of the pathogen *Bacillus thuringiensis* (Allsopp *et al.*, 1996a), have been considered, but, at present, are not viable. Priority areas and strategies for the development of canegrub IPM are:

1. Improved efficiency of current controls:
Identify the causes of field failures of suSCon® Blue (14% chlorpyrifos) and Mocap® (10% ethoprop); identify the most efficient placement of suSCon® Blue, Mocap® and Rugby® (10% cadusafos) for different canegrubs and crop environments; and improve insecticide application machinery for treating ratoon crops.
2. Development of substitute controls:
Develop other controlled-release and knockdown insecticides targeting larvae; develop microbial insecticides active against larvae.

3. System redesign:
Identify cultural practices restricting the increase of canegrubs and incorporate these into canegrowing practices and improve natural resistance and/or select genes (natural or synthetic) which convey resistance and incorporate these into breeding programs (Allsopp *et al.*, 1996b)
4. All approaches:
Devise and test useful population monitoring and sampling techniques; identify key factors determining spatial and temporal population levels; develop predictive models and incorporate them into decision-support systems; determine economic injury levels for a range of canegrowing conditions and formulate IPM strategies and decision-support systems for major canegrubs.

Canegrub research programs aim to give a system where:

1. Generally grub numbers will be kept low by resistant cultivars that reduce survival of grubs or slow their development, and by changes to farming practices such as green cane trash blankets, site management, crop rotation and fallowing.
2. Grub numbers will be monitored on an area-wide basis, through direct assessment of numbers and/or through a network of traps that assess adult numbers. These data will be used with computer models to forecast grub numbers. Growers may then implement the most appropriate response.
3. Fields that have a high grub risk will be treated at planting with a controlled-release insecticide and in the early second ratoon with a biological insecticide such as the fungus *Metarhizium anisopliae*. This should give grub control for four ratoons and minimise the development of resistance.
4. Fields with a moderate risk will be treated with insecticides as required; these will be either biological or synthetic and rotated to minimise resistance.
5. Fields with low grub risk will not be treated; controls such as resistant cultivars and crop management will maintain numbers below economic levels.

Rodents

Damage and pest control strategies

Two species of native rodent, the canefield rat (*Rattus sordidus*) and the grassland melomys (*Melomys burtoni*), damage sugarcane by chewing the stalks. Damaged stalks become infected with bacterial and fungal rots which lower sugar content. They are the second most important pests of Australian sugarcane with losses totalling $6 million in 1993. Damage is most severe in central and northern Queensland (Robertson *et al.*, 1995b). Control previously depended on aerial application of thallium sulfate baits.

Novel strategies

Management integrates cultural controls reducing the incidence of damage, strategic baiting reducing high numbers in particular fields, and population monitoring allowing predictions of damage (Robertson *et al.*, 1995b).

Cultural control is based on knowledge of rodent feeding habits and habitat preferences. It relies on the twin strategies of reducing immigration to canefields by excluding the weeds that attract rodents, and of reducing the area of nearby non-crop habitats available as refuges. Suitable habitats near canefields are made unsuitable by mowing grassy headlands, mowing or close grazing of grasslands, replacing unused grassland with closed-canopy forest, and clearing weeds from drains. Canefields should be kept weed free from planting until harvest with herbicides or with a trash blanket following green-cane harvesting.

The rodenticide Klerat® (0.005% brodifacoum) has conditional registration for use in sugarcane. It cannot legally be applied from aircraft, and is applied only to fields that are at risk from rodents. It is most easily applied when cane is short, before populations are large and damage is evident. Thus, it relies on predicting rodent numbers to define at-risk fields. A model predicts numbers based on rainfall and subsequent weed growth, and on numbers of rodents trapped during early growth of crops. These predictions are made available before the crop is too high to prevent adequate application of Klerat®.

Integration of weed control, population monitoring and strategic baiting is more complex than the previous simple strategy of aerial baiting each year. Growers must plan to control rodents in each November-January by controlling weeds and applying bait if predictions advise of likely damage; previous responses have been to visible damage and not to potential rodent numbers.

Sugarcane weevil borer

Damage and pest control strategies

The re-emergence of sugarcane weevil borer, *Rhabdoscelus obscurus* (Coleoptera: Curculionidae), as a serious pest is a case of 'lessons once learnt, now forgotten'. Introduction of preharvest burning and resistant cultivars in the 1940s was followed by low levels of weevil borer damage. Damage has increased in the past decade with widespread adoption of green-cane harvesting and trash blanketing (GCTB) (Robertson and Webster, 1995). No insecticides are registered and control strategies must be developed while retaining GCTB.

Novel strategies

Long term control depends on selection of resistant cultivars of sugarcane. In the interim, cultivars already approved have been screened for resistance; those at least partially resistant include Q117, Q152, Q160, Q161 and H56-752

(Robertson and Webster, 1995). Resistance may be a combination of antibiosis and nonpreference with a relationship between rind hardness early in the growth of cane and susceptibility to weevil borer. Softer-rinded cultivars such as Q120 and Q138 are bored more frequently than cultivars with harder rinds. These data are being used by the Bureau Sugar Experiment Stations (BSES) to select more resistant cultivars.

While a return to preharvest burning is unlikely, the amount of cane suitable for weevil-borer breeding that is left in fields after harvesting could be reduced. Improved harvesting efficiency, and use of shredder-toppers and trash mulching offer potential benefits and are currently being examined.

Rodent damage is also correlated with weevil incidence and control of rodents and reduction in rat-bitten cane will reduce the severity of weevil borer damage. This illustrates the complex nature of pest management and the need to understand not just target pests, but also the whole canegrowing system.

Soldier fly

Damage and pest control strategies

Soldier flies, larvae of *Inopus rubriceps* (Diptera: Stratiomyidae), feed on sugarcane roots, cutting root hairs and burrowing their forebodies into the roots. Larvae may reduce germination at planting, but, more commonly, larvae cause poor ratooning through buds failing to germinate. They damaged 430ha of sugarcane in 1993, but areas damaged were much greater during the 1960s (Samson, 1995). Dieldrin and BHC were used to control soldier fly, but their use ceased in 1991 and no insecticide is currently used. Present management recommendations (Samson, 1995) are:

1. Plough out heavily infested fields after an early harvest.
2. Destroy these stools by surface cultivation.
3. Maintain a bare fallow for as long as possible.
4. Plant a cover crop, such as cucurbits or a legume, to protect fields during periods vulnerable to erosion.
5. Cultivate during the flies' breeding season and surface cultivate for 1-2 months after that.
6. Delay planting until spring.
7. Plant tolerant cultivars.
8. Harvest plant crops and young ratoons mid- to late-season for best ratooning.
9. Continue monitoring fields each year.

Novel strategies

Four approaches have been or are being considered for minimising soldier fly losses (Samson, 1995). The first aims at minimising the number of larvae present at planting, as these are a source of damaging populations in subsequent ratoons. This could be achieved through bare fallowing, cover or break crops, cultivation, insecticides, or a reduced length of the crop cycle. The second targets immigrant adult flies by avoiding planting new crops near old infested ratoons through better farm design. The third aims to minimise the increase of larval numbers in successive generations through creating unfavourable physical environments, e.g. dry soil during the egg stage, plant resistance, encouraging natural enemies, and biological and chemical insecticides. The fourth aims to minimise effects of infestations on the crop through tolerant cultivars and promoting healthy crops. The challenge is to integrate these practices into a pest management program that is acceptable to growers and tailored to different canegrowing regions.

Wireworms

Damage and pest control strategies

Wireworms (Coleoptera: Elateridae) bore into the eyes or growing points of germinating setts or ratooning stubble. Damage is seen as poor or patchy germination and dead spindle leaves. In 1993, 2100ha of plantings in southern Queensland were treated with chlorpyrifos to protect them against wireworms. The insecticide must be applied at planting and growers decide to treat without knowing if wireworms are present. As a result, insecticide may be wasted on crops that would not have been damaged.

Novel strategies

Current research aims to minimise the use of insecticides by providing a simple, cheap system for growers to monitor fields before planting (Samson and Robertson, 1996). Baits of rolled oats are buried in the soil for 1 week. They are then dug up and wireworms counted. Preliminary results are encouraging, but more work is needed to establish the reliability of the technique and critical catch numbers that benefit from treatment. Problems in misidentifying other animals caught at the baits and the cost of sampling versus the 'peace-of-mind' benefits of treatment are barriers to the adoption of such techniques.

Earthpearls

Damage and pest control strategies

Pink earthpearls, nymphs of *Eumargarodes laingi* (Hemiptera: Margarodidae), feed on roots of sugarcane. Areas affected by earth pearls show poor crop growth patches that expand as the insects spread. Initial symptoms are stool stunting and leaf yellowing, eventually ending in stool death. Yield losses of 75% are common in the Bundaberg area. No insecticide is registered.

Novel strategies

Management strategies are based on a survey of infested fields and agronomic practices as described by Walker and Allsopp (1993) and on host-plant resistance studies detailed by Allsopp and McGill (1996). Infested fields should be ploughed out, fallowed for 12 months and ploughed and rotary-hoed during that period. Infested fields should be replanted with a tolerant cultivar such as Q147, Q135 or Q136 (in descending order of preference). All machinery used in infested areas should be cleaned thoroughly before moving to other fields to reduce spread of the pest.

Integrated pest management in sugarcane

Extensive use of pesticides in sugarcane contributes to the general perception within the community that our agriculture depends on chemicals. The issue of continuing widespread use of pesticides must be addressed. During the last few years the Australian sugar industry has taken the first tentative steps towards addressing this issue by considering IPM.

Strategies are in place to develop IPM systems for all major pests. The methods used differ between pests and depend on the pests' biology, susceptibility and importance. However, as IPM is more complex than chemical pest control, a major effort is still needed to educate growers in the principles of IPM and to provide technical support services that they require to make it a success. Continuing research is also needed because the technical knowledge base on which IPM programs are built is relatively weak in most areas. Integrated pest management has the potential to provide user-friendly, environmentally acceptable pest control at a realistic cost, but to be successful it requires the day-to-day involvement of professionals with an understanding of the crop/pest ecosystem. Research and extension expertise in areas such as damage assessment, pest monitoring, modelling and biological, chemical and cultural control is also needed. The industry must be prepared to pay for this. If we do not take this approach, the long term outcome will be an ever increasing financial burden of chemical control and increasing environmental concerns and constraints. These will mean continual high losses to pests.

Only through developing IPM strategies can we get away from the treadmill of continually replacing one insecticide with another, and build a more sustainable system. We need to consider pests as indicators of poorly designed and malfunctioning production systems. Our response must be and is to concentrate on redesigning the system and improving management rather than attempting to eliminate or directly control pests.

New Approaches to Controlling Diseases in Sugarcane

Important diseases of sugarcane and current control measures

Diseases cause significant losses of production of sugarcane in Australia, and elsewhere. The main disease control measures are the use of disease-free planting material combined with suitable farm hygiene and the use of resistant cultivars in districts where the disease risk is high. A hidden but important impact of disease is that many useful cultivars are lost from plant breeding programs through susceptibility to disease. In Australia, disease control in sugarcane is greatly aided by restricting the movement of cane between quarantine districts. Some important diseases of sugarcane, e.g. smut, are not established in Australia and national quarantine procedures are therefore necessary. Diseases of importance in Australia are briefly described below.

Viruses

The major virus-incited disease of sugarcane in Australia is Fiji disease. This virus has a double-stranded RNA genome. Currently, control is maintained by resistant cultivars, removal of diseased cane and use of disease-free planting material. The distinctive symptoms of Fiji disease are leaf galls and stunting. Chlorotic streak is believed to be virus-incited but the causal agent has not been identified. Chlorotic streak is eliminated from planting material by hot water treatment. Mosaic disease is of less importance and is caused by the sugarcane mosaic potyvirus which has a single-stranded RNA genome. Mosaic disease is controlled by rejecting susceptible cultivars.

Phytopathogenic bacteria

The most important bacterium-incited disease in Australia is ratoon stunting disease caused by the gram-positive and nutritionally fastidious bacterium *Clavibacter xyli* subsp. *xyli*. This coryneform bacterium is a xylem invading organism and causes stunting of cane, particularly in older ratoon crops and in dry conditions. The disease is controlled by strict hygiene of farm machinery and hot water treatment of planting material. Planting material can be analysed for the presence of the bacterium using microscopic and immunological tests and these tests are used in certification of planting material. Leaf scald, caused by the gram-negative bacterium *Xanthomonas albilineans*, is widely distributed in Australia but is effectively controlled by rejecting susceptible germplasm before cultivar release. Other bacterial-incited diseases of minor importance are top rot or red stripe caused by *Pseudomonas rubrilineans* and bacterial mottle caused by *Erwinia chrysanthemi*.

Phytopathogenic fungi and oomycetes

Root diseases are important in sugarcane. Pachymetra root rot is of major importance in northern, wetter regions of Australia. This disease is primarily caused by *Pachymetra chaunorhiza*, an oomycete pathogen which is unique to Australia. Control of pachymetra root rot is by use of resistant cultivars and by varietal rotation to reduce build up of spore inoculum. Root rots of sugarcane can be associated with other oomycete pathogens, e.g. *Pythium* spp., and these can cause disease in pot trials. The role played by soil pathogens, beneficial soil microorganisms and other soil biota in yield decline under intensive farming is contentious. An important fungal pathogen is *Ceratocystis paradoxa,* cause of pineapple disease, and it is effectively controlled by fungicide treatment. Sugarcane common rust caused by *Puccinia melanocephala* is controlled by resistant cultivars and red rot disease caused by *Colletotrichum falcatum,* although important in many countries, is of little significance in Australia.

Genetic engineering for resistance

Molecular biology provides tools for the improved detection of pathogens, for DNA-marker facilitated breeding and for the genetic engineering of disease resistance. The applications of DNA-based diagnostics using the highly sensitive polymerase chain reaction have so far been restricted mainly to viruses (e.g. Smith and Van de Velde, 1994; Braithwaite *et al.*, 1995) and will not be discussed here. Applications of DNA markers in breeding are discussed by Berding *et al.* (1997). This review will focus mainly on genetic engineering approaches to disease resistance. Since the first report of the genetic transformation of sugarcane (Bower and Birch, 1992) it has been possible using particle bombardment to transfer genes from a wide variety of organisms into sugarcane. Some of the current strategies that may be applied to sugarcane are described below.

Strategies for the control of viral pathogens

The first report of genetically engineering resistance to a virus was published in 1986 when transgenic tobacco producing the coat protein of the tobacco mosaic virus (TMV) was shown to be resistant to infection by TMV (Abel *et al.*, 1986). Strategies for engineering disease resistance to viruses using genes derived from the virus itself are now well proven in other plant disease systems and field assessments have been reported (Grumet, 1995). The genes that have been expressed in transgenic plants to produce resistant plants include the viral coat protein, the viral replicase, antisense sequences to the viral genome, satellite RNA, defective interfering RNA and defective viral movement protein. The genomes of several viruses of sugarcane have been cloned and at least partly sequenced (Smith *et al.*, 1996). Most successful reports of engineering resistance in other plants have been to single-stranded RNA viruses and therefore approaches using viral genes such as that encoding the coat protein would be highly feasible for the control of sugarcane mosaic virus. The coat protein of sugarcane mosaic virus has been successfully expressed in protoplasts of sugarcane (Smith *et al.*, 1992) and transgenic sugarcane plants expressing this gene are currently under evaluation (Smith *et al.*, 1996). The double-stranded RNA *Fijivirus* may also be amenable to such strategies but there is less precedent for success with this type of virus.

There have been some concerns about the deployment of genes derived from viral pathogens and the possibility that they may recombine with other viruses. Such recombination has been detected in the laboratory (Greene and Allison, 1994) and this threat may stimulate the application of highly defective viral genes or alternative, non-pathogen based genes (e.g. Tavladoracki *et al.*, 1993). In view of the importance of viral diseases in sugarcane and the high rate of success of control of viral diseases in other crops it is likely that reports of the successful genetic engineering of resistance in sugarcane will appear soon.

Strategies for the control of bacterial and fungal pathogens

Most current strategies for engineering resistance to bacterial and fungal pathogens are based on an understanding of the molecular basis of plant-pathogen interactions. With exceptions such as leaf scald disease, little work has been undertaken at the molecular level on sugarcane disease systems. It seems likely that strategies developed in other crop and model systems may be transferred to sugarcane. These strategies and current work are outlined below.

Genes encoding anti-microbial proteins

The simplest strategy for engineering resistance to fungi and bacteria is to obtain transgenic plants with high level constitutive expression of one or more genes encoding anti-microbial proteins. Anti-microbial proteins have been found in many plants and are obvious candidates for this strategy. Plant genes that have been used successfully to control at least one fungal or oomycete pathogen in a transgenic plant encode plant chitinase (Broglie *et al.*, 1991) alone and in combination with beta-1,3-glucanase (Zhu *et al.*, 1994), osmotin (Liu *et al.*, 1994), PR1a (Alexander *et al.*, 1993), ribosome-inhibiting protein (Logemann *et al.*, 1992) and plant defensins (Terras *et al.*, 1995). Proteins inhibiting *in vitro* growth of sugarcane pathogens, e.g. *Ceratocystis paradoxa* and *Pythium graminicola,* have been reported (Harrison *et al.,* 1996). So far, proteins active against important pathogens, *C. xyli* sub.sp. *xyli* and *P. chaunorhiza*, have not been identified and this is an important priority.

Inactivation of pathogenicity factors

The expression of genes whose products are able to inactivate pathogenicity factors is a possible strategy for engineering resistance. Study of the sugarcane pathogen *X. albilineans* has demonstrated that this pathogen produces a toxin termed albicidin which induces the chlorosis associated with leaf scald disease (Birch and Patil, 1987). Recently, a gene which encodes an albicidin-degrading enzyme was isolated from the bacterium *Pantoea dispersa* and was expressed in transgenic sugarcane (Zhang and Birch, 1996). Preliminary infection tests reported by Zhang and Birch (1996) suggest that the transgenic plants show highly decreased disease symptoms. This significant piece of work represents the first report of engineering sugarcane for increased resistance to disease and may permit the use of plant cultivars that would currently be rejected by Australian improvement programs because of leaf scald susceptibility.

The pioneering work of Birch and colleagues on leaf scald disease emphasises the need for fundamental studies of interactions of sugarcane with their pathogens. Without such information it may not be possible to apply this strategy to other diseases. The most important bacterial disease of sugarcane is ratoon stunting disease (rsd). However, the disease organism *C. xyli* subsp. *xyli* is extremely difficult to work with experimentally when compared to *X. albilineans*. Recently, genetic transformation of the rsd-pathogen has been achieved (S. Brumbley and R. Birch, *pers. comm.*) which should permit a genetic analysis of its pathogenicity in the near future.

Synthetic hypersensitivity

This strategy depends upon the identification of plant promoters that are expressed exclusively during microbial infection. Such promoters can be fused to a gene encoding a gene product which will kill the plant cell when the gene is expressed. The principle of this strategy has been demonstrated in transgenic potato by Strittmater *et al.* (1995). The isolation of tightly controlled plant promoters associated with the early stages of compatible plant-pathogen interactions is obviously important in this strategy.

Heterologous expression of natural disease resistance genes

In the past, plant breeders of most crops have manipulated natural resistance genes in crop improvement programs. Major breakthroughs in the understanding of disease resistance have occurred in the past two years by the cloning of several plant genes for resistance to fungal, bacterial and viral pathogens (Staskawicz *et al.*, 1995). Disease resistance (*R*) genes appear to encode receptors that have a high level of specificity to avirulence determinants of particular pathogen races. Because of this specificity it is possible that *R* genes isolated from one species may not provide a broad spectrum of resistance when transferred to another plant species. Therefore, it seems most likely that the logical use of *R* genes for genetically engineering resistance may not be possible until we know more about how they function.

Expression of regulators and elicitors of the defence response

In recent years it has been shown that plant defences can be induced by endogenous low molecular weight and small polypeptide signals. Increases in cellular hydrogen peroxide concentrations can activate defences and could be effected by expression of genes encoding enzymes such as glucose oxidase. Expression of a small signal polypeptide termed systemin in transgenic tomato leads to the constitutive expression of protease inhibitors believed to provide protection against insects (McGurl *et al.*, 1994). If similar polypeptide signals exist for defence against microbes in sugarcane then their genes would be attractive novel resistance genes.

Novel phytoalexins

Phytoalexins are low molecular weight antibiotic compounds produced by plants during resistant reactions to pathogens. Most plants produce only a limited number of the phytoalexins identified in the plant kingdom. Thus it is possible that if one could produce transgenic plants carrying genes encoding enzymes capable of modifying existing phytoalexins into ones with different specificity or increased toxicity then plant resistance may improve. The principle has been demonstrated directly by transfer of genes from grapevine encoding a stilbene synthase to transgenic tobacco (Hain *et al.*, 1993). These transgenic tobacco plants produced the novel phytoalexin resveratrol and showed enhanced resistance to a fungal pathogen. The manipulation of phytoalexins may be a fruitful approach to controlling fungal pathogens in sugarcane once a greater understanding of phenolic metabolism is obtained in this species.

Generic approaches to disease control in sugarcane

From the above it is evident that there are many options for genetic engineering of disease resistance in sugarcane. Sugarcane has many significant microbial pathogens and it is likely that multiple strategies will need to be used simultaneously to obtain control that is broad enough to control several diseases at once. The wide variety of diseases means that research should focus on generic technologies capable of providing resistance to several pathogens. Most of the strategies described above will require more information on how to express foreign genes at the right time, in the right place and in sufficient quantity for them to be effective. Other hurdles that need to be crossed concern commercial and legal constraints to technologies that are being used. These and other aspects concerning transgenic sugarcane are discussed by Birch (1997).

Conclusions

The concepts of integrated pest management can be extended to management of genetically modified sugarcane plants for both pest and disease control. In other crops, genetically modified plants are already being used in the field for pest control. The Bt gene (Perlak *et al.*, 1990) has been most widely exploited because its spectrum of activity includes major pests of cotton which currently receive a high level of pesticide application. A major consideration in deployment of the Bt gene is that Bt tolerant pest genotypes will be selected if other management systems are not used. These may involve less frequent spraying and the maintenance of non-Bt carrying plants to reduce selection pressure and maintain genetic heterogeneity in the insect population. These management methods are simply an extension of IPM which incorporates the genetically modified host system. They illustrate, however, potential problems associated with reliance on only one control mechanism. Development of wide-spread, integrated controls must be a high priority for the sugarcane industry.

As we move beyond the year 2000 it is evident that more sophisticated and integrated management practices will be needed to maximise sustainable yields from an increasingly more sophisticated base of sugarcane germplasm.

References

Abel, P.P., Nelson, R.S., De, B., Hoffmann, N., Rogers, S.G., Fraley, R.T. and Beachy, R.N. (1986) Delay of disease development in transgenic plants that express the tobacco mosaic virus. *Science* 232, 738-743.

Alexander, D., Goodman, R.M., Gut-Rella, M., Glascock, C., Weymann, K., Friedrich, L., Maddox, D., Ahl-Goy, P., Luntz, T., Ward, E. and Ryals, J. (1993) Increased tolerance to two oomycete pathogens in transgenic tobacco expressing pathogenesis-related protein 1a. *Proceedings of the National Academy of Sciences, USA* 90, 7327-7331.

Allsopp, P.G. (1993) Evidence for sex attraction in three species of Australian canegrub beetles (Coleoptera: Scarabaeidae: Melolonthinae). *Coleopterists Bulletin* 47, 51-52.

Allsopp, P.G., Chilcott, C.N. and McGhie, T.K. (1996a) Activity of proteins from two New Zealand strains of *Bacillus thuringiensis* against larvae of *Antitrogus consanguineus* (Blackburn) (Coleoptera: Scarabaeidae). *Australian Journal of Entomology* 35, 107-112.

Allsopp, P.G., McGhie, T.K., Cox, M.C. and Smith, G.R. (1996b) Redesigning sugarcane for resistance to Australian canegrubs: a potential IPM component. *Integrated Pest Mangement Reviews* 1, 79-90.

Allsopp, P.G. and McGill, N.G. (1996) Host-plant resistance: a key component in minimising losses from pink ground pearls. *Proceedings of the Australian Society of Sugar Cane Technologists* 18, 100-105.

Berding, N., Moore, P.H. and Smith, G.R. (1997) Advances in breeding technology for sugarcane. In: Keating, B.A. and Wilson, J.R. (eds), *Intensive Sugarcane*

Production: Meeting the Challenges Beyond 2000. CAB International, Wallingford, UK, pp. 189-220.

Birch, R.G. (1997) Transgenic sugarcane: opportunities and limitations. In: Keating, B.A. and Wilson, J.R. (eds), *Intensive Sugarcane Production: Meeting the Challenges Beyond 2000.* CAB International, Wallingford, UK, pp.125-140.

Birch, R.G. and Patil, S.S. (1987) Evidence that an albicidin-like phytotoxin induces chlorosis in sugarcane leaf scald disease by blocking plastid DNA replication. *Physiological and Molecular Plant Pathology* 30, 207-214.

Bower, R. and Birch, R.G. (1992) Transgenic sugarcane plants via microprojectile bombardment. *The Plant Journal* 2, 409-416.

Braithwaite, K.S., Egeskov, N.M. and Smith, G.R. (1995) Detection of sugarcane bacilliform virus using the polymerase chain reaction. *Plant Disease* 79, 792-796.

Broglie, K., Chet, I., Holliday, M., Cressman, R., Biddle, P., Knowlton, S., Mauvais, C.J. and Broglie, R. (1991) Transgenic plants with enhanced resistance to the fungal pathogen *Rhizoctonia solani. Science* 254, 1194-1197.

Greene, A.E. and Allison, R.F. (1994) Recombination between viral RNA and transgenic plant transcripts. *Science* 263, 1423-1425.

Grumet, R. (1995) Development of virus resistant plants via genetic engineering. *Plant Breeding Reviews* 12, 47-79.

Hain, R., Reif, H-J., Krause, E., Langebartels, R., Kindl, H., Vornam, B., Wiese, W., Schmelzer, E., Schrier, P.H., Stocker, R.H. and Stenzel, K. (1993) Disease resistance results from foreign phytoalexin expression in a novel plant. *Nature* 361, 153-156.

Harrison, S.J., Marcus, J.P., Goulter, K.C., Brumbley, S., Green, J.L., Maclean, D.J.and Manners, J.M. (1996) Antimicrobial proteins: New options for disease control in sugarcane. In: Wilson, J.R., Hogarth, D.M., Campbell, J.A. and Garside, A.L. (eds), *Sugarcane: Research Towards Efficient and Sustainable Production.* CSIRO Division of Tropical Crops and Pastures, Brisbane, Australia, pp. 135-137.

Liu, D., Kashchandra, G., Raghothama, G., Hasegawa, P.M. and Bressan, R.A. (1994) Osmotin overexpression in potato delays development of disease symptoms. *Proceedings of the National Academy of Sciences USA* 91, 1888-1892.

Logemann, J., Jach, G., Tommerup, H., Mundy, J. and Schell, J. (1992) Expression of a ribosome inactivating protein (RIP) leads to fungal resistance in transgenic tobacco plants. *Bio-Technology* 10, 305-308.

McGurl, B., Orozco-Cardenas, M., Pearce, G. and Ryan, C.A. (1994) Overexpression of the prosystemin gene in transgenic tomato plants generates a systemic signal that constitutively induces proteinase inhibitor synthesis. *Proceedings of the National Academy of Sciences USA* 91, 9799-9802.

Perlak, F.J., Deaton, R.W., Armstrong, T.A., Fuchs, R.L., Sims, S.R., Greenplate, J.T. and Fischhoff, B. (1990) Insect resistant cotton plants *Bio-Technology* 8, 939-945.

Robertson, L.N., Allsopp, P.G., Chandler, K.J. and Mullins, R.T. (1995a) Integrated management of canegrubs in Australia: current situation and future research directions. *Australian Journal of Agricultural Research* 46, 1-16.

Robertson, L.N., Story, P.G. and Wilson, J. (1995b) Integrated pest management for rodents in sugarcane. *Proceedings of the Australian Society of Sugar Cane Technologists* 17, 70-74.

Robertson, L. N. and Webster, D.E. (1995) Strategies for managing cane weevil borer. *Proceedings of the Australian Society of Sugar Cane Technologists* 17, 83-87.

Samson, P.R. (1995) Management of soldier flies in sugarcane. *Proceedings of the Australian Society of Sugar Cane Technologists* 17, 75-82.

Samson, P.R. and Robertson, L.N. (1996) Development of a decision-support plan for wireworm management. *Proceedings of the Australian Society of Sugar Cane Technologists* 18, 62-70.

Smith, G.R., Ford, R., Frenkel, M.J., Shukla, D.D. and Dale, J.L. (1992) Transient expression of the coat protein of sugarcane mosaic virus in sugarcane protoplasts and expression in *Escherichia coli*. *Archives of Virology* 125, 15-23.

Smith, G.R., Joyce, P.A., Handley, J.A., Sithisarn, P., Maugeri, M.M., Bernard, M.J., Berding, N., Dale, J.L. and Harding, R.M. (1996) Genetically engineering resistance to sugarcane mosaic and Fiji disease viruses in sugarcane. In: Wilson, J.R., Hogarth, D.M., Campbell, J.A. and Garside, A.L. (eds), *Sugarcane: Research Towards Efficient and Sustainable Production*. CSIRO Division of Tropical Crops and Pastures, Brisbane, Australia, pp. 138-140.

Smith, G.R. and Van de Velde, R. (1994) Detection of sugarcane mosaic virus and Fiji disease virus in diseased sugarcane using the polymerase chain reaction. *Plant Disease* 78, 557-561.

Staskawicz, B.J., Ausubel, F.M., Baker, B.J., Ellis, J.G. and Jones, J.D.G. (1995) Molecular genetics of plant disease resistance. *Science* 268, 661-667.

Strittmater, G., Janssens, J., Opsomer, C. and Botterman, J. (1995) Inhibition of fungal disease development in plants by engineering controlled cell death. *Bio-Technology* 13, 1085-1089.

Tavladoracki, P., Benvenuto, E., Trinca, S., de Martinis, D., Cattaneo, A. and Galeffi, P. (1993) Transgenic plants expressing a functional single-chain Fv antibody are specifically protected from virus attack. *Nature* 366, 469-472.

Terras, F.R.G., Eggermont, K., Kovaleva, V., Raikhel, N.V., Osborn, R.W., Kester, A., Rees, S.B., Vanderleyden, J., Cammue, B.P.A. and Broekaert, W.F. (1995) Small cysteine-rich antifungal proteins from radish *(Raphanus sativus* L.): their role in defence. *The Plant Cell* 7, 573-588.

Walker, P.W. and Allsopp, P.G. (1993) Factors influencing populations of *Eumargarodes laingi* and *Promargarodes* spp. (Hemiptera: Margarodidae) in Australian sugarcane. *Environmental Entomology* 22, 362-367.

Zhang, L. and Birch, R.G. (1996) Genetic engineering of sugarcane for leaf scald phytotoxin and disease resistance. In: Cock, J.H. and Brekelbaum, T. (eds), *Proceedings of the International Society of Sugar Cane Technologists XXII Congress, Cartegna, Colombia,* Vol. 2, pp. 397-402.

Zhu, Q., Maher, E., Masoud, S., Dixon, R.A. and Lamb, C.L. (1994) Enhanced protection against fungal attack by constitutive co-expression of chitinase and glucanase genes in transgenic tobacco. *Bio-Technology* 12, 807-812.

Chapter twelve:

Advances in Breeding Technology for Sugarcane

N. Berding[1], P.H. Moore[2] and G.R. Smith[3]

1. *Bureau of Sugar Experiment Stations (BSES), P.O. Box 122, Gordonvale, Qld 4865, Australia*
2. *USDA-ARS, P.O. Box 1057, Aiea, HI 96701, USA*
3. *BSES, P.O. Box 86, Indooroopilly, Qld 4068, Australia*

Abstract

Significant progress has been made in the genetic improvement of sugarcane. This has been by increasing sucrose yield directly or by resolving constraints to sustained or increased productivity. Current crop performance can be contrasted with that of the first interspecific hybrids released by the pioneering Dutch breeders in Java 70 years ago. Alternatively, current crop performance can be gauged against the generally accepted highest commercial yield, recorded on Oahu. Sustaining such progress requires continuing and escalating innovative applications of current and emerging technologies. In this paper, we detail technological applications to sugarcane improvement that have been documented in the past decade. These are distinguished in two respects. Most of these applications have appeared since the start of the decade. They also draw on a wide range of technologies. A general expectation that biotechnology applications may dominate this review was not fulfilled, despite such applications making valuable contributions to genomic and cytological analysis. More mundane, but still exciting, productive applications abounded. Many of the applications have improved our understanding of the *Saccharum* complex, or helped us manipulate the germplasm more effectively, or introduced greater efficiency or objectivity to assessment processes. We conclude that sugarcane improvement is a dynamic, healthy, and evolving scientific area. This is exciting and encouraging. We expect further significant progress in sugarcane improvement in the next decade. This will come from innovative applications of current and emerging technologies, with some biotechnology applications yielding significant returns.

In: Keating, B.A. and Wilson, J.R. (eds) *Intensive Sugarcane Production: Meeting the Challenges Beyond 2000*. CAB International, Wallingford, UK, 1997.

Introduction

In this paper we especially focus on advances in sugarcane breeding technologies in the past decade, building on significant reviews in 1987 (Anonymous, 1987; Berding and Skinner, 1987; Heinz, 1987a). Biotechnology was seen then as a highly important, potentially dominating, emerging technology, that threatened to cause an imbalance in resource allocation between biotechnology and traditional breeding, a concern still expressed a decade later (Duvick, 1996). More reasonable expectations for biotechnology were outlined by Law (1995). Application of molecular based approaches (Moore, 1991; Moore and Irvine, 1991) in assisting conventional improvement of sugarcane, such as genome mapping, marker-assisted selection, and other aspects are described later. Progress in genetic transformation of sugarcane is described in another chapter (Birch, 1997), prospects for molecular modification of metabolism were reviewed by Lingle (1996) and Moore *et al.* (1997), application to sugarcane diseases have been detailed elsewhere (Autrey *et al.*, 1996b; Allsopp and Manners, 1997). Consequently, discussion of biotechnology advances in this chapter is confined to several key areas.

Despite the promise of biotechnology, major advances since 1987 have arisen from a range of less glamorous technologies yielding significant returns in sugarcane improvement. These include: greater knowledge of the germplasm base, improved hybridisation through cytogenetics and control of flowering, more efficient selection procedures based on family selection, best linear unbiased prediction, improved understanding of genotype × environment (G × E) interactions, managed selection environments, and host-plant resistance. Technical advances, such as better seed storage, quicker evaluation procedures (e.g. near infra-red spectroscopy) and computerised data systems for yield, quality and other attributes, labelling, etc., have greatly raised the efficiency of breeding programs. The contributions of these diverse technologies are discussed in this paper.

Germplasm Base and Usage

Advances in characterisation and usage

Commercial sugarcane clones are usually interspecific hybrids between *Saccharum officinarum* and *S. spontaneum*, with limited infusion of *S. sinense* and *S. robustum.* Sugarcane improvement could involve the wider interbreeding group, viz. the '*Saccharum* complex', which includes *Erianthus, Miscanthus, Narenga, Ripidium*, and *Sclerostachya,* but there has been little exploitation of this germplasm in the past decade. Taxonomic relationships within the *Saccharum* complex have been neither carefully studied nor well defined (Webster and Shaw, 1995). Germplasm collection activities since 1987 have been restricted, poorly documented, and with little, if any, targeting of collection for specific traits. There now appears to be a regrettable unwillingness to make material available internationally.

Characterisation of germplasm collections up to 1987 was reviewed by Berding and Roach (1987). More recently, a large population of *S. officinarum* clones in the USDA-ARS World Sugarcane Collection in Florida was characterised (P.H. Moore, *pers. comm.*). These data are accessible through the USDA-ARS germplasm resource information network. From the same collection, Tai *et al.* (1996) characterised 125 *S. spontaneum* clones for four juice-quality and five morphological characters. There is a need for some consistency in descriptors use, and Gallacher (1994) has, for the first time for *Saccharum* spp. hybrid germplasm, developed an optimised morphological descriptor set. His set comprised 24 vegetative macromorphological descriptors selected by tree classification and a regularised discriminant function. Whilst many of these are not in the internationally accepted descriptor set, the optimised set has wide utility because of its ease of use.

Morphological descriptors recently have been overshadowed by developments in biochemical descriptors. They have intrinsic simplicity, as theoretically most can be regarded as single gene products, be highly heritable, and be reliable and provide maximum discrimination. Examples are available for use of flavonoids (Lee, 1995a), isozymes (Eksomtramage *et al.*, 1992; Gallacher *et al.*, 1995), restriction fragment length polymorphisms (RFLP) (Burnquist, 1991), and randomly amplified polymorphic DNA molecular markers (Harvey *et al.*, 1994). The potential power of recent ever-expanding developments in molecular marker technologies for discriminating among sugarcane clones is obvious, although there is perhaps now a need for a standardisation of their application to germplasm management. Not so obvious is the potential of markers for use in management of germplasm collections. Two examples are:

1. Marker data could be accumulated in a database when clones initially entered a collection, then later be used by breeders to verify the trueness to label of clones, or for genetic distance analyses to facilitate subdivision of a collection into homogeneous sub-sets, and consequent optimisation of cross-pollination combinations.
2. Collaborative experiments involving distribution of germplasm from a central source across quarantine boundaries is now near impossible but marker analysis could be used to verify identity of locally available clones for such experimentation.

Past progress in using non-*S. officinarum* germplasm to improve sugarcane performance largely centred on introgression with *S. spontaneum*, as summarised by Berding and Roach (1987), Hsu *et al.* (1988), and Burner and Legendre (1993b). Incorporation of *S. robustum* germplasm has been little exploited in *Saccharum* spp. hybrids, partly because the advantages of $2n+n$ chromosome transmission are not available. Hemaprabha and Ram (1993) studied a range of *S. robustum* derived progeny and concluded that additional backcrossing to *Saccharum* spp. hybrids would be necessary to improve sucrose content, and that intercrossing diverse I_1 (first introgression) progeny would improve cane yield. Lately, interest in use of *Erianthus arundinaceus* has expanded, as reflected in publication of molecular cytogenetic techniques to diagnose the products of introgression breeding (D'Hont *et al.*, 1995; Lee, 1995a; Besse *et al.*, 1996a; Besse *et al.*, 1996b). The potential to improve juice quality may arise in *Saccharum* hybrid crosses with *Miscanthus* spp. or *Miscanthidium* (Tai *et al.,* 1991a).

In Queensland recent success has arisen from a potent I_3 clone derived from a cross between POJ2878 and the *S. spontaneum* clone Mandalay. The I_3 clone is a combination of Indonesian (*2n*=112) and Burmese (*2n*=96) *S. spontaneum.* This parent has appeared in nine 'Q' cultivars of the 33 released from Q138 to Q170. Indications of potentially useful broadening of the genetic base from introgression of *S. spontaneum* have been evident for juice quality (Tai *et al.,* 1992), disease resistance, e.g. smut (Burner *et al.*, 1993), sugar yield (Srivastava *et al.*, 1994), general agronomic characters (Jackson and Roach, 1994), and ratooning and stress tolerance (Jackson, 1994). The latter study indicated a negative association between ratooning ability and sugar content. Srivastava *et al.* (1994) found increasing sugar yield with increasing number of generations from *S. spontaneum.* Because of its earlier flowering, crossing of *S. spontaneum* germplasm with *Saccharum* spp. hybrid material is frequently difficult. On average even I_1 progeny (Tai *et al.*, 1991b) flowered c. 67 days earlier than *Saccharum* spp. hybrid parents. Photoperiodic manipulation of flowering, or pollen storage, is necessary to effect required crosses.

Despite the extensive efforts in germplasm base-broadening in different countries, a general assessment would be that recent use of *S. spontaneum* germplasm, or for that matter any other taxa within the *Saccharum* complex, has been relatively unsuccessful by comparison with the very early nobilisation programs of Indonesia and India. However, the various crossing studies undertaken are important in understanding base-broadening breeding, and assisting in planning strategies for more efficient use of the germplasm resources.

Why has utilisation of basic *Saccharum* germplasm generally failed? Perhaps it is because there appears to be no dedicated commitment to develop material in a base broadening program to a high level before incorporation into the contemporary *Saccharum* spp. hybrid germplasm pool. Base broadening activities in another clonal crop, e.g. potato, that is taxonomically and cytologically complex have been more focused and rigorous in evaluation of germplasm, and successful for commercial potato improvement (Peloquin and Ortiz, 1994). Are there lessons here for better utilising basic germplasm of the *Saccharum* complex in, i) a focused active research effort not a part time adjunct activity, and ii) a strong collaborative network set up between the diverse research providers and the commercial client? The sugarcane world appears to lack this approach but advances in electronic communications when widely available in sugarcane institutions world-wide would provide a major impetus.

Genome analysis

The cytogenetics of the *Saccharum* complex were reviewed most recently by Heinz (1991). Whilst we now have considerable understanding of the taxonomy and evolution within the complex, and interactions within commercial hybrids, this understanding is far from complete.

Burner (1991) studied meiotic stability in the *Saccharum* complex using clones of *S. spontaneum*, *Erianthus*, *Miscanthus sinensis*, *S. robustum*, and *Narenga porphyrocoma*. Bivalent pairing predominated in all clones, and meiotic irregularities were associated with taxonomic grouping and ploidy level. Clones of *Erianthus*, *Miscanthus*, and *Narenga* appeared euploid, with fewer irregularities than *Saccharum*. Later, Burner and Legendre (1993a) evaluated meiotic stability in progeny from crosses between elite *Saccharum* spp. hybrids and *S. spontaneum* clones. Chromosome transmission was primarily n+n, and number increased with generation, from 2n=60.5 in *S. spontaneum* to 108 in the BC_3. Chromosomes paired primarily as bivalents, although variable numbers of univalents through multivalents were observed. Interestingly, they suggested that the hybrids and *S. spontaneum* clones share a common endosperm balance number, unlike *S. officinarum* and *S. spontaneum*. Because transmission is n+n rather than 2n+n, they concluded that additional cycles of recombination may be necessary to recover desirable recombinants. This conclusion is doubtful following later recombinant studies of D'Hont *et al.* (1996a).

An exciting application of new technology is computer-aided imaging to quantitatively karyotype chromosomes of *S. spontaneum* clones (Fukui *et al.*, 1994). This technique resolved that one clone, a haploid derivative (2n=32) of SES208, was a tetraploid, confirming independent findings from molecular mapping of SES208. Combined with cytological developments using *in situ* hybridisation (D'Hont *et al.*, 1995) one can anticipate marked advance in cytological understanding of the *Saccharum* complex.

Molecular applications in sugarcane genome analysis have emerged rapidly (Moore, 1996; D'Hont *et al.*, 1996b), and include surveying cytoplasmic and nuclear polymorphisms, genomic analysis through genetic mapping, comparative mapping with related species, identification of parental chromosomes in introgression breeding, as well as marking desired traits for selection.

Single dose restriction fragments have allowed mapping of a cytogenetically complex group such as *Saccharum* (Sorrells, 1992; Wu *et al.*, 1992). Their application to a population derived from *S. spontaneum* clone SES208 helped clarify its cytological origin, and its relationship to other important Gramineae species (da Silva *et al.*, 1993). The same population also was analysed by polymerase chain reaction based markers (Al-Janabi *et al.*, 1993). These two marker analyses were combined to provide a single genetic linkage map, which revealed the clone displayed polysomic inheritance, a genetic trait of an autopolyploid genome (da Silva *et al.*, 1995). Recent work with this population revealed random chromosome pairing and assortment. Further work on progeny from a cross between a 2n=80 *S. officinarum* clone and a 2n=80 *S. robustum* clone (Al-Janabi *et al.*, 1994a) found, in contrast to *S. spontaneum,* complete polysomic analysis was not observed.

D'Hont *et al.* (1994) mapped a selfed progeny of a *Saccharum* spp. hybrid, an aneuploid, using markers specific to *S. spontaneum* and *S. officinarum*. Using 53 maize probes (covering the whole maize genome) they revealed 348 RFLP bands which mapped to 25 linkage groups. Eighteen of these groups involved markers specific to *S. spontaneum*, representing seven linkage groups. Co-segregation of markers revealed considerable synteny between *S. spontaneum* and maize.

Restriction fragment length polymorphism analysis on basic and commercial germplasm separated *S. officinarum, S. robustum,* and *S. spontaneum* (Lu *et al.*, 1994b). Data for the *S. spontaneum* clones were related to geographic origin. Data for *S. barberi* and *S. sinense* supported current hypotheses for these secondary species. Genomic mapping across genera within the tribe Andropogoneae, e.g. using maize genomic probes (Grivet *et al.*, 1994), can now be used to reveal similarities or differences in genome organisation between sugarcane and other genera. Restriction fragment length polymorphisms have been used widely to characterise strong molecular differentiation between *Saccharum* and *Erianthus*. Numerous *Erianthus*-specific RFLPs and markers have been identified from intergeneric hybrids (D'Hont *et al.*, 1995) or *Erianthus* spp. (Besse *et al.*, 1996b) which also provide new information on species relationships and evolution within this genus, as well as relationships between *Erianthus* and *Saccharum*, and these genera and sorghum and maize. Within *Erianthus* (Besse *et al.*, 1996b) the greatest genetic distance was between the $2n$=20 species and the other species, and the lowest level of variability was among the *E. arundinaceus* clones from Indonesia. The latter species was distinct from all other (Indian) species. In the sample studied, variability within *S. spontaneum* was greater than in *S. officinarum*. The authors concluded that clones of $2n$=80 *S. spontaneum* were closely related to *S. officinarum*.

Fluorescent *in situ* hybridisation is an exciting new development in molecular-assisted cytology, allowing visualisation of chromosomes, intact or fragmented, in cytological preparations, as shown by D'Hont *et al.* (1995) in diagnosing *S. officinarum* × *E. arundinaceus* hybrids. Jenkin *et al.* (1995) applied this technology to visualise hybridisation sites of an rDNA probe coding for ribosomal genes on chromosomes of *Saccharum* and *Erianthus*. D'Hont *et al.* (1996a) demonstrated the technique's power by distinguishing the chromosome contribution of *S. officinarum* and *S. spontaneum* in an interspecific I_1 hybrid between these species as well as in a commercial hybrid clone. About 10% of the chromosomes of the commercial clone ($2n$=107-115) were from *S. spontaneum*. For the first time, this technique allowed verification of recombination between the two genomes. The rDNA sites also were located, verifying differences between the species for basic chromosome number and chromosome structure. This research provides a first bridge between physical and genetical mapping in sugarcane.

There is much concern for the narrow cytoplasmic base of *Saccharum* spp. hybrids, due to the recurrent use of a small number of *S. officinarum* parental clones. A study of cytoplasmic diversity in *Saccharum* complex germplasm, including *Saccharum* spp. hybrids, revealed no variability in the chloroplast genome (D'Hont *et al.*, 1993). There was no mitochondrial variation in *S. officinarum* but there was among the species. This variability in the mitochondrial genome within *S. spontaneum*, *Erianthus* and *Miscanthus* offers potential cytoplasmic germplasm sources for future crop improvement. Successful use of these sources may prove difficult because of the importance of $2n+n$ chromosome transmission.

Phylogenetic analysis of eight genera and 19 species from Andropogoneae was conducted on chloroplast DNA using chloroplast DNA probes that span the complete rice chloroplast genome (Sobral *et al.*, 1994). The survey placed the genotypes into nine groups of which seven were within the *Saccharinae*. The maternal lineage of the clones from *Narenga*, *Miscanthus*, *Sclerostachya*, and *Saccharum* formed a monophyletic group, which displayed little variation. Five of the 6 *Erianthus* spp. and *Eccoilopus longisetosus* formed a separate 'Old World' group substantially different from members of the *Saccharum* complex. The *Saccharum* genus was separated into two groups by a single mutation, with *S. spontaneum* separated from all the remaining *Saccharum* spp. and the *Saccharum* spp. hybrids. Al-Janabi *et al.* (1994b) found that the *Erianthus/Eccoilopus* was nearly as divergent from the *Saccharum* complex as from *Sorghum*. This research also found no variation for the chloroplast and mitochondrial loci surveyed in commercial *Saccharum* spp. hybrids. This lack of genetic variability in the organellar DNA at the commercial level must be cause for concern.

The above studies reveal the power of genomic analysis by sequence or marker technology. Two points of caution need to be made. Firstly, sweeping, taxonomic statements often issue from studies based on limited sampling, whether it be sampling of clones, loci, restriction enzymes, or probes. Secondly, most studies use a canvass of interesting clones from the World Collections, or lower level germplasm collections, and tacitly assume each specimen is true to label. Given the general lack of taxonomic expertise available, the apparent paucity of taxonomic verification of materials used is of great concern and could easily be very damaging to the future of *Saccharum* complex utilisation. Reports of mislabelled entries shipped from the World Collection at Miami in recent years substantiate this point.

The molecular applications described above did not exist 10 years ago. They represent some of the most exciting contributions of molecular biology to sugarcane improvement to date. Doubtless they herald an era of greatly increased understanding of sugarcane germplasm. Taxonomy, cytogenetics, collection, characterisation, use, and maintenance of *Saccharum* spp. germplasm will undergo a revolution through this technology within the coming decade.

Commercial (*Saccharum* spp. hybrid) germplasm

Evaluation data on commercial germplasm are a routine output of selection processes, but are rarely compiled formally and widely distributed. Staff at the Copersucar Technology Centre, Brazil, have made their evaluation data for *Saccharum* spp. available informally, and Indian researchers have released a database on 786 clones, mainly 'Co' (Nair and Sreenivasan, 1995).

A major concern in a clonal crop such as sugarcane is the fear of limited genetic diversity among cultivars, often unrealised until the advent of a new pest or disease. Until genomic analysis using molecular markers recently became available (e.g. Harvey *et al.*, 1994), a rigorous analysis of genetic relationships among clones was simply impossible. In a study of 20 *Saccharum* spp. hybrids, these authors found that similarity ranged from 71.7 to 89.7%, indicating very limited genetic diversity. Similar analysis on 40 commercial clones (Lu *et al.*, 1994a) gave an average genetic similarity between clones of 61%. They concluded the prime source of variability was due to *S. spontaneum.*

Conservation of the *Saccharum* complex germplasm is of vital importance. Berding and Roach (1987) observed that the field-based Miami World Collection was not optimally located in a stress-free environment, and that the preservation of clones rather than genes was a preoccupation. This situation has not really changed during the decade. However, criticism must be tempered because the major collections are funded by the USA and India and not, as is desirable, the collective World sugarcane community. The World Collections are marginally resourced which must stifle technological innovation for improved conservation. Their current status is described by Comstock *et al.* (1996) and Alexander and Viswanathan (1996), and an action plan agreed by the International Society for Sugar Cane Technologists for upgrading their status, including the concept of a core collection, is detailed by Egan (1996). An important development (see Engelmann and Engels, 1996a) would be the strengthening of links to the International Plant Genetic Resources Institute. Introduction of the concepts of networks (Engelmann and Engels, 1996b) and the use of core collections also would be important. The latter are an important concept in modern germplasm conservation, whereby research and utilisation are focused on a sub-set of a much larger collection, with the sub-set reflecting the variation present in the total collection.

The present basic clonal World Collections ideally need to be re-located to stress-free field environments. Alternatively, or additionally, *in vitro* conservation techniques to avoid field stresses and reduce maintenance costs need to be more widely used for sugarcane. Technology is developed for low temperature reduced-growth storage (e.g. Taylor and Dukic, 1993) and to cryopreserve dehydrated and encapsulated apices (Paulet *et al.*, 1993). If the concept of preserving genes rather than clones is favoured in future, then storage of true, or sexual seed is advocated (e.g. Tai *et al.*, 1994). Flowering of many entries in any collection of *Saccharum* complex germplasm is difficult to achieve which may prove a problem for such an approach.

Germplasm movement, quarantine and intellectual property

Ease of movement of germplasm is essential for sugarcane improvement, and in Australia (as elsewhere) introduced germplasm has in past decades made a significant contribution (Hogarth and Berding, 1996). Recently, there has been an increasing awareness in many countries of the risks associated with germplasm introduction. The Australian Quarantine and Inspection Service (AQIS) imposed a moratorium on the importation of clonally propagated grasses due to concern about introduction of unidentified viruses in vegetative material. Sugarcane has been exempted for the present, but the need to screen imported germplasm for both identified and unidentified viruses was acknowledged.

Threat of introduced diseases has caused three major disruptions to Australia's BSES sugarcane quarantine in the past four years preventing release of introduced material and creating a quarantine impasse affecting the sugarcane improvement program (Hogarth and Berding, 1996). Firstly, sugarcane bacilliform badnavirus (SCBV); secondly, 'yellow-red leaf fleck' (Croft and Smith, 1996) in clones from the Indian Council for Agricultural Research World Sugarcane Collection at Kannur, India, the causal agent remains unknown; and thirdly, yellow leaf syndrome (YLS) in germplasm imported from Florida (Whittle, 1994). Similar risks exist for germplasm introductions to other countries, and to prevent a complete breakdown of germplasm movement there is a need to invoke the concept of pest risk analysis (PRA) which weighs risk in cost to the industry against the potential benefits. Unfortunately in most cases there are insufficient data for such analysis. There is an identified need for more standardised plant management across international quarantine facilities (Croft *et al.*, 1996), particularly to optimise conditions needed to ensure disease expression (Dodman, 1996) if present.

Many diseases are easy to screen visually (e.g. leaf scald, smut and Fiji disease) but others are not (e.g. SCBV and sugarcane mild mosaic virus) and hence new serological and molecular-assisted tests have been developed to detect and identify sugarcane pathogens (Braithwaite and Smith, 1996). Many of these tests, however, are not yet suited to routine disease screening.

Before the Australian sugar industry can again confidently access vital foreign *Saccharum* germplasm, several issues need to be addressed:

1. Research needs to determine the importance of diseases like SCBV and YLS to sugarcane production and commercial viability.
2. Effective new tests are needed for routine assessment of quarantine samples, especially for viral diseases.
3. What significance should be attached to unidentified pathogens, mostly viral, in germplasm? Current information is insufficient to make considered decisions, but in Australia there is sufficient concern for AQIS to ban import of vegetative material of clonally propagated grasses, with sugarcane fortunately currently still excluded from this ban.
4. If PRAs reveal that material can flow through quarantine, then refinement (e.g. thermotherapy) or development of pathogen elimination techniques becomes an imperative.

Exchange of germplasm increasingly raises the question of intellectual property or plant breeding rights to protect organisations' investments in development of cultivars. In sugarcane, free exchange has been traditional between parties or countries for crop improvement programs. Modern thinking is changing, and in Australia BSES is applying under the Australian *Plant Breeders Right Act 1994* for protective rights over 'Q' cultivars. Their future use could require a royalty payment. Genetic engineering of sugarcane adds impetus to commercial protection of germplasm, particularly because application of biotechnology usually involves use of outside, patented biotechnology techniques requiring reimbursement to the owners. As well, because of special biotechnology expertise, organisations not funded by the industry now may wish to participate in sugarcane improvement and understandably wish to recover costs, if not be recompensed for their success. Protection of their innovation through plant breeding rights, and exclusive access to royalty payments via these, must be basic expectations of such investors. Molecular markers will find an important application for definitive identification of germplasm when ownership is contested (Duvick, 1996).

Hybridisation: Generation of Genetic Variation

Advances in basic breeding processes

The need for control of flowing will not be replaced by emerging technologies such as somatic fusion (Glimelius, 1993) or genetic transformation (Bower and Birch, 1992) in the foreseeable future. Reliable flowering in sugarcane germplasm is an acknowledged problem in breeding, especially in temperate and sub-tropical regions. Controlled environment and field enhancement of flowering have been reviewed by Moore and Nuss (1987) and Berding (1995). The latter study showed that adequate soil moisture was not a sufficient condition for flowering when temperatures were extreme. Photoperiod facilities have proven invaluable for enhancement and control of flowering and are now commissioned in many countries. Based on field and photoperiod facility observations, Berding and Moore (1996) have hypothesised that the level of suppression of flowering is linked to the number of days in the inductive window with maximum temperatures >32°C. Rigorous testing of this hypothesis will commence in the CSIRO phytotron in Canberra in 1997, and if validated will allow optimised seasonal timing of induction and optimised selection of the most suitable field sites.

Hybridisation procedures (Heinz and Tew, 1987) have seen only a few major technique developments in the past decade. Many attempts to improve the efficacy of the widely used Hawaiian acid crossing solution for preservation of severed flowering stalks during crossing (e.g. Ramdoyal *et al.*, 1995) have yielded little significant improvement. Research resources may better be directed elsewhere. The environmental conditions under which cross pollination is conducted is, we believe, an area worth further research to influence anther dehiscence, pollination, and fertilisation. Improvements in environmental control technology now allow more critical examination of this important issue. Pollen preservation also would help overcome problems caused by temporal differences in flowering. Tai (1988, 1993) has successfully used dried pollen stored at low temperature of several *Saccharum* species and hybrids, to produce seedlings in crosses. This potentially is an important advance.

New biochemical and molecular markers (Zheng *et al.,* 1993; D'Hont *et al.*, 1995; Lee, 1995a; Besse *et al.*, 1996a) now provide means to unequivocally check contamination in crosses, especially for critical research.

Efficient generation of crosses

Accurate prediction of performance of possible parental combinations would yield tremendous returns in any crop improvement program. In an early effort, Hogarth and Skinner (1987) developed a computerised algorithm for selection of sugarcane crosses using phenotypic evaluation data for parental clones as well as progeny. More recently, Hsu *et al.* (1991) and Wang *et al.* (1993) have described a computer system that linked parental clone, census, and selection data to a data base of desired crosses. However, crosses were compiled by breeder consensus, not mathematical estimation of parental breeding value or cross value. Chang and Milligan (1992b) seeking to identify an easy, reliable cross appraisal statistic found that best linear unbiased predictors (BLUPs) accurately predicted the potential of a cross to produce elite progeny. In other work, Chang and Milligan (1992a) found that the estimation of the sum of ranks based on family mean values for two individual traits was easier than the calculation of BLUPs, and gave equally accurate prediction of family potential. Stringer *et al.* (1996) determined the BLUP method was more effective than the empirical algorithm currently used by BSES (Hogarth and Skinner, 1986, 1987) for estimating the potential performance of parental combinations. Less information is required, making a reduction of generation interval a real possibility. Application refinements of BLUPs to sugarcane may see further efficiency increases.

Reduction of generation interval is an important aim in most crop breeding programs, with remarkable reduction in some crops, e.g. maize from 15 to 5 years. In sugarcane, scientists are concerned with two generation intervals. The first, seedling to seedling, potentially has been reduced in Australia (Meringa) and Hawaii to 7 years. In an environment unconducive to free flowering, this time is a minimum, and difficult to reduce unless clones are selected earlier than the final yield trial on the experiment station. The second is from seedling to commercial release, as short as 8 years in Queensland, but more typically 12 to 15 years. As real advantages accrue from reducing this interval, research is needed perhaps in using *in vitro* propagation of meristems early in the selection process for potentially elite clones to accelerate bulking up of clean propagation material.

Genetic transformation and somatic hybridisation

Genetic transformation offers significant future opportunities for specifically targeted transfer of new genes into crops or modification of gene action. Recent activities in sugarcane are summarised by Moore (1996). Only a few specific aspects will be mentioned here as gene transformation technologies (Birch 1997), metabolic modification (Moore *et al.*, 1997; Smith, 1997), and disease control (Allsopp and Manners, 1997) are detailed in other chapters.

Some aspects of this new technology are of concern for the eventual production of effective commercial cultivars. Somoclonal variation arising from tissue culture to produce embryogenic callus have been noted in sugarcane. If this occurs as a general consequence of the transformation procedure, then subsequent selection effort to regain agronomic desirability in transformed individuals may cancel much of the perceived advantage of the process. Genetic modification to increase pest or disease resistance will be in effect single gene solutions susceptible to breakdown under intense selection pressure. Widespread use of this approach may therefore enhance genetic vulnerability of the crop. Sugarcane improvement using gene technologies is in its infancy but its promise and interaction with breeding programs will become much clearer in the coming decade.

Somatic hybridisation has appeal for base broadening activities because problems often inherent in interspecific or intergeneric hybridisation are circumvented, e.g. synchronisation of flowering, and overcoming impediments to pollination and fertilisation. However, its development would not be simple in the *Saccharum* complex. Major problems include the impasse of regeneration from protoplasts and the questionable effect of elevated chromosome numbers.

Assessment of Selection Concepts

Pre-1987, selection strategies already were highly developed, but nevertheless in the past decade, technological improvements have given rise to new developments. These include computer-generated bar-coded labels, automation and computer-linking of laboratory analyses, e.g. for quality attributes, and at-line near infra-red spectroscopic (NIS) applications (Berding and Brotherton, 1996b). In the field, row weights can be captured directly by weighing the product stream from a harvester using truck-mounted electronic weigh bins (Hogarth and Mullins, 1989) capable of weighing 350-450 plots per day and new technology continues to improve cost effectiveness and efficiency. In whole-stalk harvested industries, plot yields are collected via hydraulic grabs linked to electronic load cells. Multiple technological developments have taken selection for sucrose yield from being an indirect process, reliant on correlated responses (Gravois *et al.*, 1991a), to a direct objective process applicable to progeny plots from seed (Cox *et al.*, 1996) to final clonal evaluation trials. The full impact of this objectivity in selection will be substantial.

Selection criteria

Breeders are under constant pressure to extend the list of selection criteria and this can be done using a combination of genetic theory and available technology. Cultivar improvement is an exercise in probability, the more non-yield criteria addressed, then the lower the probability of selecting a higher yield clone that also satisfies the non-yield criteria. Solutions to this gain vs selection load question (Berding and Skinner, 1987) have advanced little in the past decade. Continued addition of selection criteria should be accepted as the solution of last resort.

Developments in experimental design and analysis since 1987 have made little impact on selection efficiency, with the exception of analysis of genotype × environment interaction covered later. This inactivity in exploring new experimental designs and analysis should not signal that technology cannot be improved, e.g. as most programs still use unreplicated plots in early selection stages, could large gains result from using replicated smaller plots requiring the same resources, or replicated plots of current plot size?

Yield and quality

Harvest index is an overlooked, or misunderstood issue in sugarcane breeding. Sugarcane is unique among field crops in that the structural part of the plant is the harvested product, and in terms of harvest index, requires special consideration of selection objectives for cane or sucrose yield (Moore, 1989; Evans, 1993).

Record cane yields, as in north Queensland with wet years and excessive use of nitrogenous fertiliser, have resulted in heavily lodged crops with excessive sucker growth, and low sucrose content. There are now calls for cultivars that have higher sucrose content, even if they produce lower cane yield on highly fertile soils, viz. a slower growing, erect ideotype. The canopy would not open with lodging, there would be less suckering, and less dilution of the mature stalk sucrose by many low-sucrose suckers. Gravois *et al.* (1991b) using additive genetic correlation and path-coefficient analyses have assessed the genetic relationship between these characters and their implications for selection.

Ratooning is vitally important in sugarcane and ratoon yield is routinely assessed during selection. Prediction of ratoon performance from early crops within the cycle, or definition of associated traits, would be highly advantageous. Studies by Milligan *et al.* (1996) found that broad sense heritabilities for ratoon performance were low, genetic coefficients of variation were large for sucrose and cane yield, and that although second ratoon yield could be predicted by first ratoon crop yield, the best improvement for second ratoon performance would result from actual second ratoon selection. There is an obvious need to understand better the critical plant processes before ratoon yields can be predicted from early crops.

Sucrose content is a prime selection criterion, traditionally expressed on a fresh weight basis. Muchow *et al.* (1997) report that maximum stalk sucrose concentration has a relatively narrow range of 0.45-0.55g/g DW in existing cultivars. Tentative examination of NIS data (N. Berding, unpublished data), in which all major components (solubles, insolubles, and water) were analysed independently, revealed significant variation in these components. Recently, Legendre (1995) reviewed progress from recurrent selection for high sucrose content using 39 commercial clones as parents and showed a 31-39% increase over five breeding cycles, with progress lessening with each successive cycle. Improvements in early sucrose in Taiwan between 1959 and 1994 reviewed by Lo and Chen (1995) showed 4% units improvement in early season sucrose and 2% units increase at peak maturity. The procedure involved use of high sucrose germplasm in intensive crossing, early selection of seedlings, early season analysis for sucrose, and use of high sucrose clones as standards. Selection for high early season sucrose has received attention elsewhere (Domaingue *et al.*, 1990; Cox *et al.*, 1994). More recently, Cuenya and Mariotti (1996) considered that selection for high sucrose content and early maturity would yield significant progress.

Stringent recurrent selection was the reason for the success with sucrose in Louisiana and Taiwan. No other examples of similar success are available. Can such progress continue? Is genetic variation for high sucrose near exhaustion? There is little rigorous evidence to support this, especially given the small sample of *S. officinarum* and *S. spontaneum* germplasm incorporated in commercial hybrids (Roach, 1989). There must be ample opportunity to 'reinvent' modern sugarcane with even higher sucrose content despite the relatively poor success of contemporary nobilisation efforts. An effort in 1994 to do this in Australia by introducing 200 high-sucrose *S. officinarum* clones from the Indian World Collection failed because of the quarantine problems described earlier. As the sucrose content of contemporary *Saccharum* spp. hybrids exceeds that found in the contributing species (T.V. Sreenivasan, *pers. comm.*), this synergy is exploitable, and should be the focus of intense research in the coming decade. Advances in analytical techniques (by automated conventional analysis or NIS) now allow objective selection for sucrose content at all selection stages.

Adding to these approaches is genetic transformation to achieve metabolic modification of sucrose production, transport and storage. Such research is under way (Moore *et al.*, 1997). The future potential is awesome but new cultivars are unlikely within the next decade.

Disease and insect resistance

Breeding for disease resistance consumes significant resources but recent advances appear limited. This is in contrast to the severe impact of new diseases on germplasm movement. Research on inheritance of resistance to fungal and bacterial diseases features prominently in the last 10 years, e.g. smut and common rust (Wu *et al.*, 1988; Chao *et al.*, 1990; Comstock *et al.*, 1992; Hogarth *et al.*, 1993b; Balance *et al.*, 1996). Sufficient estimates are now available to allow efficient planning of breeding strategies for sugarcane improvement.

Ratoon stunting disease (RSD) caused by a xylem-limited bacterium is controlled in some industries by heat treatment and phytosanitary precautions, hence resistance is not a selection objective. Elsewhere, losses to RSD occur despite these precautions. Its incidence in basic germplasm of the *Saccharum* complex and hybrids exposed to natural infection, and the feasibility of breeding for resistance were examined by Roach (1992). A new tissue-blot enzyme immunoassay (Davis *et al.,* 1994) should speed screening for RSD resistance.

Viral diseases received some attention and are increasingly assuming greater significance for sugarcane improvement. Hogarth *et al.* (1993a) concluded for Fiji disease that use of resistant parents would produce resistant progeny, whilst Matsuoka *et al.* (1990) and de Haas and Bond (1993) examined breeding strategies for resistance to sugarcane mosaic virus, considered the most important and widespread of viral diseases affecting sugarcane.

Insect control in sugarcane in recent years has increasingly focused on host-plant resistance. Why is this? In agriculture in general there is increasing environmental sensitivity to use of agricultural chemicals. Integrated pest management programs have developed (see Allsopp and Manners, 1997) in which host-plant resistance is an important component. In general only recently have entomologists and plant geneticists been in close collaboration (de Ponti and Mollema, 1992). This is true in the Australian industry, as only recently has research started on host-plant resistance to white grubs, the larvae of melolonthine scarabs (Allsopp *et al.*, 1995; 1996), pink ground pearls *Eumargarodes laingi* Jakubski (Hemiptera: Margarodidae) (Allsopp and McGill, 1996), and sugarcane weevil borer *Rhabdoscelus obscurus* Boisduval (Coleoptera: Curculionidae) (Berding, 1996). World-wide, most literature on breeding for host-plant resistance to insects in sugarcane involves moth borers, *Chilo* spp. (Ashraf and Fatima, 1990), but there is little information on the inheritance of resistance.

Breeding for resistance to eldana borer (*Eldana saccharina* Walker (Lepidoptera: Pyralidae)) has been more systematic, innovative, and productive (Bond, 1988; Nuss, 1991). However, heritability has been poor, requiring repeated evaluation (Rutherford *et al.*, 1993). The pioneering use of NIS by the South African group to assist selection is the most exciting recent development in sugarcane resistance breeding.

Possible future strategies for increasing host-plant resistance reviewed by Allsopp *et al.* (1996) covered introduction of genes for specific antimetabolites such as proteinase inhibitors, lectins, and avidin. Subsequent assessment of stability and location of expression and biological effectiveness will be awaited with great interest.

Less traditional selection criteria may be introduced in the future as the sugarcane production system evolves, e.g. biomass to maximise total carbohydrate output, sustainability (if and when specific selection traits are identified), flood tolerance (Deren *et al.*, 1991), cold tolerance (Tai and Miller, 1993), pith influence on yield (Deren, 1992), and pith, tube (piping), and stalk density on sucrose content and stalk weight (Gravois *et al.*, 1990). Lodging is of widespread concern. Amaya *et al.* (1996) concluded that selection for lodging resistance would result in major gains. Clones with more erect habit had a greater vascular bundle density in the rind. Fertiliser use efficiency has received little attention but may become increasingly important. Efficiency can be defined as the sucrose produced per unit of nitrogen taken up, and sugarcane clones do differ in leaf nitrogen levels with associated major differences in productivity. Much research has been invested on N fertilisation of sugarcane, yet we have a minimal understanding of the genetic variation available for N uptake and utilisation. This approach is much needed and seems preferable at present to possible transformation approaches introducing transporter genes aimed at enhancing uptake of N (or other nutrients). South African research (Stevenson *et al.*, 1992; Allison and Haslam, 1993) has raised the possibility of seeking sugarcane clones that respond better to higher fertility through increased nutrient input.

Developments in screening techniques

Managed environments for disease screening are replacing natural infection because of the latter's variability and unpredictablity, e.g. for common rust (Autrey *et al.*, 1996a) and pachymetra root rot (Berding and Croft, 1996; Croft and Berding, 1996). There are real advantages in this technique in which the pathogen or inoculated plants, or both, are manipulated to minimise environmental interactions with the components. With advanced selections, there should be a high correlation with field reaction for the pathogen, and with unselected material, clones deemed resistant by the screen should be resistant in the field. The efficiency resulting from reliable, repeatable ratings is incalculable.

Near infra-red spectroscopy for analysis of high-moisture materials (rather than dried materials) is an emerging area within the technology. For material of this nature, such as fibrated sugarcane, the difficulties (Berding and Brotherton, 1996a), current (Schäffler *et al.*, 1993) and future potential applications (Meyer, 1996) were reviewed. Great advance in routine analysis of a range of quality components on clean, whole-stalk samples will follow the development of an at-line NIS application (Berding and Brotherton, 1995, 1996b), suitable for high volume clonal assessment. An exciting extension of this technology to analysis of sugarcane entering the milling train, and in product streams within the factory, primarily for process and quality control purposes is under research. Another exciting application for NIS is as an indirect screen for host resistance to eldana borer using NIS spectra for stalk surface wax (Rutherford *et al.*, 1993; Rutherford and van Staden, 1996). They suggested that screening of germplasm for eldana resistance using NIS without prior knowledge of the specific chemical constituent involved was possible. A further advance being explored by South African researchers is the use of a fibre-optic coupled remote reflectance probe to scan stalk surfaces *ex situ* for rating clonal resistance to eldana borer. Miniaturisation and robustness will eventually enable *in situ* usage. Undoubtedly, NIS will become very extensively used in sugarcane improvement (Meyer, 1996).

Family selection based on mean family performance and within family variance is replacing traditional individual clone selection as the initial screen in many breeding programs. Early assessments of this process were given by Bond (1989) and Tai and Miller (1989). In Australia, family selection at the first selection stage involves c. 20 seedlings of a family planted to a plot giving a single-row c. 10m long which is harvested *in toto* using the in-field weighing machine. Mechanisation allows plots of 200 to 300 families to be harvested within several days. Further, families could be 'replicated' within and across sites to reduce experimental error, and measure family × environment interaction. Sucrose yield is the common selection criterion. Clones are selected visually from within elite families. Mostly only clones from the best families (≈33%) are selected (Bull, 1993). Bull *et al.* (1992) found similar gains in sucrose yield for family and clonal selection but the former was more efficient. Assessment using at least two sites was needed because of G × E interactions. Jackson *et al.* (1996) assessed family versus individual selection in early selection stages. Family selection was well suited where error variation for individual clones was high, and where genetic correlations among characters were low. An unanswered question concerns the possible loss of superior clones from families rejected for low performance but which may have high variance.

Inter-genotypic competition in clonal evaluation in small plots is a well-known problem that has not been adequately addressed in the past decade. Breeders should consider the detailed analysis of detrimental effects by Bos and Cagliari (1995).

Genotype × environment interactions (G × E)

Multi-environment testing is time consuming and costly. Research has aimed to optimise G × E testing by identifying a few 'key' trial sites to represent a region (Ivory *et al.*, 1991) and maximise genetic gain for a given resource level. This approach depends on identifying the specific causes of G × E (e.g. van Eeuwijk and Elgersma, 1993; Bull *et al.*, 1996; Jackson *et al.*, 1995). Once contributors to G × E interaction are identified, they can be (i) avoided when locating trials, (ii) ameliorated in core trials, or (iii) manipulated directly in managed-environment selection trials (DeLacy *et al.*, 1996), depending on the factor and production system. Success with multi-environment testing obviously depends on the small sample of test environments adequately representing the wider range of production environments. In sugarcane, family × environment (F × E) or clone × environment (C × E) interaction may equal or exceed the main effect for family or clone (Jackson *et al.*, 1991; Bull *et al.*, 1992; Mirzawan *et al.*, 1994), and can complicate selection (Bull and Hogarth, 1990). Thus in Australia, selection is regionally based, with multi-environment trials within each region. Mechanisation of yield and quality analysis now allows initial testing to move from central research stations to multi-sites, reducing the impact of F × E and C × E effects. Analyses of southern Queensland trials (Mirzawan *et al.*, 1993) indicated large C × E effects, and suggested greater advantage to using more sites than testing ratoon performance within site for obtaining both broadly and specifically adapted clones. From trials in northern Queensland, Jackson and Hogarth (1992) similarly concluded that crop × year variation was small relative to G × E, but from other research (Jackson, 1992) recommended that multi-site testing using plant crops only be confined to early selection and not used for advanced clonal selection. Second ratoon assessment is used only for final selection. Factors such as water, nitrogen and micro-nutrients have been implicated in G × E (Jackson *et al.*, 1995; Bull *et al.*, 1996)

Research on procedures for classifying genotypic responses and environments, e.g. Bull *et al.* (1994a, b), is necessary for understanding components of G × E, and is important new knowledge gained within the last decade, the impact of which is still to be realised at the practical level. Kang *et al.* (1987) believed another aspect of G × E to be explored is the use of a stability-variance parameter to determine the relative contributions of individual genotypes to G × E.

Important in analysis of G × E is that G × year and G × crop (plant or ratoon year) are unavoidably confounded. Jones *et al.* (1993) developed a technique to minimise this using adjustments on the basis of neighbouring farm yields as a measure of the year effect. Since such adjustments resulted in only minor clonal rank changes, they concluded the effect of confounding the interactions was small for clonal selection.

Selection achievements

To assess improvement in sugarcane yield, use of a soundly based record commercial yield, e.g. 24.2t sugar ha^{-1} (Heinz, 1987b), appears a better basis than calculated theoretical potential yield. Moore (1989) found average Hawaiian industry yield was just over 50% of this record, comparing well with other major crops. Average yield (1991-5) for the Queensland industry was similar, viz. 12.0t sugar ha^{-1}. However, the maximum average yield in the Burdekin region was 17.4t sugar ha^{-1}. These values suggest sugarcane as a crop is well advanced in level of improvement. While Hawaiian production doubled from 1908 to 1984 (Plucknett and Smith, 1986), Berding and Skinner (1987) and Moore (1989) highlighted that the long term rate of improvement for sugarcane was low relative to other crops. A comprehensive overview of trends and limits in yield increases from crop improvement is given by Evans (1993).

Estimation of commercial genetic gain is vital for assessing economic value of crop improvement programs but is often difficult to establish. Reliable yield comparison of old and new cultivars grown at the same time over large areas is a daunting task, given the large year-to-year variability, often strong G × E, and few suitable genotypic records. The G × E experienced commercially must be adequately and appropriately sampled in regional variety trials.

Yield productivity gains for Queensland sugarcane (1961-96) show interesting trends (Fig. 12.1). Cane yield increased to 1978, drifted lower to 1991, and then rapidly increased to 1995. Sucrose content (CCS) showed a long term decline (possibly associated with mechanical harvesting and increased N fertiliser), with a steep rise after 1991. As a consequence, sucrose yield was static for 20 years prior to 1991. We believe the increase in all measures in the 1990s is largely due to increased breeding activities over the whole industry, e.g. see evaluation by Cox and Hansen (1995) for the central and southern regions. Correlations for yield attributes between breeding trials and agricultural performance were variable between regions. Bundaberg was good and Tully poor, the latter perhaps being associated with lodging and stool tipping. The Australian evaluations suggest that some proportion of the genetic gains is being realised, and that the static (or negative) trend in productivity from 1971 to 1989 was likely due to production factors such as: older ratoons kept; 200% increase in area per grower; industry expansion onto marginal land; increased losses due to diseases and pests; more extraneous matter in the mill supply. These negative factors appear to have outweighed the positive factors of increased irrigation, better nutrition and reduced harvesting to processing delays.

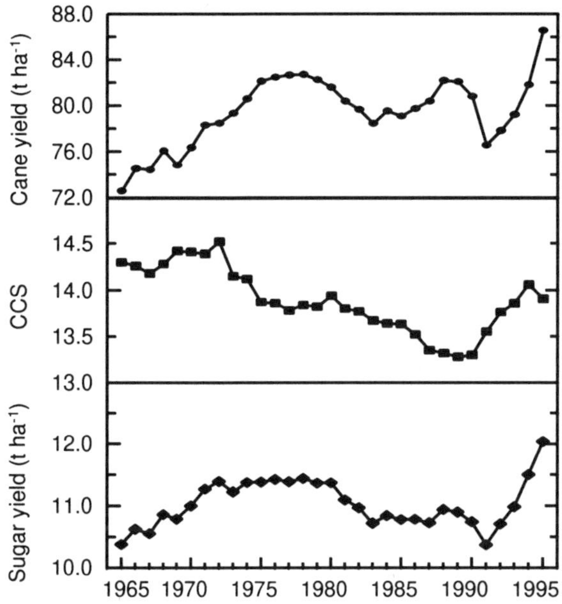

Fig. 12.1. Trends for cane yield (●, t ha^{-1}), commercial cane sugar - CCS (■, % F.W.), and sucrose yield (◆, t ha^{-1}), expressed as five-year moving means, for the Queensland industry 1961-1995.

In Hawaii, sucrose yield, as moving five-year means from 1912 to 1994, showed a long term uptrend to 1988 of c. 0.1t ha^{-1} (Fig. 12.2), but a decline from the 1988 peak of 13.6t ha^{-1} to 12.0t ha^{-1} in 1994. Possible reasons were YLS, lower fertiliser use, or perhaps a smaller crop improvement program in a declining industry.

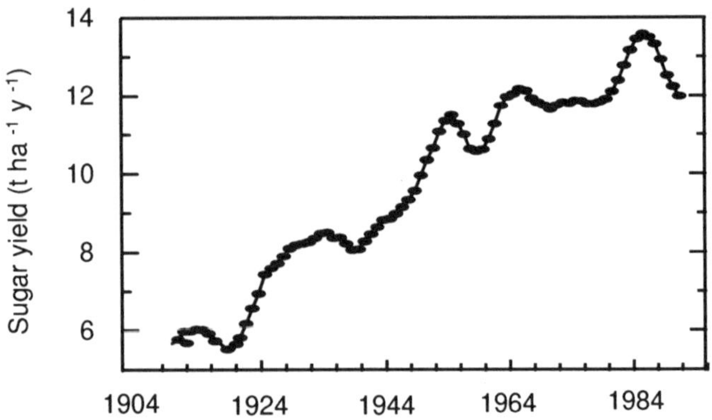

Fig. 12.2. Sucrose yield (t ha^{-1} y^{-1}) for the Hawaiian industry from 1908 to 1994 expressed as a 5-year moving mean.

Meeting or leading industry change

Major changes in industry procedures often occur faster than breeding programs can be redirected to produce cultivars suited to the new system. Anticipation of industry requirements 10 years in advance is a perennial problem facing sugarcane breeders (Irvine, 1994). Here, we present three examples of such change, viz. green cane trash blanketing, reduced row spacing, and harvester design.

Green cane trash blanketing was introduced in the 1980s into north Queensland to reduce operational costs, and was facilitated by development of new harvesters. Overnight, the crop's ecology was changed. Ratoon crops are grown under essentially zero-till. The trash blanket on ratoons lowers ground temperatures, raises soil moisture levels, and alters nutrient availability. Most of the crop is still produced by clones selected under a burnt regime. Anecdotal evidence suggests some clones are unsuited to trash-blanketing, but data are scarce. Management of clonal trials was altered once the new technology was firmly established to select cultivars under commercial agronomic practices. However, 10 years later, crop improvement programs still have not fully delivered highly adapted cultivars.

Conventional row spacings in Australia have increased to c. 1.5m largely to accommodate mechanisation. However, by reducing row spacings down to 0.5m, Bull and Bull (1996) demonstrated crop yield increases up to 100%. Particularly important in their work was the much stronger performance of 'unselected' clones compared to established cultivars selected under the normal 1.5m row spacing. Consequently, to take maximum advantage of reduced row spacing in the long term, a joint industry and research decision is needed. Some selection plots should be changed from 1.5m to narrow row spacing to exploit genetic variability for the capacity to respond to high planting densities (Bull and Bull, 1996) and select clones with the most suited agronomic and developmental traits. Inevitably there is a significant delay in finding the best new cultivars.

Changes in harvester design over the past decade influencing cutting rate and material delivered have increased losses and extraneous matter (Linedale and Ridge, 1996), perhaps costing the industry up to A$75M per year. Clonal variation for cane loss and extraneous matter was shown (Ridge and Dick, 1988). The question for breeders is whether new plant ideotypes with different architecture are needed to reduce harvesting inefficiencies? This will take time and perhaps slow genetic progress - or is the solution purely a matter for harvester design?

Future selection options

Technological advances in computers, mechanisation, and near infra-red spectroscopy were easily and readily adopted for crop improvement. Biotechnology, however, offers molecular-assisted selection procedures which are inherently different from traditional procedures. Lee (1995b) saw DNA markers as the fundamental link between crop improvement and the 'new' plant biology. He detailed four major application of DNA markers: (i) assessment of genetic diversity and merit, (ii) characterisation of plant genomes, (iii) analysis of complex traits and processes, and (iv) marker-assisted selection. Other possibilities include tagging genes of agronomic importance, determination of quantitative trait loci, comparative gene mapping, and use in alien gene transfer. The apparent slow application of molecular markers to sugarcane (Moore, 1996) is due to complex polyploidy, identification of traits coded by major genes is difficult, and most economically important traits are multi-genic. Unless automated, marker technology is time consuming and costly, as is associated phenotypic evaluation. The most promising application is bulk segregant analysis, an alternative approach for marker analysis. This technique has been used to detect polymorphisms in bulked DNA from individuals of extreme fibre values within a family (Msomi and Botha, 1994).

Albert and Schenck (1996) used molecular markers to detect DNA of a foreign genome (smut) in sugarcane. This is potentially a great advantage because this capacity for detection is well ahead of symptom expression in inoculated screening trials, which may take several years. The result is independent of pathogen × environment or pathogen × clone × environment interaction. The procedure also has possibilities for phytosanitary applications, e.g. checking planting sources.

Haploid development is potentially a potent tool for crop improvement. In *S. spontaneum*, haploids were produced by anther culture (Moore and Fitch, 1990), providing clones, and subsequent crosses, valuable for genomic analysis with molecular markers (Al-Janabi *et al.*, 1993; da Silva *et al.*, 1993; and da Silva *et al.*, 1995). Unfortunately, similar procedures for *S. officinarum* have been unsuccessful because the anthers are no longer sterile when at the correct developmental stage. An alternative path might involve the development of haploids through crossing, taking advantages of meiotic mutants or through stimulation of embryo development without fertilisation by use of wide crossing. Reducing genome complexity also is a powerful selection tool, and should be potentially very valuable in the coming decade for use in selection for higher sucrose in *S. officinarum*.

Conclusions

This review clearly shows there have been considerable gains in sugarcane improvement over the last decade. Many of these have come from application of 'traditional technologies' not being utilised at the start of the period. Many of these technologies will continue to contribute during the next decade.

The decade has been one of increasing objectivity in a wide array of our crop improvement activities. Without the development and application of these technologies, the momentum of improvement could not have been sustained. The advances may compensate partly for continued erosion of effective program size, which is a consequence of escalating demands on improvement programs.

We stand on the threshold of a potential revolution in sugarcane improvement. This is available through applications of biotechnology, encompassing the fields of genetic transformation and molecular marker technology. The basics for genetic transformation are in place, but further development is required for practical benefits to be delivered. Application of molecular marker technology in the later part of the period gives us an exciting glimpse of the future. Analysis of genomic structure, identification of chromosomes, whole or part, measurement of genetic variability, and description of synteny over distant Gramineae taxa are but part of this. These preliminary applications build on the theme of increasing objectivity. Can biotechnology cost effectively deliver the second requirement, by providing practical tools that advance sugarcane improvement?

Most importantly, we must not forget that biotechnology is a tool and not a replacement for traditional crop improvement. There are many examples of crop improvement attaining major objectives through application of the empirical art and experience of crop improvement traditionalists. Often these gains were made with minimal trimmings of science or technology. Nevertheless, the benefits of the latter are obvious in this review. The benefits to be gained from a multi-disciplinary and collaborative approach to sugarcane improvement should be obvious. The next decade can only be anticipated with excitement.

Two major deficiencies require attention for sugarcane improvement to continue effectively. The quarantine impasse currently being experienced by many programs requires resolution. This might be achieved by single or combined use of definitive diagnostic techniques, development of therapeutic treatments, or assessment of the possible economic impact of newly prominent diseases.

The world sugarcane industries collectively must provide adequate funding and support for maintenance and management of world sugarcane germplasm resources. The sugar industry is perhaps 20 years behind the mainstream germplasm conservation technologies used in other major crops. Scientific management of germplasm would serve as a base on which to build programs of broadly distributed enhancement breeding.

Acknowledgments

Dr J.K. Bull, BSES Bundaberg, is thanked for discussions on genotype × environment interaction and family selection. We acknowledge with appreciation the editorial assistance of Dr J.R. Wilson in finalising this chapter. Thanks are due Ms Ann Marsteller, Hawaii Agriculture Research Centre, Aiea, and Ms Helen Laurelin and Ms Sonya Law, BSES Indooroopilly, for helpful provision of library resources. Finally, I (N.B.) acknowledge my wife, Barbara, for her tolerance of my nocturnal activities while writing this chapter, and her continued support.

References

Albert, H.H. and Schenck, S. (1996) PCR amplification from the bE mating-type gene as a sensitive assay for the presence of sugarcane smut (*Ustilago scitaminea*) DNA. *Plant Disease*, In press.

Alexander, K.C. and Viswanathan, R. (1996) Conservation of sugarcane germplasm in India given occurrences of new viral diseases. In: Croft, B.J., Piggin, C.M., Wallis, E.S. and Hogarth, D.M. (eds), *Sugarcane Germplasm Conservation and Exchange*. Proceedings No. 67, Australian Centre International Agricultural Research, Canberra, pp. 19-21.

Al-Janabi, S.M., Honeycutt, R.J., McClelland, M. and Sobral, B.W.S. (1993) A genetic linkage map of *Saccharum spontaneum* L. 'SES 208'. *Genetics* 134, 1249-1260.

Al-Janabi, S.M., Honeycutt, R.J. and Sobral, B.W.S. (1994a) Chromosome assortment in *Saccharum*. *Theoretical Applied Genetics* 89, 959-963.

Al-Janabi, S.M., McClelland, M., Petersen, C. and Sobral, B.W.S. (1994b) Phylogenetic analysis of organellar DNA sequences in the Andropogoneae: Saccharinae. *Theoretical Applied Genetics* 88, 933-944.

Allison, J.C.S. and Haslam, R.J. (1993) Theoretical assessment of potential for increased productivity of sugarcane through increased nitrogen fertilization. *Proceedings South African Sugar Technologists' Association* 67, 57-59.

Allsopp, P.G. and Manners, J.M. (1997) Novel approaches for managing pests and diseases in sugarcane. In: Keating, B.A. and Wilson, J.R. (eds) *Intensive Sugarcane Production: Meeting the Challenges beyond 2000*. CAB International, Wallingford, UK, pp. 173-188.

Allsopp, P.G. and McGill, N.G. (1996) Host-plant resistance: A key component in minimising losses from pink ground pearls. *Proceedings of the Australian Society Sugar Cane Technologists* 18, 100-105.

Allsopp, P.G., McGhie, T.K., Cox, M.C. and Smith, G.R. (1996) Redesigning sugarcane for resistance to Australian canegrubs: a potential IPM component. *Integrated Pest Management Reviews* 1, 79-90.

Allsopp, P.G., McGhie, T.K., Smith, G.R., Ford, R. and Cox, M.C. (1995) Progress in the development of cane varieties with resistance to canegrubs. *Proceedings Australian Society Sugar Cane Technologists* 17, 97-105.

Amaya, A., Gomez, A.L., Buitrago, J.T., Moreno, C.A. and Cassalett, C. (1996) Characterization of lodging in sugarcane. *Proceedings International Society Sugar Cane Technologists* 22, 321-327.

Anonymous (1987) *Copersucar International Sugarcane Breeding Workshop*. Copersucar, Brazil.

Ashraf, M. and Fatima, B. (1990) Breeding for resistance to *Chilo* spp. in sugarcane. *Insect Science Applications* 11, 683-687.

Autrey, L.J.C., Moutia, Y. and Saumtally, S. (1996a) Screening of seedlings against rust, inoculation treatment and assessment by video image analysis. *Sugar Cane* 1, 4-10.

Autrey, L.J.C., Saumtally, S. and Dookun, A. (1996b) Application of biotechnology to sugarcane pathogens. *Proceedings International Society Sugar Cane Technologists* 22, 15-22.

Balance, M.C., Milanés Ramos, N. and Mesa López, J.M. (1996) Diallelic analysis of resistance of five sugarcane parents to *Ustilago scitaminea* Sydow. *Proceedings International Society Sugar Cane Technologists* 22, 340-345.

Berding, N. (1995) Improving flowering of sugarcane for breeding: Progress and prospects. *Proceedings Australian Society Sugar Cane Technologists* 17, 162-171.

Berding, N. (1996) Sugarcane weevil borer resistance: Breeding strategy development using survey data. *Proceedings Australian Society Sugar Cane Technologists* 18, 90-99.

Berding, N. and Brotherton, G.A. (1995) Analysis of fibrated sugarcane by near infrared reflectance spectroscopy. In: Batten, G.D., Flinn, P.C., Welsh, L.A. and Blakeney, A.B. (eds), *Leaping Ahead with Near Infrared Spectroscopy*. Royal Australian Chemical Institute, Near Infrared Spectroscopy Group, Melbourne, Australia, pp. 199-203.

Berding, N. and Brotherton, G.A. (1996a) Analysis of high-moisture material - fibrated sugarcane. *Proceedings 7th International Conference on Near Infrared Spectroscopy,* 648-654.

Berding, N. and Brotherton, G.A. (1996b) Analysis of samples from sugarcane evaluation trials by near infra-red spectroscopy using a new at-line, large cassette presentation module. In: Wilson, J.R., Hogarth, D.M., Campbell, J.A. and Garside, A.L. (eds), *Sugarcane: Research towards Efficient and Sustainable Production*. CSIRO Division of Tropical Crops and Pastures, Brisbane, Australia, pp. 57-58.

Berding, N. and Croft, B.J. (1997) Breeding for resistance to Pachymetra root rot of sugarcane: Heritability and efficiency of selection. In preparation.

Berding, N. and Moore, P.H. (1996) Towards optimized induction of flowering in sugarcane. In: Wilson, J.R., Hogarth, D.M., Campbell, J.A. and Garside, A.L. (eds), *Sugarcane: Research towards Efficient and Sustainable Production*. CSIRO Division of Tropical Crops and Pastures, Brisbane, Australia, 44-46.

Berding, N. and Roach, B.T. (1987) Germplasm collection, maintenance, and use. In: Heinz, D.J (ed.), *Sugarcane Improvement through Breeding*. Elsevier, Amsterdam, pp. 143-210.

Berding, N. and Skinner, J.C. (1987) Traditional breeding methods. In: Anonymous (ed.), *Copersucar International Sugarcane Breeding Workshop*. Copersucar, Brazil, pp. 269-320.

Besse, P., McIntyre, C.L. and Berding, N. (1996a) Ribosomal DNA variations in *Erianthus*, a wild sugarcane relative (Andropogoneae-Saccharinae). *Theoretical Applied Genetics* 92, 733-743.

Besse, P., McIntyre, C.L. and Berding, N. (1996b) Characterisation of *Erianthus* sect. *Ripidium* and *Saccharum* germplasm (Andropogoneae - Saccharinae) using RFLP markers. *Euphytica*, In press.

Birch, R.G. (1997) Transgenic sugarcane: opportunities and limitations. In: Keating, B.A. and Wilson, J.R. (eds) *Intensive Sugarcane Production: Meeting the Challenges beyond 2000.* CAB International, Wallingford, UK, pp. 125-140.

Bond, R.S. (1988) Progress in selecting for eldana resistance. *Proceedings South African Sugar Cane Technologists' Association* 62, 129-133.

Bond, R.S. (1989) Observations on family selection in the Mount Edgecombe sugarcane breeding programme. *Proceedings South African Sugar Cane Technologists' Association* 63, 132-135.

Bos, I. and Caligari, P. (1995) *Selection Methods in Plant Breeding*. Chapman and Hall, London, UK.

Bower, R. and Birch, R.G. (1992) Transgenic sugarcane plants via microprojectile bombardment. *Plant Journal* 2, 409-416.

Braithwaite, K.S., and Smith, G.R. (1996) Molecular tools for indexing in sugarcane. In: Croft, B.J., Piggin, C.M., Wallis, E.S. and Hogarth, D.M. (eds), *Sugarcane Germplasm Conservation and Exchange*. Proceedings No. 67, Australian Centre International Agricultural Research, Canberra, Australia, pp. 75-78.

Bull, J.K. (1993) Methods of analyzing genotype × environment data to enhance interpretation. *Ph.D. Thesis*, University of Queensland, Brisbane, Australia.

Bull, J.K., Basford, K.E., Cooper, M. and DeLacy, I.H. (1994a) Enhanced interpretation of pattern analyses of environments: the use of blocks. *Field Crops Research* 37, 25-32.

Bull, J.K., Bull, T.A. and Cooper, M. (1996) The importance of water and nitrogen in generating clone by environment interaction in sugarcane: A preliminary investigation based on plant crop results. In: Wilson, J.R., Hogarth, D.M., Campbell, J.A. and Garside, A.L. (eds), *Sugarcane: Research towards Efficient and Sustainable Production*. CSIRO Division of Tropical Crops and Pastures, Brisbane, Australia, pp. 49-51.

Bull, T.A. and Bull, J.K. (1996) Increasing sugarcane yields through higher planting density - preliminary results. In: Wilson, J.R., Hogarth, D.M., Campbell, J.A. and Garside, A.L. (eds), *Sugarcane: Research towards Efficient and Sustainable Production*. CSIRO Division of Tropical Crops and Pastures, Brisbane, Australia, 166-168.

Bull, J.K., Cooper, M. and Basford, K.E. (1994b) A procedure for investigating the number of genotypes required to provide a stable classification of environments. *Field Crops Research* 38, 47-56.

Bull, J.K. and Hogarth, D.M. (1990) The implications of genotype × environment interactions for evaluation of sugarcane families. II. Alternative methods of analysis. In: Kang, M.S. (ed.), *Genotype-by-Environment Interaction and Plant Breeding*. Louisiana State University, Baton Rouge, Louisiana, USA, pp. 347-363.

Bull, J.K., Hogarth, D.M. and Basford, K.E. (1992) Impact of genotype × environment interaction on response to selection in sugarcane. *Australian Journal Experimental Agriculture* 32, 731-737.

Burner, D.M. (1991) Cytogenetic analysis of sugarcane relatives (Andropogoneae: Saccharinae). *Euphytica* 54, 125-133.

Burner, D.M., Grisham, M.P. and Legendre, B.L. (1993) Resistance of sugarcane relatives injected with *Ustilago scitaminea*. *Plant Disease* 77, 1221-1223.

Burner, D.M. and Legendre, B.L. (1993a) Chromosome transmission and meiotic stability of sugarcane (*Saccharum* spp.) hybrid derivatives. *Crop Science* 33, 600-606.

Burner, D.M. and Legendre, B.L. (1993b) Sugarcane genome amplification for the subtropics: a twenty year effort. *Sugar Cane* 5-10.

Burnquist, W.L. (1991) Development and application of restriction fragment length polymorphism technology in sugarcane (*Saccharum* spp.) breeding. *Ph.D. Thesis*, Cornell University, N.Y., USA.

Chang, Y.-S. and Milligan S.B. (1992a) Estimating the potential of sugarcane families to produce elite genotypes using bivariate prediction methods. *Theoretical Applied Genetics* 84, 633-639.

Chang, Y.-S. and Milligan S.B. (1992b) Estimating the potential of sugarcane families to produce elite genotypes using univariate cross prediction methods. *Theoretical Applied Genetics* 84, 662-671.

Chao, C.P., Hoy, J.W., Saxton, A.M. and Martin, F.A. (1990) Heritability and repeatability of clone reaction to sugarcane smut in Louisiana. *Phytopathology* 80, 622-626.

Comstock, J.C., Schnell, R.J. and Miller, J.D. (1996) Current status of the world sugarcane germplasm collection in Florida. In: Croft, B.J., Piggin, C.M., Wallis, E.S. and Hogarth, D.M. (eds), *Sugarcane Germplasm Conservation and Exchange.* Proceedings No. 67, Australian Centre International Agricultural Research, Canberra, Australia, pp. 17-18.

Comstock, J.C., Wu, K.K. and Schnell, R.J. (1992) Heritability of resistance to sugar cane rust. *Sugar Cane* 7-10.

Cox, M.C. and Hansen, P.B. (1995) Productivity trends in southern and central regions and the impact of new varieties. *Proceedings Australian Society Sugar Cane Technologists* 17, 1-7.

Cox, M.C., Hogarth, D.M. and Hansen, P.B. (1994) Breeding for high early season sugar content in a sugarcane (*Saccharum* spp. hybrids) improvement program. *Australian Journal Agricultural Research* 45, 1569-1575.

Cox, M.C., McRae, T.A., Bull, J.K. and Hogarth, D.M. (1996) Family selection improves the efficiency and effectiveness of as sugarcane improvement program. In: Wilson, J.R., Hogarth, D.M., Campbell, J.A. and Garside, A.L. (eds), *Sugarcane: Research towards Efficient and Sustainable Production.* CSIRO Division of Tropical Crops and Pastures, Brisbane, Australia, pp. 42-43.

Croft, B.J. and Berding, N. (1997) Breeding strategies for Pachymetra root rot resistance of sugarcane. In preparation

Croft, B.J., Piggin, C.M., Wallis, E.S. and Hogarth, D.M. (1996) *Sugarcane Germplasm Conservation and Exchange.* Proceedings No. 67, Australian Centre International Agricultural Research, Canberra, Australia.

Croft, B.J. and Smith, G.R. (1996) Major diseases affecting sugarcane production in Australia and recent experiences with sugarcane diseases in quarantine. In: Croft, B.J., Piggin, C.M., Wallis, E.S. and Hogarth, D. (eds), *Sugarcane germplasm conservation and exchange.* Proceedings No. 67, Australian Centre International Agricultural Research, Canberra, Australia, pp. 55-58.

Cuenya, M.I. and Mariotti, J.A. (1996) Breeding sugarcane for early high-sugar content in subtropical climates. *Proceedings International Society Sugar Cane Technologists* 22, 316-320.

da Silva, J., Honeycutt, R.J., Burnquist, W., Al-Janabi, S.M., Sorrells, M.E., Tanksely, S.D. and Sobral, B.W.S. (1995) *Saccharum spontaneum* L. `SES208' genetic linkage map combining RFLP- and PCR-based markers. *Molecular Breeding* 1, 165-179.

da Silva, J., Sorrells, M.E., Burnquist, W.L. and Tanksley, S.D. (1993) RFLP linkage map and genome analysis of *Saccharum spontaneum. Genome* 36, 782-791.

Davis, M.J., Dean, J.L., Miler, J.D. and Shine Jr, J.M. (1994) A method to screen for resistance to ratoon stunting disease of sugarcane. *Sugar Cane* 9-16.

de Haas, A.O. and Bond, R.S. (1993) Selecting mosaic resistant sugarcane varieties for the Natal midlands. *Proceedings South African Sugar Technologists' Association* 67, 94-97.

DeLacy, I.H., Basford, K.E., Cooper, M., Bull, J.K. and McLaren, C.G. (1996) Analysis of multi-environment trials - an historical perspective. In: Cooper, M. and Hammer, G.L. (eds), *Plant Adaptation and Crop Improvement.* CAB International, Wallingford, UK, pp. 39-124.

de Ponti, O.M.B. and Mollema, C. (1992) Emerging breeding strategies for insect resistance. In: Stalker, H.T. and Murphy, J.P. (eds), *Plant Breeding in the 1990s*. CAB International, Wallingford, UK, pp. 323-346.

Deren, C.W. (1992) Stability and heritability of pith in sugarcane and its influence on yield. *Plant Breeding* 109, 242-247.

Deren, C.W., Snyder, G.H., Miller, J.D. and Porter, P.S. (1991) Screening for heritability and flood-tolerance in the Florida (CP) sugarcane breeding population. *Euphytica* 56, 155-160.

D'Hont, A., Grivet, L., Lu, Y.H., Roques, D., Feldman, P., Rao, P.S., Daugrois, J.H., Dufor, P., Berding, N., Walker, D.I.T., Harmon, P. and Glaszmann, J.C. (1996b) The genome of modern sugarcane varieties. *Proceedings International Society Sugar Cane Technologists* 22, 363-367.

D'Hont, A., Lu, Y.-H., González de Leon, D., Grivet, L., Feldmann, P., Lanaud, C. and Glaszmann, J.C. (1994) A molecular approach to unravelling the genetics of sugarcane, a complex polyploid of the Andropogoneae tribe. *Genome* 37, 222-230.

D'Hont, A., Lu, Y.-H., Feldmann, P. and Glaszmann, J.C. (1993) Cytoplasmic diversity in sugarcane revealed by heterologous probes. *Sugar Cane* 12-15.

D'Hont, A., Rao, P.S., Feldmann, P., Grivet, L., Islam-Faridi, N., Taylor, P. and Glaszmann, J.C. (1995) Identification and characterisation of sugarcane intergeneric hybrids, *Saccharum officinarum* x *Erianthus arundinaceus*, with molecular markers and DNA in situ hybridisation. *Theoretical and Applied Genetics* 91, 320-326.

D'Hont, A., Grivet, L., Feldmann, P., Rao, S., Berding, N. and Glaszmann, J.C. (1996a) Characterisation of the double genomic structure of modern sugarcane cultivars (*Saccharum* spp.) by molecular cytogenetics. *Molecular General Genetics* 250, 405-413.

Dodman, R.L. (1996) Optimizing disease expression in quarantine. In: Croft, B.J., Piggin, C.M., Wallis, E.S. and Hogarth, D.M. (eds), *Sugarcane Germplasm Conservation and Exchange*. Proceedings No. 67, Australian Centre International Agricultural Research, Canberra, Australia, pp. 64-66.

Domaingue, R., Mamet, L. and Mungur, H. (1990) Selection for varieties suitable for early harvest in Mauritius. *Proceedings South African Sugar Technologists' Association* 64, 30-33.

Duvick, D.N. (1996) Plant breeding, an evolutionary concept. *Crop Science* 36, 539-548.

Egan, B.T. (1996) Role of the ISSCT in promoting sugarcane germplasm collections and their maintenance. In: Croft, B.J., Piggin, C.M., Wallis, E.S. and Hogarth, D.M. (eds), *Sugarcane Germplasm Conservation and Exchange*. Proceedings No. 67, Australian Centre International Agricultural Research, Canberra, Australia, pp. 12-13.

Eksomtramage, T., Paulet, F., Noyer, J.L., Feldmann, P. and Glaszmann, J.C. (1992) Utility of isozymes in sugarcane breeding. *Sugar Cane*, 14-21.

Engelmann, F. and Engels, F. (1996a) The role of IPGRI in germplasm conservation. In: Croft, B.J., Piggin, C.M., Wallis, E.S. and Hogarth, D.M. (eds), *Sugarcane Germplasm Conservation and Exchange*. Proceedings No. 67, Australian Centre International Agricultural Research, Canberra, Australia, pp. 14-16.

Engelmann, F. and Engels, F. (1996b) Networking efforts in germplasm preservation. *Proceedings International Society Sugar Cane Technologists* 22, 245-249.

Evans, L.T. (1993) *Crop evolution, adaptation and yield*. Cambridge University Press, Cambridge, UK.

Fukui, K., Ohimido, N., Ha, S. and Moore, P.H. (1994) Analysis and utility of chromosome information. 67. Complete identification of wild sugarcane chromosomes. *Japanese Journal Breeding* 44 (Suppl. 2), 29.

Gallacher, D.J. (1994) Development of a minimum descriptor set for individuals of *Saccharum* spp. hybrid germplasm. *Ph.D. Thesis*. James Cook University, Townsville.

Gallacher, D.J., Lee, D.J. and Berding, N. (1995) Use of isozyme phenotypes for rapid discrimination among sugarcane clones. *Australian Journal Agricultural Research* 46, 601-609.

Glimelius, K. (1993) Evolution and crop improvement. In: Imrie, B.C. and Hacker, J.B. (eds), *Focused Plant Improvement: Towards Responsible and Sustainable Agriculture*. Proceedings Tenth Australian Plant Breeding Conference, April 1993. Vol. 1, pp. 32-36.

Gravois, K.A., Milligan, S.B. and Martin, F.A. (1990) The role of pith, tube, and stalk density in determining sugarcane sucrose content and stalk weight. *Theoretical Applied Genetics* 79, 273-277.

Gravois, K.A., Milligan, S.B. and Martin, F.A. (1991a) Indirect selection for increased sucrose yield in early sugarcane testing stages. *Field Crops Research* 26, 67-73.

Gravois, K.A., Milligan, S.B. and Martin, F.A. (1991b) Additive genetic effects for sugarcane yield components and implications for hybridization. *Tropical Agriculture (Trinidad)* 68, 376-380.

Grivet, L., D'Hont, A., Dufour, P., Hamon, P., Roques, D. and Glaszmann, J.C. (1994) Comparative genome mapping of sugar cane with other species within the Andropogoneae tribe. *Heredity* 73, 500-508.

Harvey, M., Huckett, B.I. and Botha, F.C. (1994) Use of polymerase chain reaction (PCR) and random amplification of polymorphic DNAs (RAPDs) for the determination of genetic distances between 21 sugarcane varieties. *Proceedings South African Sugar Technologists' Association* 68, 36-40.

Heinz, D. J (1987a) *Sugarcane Improvement through Breeding*. Elsevier, Amsterdam, The Netherlands.

Heinz, D.J (1987b) Sugarcane improvement: Current productivity and future opportunities. In: Anonymous (ed.), *Copersucar International Sugarcane Breeding Workshop*. Copersucar, Brazil, pp. 55-70.

Heinz, D.J (1991) Sugarcane cytogenetics. In: Tsuchiya, T. and Gupta, P.K. (eds), *Chromosome Engineering in Plants: Genetics, Breeding, Evolution*. Elsevier, Amsterdam, The Netherlands, pp. 279-293.

Heinz, D.J and Tew, T.L. (1987) Hybridization procedures. In: Heinz, D.J (ed.), *Sugarcane Improvement through Breeding*. Elsevier, Amsterdam, The Netherlands, pp. 313-342.

Hemaprabha, G. and Ram, B. (1993) Genetic variability in nobilization stages of *Saccharum* robustum Brandes et Jeswiet ex Grassl. *Sugar Cane* 6, 6-9.

Hogarth, D.M. and Berding, N. (1996) Importance of international germplasm exchange to the Australian sugar industry. In: Croft, B.J., Piggin, C.M., Wallis, E.S. and Hogarth, D.M (eds), *Sugarcane germplasm conservation and exchange*. Proceedings No. 67, Australian Centre International Agricultural Research, Canberra, pp. 2-5.

Hogarth, D.M. and Mullins, R.T. (1989) Changes in the BSES plant improvement program. *Proceedings International Society Sugar Cane Technologists* 20, 956-961.

Hogarth, D.M., Reimers, J.F., Ryan, C.C. and Taylor, P.W.J. (1993a) Quantitative inheritance of Fiji disease resistance in sugarcane. *Field Crops Research* 34, 175-186.

Hogarth, D.M., Ryan, C.C. and Taylor, P.W.J. (1993b) Quantitative inheritance of rust resistance in sugarcane. *Field Crops Research* 34, 187-193.

Hogarth, D.M. and Skinner, J.C. (1986) Computerization of cane breeding records. *Proceedings International Society Sugar Cane Technologists* 19, 478-491.

Hogarth, D.M. and Skinner, J.C. (1987) Computerization of parental selection. In: Anonymous (ed.), *Copersucar International Sugarcane Breeding Workshop*. Copersucar, Brazil, pp. 87-101.

Hsu, S.-Y., Lin, C.-J., and Lo, C.C. (1991) Personal computer as an aid in sugarcane crossing program. *Report Taiwan Sugar Research Institute* 132, 1-11.

Hsu, S-Y, Lo, C.C. and Shih, S.C. (1988) Use of a *Saccharum spontaneum* derivative in sugarcane crossing. Report Taiwan Sugar Research Institute, 1-8.

Irvine, J.E. (1994) Changes in sugarcane breeding. *Sugar Journal* 57, 17-18.

Ivory, D.A., Kaewmeechai, S., DeLacy, I.H. and Basford, K.E. (1991) Analysis of the environmental component of genotype x environment interaction in crop adaptation evaluation. *Field Crops Research* 28, 71-84.

Jackson, P.A. (1992) Genotype x environment interactions in sugarcane. II. Use of performance in plant cane as an indirect selection criterion for performance in ratoon crops. *Australian Journal Agricultural Research* 43, 1461-1470.

Jackson, P. (1994) Genetic relationships between attributes in sugarcane clones closely related to *Saccharum spontaneum*. *Euphytica* 79, 101-108.

Jackson, P.A., Bull, J.K. and McRae, T.A. (1996) The role of family selection in sugarcane breeding programs and the effect of genotype x environment interactions. *Proceedings International Society of Sugar Cane Technologists* 22, 261-270.

Jackson, P.A. and Hogarth, D.M. (1992) Genotype x environment interactions in sugarcane. I. Patterns of response across sites and crop-years in North Queensland. *Australian Journal Agricultural Science* 43, 1447-1459.

Jackson, P.A., Horsley, D., Foreman, J., Hogarth, D.M. and Wood, A.W. (1991) Genotype x environment (GE) interactions in sugarcane variety trials in the Herbert. *Proceedings Australian Society Sugar Cane Technologists* 13, 103-109.

Jackson, P.A., McRae, T.A. and Hogarth, D.M. (1995) Selection of sugarcane families across variable environments II. Patterns of response and association with environmental factors. *Field Crops Research* 43, 119-130.

Jackson, P.A. and Roach, B.T. (1994) Performance of sugar-cane progeny from crosses between clones derived from diverse *S. spontaneum* sources. *Tropical Agriculture (Trinidad)* 71, 57-61.

Jenkin, M.J., Reader, S.M., Purdie, K.A. and Miller, T.E. (1995) Detection of rDNA sites in sugarcane by FISH. *Chromosome Research* 3, 444-445.

Jones, P.N., Ferraris, R. and Chapman, L.S. (1993) A technique for minimizing confounding of genotype x year and genotype x crop type effects in sugarcane. *Euphytica* 67, 199-204.

Kang, M.S., Glaz, B. and Miller, J.D. (1987) Interrelationships among stabilities of important agronomic traits in sugarcane. *Theoretical Applied Genetics* 74, 310-316.

Law, C.N. (1995) Genetic manipulation in plant breeding - prospects and limitations. *Euphytica* 85, 1-12.

Lee, D.J. (1995a) Enhancement of *Saccharum* spp. hybrid material by introgression with *Erianthus arundinaceus* germplasm. *Ph.D. Thesis*. James Cook University, Townsville.

Lee, M. (1995b) DNA markers and plant breeding programs. In: Sparks, D.L. (ed.), *Advances in Agronomy*. Academic Press, San Diego, USA, pp. 265-344.

Legendre, B.L. (1995) Potential for increasing sucrose content of sugarcane: An assessment of recurrent selection in Louisiana. *Sugar Cane* 3, 4-8.

Linedale, A.I. and Ridge, D.R. (1996) A successful campaign to minimise harvesting losses within the Queensland sugar industry. *Proceedings Australian Society of Sugar Cane Technologists* 18, 1-5.

Lingle, S.E. (1996) Teaching an old dog new tricks: Molecular methods in sugarcane physiology. *Proceedings International Society Sugar Cane Technologists* 22, 374-379.

Lo, C.C. and Chen, Y.-H. (1995) Breeding for sucrose content improvement in sugarcane varieties. *Taiwan Sugar* 42, 10-16.

Lu, Y.-H., D'Hont, A., Paulet, F., Grivet, L., Arnaud, M. and Glaszmann, J.C. (1994a) Molecular diversity and genome structure in modern sugarcane varieties. *Euphytica* 78, 217-226.

Lu, Y.-H., D'Hont, A., Walker, D.I.T., Rao, P.S., Feldmann, P., and Glaszmann, J.C. (1994b) Relationships among ancestral species of sugarcane revealed with RFLP using single copy maize nuclear probes. *Euphytica* 78, 7-18.

Matsuoka, S., Masuda, Y. and Arizono, H. (1990) Breeding procedures for resistancc to sugar cane mosaic virus in Brazil. *Sugar Cane* 12-16.

Meyer, J.H. (1996) Reaping the benefits of near infra-red spectroscopy in the South African sugar industry. In: Wilson, J.R., Hogarth, D.M., Campbell, J.A. and Garside, A.L. (eds), *Sugarcane: Research towards Efficient and Sustainable Production.* CSIRO Division of Tropical Crops and Pastures, Brisbane, Australia, 183-186.

Milligan, S.B., Gravois, K.A. and Martin, F.A. (1996) Inheritance of sugarcane ratooning ability and the relationship of younger crop traits to older crop traits. *Crop Science* 36, 45-50.

Mirzawan, P.D.N, Cooper, M., DeLacy, I.H. and Hogarth, D.M. (1994) Retrospective analysis of the relationships among test environments of the Southern Queensland sugarcane breeding programme. *Theoretical and Applied Genetics* 88, 707-716.

Mirzawan, P.D.N, Cooper, M. and Hogarth, D.M. (1993) The impact of genotype x environment interactions for sugar yield on the use of indirect selection in southern Queensland. *Australian Journal Experimental Agriculture* 33, 629-638.

Moore, P.H. (1989) Physiological basis for varietal improvement in sugarcane. In: Mohan Naidu, K., Sreenivasan, T.V. and Premachandran, M.N. (eds), *Sugarcane Varietal Improvement.* Sugarcane Breeding Institute, Coimbatore, India, pp. 19-55.

Moore, P.H. (1991) Prospects for strengthening the conventional breeding of sugarcane with molecular-based approaches. *Hawaiian Planters' Record* 61, 1-23.

Moore, P.H. (1996) Progress in sugarcane molecular biology. *Proceedings International Society Sugar Cane Technologists* 22, 353-362.

Moore, P.H., Botha, F.C., Furbank, R.T. and Grof, C.P.L. (1997) Potential for overcoming physio-biochemical limits to sucrose accumulation. In: Keating, B.A. and Wilson, J.R. (eds) *Intensive Sugarcane Production: Meeting the Challenges beyond 2000.* CAB International, Wallingford, UK, pp. 141-155.

Moore, P.H. and Fitch, M.M.M. (1990) Sugarcane (*Saccharum* spp.): Anther culture studies. In: Bajaj, Y.P.S. (ed.) *Haploids in Crop Improvement I.* Springer-Verlag, New York, USA, pp. 480-497.

Moore, P.H. and Irvine, J.E. (1991) Genomic mapping of sugarcane and its potential contribution to improvement and to selection of new varieties. *Proceedings South African Sugar Cane Technologists' Association* 65, 96-102.

Moore, P.H. and Nuss, K.J. (1987) Flowering and flower synchronization. In: Heinz, D.J (ed.), *Sugarcane Improvement through Breeding.* Elsevier, Amsterdam, The Netherlands, pp. 273-311.

Msomi, N. and Botha, F.C. (1994) Identification of molecular markers linked to fibre using bulk segregant analysis. *Proceedings South African Sugar Technologists' Association* 68, 41-45.

Muchow, R.C., Robertson, M.J., Wood, A.W. and Keating, B.A. (1997) Assessing limits to sugarcane yield. *Proceedings Australian Society of Sugar Cane Technologists* 19, In press.

Nair, N.V. and Sreenivasan, T.V. (1995) *Sugarcane database: Vol I. Database on Co varieties. Part 1. Yield, quality and flowering*. Sugarcane Breeding Institute, Coimbatore, India.

Nuss, K.J. (1991) Screening sugarcane varieties for resistance to eldana borer. *Proceedings South African Sugar Cane Technologists' Association* 65, 92-95.

Paulet, F., Engelmann, F. and Glaszmann, J.C. (1993) Cryopreservation of apices of *in vitro* plantlets of sugarcane (*Saccharum* spp. hybrids) using encapsulation/dehydration. *Plant Cell Reports* 12, 525-529.

Peloquin, S.J. and Ortiz, R. (1994) Techniques for introgressing unadapted germplasm to breeding populations. In: Stalker, H.T. and Murphy, J.P. (eds), *Plant Breeding in the 1990s*. CAB International, Wallingford, UK, pp. 485-507.

Plucknett, D.L. and Smith, N.J.H. (1986) Sustaining agricultural yields. *BioScience* 36, 40-45.

Ramdoyal, K., Badaloo, G. and Mangar, M. (1995) Effect on seed setting of potassium metabisulphite in sugarcane crossing solution. *Sugar Cane* 3-7.

Ridge, D.R. and Dick, R.G. (1988) Current research on green cane harvesting and dirt rejection by harvesters. *Proceedings Australian Society of Sugar Cane Technologists* 10, 19-25.

Roach, B.T. (1989) Origin and improvement of the genetic base for sugarcane. *Proceedings Australian Society Sugar Cane Technologists* 11, 492-503.

Roach, B.T. (1992) Susceptibility to ratoon stunting disease in the *Saccharum* complex and feasibility of breeding for resistance. *Sugar Cane* 1-11.

Rutherford, R.S., Meyer, J.H., Smith, G.S. and van Staden, J. (1993) Resistance to *Eldana saccharina* (Lepidoptera: Pyralidae) in sugarcane and some phytochemical correlations. *Proceedings South African Sugarcane Technologists' Association* 67, 82-87.

Rutherford, R.S. and van Staden, J. (1996) Towards a rapid near-infrared technique for prediction of resistance to sugarcane borer *Eldana saccharina* Walker (Lepidoptera: Pyralidae) using stalk surface wax. *Journal Chemical Ecology* 22, 681-694.

Schäffler, K.J., Dunsmore, A.N. and Meyer, J.H. (1993) Rapid analysis of sugar products by near infrared spectroscopy. *Proceeding South African Sugar Cane Technologists' Association* 67, 222-229.

Smith, F.W. (1997) Nutrient uptake and metabolism: prospects for molecular modification to improve efficiency. In: Keating, B.A. and Wilson, J.R. (eds) *Intensive Sugarcane Production: Meeting the Challenges beyond 2000.* CAB International, Wallingford, UK, pp. 157-172.

Sobral, B.W.S., Braga, D.P.V., LaHood, E.S. and Keim, P. (1994) Phylogenetic analysis of chloroplast restriction enzyme site mutations in the *Saccharinae* Griseb. subtribe of the *Andropogoneae* Dumort. tribe. *Theoretical Applied Genetics* 87, 843-853.

Sorrells, M.E. (1992) Development and application of RFLPs in polyploids. *Crop Science* 32, 1086-1091.

Srivastava, B.L., Cooper, M. and Mullins, R.T. (1994) Quantitative analysis of the effect of selection history on sugar yield adaptation of sugarcane clones. *Theoretical Applied Genetics* 87, 627-640.

Stevenson, D.W.A., van der Merwe, A., Benninga, W. and Allison, J.C S. (1992) Response of different sugarcane varieties to greater than normal applications of nitrogen. *Proceedings South African Sugar Technologists' Association* 66, 50-53

Stringer, J.K., McRae, T.A. and Cox, M.C. (1996) Best linear unbiased prediction as a method of estimating breeding value in sugarcane. In: Wilson, J.R., Hogarth, D.M., Campbell, J.A. and Garside, A.L. (eds), *Sugarcane: Research towards Efficient and Sustainable Production.* CSIRO Division of Tropical Crops and Pastures, Brisbane, Australia, 39-41.

Tai, P.Y.P. (1988) Long-term storage of *Saccharum spontaneum* L. pollen at low temperature. *Sugar Cane* 12-16.

Tai, P.Y.P. (1993) Low temperature preservation of F_1 pollen in crosses between noble or commercial sugarcane and *Saccharum spontaneum* L. *Sugar Cane* 5, 8-11.

Tai, P.Y.P., Gan, H., He, H. and Miller, J.D. (1991a) Phenotypic characteristics of F_2 and BC_1 progenies from sugarcane intergeneric crosses. *Journal American Society Sugar Cane Technologists* 11, 38-47.

Tai, P.Y.P, He, H., Gan, H. and Miller, J.D. (1991b) Flowering of hybrids from commercial sugarcane x *Saccharum spontaneum* crosses. *Journal American Society Sugar Cane Technologists* 11, 69-74.

Tai, P.Y.P, He, H., Gan, H. and Miller, J.D. (1992) Variation for juice quality and fibre content in crosses between commercial sugarcane and *Saccharum spontaneum. Journal American Society Sugar Cane Technologists* 12, 47-57.

Tai, P.Y.P and Miller, J.D. (1989) Family performance at early stages of selection and frequency of superior clones from crosses among Canal Point cultivars of sugarcane. *Journal American Society Sugar Cane Technologists* 9, 62-70.

Tai, P.Y.P. and Miller, J.D. (1993) Recurrent selection for cold tolerance of sugarcane. *Journal American Society Sugar Cane Technologists* 13, 40-45.

Tai, P.Y.P, Miller, J.D. and Legendre, B.L. (1994) Preservation of *Saccharum spontaneum* germplasm through storage of true seed. *Sugar Cane* 6, 3-8.

Tai, P.Y.P, Miller, J.D. and Legendre, B.L. (1996) Evaluation of the World Collection of *Saccharum spontaneum* L. *Proceedings International Society Sugar Cane Technologists* 22, 250-260.

Taylor, P.W.J. and Dukic, S. (1993) Development of an *in vitro* culture technique for conservation of *Saccharum* spp. hybrid germplasm. *Plant Cell Tissue Organ Culture* 34, 217-222.

van Eeuwijk, F.A. and Elgersma, A. (1993) Incorporating environmental information in an analysis of genotype by environment interaction for seed yield of perennial ryegrass. *Heredity* 70, 447-457.

Wang, T.-H., Hour, A.-L., Hsu, S.-Y. and Lo, C.C. (1993) Data base management system for sugarcane breeding program. *Taiwan Sugar* 40, 8-15 and 20-21.

Webster, R.D. and Shaw, R.B. (1995) Taxonomy of the native North American species of *Saccharum* (Poaceae: Andropogoneae). *Annals Missouri Botanical Gardens* Submitted. (cited in Burner and Webster, 1995).

Whittle, P. (1994) Yellow leaf syndrome maybe in Australia. *BSES Bulletin* 48, 14-15.

Wu, K.K., Burnquist, W., Sorrells, M.E., Tew, T.L., Moore, T.L. and Tanksley, S.D. (1992) The detection and estimation of linkage in polyploids using single-dose restriction fragments. *Theoretical Applied Genetics* 83, 294-300.

Wu, K.K., Heinz, D.J and Hogarth, D.M. (1988) Association and heritability of sugarcane smut resistance to races A and B. *Theoretical Applied Genetics* 75, 754-760.

Zheng, D., Lin, Y. and Tai, P.Y.P. (1993) Electrophoretic analysis of intergeneric hybrids of Saccharum x related genera. *Journal American Society Sugar Cane Technologists* 13, 46-53.

Chapter thirteen:

Nitrogen Management in Intensive Agriculture: Sugarcane in Australia

B.A. Keating[1], K. Verburg[2], N.I. Huth[1] and M.J. Robertson[1]

1. *CSIRO Tropical Agriculture and CRC for Sustainable Sugar Production, 306 Carmody Road, St Lucia, Qld 4067, Australia*
2. *CSIRO Soils, 306 Carmody Road, St Lucia, Qld 4067, Australia*

Abstract

The substantial increase in N fertiliser used in the Australian sugar industry over the last 40 years is mirrored in intensive cropping industries world-wide. Whilst enabling higher production, downsides of increasing N fertiliser input have included acidification, contamination of ground and surface water and enhanced greenhouse gas emission.

Issues of N requirement, N loss and opportunities for improved N management are explored in this paper with the aid of a crop-soil simulation model (APSIM-Sugarcane) which encapsulates what we know of N supply and utilisation in sugarcane systems. While the study focuses on sugarcane in Australia, the principles established are relevant to other intensive cropping systems in other parts of the world.

The consistent theme through the analysis is the way in which decisions on N fertilisation rate affect the balance between production, profit and environmental impact. Optimum N rates are identified and related to other factors influencing production. Response curves are estimated to be very flat once the optimum is reached and there is little direct economic incentive to better match N fertiliser management to soil and climate condition and crop requirement. There are, however, major gains to be made in reducing the potential for N leaching loss by eliminating the worst cases of over-fertilisation. We conclude that changes in N fertilisation practice for sugarcane in Australia are more likely to come from an environmental imperative than an economic one.

In: Keating, B.A. and Wilson, J.R. (eds) *Intensive Sugarcane Production: Meeting the Challenges Beyond 2000*. CAB International, Wallingford, UK, 1997.

Introduction

In common with many intensive agricultural industries world-wide, N fertiliser use within the Australian sugar industry has risen substantially over the last 40 years. Rates used, often exceeding the industry standard recommendations, are high by world standards. While rapid expansion of N fertiliser use has enabled world food production to stay somewhat in step with population growth, downsides have included soil acidification, nitrate loss to ground or surface waters and elevated greenhouse gas emission.

The challenge facing intensive agriculture is to apply sufficient N fertiliser to achieve the potential productivity of a crop set by cultivar, soils and climate, but to avoid loss of N that can damage the resource base, or the surrounding environment. Meeting this challenge requires an understanding of the complex N cycle, and its interactions with weather, soil, crop and management. In this paper models are used to integrate the complexity and variability of the N cycle and its interactions with the crop management system.

These models embody concepts of N demand to achieve a yield potential and N supply from a range of sources, that are subject to a range of loss processes. N fertilisation is initially considered in the context of these concepts of demand and supply. With this background, a crop-soil simulation model is used to explore the way in which decisions on N fertilisation affect the balance between production, profit and environmental impact.

N Demand and Supply

Nitrogen is vital for most plant metabolic processes (Engels and Marschner, 1995) and is present principally as proteins, amino acids and amides, and sometimes as small quantities of nitrate and ammonium. The chlorophyll molecule contains nitrogen, and hence one of the first impacts of N deficiency is chlorosis or loss of green colour from the photosynthetic tissues. The N requirement in a range of high yielding situations is presented in Table 13.1. Note that the N requirements of sugarcane are similar to those of wheat and maize, despite the total biomass of the sugarcane crop being about four times greater than that of wheat and maize (e.g. 80t ha^{-1} d.w. for sugarcane compared to 20t ha^{-1} d.w. for wheat and maize)

In intensive cropping systems, the key repositories for N are soil organic matter, soil mineral N (e.g. nitrate and ammonium) and crop biomass. Typically, mineral N in the soil or the N in plant parts represents only 1-4% of the total N in the plant-soil system. A range of biological and physical processes result in gain, loss or transformation of N in the plant or soil. Whilst involving only a small quantity of the total system N at any one time, these processes have a major impact on the performance of the biological system. Key processes include ammonification, nitrification, immobilisation, denitrification, volatilisation, leaching, crop uptake and symbiotic and non-symbiotic fixation.

Table 13.1. Quantities of N taken up and removed in harvested products for a range of high yielding crops. [Yields either per crop or year. Sugarcane, sugarbeet, potato and oil palm are on the basis of fresh weight of harvested product]

Crop	Yield t ha^{-1}	N uptake kg ha^{-1}	N removal kg ha^{-1}	Reference
Wheat	10.5	260	200	Neeteson, 1995
Maize	12.0	275	170	Muchow and Davis, 1988
Potato	45	253	153	Rauschkolb and Hornsby, 1994
Sugarbeet	56	253	112	Rauschkolb and Hornsby, 1994
Sugarcane	200	240	140	Muchow *et al.*,1996 Catchpoole and Keating, 1995
Rubber	2.2	240	22	Chew and Pushparajah, 1995
Oil Palm	35	190	116	Chew and Pushparajah, 1995
Cocoa	2.2	170	62	Chew and Pushparajah, 1995
Coconut	4.6	130	54	Chew and Pushparajah, 1995
Fruit trees	-	48	9 - 34	Sanchez *et al.*, 1995
Grapes	-	84	35	Sanchez *et al.*, 1995

In the absence of fertilisation, plants incapable of symbiotic N fixation rely on net mineralisation from soil organic matter and N input from atmospheric sources. Many soils, when first cultivated, are capable of net mineralisation rates in the order of 200kg N ha^{-1} y^{-1} (Dalal and Mayer, 1987) but these rates often decline to < 50kg N ha^{-1} y^{-1} as organic matter declines. Quantity of N mineralised varies with soil and climate, and reflects a dynamic equilibrium between gain and loss of C and N. Ultimately, steady state conditions prevail, whereby organic N pools are stable and N removal in harvested products balances inputs from atmospheric sources such as rainfall, dry deposition of ammonia or non-symbiotic N fixation. Inputs from these latter sources of 20-40kg N ha^{-1} y^{-1} are not unreasonable (Meisinger and Randall, 1991).

N Fertilisation and Intensive Crop Production

While wheat has been grown continuously without N fertiliser in trials at Rothamsted, UK, for over 150 years, yields are < 20% of those from N fertilised plots (Jenkinson, 1991). Agricultural intensification to meet a growing world population has gone hand in hand with an increasing use of N fertiliser, considerably outstripping growth in use of other nutrients (Fig. 13.1). The drop in fertiliser usage beyond 1989 has been associated with political changes in

Eastern Europe. In Europe and the US, while crop yield increases have been large, they have not kept pace with N fertiliser inputs and N fertiliser use efficiency has declined (Sinclair, 1997). In contrast, sub-Saharan Africa has lagged behind the rest of the world in both fertiliser use and food production (Stoorvogel *et al.*, 1993).

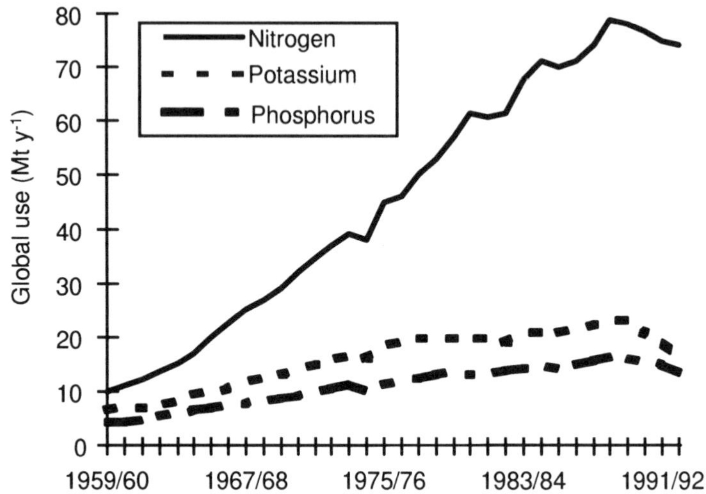

Fig. 13.1. Historical patterns of global N, P and K fertiliser use [after Bumb ,1995]

Agricultural systems vary in the extent to which N fertiliser input replaces N removal in the harvested product (Table 13.2). There is a clear trend for crops with higher gross value per unit area to apply N fertiliser far in excess of what is removed in the harvested product. At the other extreme, some dryland cereal production systems are unsustainable because of nutrient depletion.

N Impact on the Resource Base and the Environment

While N fertilisation has increased productivity in agricultural systems around the world, it has not been without cost. This has principally been in terms of soil acidification, negative effects of N loss to ground or surface water, and emission of the greenhouse gas, nitrous oxide.

Soil acidification

Soil acidification is a complex process influenced by soil, climate and management factors. Cycling of N in the soil-crop system would have neutral effect on soil pH if either the system was truly closed or if the quantity and form of N gain balanced the quantity and form of N loss (Helyar, 1976). This rarely occurs and the form and amount of fertiliser addition, N transformation, N fixation, N uptake by crops and N loss all interact with other chemical equilibria in the soil to modify pH. Generally this modification is towards acidification in response to N leaching losses and product removal, at a rate strongly influenced by the buffering capacity of the soil.

Table 13.2. Nitrogen inputs and removal in harvested products from a range of agricultural systems.

System	N inputs* (kg ha^{-1}crop^{-1})	N removal (kg ha^{-1}crop^{-1})	Ratio of inputs to removal
Subsistence maize in Kenya[1]	20	20	1.0
Nth-east Australia grains[2]	20	50	0.4
Sth Australian wheat/sheep[3]	50	40	1.2
Australian sugarcane (burn)[4]	180	120#	1.5
Australian sugarcane (green)[4]	180	70	2.6
Wheat in UK[5]	200	130	1.5
Maize in US[6]	180	100	1.8
Intensive vegetables[7]	158-242	21-111	2-8
Grapes[8]	69	18	3.8
Perennial fruit trees[8]	131-174	9-34	5-5

* All systems assumed to receive 20kg N ha^{-1}crop^{-1} from rainfall, dry deposition and non-symbiotic fixation.

\# Includes N lost from burning of trash and tops.

1. B.A. Keating (unpublished data)
2. Unfertilised grain cropping on initially fertile vertisols. Based on 2t grain ha^{-1} at 2.5%N
3. See McCown *et al.* (1988). Assumes 60kg N ha^{-1} fixed during a 3 year pasture phase exploited over a 2 year cropping phase.
4. Keating *et al.* (1994a), Vallis and Keating (1994), Catchpoole and Keating (1995)
5. Addiscott *et al.* (1991)
6. Schepers and Fox (1989), Bock and Hergert (1991)
7. Tomatoes drip irrigated at Bundaberg, K.L. Weier and B.A. Keating (unpublished data)
8. Sanchez *et al.* (1995), Miller and Smith (1976)

Nitrate losses to surface and groundwaters

Water moving through soil also moves dissolved ions, with nitrate often the major constituent. This nitrate finds its way to ground and surface water, estuaries and ultimately the marine environment, and together with other nutrients, most notably phosphate, can lead to eutrophication and ecological imbalance.

Of equal concern and greater public notoriety has been the problem of nitrate in groundwater used as sources of potable water. The draft NHMRC drinking water guideline (NHMRC 1994) for Australia states that nitrate should not exceed 50mg NO_3 l^{-1} for infants under 3 months of age or 100mg NO_3 l^{-1} for all other users. The WHO and European Union set standards of 45 to 50mg NO_3 l^{-1} respectively. Addiscott *et al.* (1991) reviewed the 'nitrate problem' and its relationship to farming and N fertilisation with an emphasis on Europe, whilst Madison and Brunett (1985) reported on extensive contamination of shallow aquifers below agricultural districts in the USA.

In Queensland, 2% of all bores sampled have nitrate concentration > 50mg NO_3 l^{-1} (Keating *et al.*, 1997a). Some of this was in groundwater associated with basaltic geological structures of the Great Dividing Range and may be of natural origin. The bulk of the remainder was in alluvial aquifers along the coast supporting intensive horticultural and sugarcane production.

Greenhouse gas emissions

Nitrous oxide is released from agricultural soils through nitrification of ammonium or denitrification of nitrate; the latter is the dominant process in wet agricultural environments. Nitrous oxide is a significant contributor to global warming, with a 'Greenhouse Warming Potential' 290 times that of the equivalent weight of CO_2 (Galbally *et al.*, 1992). Fertiliser addition that increases these N transformation and loss processes has resulted in enhanced nitrous oxide emissions to the atmosphere (Duxbury, 1994; Mosier, 1994).

Nitrogen Management of Sugarcane in Australia

Nitrogen usage by industry

In Australia, canegrowers spend c. A$70M each year on N fertilisers. Do they get a good return on this investment; where do the 80,000t N applied each year end up; what are the needs and opportunities for improved N management ?

Wood *et al.* (1997) detail the current recommendations and practice for N fertilisation of sugarcane in Australia (see also Calcino, 1994). These recommendations can be summarised as 120kg N ha^{-1} for plant crops and 160kg N ha^{-1} for ratoon crops, adjusted upwards for the higher yield potential of the Burdekin River District and downwards for 'rich' soils. While responses to N between field trials have been variable (Chapman, 1982; Keating *et al.*, 1994a), the large numbers of trials involved provides some confidence in these general recommendations. They are set high enough to be on the plateau of yield in most situations, but not so high as to depress commercial cane sugar (CCS) upon which growers' payments depend. A significant point in relation to the thrust of this paper is that growers appear to chose to 'err on the high side' of these recommendations, sometimes to a significant degree. N fertiliser use has expanded more rapidly than has cane yield, e.g. for the Herbert (Fig. 13.2a), and data for a number of regions indicate that an N rate higher than current recommendations is used on a significant proportion of farms (Fig. 13.2b).

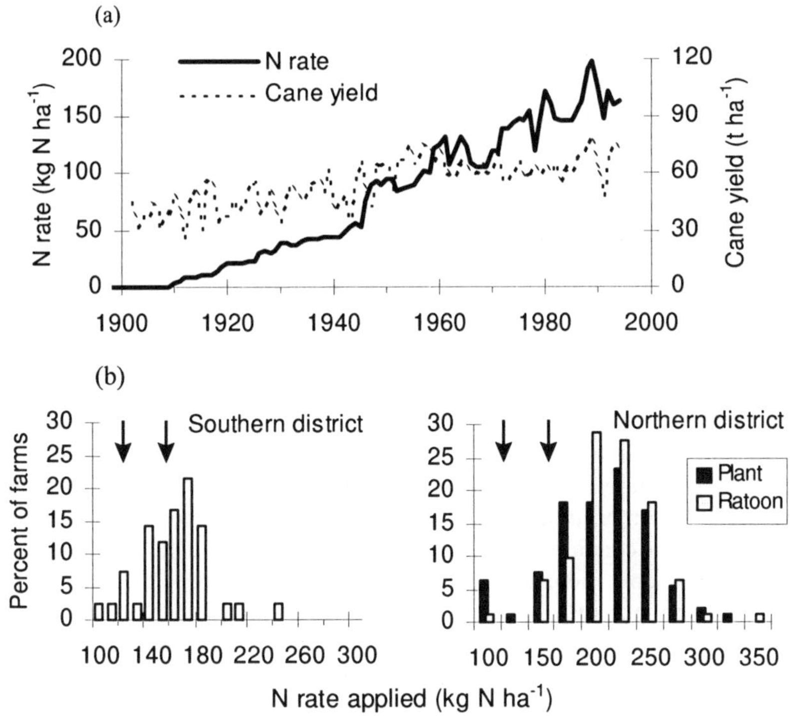

Fig. 13.2. Examples of N fertiliser use in the Australian sugar industry: (a) historical N inputs into the Herbert River district compared with cane yields (after Johnson, 1997) and (b) variation in N rate applied for farms in a southern (T. Wilcox, unpublished data) and northern (Herbert Cane Protection and Productivity Board, unpublished data) cane growing district of Queensland. [Arrows indicate recommended rates]

In the next section, we explore the questions of cane response to N fertiliser, to understand better what might motivate growers to 'err on the high side' of N fertilisation rates, and to explore some of the non-production implications of such practices over the long-term.

Analysis of response of sugarcane to N fertilisation

Because of the variability in weather, soil and management practice, we have integrated in a model framework current knowledge of sugarcane growth, response to N and N behaviour in the soil. Previous experience (Keating *et al.*, 1991; Probert *et al.*, 1995) has shown that the combination of experimentation and model development/application provides valuable insights into system performance with respect to N management.

The model framework

The APSIM systems model (Agricultural Production systems SIMulator; McCown *et al.*, 1996) has a structure in which crops and major soil processes are dealt with in separate modules. Modules can be chosen that reflect the particular system under study. The soil-water-solute model SWIMv2 (Verburg *et al.*, 1997), which is based on the Richards' and advection-dispersion equations, has been interfaced with APSIM (Huth *et al.*, 1996) to allow a detailed treatment of the movement of water and solutes in the soil-plant-atmosphere continuum. Performance of SWIMv2 with respect to modelling water and solute flux has been reported by Verburg (1997).

Reports of crop model performance can be found in Muchow *et al.* (1997), Robertson *et al.* (1997) and Keating *et al.* (1994b, 1997b). Performance of the APSIM-SWIM combination with respect to both water and solute flux and N uptake by sugarcane crops has been reported by Verburg *et al.* (1996). Figure 13.3 provides an example of model performance at a site in north Queensland. The example demonstrates the performance of the crop module in N sufficient and N deficient conditions, as well as the performance of the soil N module in terms of N supply from fertiliser and soil organic sources over two years of cropping. The lack of response to N in the first year of an N trial is common, particularly for plant crops. The response is strong in year 2 because soil organic N pools have been depleted. These soil organic matter processes are addressed within the modelling framework and appear adequate for simulation of the observed phenomena.

Simulation methods

APSIM was configured with the soil-water-solute module SWIMv2, a sugar crop module (Keating *et al.,* 1997b), a surface residue module (Probert *et al.,* 1997) and a soil N module (SOILN, Probert *et al.,* 1997). The climate used was that of Ingham (18.65°S, 146.17°E) in north Queensland and daily rainfall data were used from 1895 to 1994. Procedures used in estimating temperature and radiation are outlined by Muchow *et al.* (1997). The soil simulated was 1.2m deep, freely draining and had a plant available water capacity of 150mm. The standard profile contained 1% organic C with a C:N ratio of 14 in the surface layers. Initial mineral N and organic C levels were varied to establish three soil states of differing N status. Growth and N balance of a first ratoon crop under either rainfed or full irrigation conditions were simulated over the climate record. The soil and crop status was re-initialised each year to eliminate carry-over effects via the soil, and to focus on the impact of climate alone. Effects of N fertilisation from 0 to 300kg N ha^{-1} y^{-1} were examined.

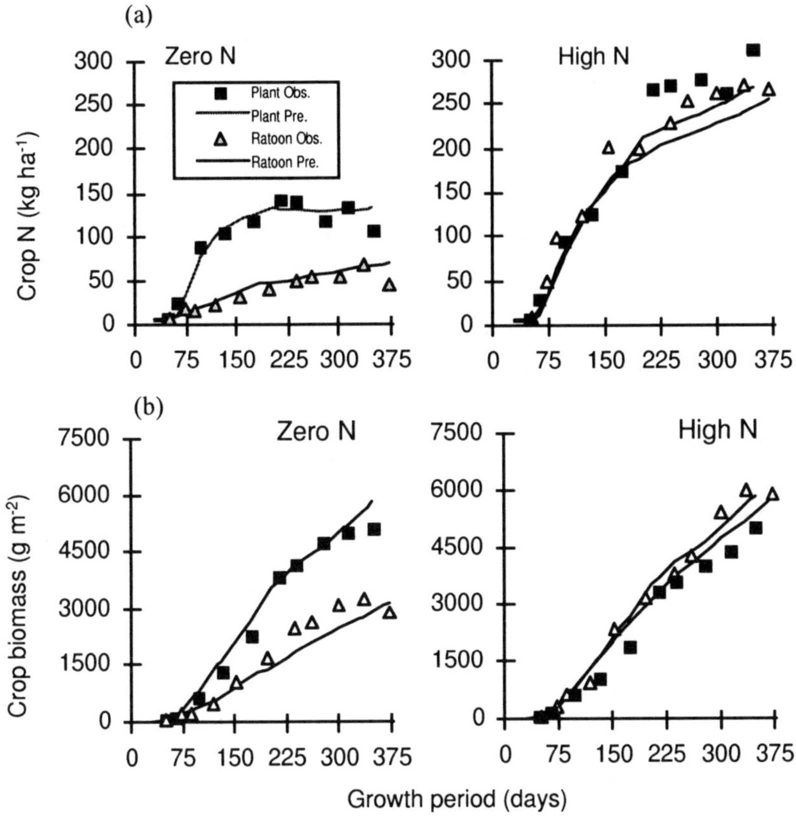

Fig. 13.3. Comparison of observed and predicted time trends for (a) N content and (b) biomass of N deficient (zero N) and N sufficient (high N) plant and ratoon crops of sugarcane grown in the Herbert River district in 1992-1993. [Data from Muchow *et al.*, 1996]

A number of assumptions are implicit in the simulations. Firstly, no volatilisation of N was simulated and hence the simulations are for a situation akin to incorporated fertiliser in a soil with moderate cation exchange capacity (Biggs *et al.*, 1996). Secondly, we did not consider the effects of N supply on CCS, an important component of the cane payment system in Australia. While effects are possible, they have been inconsistent and the crop models are not capable at present of reliable simulation of this character. Finally, the simulated crops were not limited by pests, disease, weeds, cyclone damage, flooding or waterlogging. Hence we may expect the yields simulated to exceed actual district averages.

Methods of output analysis

Modified Mitscherlich functions (eq. 1) were fitted to simulated cane yield responses to N fertiliser. The modification involved adding an exponent (d) to the fertiliser rate, allowing for a sigmoidal component of the N response curve (Barrow and Mendoza, 1990).

$Y = Yo - b * \exp(-c*N^d)$ Eq. (1)
where Y = cane yield ($t\ ha^{-1}$), N = N fertilisation rate ($kg\ N\ ha^{-1}$),
Yo = yield in the absence of N fertiliser, and b, c and d are fitted constants.

Gross margins (GM) were calculated as:
$GM = Y * (P_y - HC) - N * P_n$ Eq. (2)
where P_y = cane price ($\$\ t^{-1}$), P_n = N fertiliser price ($\$\ kg^{-1}$ of N) and
HC = harvest costs ($\$\ t^{-1}$).

The differential of equation (2) was solved for ${}^{dGM}/_{dN} = 0$ and an economic optimum N rate calculated (N_{opt}). We have valued cane at \$30 t^{-1}, fertiliser at \$1 kg^{-1} of N and harvesting costs at \$6.50 t^{-1} cane.

Simulated response to N

Average responses over the climate record

The average (over 100y) rainfed cane yield of 101t ha^{-1} simulated for non-limiting fertility (Fig. 13.4) was consistent with local experience and so was that with full irrigation (164t ha^{-1}) (Muchow *et al.*, 1996, 1997). Marginal returns from N fertilisation were calculated (Fig. 13.4b) and the economic optimum N rate (N_{opt}) was, as expected, different for different production levels (rainfed vs irrigated). Note, as N rate exceeds N_{opt}, production reaches a plateau and economic returns fall slightly, but losses such as leaching increase substantially (Fig. 13.4c). The higher leaching under the rainfed system, whilst somewhat counter intuitive, arises from both the lower crop growth and crop demand in that system and the fact that an efficient irrigation practice was assumed in the simulation.

The effect of initial soil fertility on cane response to N fertiliser is examined in Fig. 13.5. Without added fertiliser, the three states of N fertility analysed supported average cane yield under full irrigation ranging from 36 to 133t ha^{-1} (Fig. 13.5a). These initial soil fertility levels were chosen to span the range typical of cane production zones (Wood *et al.*, 1997). Clearly, more fertile soils will be less responsive to input of fertiliser N, a lower N rate will be economically optimal (Fig. 13.5b), and higher leaching loss will be likely as N rate exceeds the economic optimum (Fig. 13.5c).

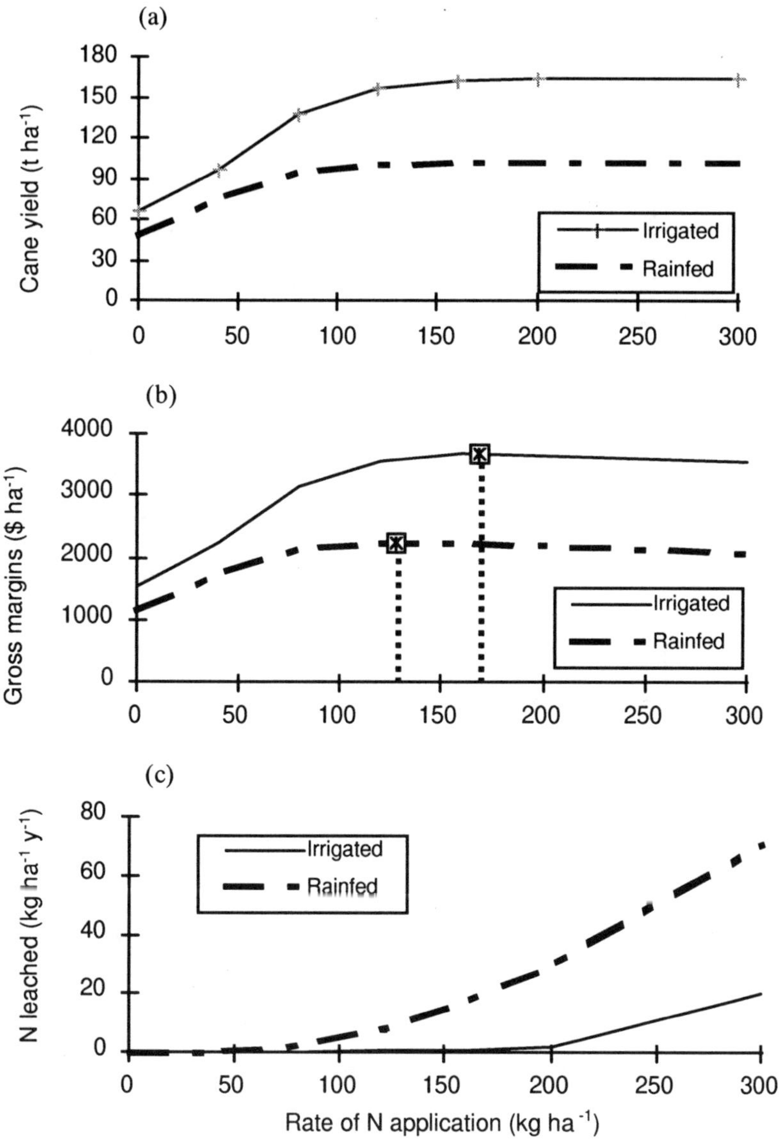

Fig. 13.4. Simulated response of a sugarcane crop to N fertiliser averaged over 100 years of Ingham weather under either rainfed or fully irrigated conditions: (a) cane yields, (b) gross margins (broken vertical line indicates N_{opt}), and (c) N leaching.

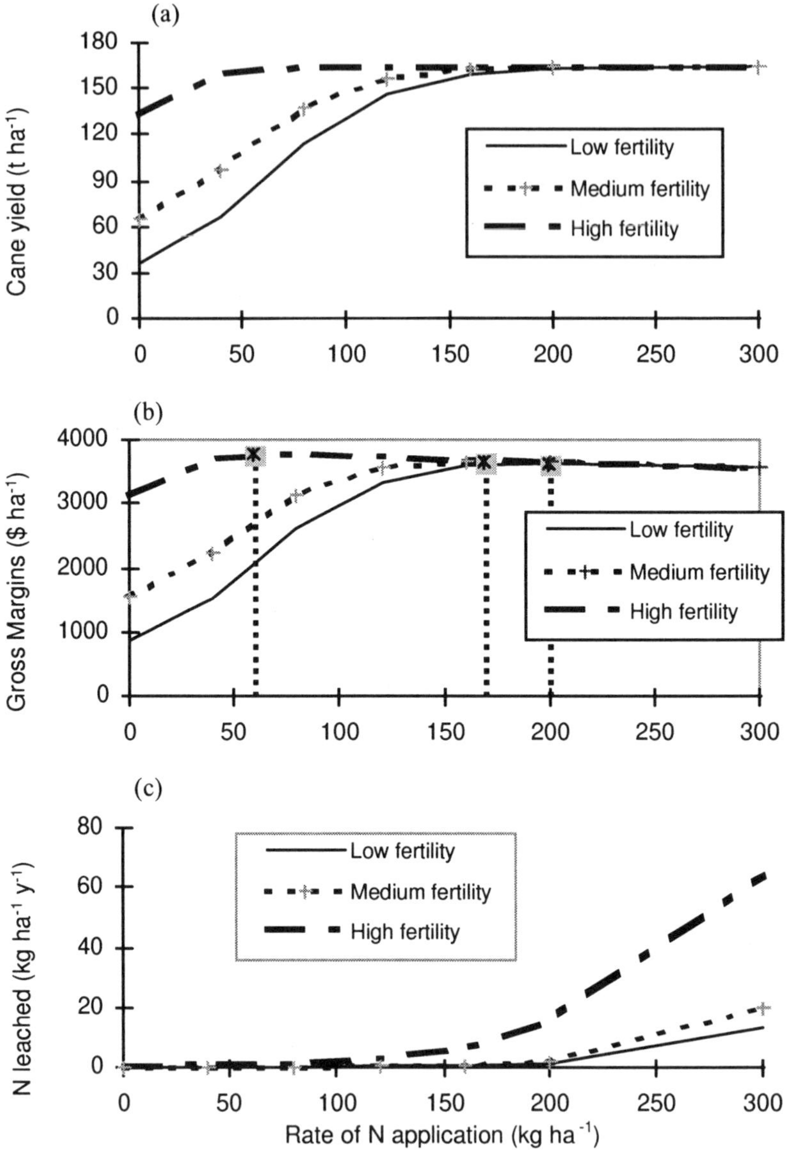

Fig. 13.5. Simulated response of sugarcane to N fertiliser averaged over 100 years at Ingham under three levels of soil fertility. (High: 1.0% C in surface soil, 110 kg min-N ha^{-1} in profile at ratooning; Medium: 1.0% C , 25kg min-N ha^{-1}; Low: 0.6% C, 25kg min-N ha^{-1}). (a) Cane yields, (b) Gross margin [vertical broken lines indicate N_{opt}], and (c) N leaching.

Variability in N response

Variability in response, summarised in terms of variations in N_{opt}, was greater for the rainfed treatment (Fig. 13.6a) than under full irrigation (Fig. 13.6b). These ranged from 40 to 200kg ha^{-1} in the rainfed system and 150 to 200kg ha^{-1} under full irrigation. Because of a close association between crop yield and N uptake, variation in yield dominated variability in N response, and a strong relationship is simulated between optimum N rate and yield when N was non-limiting (Fig. 13.7).

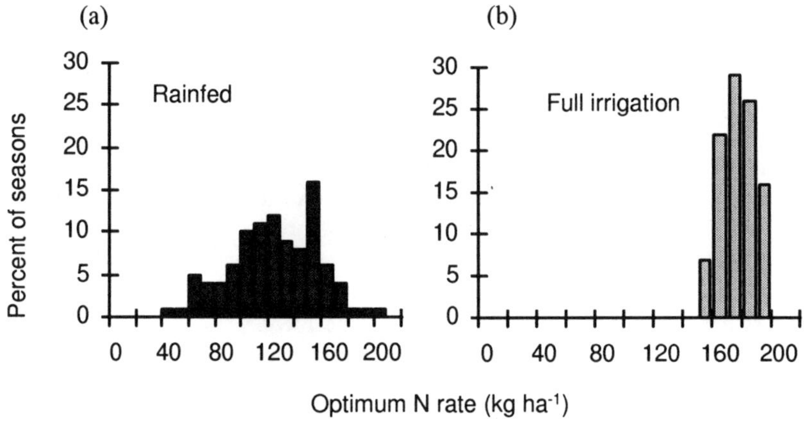

Fig. 13.6. Variability in optimum N rate (N_{opt}) simulated for (a) rainfed and (b) irrigated sugarcane production over 100 years at Ingham, Qld.

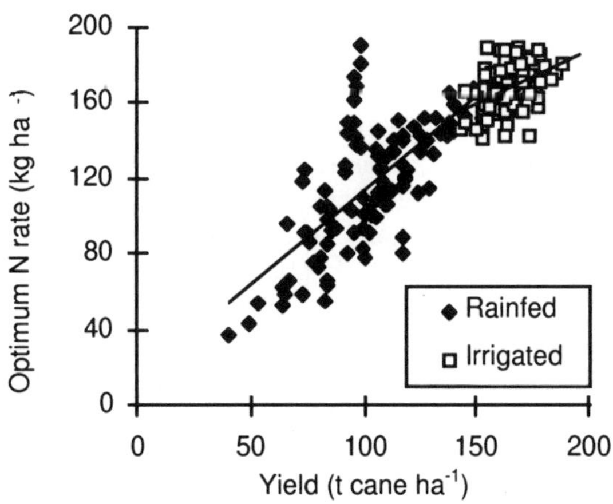

Fig. 13.7. Relationship between optimum N rate and yield when N was non-limiting, simulated at Ingham, Qld under rainfed and irrigated conditions.

Implications for N management in the sugarcane industry

There is a long history of agronomists and economists studying yield / fertiliser responses (e.g. Colwell, 1970; Dillon and Anderson, 1990; Bullock and Bullock, 1994). The bio-economic literature consistently refers to the relative insensitivity of economic returns at levels of input around the optimum (Anderson, 1975; Jardine, 1975; White, 1975). In other words, the response functions are relatively flat around the optimum and there is little to be gained by investing management and research resources in 'fine-tuning' N management. Jardine (1975) demonstrated that in many situations, one half the optimal fertiliser rate generates 90% of the gross margins. Anderson (1975) summed up the economic argument with respect to optimal levels of decision variables succinctly in the quote: *'precision is pretence and great accuracy is absurdity'*.

Our simulation studies are consistent with this view. Fertiliser N rates between 97 and 160kg ha^{-1} generated returns above 95% of the optimum (Fig. 13.4b). The higher the value of the product, relative to the costs of the fertiliser inputs, the greater the insensitivity of return to the precise level of fertiliser input. These economic forces contribute to the phenomenon that the ratio of N input to N offtake in the harvested product increases substantially as one moves from extensive to intensive agricultural systems (Table 13.2).

Use of 'supra-optimal' levels of N fertiliser by many canegrowers in Australia would appear to stem from:

1. Growers appreciating that returns are very insensitive to N rates above the economic optimum, but are more sensitive to N rates below the optimum.
2. Growers perceiving levels of risk and uncertainty that encourage the use of rates high enough to ensure sub-optimal inputs never arise.

If economic returns were the only consideration in determining N_{opt} for sugarcane, there would be little case for attempting to fine tune rates. However, increasing concern for the resource base and the environment will motivate growers to ensure that N fertilisation rates are not excessive. Given these opposing forces, we argue that more precise N fertiliser management, that better reflects soil supply and crop demand, is a goal worth striving for in the sugarcane industry. The simulation (Fig. 13.8) highlights the potential for N losses with excessive N use. N_{opt} is closely related to likely cane yield and there are opportunities to better promote this concept within the industry. For N applied efficiently (eg. no volatilisation losses), the rule of thumb of 1.4kg N t^{-1} of cane should be adequate for the first 100t of crop and 1.0kg N for each tonne above that (Fig. 13.9). The simulation study also reinforces the existing recommendations that prescribe lower N rates for more fertile soils (Fig. 13.5) and plant crops (data not shown).

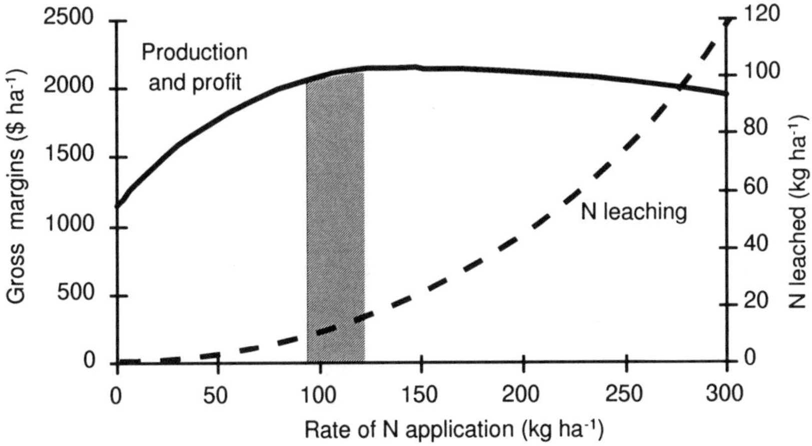

Fig. 13.8. Average gross margins and N leaching simulated in a plant and 4 ratoon cycle at Ingham, Qld over 100y of climate data. [Hatched area indicates a range of N rates that achieve both optimal economic returns and acceptable N leaching losses]

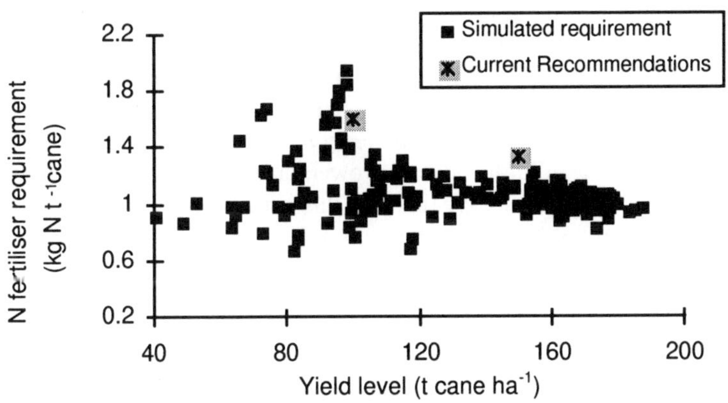

Fig. 13.9. 'Rule of thumb' measure of N fertiliser required to produce one tonne of cane. [Estimated from simulated N_{opt} and cane yield over 100 years at Ingham, Qld. Points denoted as current recommendations relate to industry recommendations (Calcino, 1994) for 100 and 150t ha^{-1} crops]

The conclusions from the simulation study differ from current industry recommendations more in degree than substance. Whilst simulated N_{opt} values are c. 20% lower than current recommendations we suspect the latter have a degree of risk avoidance built into them, and were developed before there arose an environmental dimension to the N fertilisation issue. Some growers appear to opt for their own element of insurance by using higher than recommended rates.

While N process research, and the model construction that has accompanied it, has provided results largely consistent with current recommendations, its value lies in providing a more comprehensive understanding of dynamics of N within the sugarcane crop-soil system. This understanding provides the basis for both improvements in technology and the knowledge-base of growers and advisers.

Future Directions for N Research and Management in Sugarcane

We see four directions for future work. Firstly, assessment of any negative impacts on the resource base or the environment. Such work is underway in a number of locations in the Australian sugar industry, notably for surface waters in the South Johnstone (McShane *et al.*, 1993) and Herbert (Bramley and Johnston, 1997) catchments, and for groundwaters in Bundaberg (Keating *et al.*, 1997a).

Secondly, expanding the knowledge that growers themselves have of the N cycle and the implications of excessive N fertilisation not only in terms of their production/profits, but also in terms of any negative effects on their soils or the wider environment. We see this as a priority and believe the models can contribute to this learning experience.

Thirdly, achieving better efficiency of N use through new products or methods of application. Nitrification inhibitors and slow release fertiliser formulations have offered promise for some time but to date the economics appear unattractive for widespread application. Trickle irrigation and fertigation provide a quantum leap in the control growers have over N supply but < 1% of the crop is trickle irrigated. We believe this area will grow and offers good prospects for control of N leaching loss in sensitive areas.

A fourth aspect that warrants attention is the monitoring of crop or soil N status as a basis for improved decision making. Crop monitoring has proven valuable in various crop systems, e.g. sap nitrate in cereal (Elliot *et al.*, 1993) and horticultural crops (Olsen and Lyons, 1994), and the chlorophyll meter in maize (Schepers *et al.*, 1992). Logistical difficulties with N application during later growth and the potential for such applications to impact negatively on CCS suggest there is limited scope for within-season tactical adjustments to N management of sugarcane. The insensitivity of the system to current fertiliser application because of the 'buffering' role of mineralisation of soil organic matter as a source of N for the current crop (Keating *et al.*, 1993) reinforces this judgement. Strategic assessments at a block-specific level as a basis for next year's crop hold greater promise. Soil testing forms part of the South African fertiliser recommendation scheme (Moberly *et al.*, 1984). However, predictive tests remain unproven for sugarcane in Australia. Current research in Australia indicates that the analysis of nitrogenous compounds in sugarcane juice at the mill holds promise as a basis of identifying under- or over-fertilisation (B.A. Keating, 1997, unpublished data).

Investment in additional information specific to a particular management decision (e.g. crop or soil testing prior to fertilising a particular paddock) can be viewed as an effort to reduce uncertainty. The benefit of such an investment will depend on the cost of gathering information relative to the benefit for improved decision making. With respect to the later, the flatness of the economic response curves around the optimum has limited the attractiveness of elaborate monitoring schemes in other systems, e.g. wheat / lupins in Western Australia (D. Pannell, *pers. comm.*) and wheat in the north-east grains region, (P. Hayman, *pers. comm.*). However, the need to reduce negative consequences for the environment may mean that such investment is warranted for N fertiliser management of sugarcane.

General Prognosis for N Research and Management

Research on N in agriculture has been substantial, reflecting its importance for crop productivity, and its potential negative resource and environmental consequences. The issues are complex, involving crop and soil dimensions, with a large number of gain, loss and transformation processes operating within a variable climatic regime and over many diverse soil types. Time scales of significance range from days for an N leaching event to decades and beyond for changes in soil organic matter content.

Farmers need to manage this system with all its diversity, variability, uncertainty and complexity and they generally do this very well. They attempt to optimise management of their system within the parameters that they know about and that are of concern to them, and the constraints they operate under. This is true whether it be for subsistence maize farmers in semi-arid Africa (Keating *et al.*, 1991) or for the group that is our focus in this paper, intensive sugarcane growers on the wet tropics of Australia.

Researchers also need to confront the same issues of variability and complexity. They do this in part through controlled experiments, and as this paper demonstrates, through quantitative modelling to integrate what we know about individual processes, and address important questions of system performance.

This review has focused on short-term dynamics of N in sugarcane systems. Much longer-term periods, e.g. 10 to 100 years, are needed for issues such as acidification and shifts in soil organic matter to fully develop. Models like that deployed here also have an important role to play as a framework for the investigation of these long term effects (Vallis *et al.*, 1996).

If researchers have a role in improving farmers' **management** of nitrogen (as distinct from developing new technologies), it will most likely be in broadening the knowledge base upon which farmers optimise their management. In the context of sugarcane production systems, this will be in identifying and communicating the tradeoffs between productivity, profitability, sustainability of the resource base, and environmental impact.

Acknowledgments

The authors acknowledge the contribution of Dr M.E Probert to the simulation study and for constructive review of this manuscript. Support for experimental and modelling activities that have contributed to our understanding of the nitrogen cycle for sugarcane has come from the Sugar Research and Development Corporation and the Land and Water Resources Research and Development Corporation. We acknowledge the Bundaberg and Herbert Productivity Boards for their efforts in assessing fertiliser input levels and their assistance in making these data available for this review.

References

Addiscott, T.M. Whitmore, A.P. and Powlson, D.S. (1991) *Farming, Fertilizers and the Nitrate Problem.* CAB International, Wallingford, UK, 170pp.

Anderson, J.R. (1975) One more or less cheer for optimality. *Journal of the Australian Institute of Agricultural Science* 41, 195-197.

Barrow, N.J. and Mendoza, R.E. (1990). Equations for describing sigmoid yield responses and their application to some phosphate responses by lupins and by subterranean clover. *Fertilizer Research*, 22, 181-188.

Biggs, J.S., Vallis, I., Kokot, S. and Keating, B.A. (1996) Effect of depth of urea application on loss of nitrogen by volatilisation from acid soils. In: Wilson J.R., Hogarth, D.M., Campbell, J.A. and Garside, A.L. (eds), *Sugarcane: Research Towards Efficient and Sustainable Production.* CSIRO Division of Tropical Crops and Pastures, Brisbane, Australia, pp. 198-199.

Bock, B.R. and Hergert, G.W. (1991) Fertilizer nitrogen management. In: Follett, R.F., Keeney, D.R. and Cruse, R.M. (eds), *Managing Nitrogen for Groundwater Quality and Farm Profitability.* Soil Science Society of America, Madison, USA, pp.139-164.

Bramley, R.G.V. and Johnson, A.K.L. (1997). Land use impacts on nutrient loading in the Herbert River. In: Hunter, H.M., Eyles, A.G. and Rayment, G.E. (eds), *Downstream Effects of Land Use.* Department Natural Resources, Brisbane, Australia, pp. 93-96.

Bullock, D.S. and Bullock, D.G. (1994) Calculation of optimal nitrogen fertilizer rate. *Agronomy Journal* 86, 921-923.

Bumb, B.L. (1995). World nitrogen supply and demand: An overview. In: Bacon, P.E. (ed.), *Nitrogen Fertilization in the Environment.* Marcel Dekker, New York, USA, pp. 1-41.

Calcino, D. (1994) *Australian Sugarcane Nutrition Manual.* BSES and SRDC, Brisbane, Australia, 60pp.

Catchpoole, V.R. and Keating, B.A. (1995) Sugarcane yield and nitrogen uptake in relation to profiles of mineral-nitrogen in the soil. *Proceedings of the Australian Society of Sugar Cane Technologists* 17, 187-192.

Chapman, L.S. (1982) Estimating sugar yield responses from N, P and K fertilizers in Queensland. *Proceedings of Australian Society of Sugarcane Technologists* 4, 147-153.

Chew, P.S. and Pushparajah, E. (1995) Nitrogen management and fertilization of tropical plantation tree crops. In: Bacon, P.E. (ed.), *Nitrogen Fertilization in the Environment*, Marcel Dekker, New York, USA, pp. 225-294.

Colwell, J.D. (1970) Precision in the estimation of the fertilizer requirements of wheat: Need, feasibility and economic desirability. *Journal of the Australian Institute of Agricultural Science* 36, 273-278.

Dalal, R.C. and Mayer, R.J. (1987) Long-term trends in fertility of soils under continuous cultivation and cereal cropping in southern Queendland. VII. Dynamics of nitrogen mineralisation potentials and microbial biomass. *Australian Journal Soil Research* 25, 83-93.

Dillon, J.L. and Anderson, J.R. (1990) *The Analysis of Response in Crop and Livestock Production.* 3rd Edition, Pergamon Press, Oxford, UK, 250pp.

Duxbury, J.M. (1994) The significance of agricultural sources of greenhouse gases. *Fertilizer Research* 38, 151-163.

Elliot, D.E., Pelham, S.D. and Reuter, D.J. (1993) Synchronising diagnosis and correction of nitrogen deficiency in barley grown in semi-arid environments. In: Barrow, N.J. (ed.), *Plant Nutrition - from Genetic Engineering to Field Practice.* Kluwer Academic Publishers, Dordrecht, Netherlands, pp. 475-478.

Engels, C. and Marschner, H. (1995) Plant uptake and utlization of nitrogen. In: Bacon, P.E. (ed.), *Nitrogen Fertilization in the Environment.* Marcel Dekker, New York, USA, pp. 41-82.

Galbally, I.E., Fraser, P.J., Meyer, C.P. and Griffith, D.W.T. (1992) Biosphere-atmosphere exchange of trace gases over Australia. In: Gifford, R.M. and Barson, M.M. (eds), *Australia's Renewable Resources: Sustainability and Global Change.* Bureau of Rural Resources, Proceedings No. 14, Canberra, Australia, pp. 117-149.

Helyar, K.R. (1976) Nitrogen cycling and soil acidification. *Journal of the Australian Institute of Agricultural Science* 42, 217-221.

Huth, N.I., Keating, B.A., Bristow, K.L., Verburg, K. and Ross, P.J. (1996) SWIMv2 in APSIM: An integrated plant, soil, water and solute modelling framework. *Proceedings 8th Australian Agronomy Conference*, 1996, Toowoomba, Australia, pp. 667.

Jardine, R. (1975) Two cheers for optimality. *Journal of the Australian Institute of Agricultural Science* 41, 30-34.

Jenkinson, D.S. (1991) The Rothamsted long-term experiments: Are they still of use? *Agronomy Journal* 83, 2-10.

Johnson, A.K.L. (1997) Quantifying current and historical inputs of fertilizer N and P to rural lands in the Herbert River Catchment. In: Hunter, H.M., Eyles, A.G. and Rayment, G.E. (eds), *Downstream Effects of Land Use.* Department Natural Resources, Brisbane, Australia, pp. 273-275.

Keating, B.A., Bauld, J., Hillier, J., Ellis, R., Weier, K.L., Sunners, F. and Connell, D. (1997a) Leaching of nutrients and pesticides to Queensland groundwaters. In: Hunter, H.M., Eyles, A.G. and Rayment, G.E. (eds), *Downstream Effects of Land Use.* Department Natural Resources, Brisbane, Australia, pp. 151-164.

Keating, B.A., Godwin, D.C. and Watiki, J.M. (1991) Optimization of nitrogen inputs under climatic risk. In: R.C. Muchow and J.A. Bellamy (eds), *Climatic Risk in Crop*

Production - Models and Management for the Semi-Arid Tropics and Sub-Tropics. CAB International, Wallingford, UK, pp 329-357.

Keating, B.A., Liu, D., Inman-Bamber, G., Robertson. M.J., Muchow, R.C. and Kingston, G. (1994b) Comparison of levels of approach in modelling sugarcane potential yield. In: Robertson, M.J. (ed.), *Proceedings Workshop on Research and Modelling Approaches to Assess Sugarcane Production Opportunities and Constraints.*. ISBN 0 643 05768 4, University of Queensland., Brisbane, Australia, pp. 43-52.

Keating, B.A., Robertson, M.J., Muchow, R.C. and Huth, N.I. (1997b) Modelling sugarcane production systems.. 1. Description and validation of the sugarcane module. *Field Crops Research*, In prep.

Keating, B.A., Vallis, I., Hughes, M. and Ridge, D.R. (1993). Strategic directions for nitrogen research - a view from the south. *Proceedings of the Australian Society of Sugar Cane Technologists* 15, 276-284.

Keating, B.A., Vallis, I., Parton, W.J., Catchpoole, V.R., Muchow, R.C. and Robertson, M.J. (1994a) Modelling and its application to nitrogen research and management for sugarcane. *Proceedings of the Australian Society of Sugar Cane Technologists* 16, 131-142.

Madison, R.J. and Brunett, J.O. (1985) Overview of reoccurrence of nitrate in groundwater of the United States. *US Geological Survey, Water Supply Paper 2275.* US Government Print Office, Washington DC, USA.

McCown, R.L., Cogle, A.L., Ockwell, A.P. and Reeves, T.G. (1988). Nitrogen supply to cereals in legume ley systems under pressure. In: Wilson, J.R. (ed.), *Advances in Nitrogen Cycling in Agricultural Ecosystems.* CAB International, Wallingford, UK, pp. 292-394.

McCown, R.L., Hammer, G.L., Hargreaves, J.N.G. Holzworth, D.P. and Freebairn, D.M. (1996) APSIM: a novel software system for model development, model testing and simulation in agricultural systems research. *Agricultural Systems* 50, 255-271.

McShane, T.J., Reghenzani, J.R., Prove, B.G. and Moody, P.W. (1993). Nutrient balances and transport from sugar cane land - a preliminary report. *Proceedings of the Australian Society of Sugar Cane Technologists* 15, 269-275.

Meisinger, J.J. and Randall, G.W. (1991) Estimating nitrogen budgets for soil-crop systems. In: Follett, R.F., Keeney, D.R. and Cruse, R.M. (eds), *Managing Nitrogen for Groundwater Quality and Farm Profitability*, Soil Science Society of America, Madison, USA, pp. 85-124.

Miller, R.J. and Smith, R.B. (1976) Nitrogen balance in the southern San Joaquin Valley. *Journal Environmental Quality* 5, 274-78.

Moberly, P.K., Wood, R.A., Meyer, J.H., Leibbrandt, N.B. and McIntyre, R.K. (1984) Soil: an important factor to consider when making N recommendations. *Proceedings Nitrogen Symposium*, Soils Irrigation Research Institute, Pretoria, South Africa, pp. 10-12.

Mosier, A.R. (1994) Nitrous oxide emissions from agricultural soils. *Fertilizer Research* 37, 191-200.

Muchow, R.C. and Davis, R. (1988) Effect of nitrogen supply on the comparative productivity of maize and sorghum in a semi-arid tropical environment. *Field Crops Research* 18, 17-30.

Muchow, R.C., Robertson, M.J. and Keating, B.A. (1997) Limits to the Australian sugar industry: climatic and biological factors. In: Keating, B.A. and Wilson, J.R. (eds),

Intensive Sugarcane Production: Meeting the Challenges Beyond 2000. CAB International, Wallingford, UK, pp. 37-54.

Muchow, R.C., Robertson, M.J., Wood, A.W. and Keating, B.A. (1996) Effect of nitrogen on the time-course of sucrose accumulation in sugarcane. *Field Crops Research* 47, 143-153.

Neeteson, J.J. (1995) Nitrogen management for intensively grown arable crops and field vegetables. In: Bacon, P.E. (ed.), *Nitrogen Fertilization in the Environment*, Marcel Dekker, New York, USA, pp. 295-326.

NHMRC (1994) *Australian Drinking Water Guidelines.* National Health and Medical Research Council / Agricultural and Resource Management Council of Australia and New Zealand. Draft, June 1994, Canberra, Australia.

Olsen, J.K. and Lyons, D.J. (1994) Petiole sap nitrate is better than total nitrogen in dried leaf for indicating nitrogen status and yield responsiveness of capsicum in subtropical Australia. *Australian Journal Experimental Agriculture* 34, 835-844.

Probert, M.E., Dimes, J.P., Keating, B.A., Dahal, R.C. and Strong, W.M. (1997) APSIM's water and nitrogen modules and simulation of the dynamics of water and nitrogen in fallow systems. *Agricultural Systems* (in press).

Probert, M.E., Keating, B.A., Thompson, J.P. and Parton, W.J. (1995) Modelling water, nitrogen and crop yield for a long-term fallow management experiment. *Australian Journal Experimental Agriculture* 35, 941-950.

Rauschkolb, R.S. and Hornsby, A.G. (1994) *Nitrogen Management in Irrigated Agriculture.* Oxford Univ. Press, Oxford, UK, 247 pp.

Robertson, M.J., Inman-Bamber, N.G. and Muchow, R.C. (1997) Opportunities for optimising the use of limited water by the sugarcane crop. In: Keating, B.A. and Wilson, J.R. (eds), *Intensive Sugarcane Production: Meeting the Challenges Beyond 2000.* CAB International, Wallingford, UK, pp. 287-304.

Sanchez, E.E., Khemira, H., Sugar, D. and Righetti, T.L. (1995) Nitrogen management in orchards. In: Bacon, P.E. (ed.), *Nitrogen Fertilization in the Environment*, Marcel Dekker, New York, USA, pp. 327-380.

Schepers, J.S. and Fox, R.H. (1989) Estimation of N budgets for crops. In: Follett, R.F. (ed.), *Nitrogen Management and Groundwater Protection*, Elsevier, Amsterdam, Netherlands, pp. 221-246.

Schepers, J.S., Francis, D.D., Vigil, M.F. and Below, F.E. (1992) Comparison of corn leaf nitrogen and chlorophyll meter readings. *Communications in Soil Science and Plant Analysis* 23, 2173-2187.

Sinclair, T.R. (1997) Yield 'plateaus' in grain crops: the topography of yield increase. In: Keating, B.A. and Wilson, J.R. (eds), *Intensive Sugarcane Production: Meeting the Challenges Beyond 2000.* CAB International, Wallingford, UK, pp. 87-102.

Stoorvogel, J.J., Smaling, E.M.A. and Janssen, B.H. (1993) Calculating soil nutrient balances in Africa at different scales I. Supra-national scale. *Fertilizer Research* 35, 227-236.

Vallis, I. and Keating, B.A. (1994) Uptake and loss of fertilizer and soil nitrogen in sugarcane crops. *Proceedings of the Australian Society of Sugar Cane Technologists* 16, 105-113.

Vallis, I, Parton, W.J., Keating, B.A. and Wood, A.W. (1996) Simulation of the effects of trash and fertilizer management on soil organic matter levels and yields of sugarcane. *Soil and Tillage Research* 38, 115-132.

Verburg, K. (1997) *Methodology in Soil-Water-Solute Balance Modelling: An Evaluation of the SoilWat and SWIMv2 Models*, Report of workshop, Brisbane, Australia, 16-18 May 1995, CSIRO Australia Division of Soils Divisional Report, 87 pp. (in press).

Verburg, K., Keating, B.A., Bristow, K.L., Huth, N.I., Ross, P.J. and Catchpoole, V.R. (1996) Evaluation of nitrogen fertilizer management strategies in sugarcane using APSIM-SWIM. In: Wilson J.R., Hogarth, D.M., Campbell, J.A. and Garside, A.L. (eds), *Sugarcane: Research Towards Efficient and Sustainable Production.* CSIRO Division of Tropical Crops and Pastures, Brisbane, Australia, pp. 200-202.

Verburg, K., Ross, P.J. and Bristow, K.L. (1997) *SWIMv2 Manual.* CSIRO Division of Soils Divisional Report, CSIRO, Australia, 93 pp. (in press).

White, D.H. (1975) The search for an optimal stocking rate: Continuing the saga. *Journal of the Australian Institute of Agricultural Science* 41, 192-194.

Wood, A.W., Bramley, R.G.V., Meyer, J.H. and Johnson, A.K.L. (1997). Opportunities for improving nutrient management of sugarcane. In: Keating, B.A. and Wilson, J.R. (eds), *Intensive Sugarcane Production: Meeting the Challenges Beyond 2000.* CAB International, Wallingford, UK, pp. 243-266.

Chapter fourteen:

Opportunities for Improving Nutrient Management in the Australian Sugar Industry

A.W. Wood[1,2], R.G.V. Bramley[1,4], J.H. Meyer[3] and A.K.L. Johnson[1,5]

1. *CRC for Sustainable Sugar Production*
2. *CSR Sugar Mills, PMB, Ingham, Qld 4850, Australia*
3. *SASEX, Mt Edgecombe 4300, Natal, South Africa*
4. *CSIRO Division of Soils and 5. CSIRO Division of Tropical Crops and Pastures, Davies Laboratory, PMB, Aitkenvale, Qld 4814, Australia*

Abstract

The paper reviews nutrient management for sugarcane, particularly from an Australian perspective, and examines changes in fertilisation strategies over the last 40 years. Limitations of current approaches to nutrient management are evaluated through a critical analysis of nutrient management recommendations. Areas where knowledge is inadequate are noted and opportunities for improving the management of nutrients identified. These include the use of crop response models and precision agriculture technologies.

Introduction

Nutrient management for sugarcane, as for many intensive cropping systems, involves large inputs of nitrogen (N), phosphorus (P) and potassium (K) fertiliser. Fertilisation strategies are directed primarily at maximising production, focusing on supply to the crop rather than on managing the soil resource. Consequently, until recently, little research effort had been directed at the efficiency of fertiliser use, nutrient imbalances in the soil and crop, or the

In: Keating, B.A. and Wilson, J.R. (eds) *Intensive Sugarcane Production: Meeting the Challenges Beyond 2000*. CAB International, Wallingford, UK, 1997.

potential adverse effects of the application of high rates of a narrow range of nutrients. Examples of such effects are soil acidification, reductions in soil organic matter and cation exchange capacity (CEC), or possible environmental impacts of off-farm nutrient export.

This paper reviews current practice and research results relevant to nutrient management of sugarcane, and on the basis of that review, argues that there is a critical need to re-evaluate the way in which nutrients are supplied and managed for sugarcane production. The central case developed is for a system of nutrient management that focuses more on the physical, chemical and biological properties of soils and that embodies some of the principles of sustainable agriculture, whilst retaining a focus on increasing productivity. We argue for a shift in emphasis away from supplying large quantities of N, P and K to the crop, and instead propose an alternative philosophy which:

1. Focuses on the management of different soils to enhance their ability to store and supply a wide range of nutrients to the crop.
2. Takes into account the balance of different nutrients in the soil.
3. Emphasises the maintenance of high levels of soil organic matter.
4. Has the goal of the long term improvement of soil fertility through the enhancement of natural soil processes and nutrient cycles.

Current Nutrient Management Recommendations

The Australian sugar industry currently uses generalised fertiliser recommendations, with (apart from N), no specific recommendations for different regions, climatic conditions or soil types (Calcino, 1994). Whilst N recommendations are based on yield response functions developed in different canegrowing districts, fertiliser recommendations for other nutrients are based on soil analysis where the philosophy is to maintain soil nutrient levels between a critical value and a level above which no response to fertiliser can be expected (Kingston *et al.*, 1991). Other methods of deriving recommendations for cane growers are not usually employed, although foliar analysis was used by CSR in the 1960s for a fertiliser advisory service in north Queensland (Farquhar, 1965).

The N recommendations are based on regional yield responses from over 200 N rate trials (Chapman, 1994). Whilst large regional differences in response were observed, Chapman (1994) states that separate recommendations for different regions, apart from the Burdekin, could not be justified when mean yield responses were calculated for each district. The lack of regional differentiation may be the result of a failure to differentiate between soils, the statistical procedures used, the response functions fitted to the data or the use of relative rather than absolute yield. N fertiliser recommendations are lower for plant cane than for ratoons and are reduced in special situations, e.g. where dry conditions limit yield or where fertile alluvial soils reduce the N fertiliser requirement (Table 14.1). Recommendations are also reduced when low sugar prices prevail. Chapman (1994) justified his recommendations on economic grounds by calculating the probability of achieving returns on the money spent on N fertiliser. He did not factor into his calculations the potential cost of environmental damage arising from excessive N applications, nor did he account for the method of application or the inefficiency of N use. For plant and ratoon cane, c. 20-25% of the N is commonly applied as an NPK mixture at planting or soon after ratooning. The remainder is applied at around canopy closure, usually as urea. Recently, more growers have favoured the convenience of an NPK fertiliser for ratoons applied in 'a one shot' operation. Urea is by far the cheapest and most popular form of N fertiliser; however, when applied to the surface of green cane trash blankets, significant ammonia volatilisation losses can occur, typically c. 40% of N applied (Freney *et al.*, 1994).

Phosphorus fertiliser recommendations (Table 14.1) are based on a dilute sulphuric acid soil extraction known as the BSES-P test (Kerr and von Steiglitz, 1938). They differ according to expected crop yields but do not account for differences in soil P reaction between soils (Bramley *et al.,* 1995; Bramley and Wood, 1996). Most uncultivated soils or new cane areas are very low in P and require large initial applications of P fertiliser. Older cane areas have high soil P levels due to a long history of P application (Chapman *et al.*, 1981). The main form of P fertiliser now used is diammonium phosphate, usually as a constituent of NPK mixtures.

Potassium fertiliser recommendations are based on two soil analyses: exchangeable or readily available K and nitric acid extractable or reserve K (Table 14.1). Neither the CEC nor the proportion of the CEC occupied by K are taken into account. Most K fertiliser is muriate of potash (KCl) and is often applied as a constituent of NPK mixtures.

Calcium is usually applied as lime, gypsum or dolomite. Recommended lime rates are based on soil exchangeable Ca levels (Table 14.1), but do not consider measures of soil acidity such as soil pH, pH buffering capacity or exchangeable aluminium. Recommendations are focused solely on the crop requirement for Ca.

Magnesium is usually applied as magnesium oxide, dolomite or a blend of calcium and magnesium carbonate. Recommendations are based on soil exchangeable Mg levels (Table 14.1). As for K and Ca, neither soil CEC nor the % of the CEC occupied by Mg are taken into account.

Sulphur recommendations are based on a soil SO_4-S analysis (Table 14.1). Whilst most sugarcane soils used to have high S levels due to widespread use of ammonium sulfate and superphosphate, the change to high analysis N and P fertilisers meant that little or no S was being applied. More recently, most NPK fertiliser mixtures have been modified to include a small amount of S as NH_4 SO_4.

Table 14.1. Current nutrient management recommendations for sugarcane in Australia (Calcino, 1994)

Nitrogen

	Crop Class	N rate (kg ha^{-1}) Burdekin	N rate (kg ha^{-1}) Other regions
Sugar price < A$ 300 t^{-1}	Plant	135	120
	Ratoon & replant	210	160
Sugar price > A$ 300 t^{-1}	Plant	150	120-150
	Ratoon & replant	270	160-200
Dry land or rich land	Plant	-	80
regardless of sugar price	Ratoon & replant	-	120

Phosphorus

Soil P (mg kg^{-1})	Recommended P rate (kg ha^{-1}): Crop yield >120t ha^{-1} Plant & replant	Crop yield >120t ha^{-1} Ratoon	Crop yield <120t ha^{-1} Plant & replant	Crop yield <120t ha^{-1} Ratoon
<6	80	40	80	25
6-10	80	40	40	25
11-20	40	40	25	25
21-40	20	20	20	20
>40	20	nil	20	nil

Potassium

Soil K Level (me%)		K rate (kg ha^{-1}) Plant	K rate (kg ha^{-1}) Ratoon & replant
Low	exch. K <0.24 nitric K <0.2	100	120
Intermediate	exch. K <0.24 nitric K 0.2-0.6	80	100
High	exch. K >0.24 nitric K >0.6	K rate can be reduced in adverse economic times	

Calcium

Soil Ca level (me%)	Situation	Lime rate (t ha^{-1})	Ca supplied (kg ha^{-1})
Ca deficient <0.55	Likely response to Ca	5.0	1800
Ca marginal 0.55-1.25	Possible response to Ca	2.5	900
Ca adequate >1.25	Adequate Ca	nil	nil

Magnesium

Soil Mg level (me%)	Situation	Application rate of product	Mg supplied (kg ha^{-1})
Mg deficient <0.10	Likely response to Mg	MgO at 300kg ha^{-1} OR Dolomite at 1.6t ha^{-1} OR Ca/Mg blend at 3.2t ha^{-1}	160
Mg marginal 0.10-0.25	Possible response to Mg	MgO at 150kg ha^{-1} OR Dolomite at 0.8t ha^{-1} OR Ca/Mg blend at 2.6t ha^{-1}	80
Mg adequate >0.25	Adequate Mg	nil	nil

Sulphur

Soil sulfate-S level (mg kg^{-1})	Situation	S rate (kg ha^{-1})
0-2	Possibly S deficient	25
3-7	Possibly S marginal	10
>7	Possibly S adequate	nil

Zinc

Soil zinc level (mg kg^{-1})	Zn rate (kg ha^{-1})	Application rate of product
0-0.6	10	Zinc sulfate heptahydrate @ 44kg ha^{-1} OR zinc fortified planting mixtures at normal rate
>0.6	nil	nil

The only other nutrient for which recommendations have been developed is zinc (Table 14.1). These are based on a dilute hydrochloric acid soil extraction procedure (Reghenzani, 1993). No reliable indicators of nutrient deficiency have been developed for copper, manganese, iron, boron, molybdenum or silicon, although leaf tissue analysis is recommended for manganese and boron.

Long Term Trends in Fertiliser Use and Soil Fertility

Pulsford (1991, 1993) reviewed N and P application to coastal river catchments in Queensland over the last 80 years. These showed a rapid increase in fertiliser usage since the mid 1940s, particularly in those catchments with significant amounts of sugarcane. Johnson (1996) similarly reviewed N and P usage for sugarcane in the Herbert River catchment over the last 130 years (Fig. 14.1). Fertiliser application rates increased rapidly over the last 50 years, characterised by a series of peaks and troughs rather than by a steady increase. From 1970 to 1990, N rates increased from 120 to almost 200kg ha^{-1} and P from 20 to almost

30kg ha^{-1}. These values are consistent with a state-wide increase in average N application from 120 to 210kg ha^{-1} between 1966 and 1981 (BSES, 1985).

Wood (1984, 1985) suggested that increased fertiliser use in the Herbert over the late 1960s to the early 1980s may have been a response to a period of declining crop yields over that same period. This may also have been the case in other parts of Queensland as there was a plateauing in state sugar yields between 1972 and 1982 (BSES, 1985). The perception that cane yields can be increased by simply applying more fertiliser may still be widespread as current nutrient application rates are well in excess of recommended rates.

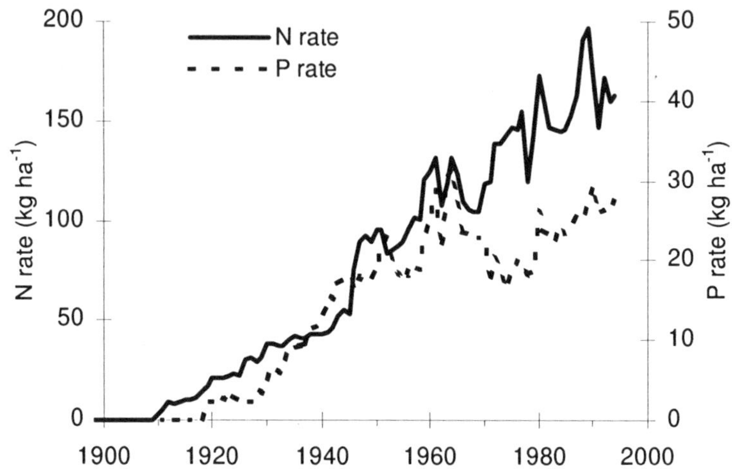

Fig. 14.1. Historical nitrogen and phosphorus fertiliser use for sugarcane in the Herbert catchment.

In the Herbert, the average N use is currently well above industry recommendations and there are large differences in fertiliser application rates between growers (Fig. 14.2). Applications of P are also high considering the large areas of the Herbert that have soil P levels well in excess of 40mg kg^{-1} using the BSES P test. Applications of K on the other hand are below recommended rates and appear to have remained fairly constant since the early 1970s (Wood, 1984).

Increases in fertiliser use have not been matched by improved soil fertility. Evidence for this has come from comparing soil in sugarcane fields and adjacent uncultivated areas in the Herbert, Burdekin and Tully districts (Wood, 1985; Bramley *et al.*, 1996a). The uncultivated sites had never grown sugarcane, whilst the sugarcane fields had been cropped for at least 20 years. The sugarcane soils had degraded both physically and chemically with higher bulk density and lower porosity, were more acid, had higher phosphorus levels, less organic matter and were lower in exchangeable Ca, Mg, Na, K and CEC.

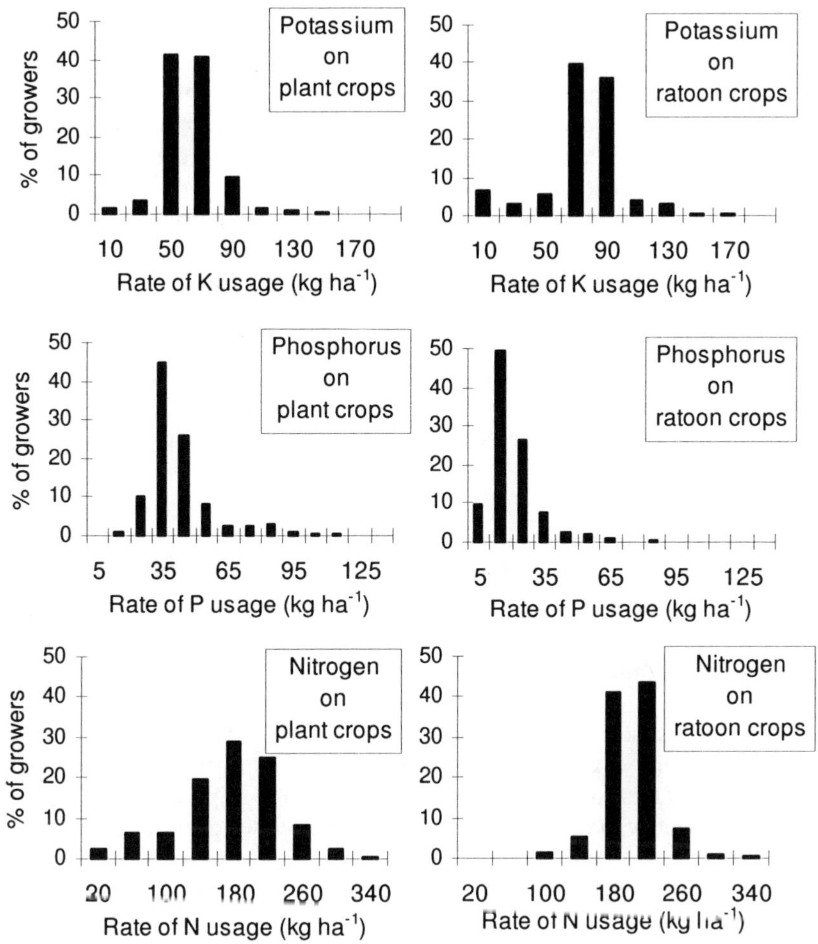

Fig. 14.2. Application rates of nitrogen, phosphorus and potassium to the 1995 sugarcane crop in the Herbert River district

Growers' Attitudes to Nutrient Management

There has been little formal economic analysis of sugarcane production systems or research into the attitudes which sugar producers have towards risk. This is unfortunate given the high input focus of most growers. However, preliminary results from a survey of 120 growers in the Herbert River District (Johnson, 1995) showed that growers consider a wide range of sources of variability to be important for nutrient management. Weather conditions, soil type and cane variety were all taken into account. Cane price effects were not rated highly and this is important given that existing industry fertiliser recommendations for N in particular are related to the price of sugar.

The survey indicated that growers considered a single NPK fertiliser recommendation for the whole Herbert River district was unsatisfactory. Only 38% of respondents followed the current industry fertiliser recommendations and 87% considered that farm-specific recommendations would significantly improve on their productivity. From the growers' perspective, fertiliser recommendations need to account for geographic variability at a higher level of spatial resolution than currently exists. Producers considered the timing of fertiliser application to be vitally important. This may reflect the widespread adoption of green cane trash blanketing in the Herbert and the critical importance of timing urea application to coincide with wet weather to minimise ammonia loss. Many producers expressed a desire for economically viable, slow release fertilisers to maximise fertiliser nutrient use efficiency and minimise nutrient losses.

Limitations, Knowledge Gaps and Opportunities for Improved Nutrient Management

Canegrower attitudes to nutrient management

A large gap exists in our understanding of the attitudes and level of understanding that sugar producers have towards nutrient management. The Herbert survey (Johnson, 1995) was an important first step, but there is a need for more dialogue with canegrowers to answer the following questions:

1. Why do only a few growers follow current industry nutrient recommendations?
2. Why have growers increased their N and P application rates so much over the last 20 years?
3. Why do only a small proportion of growers conduct routine soil testing?
4. What do growers require in terms of improved nutrient management recommendations?
5. What would the attitude be towards the development of more elaborate decision aids?

Until we have answers to some of these questions it will not be possible to develop more effective methods of nutrient management and decision aids that will be readily adopted by growers.

Focus on a narrow range of nutrients

Crude comparisons with the amount of nutrients applied as fertiliser indicate an excess of supply over that removed in the crop for N and P but a deficit situation for all other nutrients (Table 14.2) (Chapman *et al.*, 1981). In the latter case, this demonstrates that insufficient emphasis has been given to replacing those nutrients being removed by the crop in current fertiliser programs. Exceptions apply where nutrients are already present in the soil or where they are being applied in irrigation water. Thus, a major limitation of nutrient management strategies in the Australian industry has been the development of fertiliser recommendations which focus on too narrow a range of nutrients: traditionally, N, P and K and more recently S. This reflects a focus of attention on those nutrients most likely to deliver a current economic response. Little notice has been taken of other essential nutrients, or of the interactions between them.

Table 14.2. Mean nutrient removal (kg ha^{-1}) by cane crops (cane, tops, trash) in different districts for northern to southern Queensland [After Calcino, 1994]

Nutrient	Mossman to Cairns	Babinda to Tully	Herbert	Burdekin	Mackay	Bunda-berg	Maryborough to Rocky Point
Nitrogen	143	122	128	154	123	150	134
Phosphorus	18	17	15	37	18	23	20
Potassium	217	208	139	276	203	260	234
Calcium	39	27	30	55	41	51	31
Magnesium	26	18	29	57	35	52	36
Sulfur	35	25	25	47	25	48	34
Copper	0.09	0.07	0.09	0.11	0.09	0.09	0.12
Zinc	0.68	0.46	0.43	0.59	0.38	0.37	0.39
Iron	6.42	7.32	7.49	5.65	6.93	8.61	6.56
Manganese	6.79	3.91	4.89	1.90	3.64	2.66	2.89
Cane yield (t ha^{-1})	100	93	74	119	84	[illegible]	85

At least 14 elements are essential for normal cane growth and development (Anderson and Bowen, 1990). Although more than 100 years ago Liebig insisted that all nutrients removed from the soil by crops should be replaced, modern agriculture commonly replaces only 3 or 4, and in doing so usually applies them in relatively heavy concentrations. This approach to fertilisation can give rise to nutrient imbalances and deficiencies in the soil; a situation exacerbated by recent trends for fertilisers to contain fewer impurities and less trace elements.

The lack of focus on micronutrients in the Australian industry is evident from the lack of practical recommendations and understanding of micronutrient behaviour in sugarcane soils. Growers are advised not to attempt to rectify potential nutrient deficiencies until symptoms of deficiency are visible in the crop. This limited economic approach to fertilisation is dangerous as it restricts the detection and rectification of potential nutrient deficiencies and encourages the perpetuation of nutrient imbalances in the soil. It is also probable that crop yields are being reduced by sub-clinical deficiencies long before the development of deficiency symptoms in the crop.

Evaluation of sugarcane areas in north Queensland at risk from Zn deficiency (Reghenzani, 1993) illustrates the potential benefits which might accrue from broadening the focus of nutrient management recommendations. Whilst work overseas has largely defined the role of minor elements in cane growth (Anderson and Bowen, 1990), there is a need to assess their current levels of availability in different soil types and whether maintenance applications are needed.

Silicon is an element that has received little attention in Australia, but considerably more overseas. Application of calcium silicate slag has given larger increases in sugar yields than lime on highly weathered, acidic soils for reasons associated with overcoming Si deficiency, increasing soil P availability, increasing soil pH, Ca and Mg, and ameliorating Al toxicity (Fox *et al.,* 1967; Moberly and Meyer, 1975). Silicon may also be implicated in the superior crop responses to cement compared with lime in north Queensland (Ridge *et al.,* 1980). Further research into the role played by silicates in both soil nutrient availability and crop growth is warranted.

Recommendations are too broad

Most fertiliser recommendations are industry-wide with virtually no matching of rate to regional climatic and soil conditions (Calcino, 1994). Whilst this approach is simple, it is obviously imprecise because important soil and climatic factors can influence the availability of nutrients to the crop. It is therefore at odds with the philosophy of sustainable agricultural production and is likely to result in the inefficient use of fertiliser (Bramley and Wood, 1996).

Bramley *et al.* (1995) demonstrated how this approach is flawed with respect to P fertiliser use in two contrasting Herbert River soils. Bramley and Wood (1996) further showed how a fundamental knowledge of P behaviour in sugarcane soils could be used to help derive P fertiliser recommendations. Similarly, Wood *et al.* (1996) and Wood and Bramley (1996) have argued that in making recommendations for Ca, Mg and K, a knowledge of the charge characteristics of sugarcane soils is needed as CEC is currently ignored.

Given the strong support by canegrowers in the Herbert for farm-specific or soil-specific fertiliser recommendations (Johnson, 1995), there are strong reasons for improving the information available on the distribution and properties of different soil types, at a scale that is meaningful to growers. Wood and Bramley (1996) have shown how a detailed soil survey of the Herbert River District coupled with a comprehensive soil analytical database might be used for the development of more precise nutrient management recommendations for groups of soils having similar physical and chemical properties. For other sugar producing districts, there is a need either to conduct more detailed soil surveys so that soils data are available at the farm block level or to use a quicker, lower cost approach based on growers assisting with the on-farm identification of major soil types, as suggested by Keig *et al.* (1991). Alternatively, large data sets such as those of Wood and Bramley (1996) may be analysed to identify 'key' soil properties of greatest importance to sugarcane production. Maps of these properties could then be produced at a much reduced cost compared with more holistic soil surveys.

Poor focus on management of soil properties and soil fertility

Fertiliser recommendations in the Australian sugar industry were developed when maximisation of production and short-term economic gain were the key factors driving the industry. Little emphasis has been given to the development and maintenance of a high level of soil organic matter even though it is directly related to development of a biologically fertile soil and the efficient nutrient cycling. With the increasing adoption of green cane harvesting in Australia (Wood, 1991), there is a need to investigate the long term impact of trash retention on soil organic matter and nutrient supply. Preliminary predictions based on 15 years of soil sampling and cane harvest data in the Herbert suggest that trash retention could increase soil organic carbon under a sugarcane monoculture by almost 40% (Vallis *et al.*, 1996). They also suggest that, over the long term, fertiliser N inputs could be reduced by up to 40kg N ha^{-1} $year^{-1}$ due to increased N mineralisation rates. Rotations with other crops may well permit further reductions in fertiliser N use (A.L. Garside, *pers. comm.*). With retention of unburnt crop residues, Sutton *et al.* (1996) demonstrated increase in soil microbial biomass and N mineralisation rates indicative of long term changes in soil organic matter and enhanced soil biological activity and nutrient turnover.

The contribution of soil organic matter in supplying nutrients to the crop through natural mineralisation processes has not been recognised in Australian nutrient management programmes. In contrast, the South African sugar industry has identified soil organic matter as one of the most important factors to consider when making N recommendations (Moberly *et al.*, 1984). Soil types are grouped according to the amount of N mineralised from their organic matter and different N fertiliser recommendations apply to each group (Meyer and Wood, 1994). Recent N research in Australia (Keating *et al.*, 1993) has suggested that differences in soil N mineralisation potential are a major reason for the highly variable nature of crop responses in N rate trials and that sources of N in the soil play a more important role in sugarcane nutrition than fertiliser N. The most likely source of the soil N is the soil organic N pool, much of which has built up through past N fertilisation. It is therefore important to develop a reliable method of using soil properties to predict the amount of N mineralisation and likely crop response to applied N so that fertiliser is not wasted in a non-responsive situation. Research on N mineralisation in Australia is needed to decide whether a similar system to that used in the South African sugar industry is justified for our sugarcane soils.

Current fertiliser recommendations also lack precision because they do not account for the rate and magnitude of the reaction between added nutrients and soils, e.g. industry P recommendations ignore soil differences in P sorption and the rate of the reaction between sorbed P and the soil. Wood (1988) and Toreu *et al.* (1988) demonstrated substantial differences in these properties for soils of the Herbert River district and wet tropical coast. Despite these findings, reviews of BSES trial data have shown 'no strong justification for special application rates on phosphorus 'fixing soils' which have a history of adequate phosphorus fertiliser application' (BSES, 1985). This latter philosophy has recently been challenged by Bramley *et al.* (1995) and Bramley and Wood (1996).

Soil acidification under sugarcane production systems is a threat to the sustainability of the Australian Sugar Industry, yet this is not considered in nutrient management recommendations. Lime is recommended only where soil exchangeable Ca is deficient or marginal. Whilst current cane varieties are regarded as tolerant of high Al levels (Hetherington *et al.*, 1986), it appears likely that cane breeders have been selecting sugarcane varieties that are adapted to a steadily acidifying environment (Schroeder *et al.,* 1996). Unlike liming recommendations for sugarcane in South Africa or for other crops in Australia or elsewhere, which are based on exchangeable Al and clay % (Schroeder *et al.,* 1994), the Australian sugar industry ignores soil properties such as pH, pH buffering capacity, exchangeable Al or CEC. The emphasis on soil Ca means that lime recommendations do not adequately address the problem of soil acidification (Wood, 1985; Bramley *et al.,* 1996a) which occurs largely through application of N fertiliser and removal of cations in harvested cane. Furthermore, the recommendations fail to account for differences in the nature and rate of reaction between lime and different soil types.

BSES have validated their lime recommendations using trials on acid soils with 'adequate' Ca in north Queensland. These demonstrated no response to lime applications of up to 10t ha^{-1} in plant or first ratoon crops (Hurney, 1971; Haysom *et al.,* 1986). Quinan and Wood (1989, 1993) challenged these findings with long term trials which demonstrated a continuing response to lime over an 8 year period and a strong relationship between soil pH and yield response to liming on soils with initially inadequate Ca. They suggested lime effects needed to be evaluated over a long timeframe. Further research is needed to investigate whether liming recommendations can be based on more precise criteria such as soil pH, exchangeable aluminium, cation exchange capacity and pH buffering capacity and a long term research project has just commenced in the Herbert to answer this question. It will also investigate the long term effects of ameliorating soil acidity on sugarcane growth and soil chemical properties, and look at possible reasons for the depression in CCS commonly associated with liming (Kingston *et al.*, 1996). South African work suggests that higher plant N may be involved (Wood, 1993), perhaps warranting the investigation of lime application on nitrification and N mineralisation. Possibly, lime at low levels could be used to reduce N requirement, particularly with trash blanketing.

Another soil property ignored in formulating nutrient recommendations is the ability of different soils to retain nutrient cations. Calcium, Mg and K recommendations are based on the measurement of individual exchangeable cations in the soil. Soil CEC and the proportion due to each nutrient are not considered, which is surprising since Ca, Mg and K interact closely and compete for charge sites. Recommendations also ignore the relationship between soil pH and the proportion of soil CEC occupied by acidic cations (Fig. 14.3). In soils <pH4.6, more than 50% of the CEC can be occupied by acidic cations. As a consequence, the nutrient cations that might otherwise occupy this portion of the CEC are subject to loss through leaching. Similarly, soils with a very low CEC (<2.0cmol(+) kg^{-1}) require special recognition and management. Large single nutrient applications are inappropriate on such soils as they can lead to wastage of fertiliser. Also, if such soils become too acidic, Al toxicity could become a potential problem.

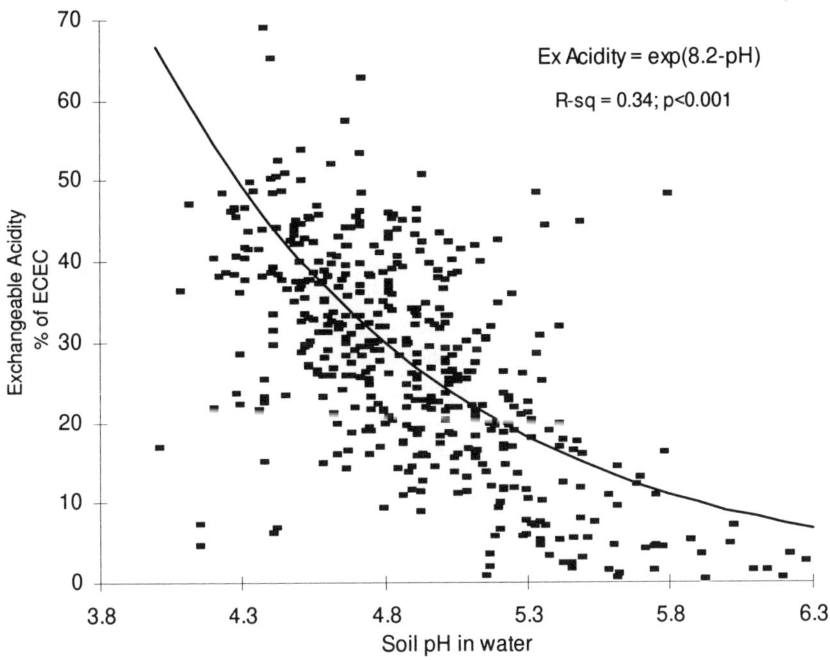

Fig. 14.3. Relationship between soil pH and the proportion of soil CEC occupied by acidic cations

Nutrient interactions or imbalances have been largely ignored

Since the sugar industry considers each nutrient in isolation, rather than using a holistic approach which looks at interrelationships between all the important nutrients, there have been very few studies of nutrient interactions or induced deficiencies caused by nutrient imbalance. Examples of interactions between soil and applied nutrients noted for sugarcane (Anderson and Bowen, 1990) include: excess P reducing the availability of Zn and Cu; high K inhibiting uptake of Ca and Mg; lime application reducing the availability of Mn, B and Zn; the effects of Ca, P and Fe on soluble Al levels; and the interactions of Mn with Si and Fe. Extensive literature exists for other crops on the importance of vesicular arbuscular mycorrhizae (VAM) but their importance in sugarcane growth is unknown.

It is dangerous to assume that by adopting recommended nutrient application rates there will be no induced deficiencies. For example, the high soil P in many old cane areas may be restricting the availability of Cu and Zn, though this would not be evident from current soil testing procedures. In Hawaii it has been suggested that excess P may be inhibiting the uptake of Zn in some sugarcane fields (HSPA, 1994). High soil P levels may also have a moderating effect on soluble Al levels through the formation of Al/P complexes which may explain the apparent tolerance of sugarcane to high exchangeable Al levels in the soil.

Advisory procedures

The limited and highly generalised information available to canegrowers in Australia to help them develop fertiliser recommendations is in contrast to some sophisticated fertiliser advisory services available overseas. There is no fertiliser advisory service provided by the industry. Growers take their own soil samples which are analysed by commercial laboratories operated mainly by fertiliser companies. Local representatives of these companies (whose training is often limited to a three-day course in soil test interpretation) then provide a fertiliser recommendation which is based on their interpretation of the soil test in relation to industry recommendations. In contrast, in many other sugarcane producing countries, a fertiliser advisory service based on both soil and leaf analysis is available to cane growers. In South Africa, soil-specific fertiliser recommendations covering a whole crop cycle based on a pre-plant soil analysis, annual analysis of leaf samples and on the use of nutrient ratio indices based on the DRIS system are provided for growers (Wood, 1987). A similar integrated system is available in Florida (Elwali and Gascho, 1984). In Hawaii, a highly sophisticated crop logging system has been used for monitoring the development of plantation sugarcane crops which included leaf analysis at 5 weekly intervals (Clements, 1980). In Mauritius, foliar analysis is widely used in association with soil analysis for evaluating nutrient requirements and for examining nutrition trends (Ng Kee Kwong *et al.*, 1988). There are opportunities for much more sophisticated advisory procedures to be adopted in Australia.

Lack of focus on nutrient use efficiency, loss processes and crop requirements

Strategic research to improve our understanding of the efficiency of nutrient use and the major processes involved began less than 10 years ago. Prior to this, research through an extensive series of agronomic trials was directed mainly at providing recommendations on fertiliser rates and methods of application and on developing appropriate soil and plant tests to assist with making recommendations. These trials were not always supported by intensive soil and plant sampling or environmental monitoring, so it was not always possible to seek explanations for the trial results in terms of nutrient uptake and losses.

Most of the recent strategic research has focused on N and the various processes making up the N cycle in sugarcane production. This has included studies of the recovery and losses of fertiliser N (Vallis and Keating, 1994), volatilisation losses (Freney *et al.*, 1994), leaching losses (Catchpoole and Keating, 1995), N fixation (Chapman *et al.*, 1992) and crop uptake (Muchow and Robertson, 1994). Some of the main findings from this research, summarised by Keating *et al.* (1993), are: (i) responses to fertiliser N are highly variable; (ii) losses of fertiliser N are large with leaching, denitrification and volatilisation all important in different circumstances; (iii) fertiliser N is not the major N source for cane crops in the short term; (iv) fertiliser N is vital for the long term maintenance of soil fertility; (v) split applications of N rarely improve cane yield, although if environmental impact is accounted for, they may reduce costs.

Whilst our knowledge of N pathways and loss processes has advanced considerably, little is known about losses, residual value of previously applied fertiliser and nutrient use efficiency for other nutrients. A limited understanding of crop requirements and the pattern of uptake for different nutrients has constrained the development of management strategies that reduce nutrient losses from the system whilst increasing sugar yields. Reductions in nutrient losses may be achieved if ways can be found to synchronise the supply of available nutrients from the soil with pattern of uptake by the crop. However, this will vary with cane variety, crop age, soil and climatic conditions, crop management and crop yield potential.

Sugarcane nutrition and crop health

It is reasonable to assume that crops have a greater resistance to pests and diseases if they are supplied with adequate and balanced nutrition and are growing in a fertile and biologically active soil where natural processes and cycles are operating. Support for this thesis comes from Anderson and Dean (1986) who found that rust severity was related to both nutrient deficiencies and excesses, and that control of rust may be possible by modifying soil and crop nutritional status. In South Africa, damage from the stalk borer *Eldana sacharina* is substantially increased where growers apply excessive amounts of N (Atkinson and Nuss, 1989). In Australia, decline in productive capacity of land which has been under sugarcane for many years may be in part related to nutritional factors (Bramley *et al.,* 1996a). Whilst the focus has so far been largely on the effects of root infection by a range of pathogenic fungi (Magarey *et al.*, 1995), changes in soil physical, chemical (Bramley *et al.,* 1996a) and biological properties under the prevailing system of sugarcane monoculture may have been detrimental to root health, resulting in reduced root function and nutrient uptake thus making crops more susceptible to attack by a range of different pathogens.

Environmental issues

Although large losses of applied N have been clearly established, investigations in Australian catchments of where lost nutrients go in the environment or their environmental impact at a regional scale have only recently commenced (McShane *et al.*, 1993; Bramley *et al.*, 1994; Bramley and Johnson, 1996). The acquisition of this type of data is essential for the development of strategies for reducing nutrient losses from sugarcane areas and for their environmental impact. Environmental issues are likely to impact markedly on nutrient management in the future and it is vital that decisions are based on fact rather than conjecture.

An Improved Philosophy for Nutrient Management

Improvements to nutrient management in the Australian sugar industry need to deliver benefits in two key areas: (i) they need to lead to the creation of a more ecologically sustainable sugar industry; and (ii) they should be economically beneficial. To do this, nutrient management systems should aim at:

1. Enhancing sustainability through a long term improvement in soil fertility and crop health.
2. Improved environmental protection to reduced reparation costs (associated with environmental damage).
3. Improvements in crop nutrition leading to higher yields and improved cane and sugar quality giving higher returns for growers and millers, reduced sugar milling and refining costs, and improved marketability of sugar.
4. Development of practices that reduce fertilisation costs and growers' costs of production (and reducing the amount of nutrients likely to be lost to the environment).

To achieve these benefits, we suggest a change of philosophy is required in the way nutrients are supplied and managed for sugarcane production. The current approach exploits soil fertility and emphasises the delivery of nutrients directly to the crop to achieve short term economic gain, possibly at the expense of the downstream environment. A new approach is required which aims to create biologically fertile soils capable of storing and supplying a wide range of nutrients to the crop through natural soil processes, and in which fertilisers are used for soil fertility maintenance.

This philosophy has driven fertiliser management in New Zealand for many years (Cornforth and Sinclair, 1984) and is now embodied in a computer-based decision support model for fertiliser management called Outlook™ (Marshall, 1995). The NZ system, which could be readily transferred to the Australian sugar industry, is based on the concept of a pool of nutrients cycling in the soil-plant-animal system (Metherell *et al.*, 1995) and an understanding of the reactions associated with this pool. It uses baseline information on soil nutrient status for a range of soils (e.g. Roberts *et al.*, 1994, 1996), information on previous fertiliser history (McCall and Thorrold, 1991), a knowledge of crop response to fertiliser addition and the effectiveness of different fertilisers used under different conditions (Edmeades *et al.,* 1991) in addition to appropriate economic analysis. For each soil, a 'biological optimum' soil test level is defined and once this has been attained, maintenance fertiliser rates are used to offset nutrient removal in products, and in the case of nutrients like P, declining availability over time. This system will by definition lead to more efficient use of added nutrients, and a minimisation of the potential for nutrient export off-site and associated environmental impact (Bramley *et al.* 1996b).

Bowden (1989) has used similar arguments to advocate moving away from making single figure fertiliser recommendations to a situation where farmers can be shown how yields and profits will vary with fertiliser rate and with different seasonal conditions. By predicting nutrient response surfaces rather than requirements, it helps farmers make better decisions on fertiliser management and shows what happens to yields and profits in certain situations, e.g. if they choose to spend more or less on fertiliser or if they experience a season which is better or worse than average. The ability to answer such 'what if' questions allows farmers to make informed decisions for their particular situation. When this information is not available, farmers may choose to deviate from single figure recommendations without knowing the consequences of their action, as currently happens in the sugar industry.

Another example is 'NPK Decide', used for fertiliser management for wheat in Western Australia (Bowden and Bennett, 1976; Burgess, 1988). The model takes into account the pool of nutrients in the soil and crop residues, the residual value of previously applied fertiliser, the potential availability of nutrients in relation to various nutrient loss processes, the nutrient uptake by the crop, likely crop yields as determined by nutrient uptake and potential crop yields, and an index of crop quality. By using this information, the program estimates how the crop would respond to various application rates of fertiliser. This response is then translated into economic terms relevant to the farm in question.

Probably the best way of handling the variation in nutrient response due to uncontrolled factors from the decision making point of view is to use some form of nutrient modelling. Angus *et al.* (1993) outlined three main uses for nutrient modelling: (i) to develop sensitivity analyses which show the effects of changing fertiliser inputs on yield and profitability; (ii) to develop fertiliser recommendations within a probability framework so that farmers can be shown the chances of obtaining particular returns from different levels of fertiliser input; and (iii) to develop tactical predictions as the season progresses to assist decisions on the need for additional fertiliser.

As part of the adoption of a new philosophy of soil improvement, there is a need to achieve a better understanding of sugarcane soils and develop a system of 'management by soil type' or 'precision agriculture' (PA). PA aims to match inputs with yields at much more detailed scales such as the paddock or cane block. It is not a new concept and has been successfully adopted by farmers in parts of the United States where it has resulted in increased crop yields, savings in fertiliser and herbicides, and reduced potential for off-farm environmental damage (Larson and Robert, 1991; Carr *et al.*, 1991). McBratney and Whelan (1995) outlined the numerous benefits of PA for the Australian cotton industry and similar benefits are likely to apply to the sugar industry.

Application of PA to provide precise fertiliser recommendations for sugarcane crops would require the following:

1. Development of a mapping system for cane yields at block level.
2. Detailed mapping of the key properties of soils in sugarcane regions.
3. Development of a spatially referenced database for storing digital information on soils, nutrient behaviour and yield variability.
4. Characterisation of soil nutrient behaviour such as: reaction of added nutrients with soils; soil hydraulic properties; charge characteristics; sorption and desorption of fertiliser nutrients; composition and balance of nutrient cations; soil pH buffering capacity; mineralisation of organic matter; and interactions between added nutrients and soil pH.
5. A reassessment of current fertiliser management practices.
6. Linkage of the information collected to a GPS-guided system for precise soil management.

Additional benefits of PA include its use as a means of predicting the risk of nutrient loss from different soils and modifications to irrigation management and weed control. Experience with other crops has shown that farmers will readily adopt this approach once they have seen the maps showing crop yield variation across a paddock and realised the opportunities for yield improvement with more precise and targeted soil management (S. Cook, *pers. comm.*).

We also believe that a more sophisticated and scientific basis for making decisions on the timing of fertiliser application should be sought. More efficient fertiliser use and reduced nutrient loss off-site should accrue, but would depend on better detailed knowledge of crop nutrient requirements over time than is currently available. Retention of nutrients in fallow land requires careful management. Garside *et al.* (1996) suggested that growing a fallow crop with high N uptake is a good strategy in high rainfall districts for capturing soil mineral N from the soil and preventing its removal by denitrification and leaching.

Conclusions

It is likely that Australia will follow the trend emerging in many countries for rural industries to be held accountable for the on-site and downstream effects of their activities and to comply with the principles of ecologically sustainable development. Some countries, such as The Netherlands and parts of the USA, have already introduced legislation governing the use of fertilisers and agricultural chemicals so that off-farm nutrient exports are subject to penalties. Accordingly, environmental issues are likely to play an increasingly important part in influencing nutrient management strategies. Whether this happens or not, there is clearly much scope for improving the management of nutrients in the Australian Sugar Industry. If the costs of rectifying possible environmental degradation were included as costs of production, farm profitability would be enhanced considerably with the adoption of more sustainable soil and nutrient management practices that improved the efficiency of fertiliser use. It is therefore timely that current methods of nutrient management be scrutinised and a new philosophy put forward for consideration.

Acknowledgments

The authors would like to thank Leonie Baker for her assistance in developing this chapter and the Herbert Cane Protection and Productivity Board for access to unpublished data on fertiliser usage by growers.

References

Anderson, D.L. and Bowen, J.E. (1990) *Sugarcane Nutrition.* Potash and Phosphate Institute, Atlanta, USA.

Anderson, D.L. and Dean, J.L. (1986) Relationship of rust severity and plant nutrients in sugarcane. *Phytopathology* 76, 581-585.

Angus, J.F., Bowden, J.W. and Keating, B.A. (1993) Modelling nutrient responses in the field. In: Barrow, N.J. (ed.), *Plant Nutrition - from Genetic Engineering to Field Practice,* Kluwer Academic Publishers, Dordrecht, The Netherlands, pp. 59-68.

Atkinson, P.R. and Nuss, K.J. (1989) Associations between host-plant nitrogen and infestations of the sugarcane borer, *Eldana saccharina* Walker (Lepidoptera:Pyralidae). *Bulletin Entomological Research* 79, 489-506.

Bowden, J.W. (1989) Predicting phosphorus fertilizer requirements in Australia. *Agricultural Science* 2, 22-25.

Bowden, J.W. and Bennett, D. (1976) The 'Decide' Model for predicting superphosphate requirements. In: *Proceedings of a Symposium 'Phosphate in Agriculture',* November 1974, Australian Institute of Agricultural Science, Victoria Branch, Melbourne, Australia, pp. 6.1 - 6.36

Bramley, R.G.V., Ellis, N., Nable, R.O. and Garside, A.L. (1996a) Changes in soil chemical properties under long term sugarcane monoculture and their possible role in sugar yield decline. *Australian Journal of Soil Research* 34, 967-984.

Bramley, R.G.V. and Johnson, A.K.L. (1996) Land use impacts on nutrient loading in the Herbert River. In: Hunter, H.M., Eyles, A.G. and Rayment, G.E. (eds), *Downstream Effects of Land Use.* Department of Natural Resources, Brisbane, Australia, pp. 93-96.

Bramley, R.G.V., Johnson, A.K.L. and Smettem, K.R.J. (1994) Quantifying the effects of different rural land use on water quality in the catchment of the Herbert River - Preliminary results. *Proceedings of the Australian Society of Sugarcane Technologists* 16, 21-29.

Bramley, R.G.V., Johnson, A.K.L. and Wood, A.W. (1996b) Nutrient loss in the Herbert River catchment and strategies for its reduction. In: Uren, N. (ed.), *Proceedings of the Australian Society of Soil Science*, National Soils Conference, Melbourne, Australia, pp. 23-24.

Bramley, R.G.V. and Wood, A.W. (1996) More precise P fertiliser recommendations to minimise potential downstream impacts of the sugar industry. In, Hunter, H.M., Eyles, A.G. and Rayment, G.E. (eds), *Downstream Effects of Land Use.* Department of Natural Resources, Brisbane, Australia, pp. 269-272.

Bramley, R.G.V., Wood, A.W. and Cristaudo, R. (1995) Improving the precision of phosphorus fertilizer recommendations for sugarcane. *Proceedings of the Australian Society of Sugarcane Technologists* 17, 179-186.

BSES (1985) Sugarcane Nutrition Review. *BSES Bulletin* 10, 4-23

Burgess, S.J. (1988) Going beyond single-figure fertilizer recommendations. *Journal of Agriculture Western Australia* 29, 12-16.

Calcino, D.V. (1994) *Australian Sugarcane Nutrition Manual.* SRDC/BSES, Indooroopilly, Qld, Australia, 60pp.

Carr, P.M., Carlson, G.R., Jacobsen, J.S., Nielsen, G.A. and Skogley, E.O. (1991) Farming soils, not fields: A strategy for increasing fertilizer profitability. *Journal of Production Agriculture* 4, 57-61.

Catchpoole, V.R. and Keating, B.A. (1995) Sugarcane yield and N uptake in relation to profiles of mineral-N in the soil. *Proceedings of the Australian Society of Sugarcane Technologists* 17, 187-192.

Chapman, L.S. (1994) Fertilizer N management in Australia. *Proceedings of the Australian Society of Sugarcane Technologists* 16, 83-92.

Chapman, L.S., Halsall, D.M. and Gibson, A.H. (1992) Biological N fixation and sugarcane. *Proceedings of the Australian Society of Sugarcane Technologists*14, 90-93.

Chapman, L.S., Haysom, M.B.C. and Chardon, C.W. (1981) Checking the fertility of Queensland's sugar land. *Proceedings of the Australian Society of Sugarcane Technologists* 3, 325-332.

Clements, H.F. (1980) *Sugarcane Crop Logging and Crop Control: Principles and Practices*. University Press of Hawaii, Honolulu, USA.

Cornforth, I.S and Sinclair, A.G. (1984) *Fertiliser and Lime Recommendations for Pastures and Crops in New Zealand* (2nd revised edition). New Zealand Ministry of Agriculture and Fisheries, Wellington, NZ.

Edmeades, D.C., Watkinson, J.H., Perrott, K.W., Sinclair, A.G., Ledgard, S.F., Rajan, S.S.S., Brown, M.W., Roberts, A.H.C., Thorrold, B.T., O'Connor, M.B., Floate, M.J.S., Risk, W.H. and Morton, J.D. (1991) Comparing the agronomic performance of soluble and slow release phosphate fertilisers: the experimental basis for RPR recommendations. *Proceedings of the New Zealand Grassland Association* 53, 181-190.

Elwali, A.M.O. and Gascho, G.J. (1984) Soil testing, foliar analysis and DRIS as guides to sugarcane fertilization. *Agronomy Journal* 76, 466-470.

Farquhar, R.H. (1965) The interpretation and use of tissue analysis within a fertilizer advisory service for sugarcane in north Queensland. *Proceedings International Society of Sugarcane Technologists* 12, 227-236.

Fox, R.L., Silva, J.A., Younge, O.R., Plucknett, D.L. and Sherman, G.D. (1967) Soil and plant silicon and silicate response by sugarcane. *Soil Science Society of America Proceedings* 31, 775-779.

Freney, J.R., Denmead, O.T., Wood, A.W. and Saffigna, P.G. (1994) Ammonia loss following urea addition to sugarcane trash blankets. *Proceedings of the Australian Society of Sugarcane Technologists* 16, 114-121.

Garside, A.L., Berthelsen, J.E., Richards, C.L. and Toovey, L.M. (1996) Fallow legumes on the wet tropical coast: some species and management options. *Proceedings of the Australian Society of Sugarcane Technologists* 18, 202-208.

Haysom, M.B., Hurney, A.P. and Nielsen, P.J. (1986) The influence of lime on the acid soils of Ingham. *Proceedings of the Australian Society of Sugarcane Technologists* 8, 55-62.

Hetherington, S.J., Asher, C.J. and Blamey, F.P.C. (1986) Tolerance of sugarcane to Al in soil and solution culture. *Proceedings of the Australian Society of Sugarcane Technologists* 8, 63-68.

HSPA (1994) *Annual Report. Hawaiian Sugar Planters Association*, Honolulu, Hawaii, USA, p. 16.

Hurney, A.P. (1971) An evaluation of liming in the Cairns-Babinda area. *Proceedings of the Queensland Society of Sugarcane Technologists* 38, 109-112.

Johnson, A.K.L. (1995) Risk perceptions and nutrient management responses in the Australian sugar industry: preliminary results from the Herbert River District. *Proceedings of the Australian Society of Sugarcane Technologists* 17, 172-178.

Johnson, A.K.L. (1996) Quantifying Current and Historical Inputs of Fertiliser N and P to Rural Lands in the Herbert River Catchment. In: Hunter, H.M., Eyles, A.G. and Rayment, G.E. (eds), *Downstream Effects of Land Use*, Department of Natural Resources, Brisbane, Australia, pp. 273-276.

Keating, B.A., Vallis, I., Hughes, M. and Ridge, D.R. (1993) Strategic directions for N research - a view from the south. *Proceedings of the Australian Society of Sugarcane Technologists* 15, 276-284.

Keating, B.A., Vallis, I., Parton, W.J., Catchpoole, V.R., Muchow, R.C. and Robertson, M.J. (1994) Modelling and its application to N management and research for sugarcane. *Proceedings of the Australian Society of Sugarcane Technologists* 16, 131-142.

Keig, G., McAlpine, J.R. and Pointing, E.M. (1991) Caneland productivity monitoring and enhancement - appropriate use of information technology. *Proceedings of the Australian Society of Sugarcane Technologists* 13, 138-145.

Kerr, H.W. and von Steiglitz, C.R. (1938) The laboratory determination of soil fertility. *BSES Technical Communication* No. 9, BSES, Brisbane, Australia.

Kingston, G., Dickenson, J. and Smith, D. (1991) Rural land use practices in coastal North Queensland. In: Yellowlees, D. (ed.), *Land Use Patterns and Nutrient Loading of the Great Barrier Reef Region.* James Cook University, Townsville, Australia, pp. 1-17.

Kingston, G., Hurney, A.P. and Aitken, R.L. (1996) Depression of ccs in sugarcane after use of liming products: economic significance and associations with soil chemistry and leaf nutrient composition. *Proceedings of the Australian Society of Sugarcane Technologists* 18, 164-173.

Larson, W.E. and Robert, P.C. (1991) Farming by soil. In: Lal, R. and Pierce, F.J. (eds), *Soil Management for Sustainability.* Soil and Water Conservation Society, Ankeny, Iowa, USA, pp. 103-112.

Magarey, R.C., Yip, H.Y., Bull, J.I. and Johnson, E.J. (1995) Recent studies into the soil biology of yield decline. *Proceedings of the Australian Society of Sugarcane Technologists* 17, 128-133.

Marshall, P.R. (1995) An introduction to Outlook™: software for evaluation of phosphate fertiliser strategies. In: Currie, L.D. and Loganathan, P. (eds), *Fertiliser Requirements of Pasture and Field Crops* Occasional Report No. 8. Fertiliser and Lime Research Centre, Massey University, New Zealand, pp. 40-46.

McBratney, A.B. and Whelan, B.M. (1995) The potential for site-specific management of cotton farming systems. Cooperative Research Centre for Sustainable Cotton Production, Narrabri, NSW, Australia.

McCall, D.G. and Thorrold, B.S. (1991) Fertiliser history is a useful predictor of soil fertility status. *Proceedings of the New Zealand Grassland Association* 53, 191-196.

McShane, T.J., Reghenzani, J.R., Prove, B.G. and Moody, P.W. (1993) Nutrient balances and transport from sugarcane land - A preliminary report. *Proceedings of the Australian Society of Sugarcane Technologists* 15, 268-275.

Metherell, A.K., McCall, D.G. and Woodward, S.J.R. (1995) OUTLOOK™: A phosphorus fertiliser decision support model for grazed pastures. In: Currie, L.D. and Loganathan, P. (eds), *Fertiliser Requirements of Pasture and Field Crops* Occasional Report No. 8. Fertiliser and Lime Research Centre, Massey University, New Zealand, pp. 24-39.

Meyer J.H. and Wood, R.A. (1994) Nitrogen management of sugarcane in South Africa. *Proceedings of the Australian Society of Sugarcane Technologists* 16, 93-104.

Moberly, P.K. and Meyer, J.H. (1975) The amelioration of acid soils in the South African Sugar Industry. *Fertilizer Society of South Africa Journal* 2, 57-66.

Moberly, P.K., Wood R.A., Meyer, J.H., Leibbrandt, N.B. and McIntyre, R.K. (1984) Soil: an important factor to consider when making N recommendations. *Proceedings Nitrogen Symposium*, Soils Irrigation Research Institute, Pretoria, South Africa, pp. 10-12.

Muchow, R.C. and Robertson, M.J. (1994) Relating crop N uptake to sugarcane yield. *Proceedings of the Australian Society of Sugarcane Technologists* 16, 122-130.

Ng Kee Kwong, K.F., Gauthier, J. and Deville, J. (1988) Foliar diagnosis of sugarcane in Mauritius - a historical review. *Sugar Cane*, 1988 Spring Supplement, 1-7.

Pulsford, J.S. (1991) Historical inputs of fertilizer nutrients on to agricultural lands of coastal north Queensland. In: Yellowlees, D. (ed.), *Land Use Patterns and Nutrient loading of the Great Barrier Reef Region.* James Cook University, Townsville, Australia, pp. 39-52.

Pulsford, J.S. (1993) Historical Nutrient Usage in Coastal Queensland River Catchments Adjacent to the Great Barrier Reef Marine Park. *Report to the Great Barrier Reef Marine Park Authority.* GBRMPA, Townsville, Australia.

Quinan, D.J. and Wood, A.W. (1989) The effects of lime applications on acid soils in the Hambledon mill area. *Proceedings of the Australian Society of Sugarcane Technologists* 11, 70-75.

Quinan, D.J. and Wood, A.W. (1993) Long term effects of lime applications on some soils in the Cairns area. *Proceedings of the Australian Society of Sugarcane Technologists* 15, 368.

Reghenzani, J.R. (1993) A survey of the nutritional status of north Queensland sugarcane soils with particular reference to zinc. *Proceedings of the Australian Society of Sugarcane Technologists* 15, 298-304.

Ridge, D.R., Hurney, A.P. and Haysom, M.B.C. (1980) Some aspects of calcium and magnesium nutrition in north Queensland. *Proceedings of the Australian Society of Sugarcane Technologists* 2, 55-61.

Roberts, A.H.C., Morton, J.D., O'Connor, M.B. and Edmeades, D.C. (1996) Building a solid foundation for pasture production in Northland: P, K, S and lime requirements. *Proceedings of the New Zealand Grassland Association* 57, 119-125.

Roberts, A.H.C., Webb, T.H., Morton, J.D., O'Connor, M.B. and Edmeades, D.C. (1994) Building a solid foundation - sulphur, phosphorus and potassium requirements for the sedimentary soils of North Canterbury. *Proceedings of the New Zealand Grassland Association* 56, 7-12.

Schroeder, B.L., Robinson, J.B., Wallace, M. and Turner, P.E.T. (1994) Soil acidification: occurrence and effects in the South African sugar industry. *Proceedings South African Sugar Technologists Association,* pp. 70-74.

Schroeder, B.L., Turner, P.E.T., Meyer, J.H. and Robinson, J.B. (1996) Advances in quantifying soil acidity and acidification rates in the South African sugar industry. In: Wilson J.R., Hogarth, D.M., Campbell, J.A. and Garside, A.L. (eds), *Sugarcane: Research Towards Efficient and Sustainable Production.* CSIRO Division of Tropical Crops and Pastures, Brisbane, Australia, pp. 256-258.

Sutton, M.R., Wood, A.W. and Saffigna, P.G. (1996) Long term effects of green cane trash retention on Herbert River soils. In: Wilson J.R., Hogarth, D.M., Campbell, J.A. and Garside, A.L. (eds), *Sugarcane: Research Towards Efficient and Sustainable Production.* CSIRO Division of Tropical Crops and Pastures, Brisbane, Australia, pp. 178-180.

Toreu, B.N., Thomas, F.G. and Gillman G.P. (1988) Phosphate - sorption characteristics of soils of the North Queensland coastal region. *Australian Journal of Soil Research* 26, 465-477.

Vallis, I. and Keating, B.A. (1994) Uptake and loss of fertiliser and soil N in sugarcane crops. *Proceedings of the Australian Society of Sugarcane Technologists* 16, 105-113.

Vallis, I., Parton, W.J., Keating, B.A. and Wood, A.W. (1996) Simulation of the effects of trash and N fertilizer management on soil organic matter levels and yields of sugarcane. *Soil and Tillage Research* 38, 115-132.

Wood, A.W. (1984) Some aspects of soil degradation under intensive sugarcane cultivation in the Herbert Valley. *Proceedings National Soils Conference*, Brisbane, Australia, p. 221.

Wood, A.W. (1985) Soil degradation and management under intensive sugarcane cultivation in north Queensland. *Soil Use and Management* 1, 120-124.

Wood, A.W. (1988) Phosphate sorption characteristics of sugarcane soils in the Ingham area. *Proceedings of the Australian Society of Sugarcane Technologists* 10, 111-117.

Wood, A.W. (1991) Management of crop residues following green harvesting of sugarcane in north Queensland. *Soil & Tillage Research* 20, 69-85.

Wood, A.W. and Bramley, R.G.V. (1996) Soil survey - a tool for better fertilizer management in the Australian Sugar Industry. In: Wilson J.R., Hogarth, D.M., Campbell, J.A. and Garside, A.L. (eds), *Sugarcane: Research Towards Efficient and Sustainable Production.* CSIRO Division of Tropical Crops and Pastures, Brisbane, Australia, pp. 189-193.

Wood, A.W., Bramley, R.G.V. and White, R.E. (1996) Non-pedological soil classification - a useful tool for on-farm management of sugarcane production. In: Uren, N. (ed.), *Proceedings of the Australian Society of Soil Science,* National Soils Conference, Melbourne, Australia, pp. 319-320.

Wood, R.A. (1987) The basis of fertilizer recommendations for the South African sugar industry. *Sugar Cane* 1987, 2, 10-15.

Wood, R.A. (1993) Calcium, magnesium and S in sugarcane production. *South African Sugar Journal* 77, 72-73.

Chapter fifteen:

New Technologies for Sugar Milling and By-Product Modification

C.J. Allen[1], M.J. Mackay[2], J.H. Aylward[2] and J.A. Campbell[2]

1. *Department of Chemistry & Chemical Engineering, James Cook University of North Queensland, Townsville, Qld 4811, Australia*
2. *CSIRO Tropical Agriculture, 306 Carmody Road, St Lucia, Qld 4067, Australia*

Abstract

The Australian sugar industry is amongst the most competitive in the world and is able to compete at low world market prices without significant government protection. It is currently expanding, with many mills substantially increasing their capacity. Once the resulting benefits of economies of scale have been realised, it is likely that the long-term outlook for the industry will return to one of reducing returns and increasing costs. A major opportunity exists to improve the long term viability of the industry by utilising all the components in sugarcane to produce not only sucrose but a range of high-value by-products and hence reduce exposure to sugar price cycles and volatility.

This paper considers the Australian industry and highlights some perceived deficiencies with the current processes. New technology is discussed which would allow the key components of sugarcane to be separated and more efficiently utilised. This technology allows for the separation of the constituent components of the cane prior to processing, may provide expanded opportunities for the utilisation of plant fibre and molasses, and increases the potential for novel applications of sucrose. By introducing a cane separator at the front end of a mill the authors propose a system that would produce high recovery of juice of higher purity, and higher quality uncontaminated by-products. These improved by-products in turn provide opportunities to add value to the milling process

The views expressed in this paper are the personal views of the authors and do not necessarily represent the views of the authors' organisation or companies

In: Keating, B.A. and Wilson, J.R. (eds), *Intensive Sugarcane Production: Meeting the Challenges Beyond 2000*. CAB International, Wallingford, UK, 1997.

through the application of new extraction processes, molecular biological modifications and innovative fermentation techniques. Research areas that could lead to significant changes in the current sugar processing system are highlighted. They include a liquid separation process that produces very low sugar colour, the evaluation of cane wax and its constituent chemicals, the application of enzymes for improved paper pulp or pretreatment of bagasse for animal feed, ethanol production through direct microbial conversion, and low joule sucrose sweeteners. A vision of the sugarcane processing factory of the future is presented.

Introduction

Sugar milling in Australia is a mature industry; 131 years old in Queensland. Like most mature industries, it is also very conservative. Despite changes towards increased capacity, reduced manning levels (automation) and improved equipment design, the fundamental processes have remained unchanged.

The sugar industry is a major export earner for Australia. In 1994, around 5Mt of raw sugar were produced from nearly 35Mt of harvested sugarcane. Approximately 85% of this sugar was exported earning around A$1.5B (CANEGROWERS, 1996). The Australian industry is highly competitive and the area under sugarcane is currently undergoing major expansion. However, in real terms, the value of sugar is declining. If the value of sugar had kept up with inflation since 1966, the current value would be around A616t^{-1}$, compared to the actual value of A320t^{-1}$. Many cost saving measures have already been implemented in milling, and economies from continuous (24h, 7 days per week) crushing and expansion will soon have been realised. To enhance future competitiveness, it is suggested that sugar mills will need to diversify into other products.

A sugarcane stalk contains many valuable components other than sucrose, and the sugar mill of the future could produce many products apart from raw sugar and molasses. In this paper, we will explore some of the exciting new developments for the processing of sugarcane at the mill and the subsequent use of sugarcane by-products, and describe how a mill of the future might operate. This paper cannot present a detailed summary of all aspects of by-product utilisation and modification. To readers with specific interests in this subject, we recommend reviews by GEPLACEA (1988), Paturau (1989) and Singh and Solomon (1995).

Current Status of Sugar Milling in Australia

Currently, the main objective of sugarcane processing in Australia is the extraction of (relatively pure) sucrose. To that end, the fundamental stages are:

1. Receive cane (billets, extraneous matter and dirt).
2. Break down the fibre to open up juice bearing (parenchyma) cells.
3. Crush the fibre and wash it with water to extract the juice.
4. Clarify the juice to remove as many impurities as possible.
5. Concentrate the juice by evaporation.
6. Crystallise - usually in a series of three boilings.
7. Separate the crystals from the mother molasses by centrifugation.
8. Dry and cool the sugar crystals.
9. Cool and dilute the final molasses.

The process is fundamentally flawed. The first steps intimately mix the sucrose with all of the other presently unwanted components present in the cane supply. The rest of the process then expends considerable capital and energy in an attempt to separate them all again. The technologies to avoid the wasteful aspects of the current system exist.

To date, mills in Australia have been interested only in producing sucrose. Whilst a number of by-product processes have been developed and commercialised around the world (GEPLACEA, 1988), in Australia the main interests have been the production of ethanol from molasses, power generation for export into the electricity grid (cogeneration) and bagasse-based animal feed. On average, in Australian mills 100t of sugarcane will yield 14.3t of raw sugar, 27.2t of bagasse, 5.2t of filter cake, 2.6t of molasses, and 50.7t of waste water. Cuban reports show similar production trends, viz. 12.6% raw sugar, 28.0% bagasse, 4.0% filter cake, 3.2% molasses and 52.2% water (GEPLACEA, 1988).

Figure 15.1 shows a simplified Australian sugar mill process, indicating that there are few end-products. At present the value of the non-sucrose by-products is relatively negligible. Currently, sugarcane is shredded in hammer mills and crushed in a multi-roller milling train. Bagasse is burnt in boilers to raise steam and generate power. Juice is clarified and filtered, producing filter cake as a by-product, before being concentrated by multiple effect evaporation. The syrup is subjected to multiple crystallisation and centrifugal separation to produce raw sugar and molasses.

The current model for Australian milling outputs (Fig. 15.1) has not changed significantly for more than 100 years. The high value of sucrose has meant that there has been only limited interest in the utility of other outputs. However, in light of increased international competition and comparatively low (and decreasing) sucrose prices (Fry, 1994) there will be an increased effort to maximise other economically viable outputs from the milling process. In Australia, molasses is currently sold at low-value for animal feed or as distillery feed-stock. Bagasse is burnt in low efficiency boilers to provide energy for mills and sometimes for export to the electricity grid. Mills have a surplus of bagasse and operate energy inefficient factories in an attempt to use up the excess. Disposal of unburnt bagasse can cause severe environmental problems associated with water run-off. Filter cake (produced by the juice clarification process) and the ash from the boilers are returned to the cane fields as fertiliser at a negative return to the mill, and waste water is usually treated and discharged.

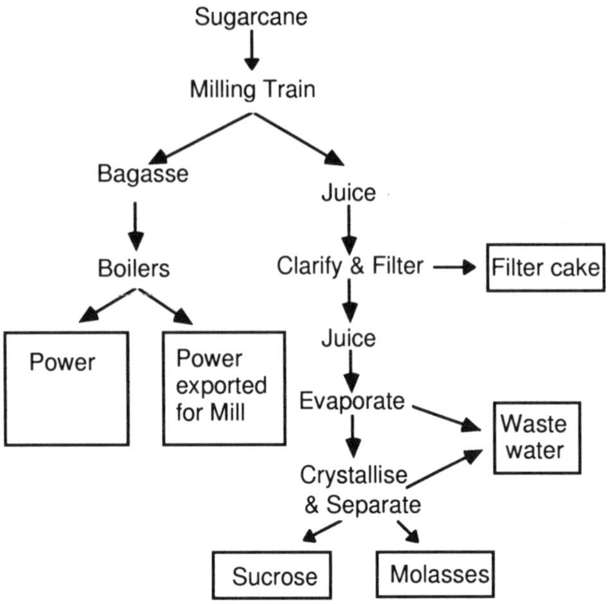

Fig. 15.1. Schematic diagram of sugarcane processing as currently practised in Australia. Valuable end-products are shown in boxes.

New technologies are emerging and have potential in four areas of sugarcane processing: (i) the removal and utilisation of impurities from cane before or immediately after milling, (ii) better utilisation of plant fibre, (iii) improved utilisation of molasses, and (iv) the use of sucrose as a substrate for novel products. The current state of our understanding and the potential for improvement are presented below.

Impurity removal and utilisation

The potential for removal of impurities from cane offers significant benefits, not only for obtaining valuable by-products such as wax and polysaccharides but also enhanced milling, clarification, and crystallisation. In addition, the removal of dirt, which enters the process adhering to the cane billets, is likely to lead to substantial maintenance savings.

Currently, the sugarcane supplied to the mill contains a variety of materials: billets of sound cane, dirt, tops (cabbage), roots, green and dead leaf material, other materials (rocks, harvester parts, dead animals, etc). Only the billets of sound cane are of any use for sugar processing. They comprise three layers. The thin outer epidermis contains wax and a number of complex polysaccharides which contribute to sugar recovery and quality problems. The rind contains long fibrous cells that provide the structural strength of the cane stalk. Rind contains most of the impurities, the unwanted colour-forming components and some sugar.

The inner section of the stalk, or pith, has less of the long fibre cells and more of the soft, shorter fibres. It also contains most of the sugar-bearing juice (parenchyma) cells. There is considerable scope to improve: (i) the quality of the raw material provided to the mills through cane cleaning procedures, (ii) the processing of cane billets at the mill through new separation technologies, and (iii) the extraction of sucrose via new filtration techniques.

Fibre utilisation

Bagasse is a rich potential source of many materials. It is a renewable resource and processes based on it would be 'greenhouse neutral' and would add significant value to what is currently a waste product. Bagasse comprises water, sucrose and other soluble solids, wax and other insoluble solids, and fibre. The fibre component (cellulose, hemicellulose and lignin) is made up of tough, hard-walled, long cylindrical cells of the rind and vascular tissues (true fibre) and soft, thin-walled, irregular-shaped parenchymatous cells of the inner stalk tissue (pith).

Bagasse can be broken down by a two-stage hydrolysis to yield glucose and xylose. A key area of current research is optimising the conversion of cellulose to glucose and xylan to xylose. If this technology can be perfected at low cost, glucose and xylose are the starting point for a range of high value chemicals, such as xylitol (Gurgel *et al.*, 1995). Xylitol is metabolised independently of insulin in the mammalian system, and has application in the preparation of foodstuffs for diabetics as well as being used as a generic low-joule sweetener (Ramalinga Setty *et al.*, 1995).

Technologies for production of chemicals such as furfural, cogeneration, and the production of medium density fibre board or paper from bagasse are already well established (Paturau, 1989). As new uses are developed for bagasse, the operational focus of mills will switch from energy inefficient to energy efficient as the fibre component becomes too valuable a resource to be wasted.

Molasses utilisation

There is considerable scope for improving the utilisation of molasses. Molasses contains around 35% sucrose, 20% water, 15% reducing sugars (mainly glucose and fructose), 15% ash (mainly inorganic salts), and small quantities of organic materials, including proteins, amino acids, nucleotides, peptides, vitamins, organic acids, phytosterols, gums, dextran, etc.

There are a number of opportunities to add value to this stream. The production of high-value chemicals by biotechnology has major potential. A number of minor components, such as phytosterols and aconitic acid, might be economically extracted. The recovery of additional sucrose or the production of a pure glucose/fructose sweetener offer very attractive possibilities.

Novel products from sucrose

To date the international demand for sucrose as an end-product has meant that there has been little economic advantage to the use of sugarcane sucrose as a feed stock for the production of valuable compounds, despite its high chemical purity. Market trends and the progress of microbial manipulation will make such production systems increasingly attractive into the next century.

Within this framework, we will now discuss relevant new technologies, and offer a view of how milling facilities could develop to cope with the needs of the future.

New Technologies for Sugar Milling and By-Product Processing

Impurity removal and utilisation

Cane cleaning

It is generally recognised that extraneous matter in the cane supply is deleterious to sugar milling operations and is a major cost to the industry as a whole. In some countries the industry is already practising cane cleaning techniques. However, in Australia the current cane payment arrangement hinders the introduction of such practices.

Extraneous matter of major concern in milling are dirt, trash and tops. Tops introduce impurities into the mill, marginally reducing sugar recovery. Attempts to reduce the level of tops in cane would need to be tempered against acceptable cane loss. Dirt in cane causes extensive wear in mills and boilers and increases mud losses and fly ash disposal costs.

With current green cane harvesting, there is limited scope to remove more trash in the field without increasing the loss of sound billets. Current losses of sound billets are estimated to account for between 2% and 8% of cane harvested - a significant but calculated loss to the industry. Recovering more billets, and therefore increasing the trash content, significantly reduces cane weight per bin and increases the transportation costs. The effect of the trash in the factory is felt in reduced sugar recovery, decreased crushing rate and lower CCS.

The 21st century mill will employ both dry cleaning systems (which remove trash) and wet cleaning systems (which remove dirt). The feed to the mill will be clean billets of cane. Dirt will be recovered and returned to the fields. The extra fibre, supplied as trash, may be used for animal feed production or will be conveyed directly to the boilers to generate energy.

Cane separation

The 21st century sugar mill will employ cane separation technology as part of its by-product strategy, probably alongside conventional crushing mills or diffusers. Figure 15.2 shows the principles of a modern cane separator. Clean cane billets are fed vertically into the separator from an accelerating conveyor. The billets, guided by two rollers, hit a wedge and are split into two halves along their longitudinal axis. Each half billet is then transferred through identical stations in the separator. Intermediate feed rolls flatten the billet halves prior to entering the pith milling station. The pith milling station removes the internal pith juice fraction by milling the soft pith material, fracturing over 98% of the juice bearing pith cells in the process. This process consumes much less energy than conventional shredding and milling. The speed of the billet half is controlled by a transporter roll which grips the rind side of the billet half while the high speed milling cutter removes the pith/juice fraction. The pith/juice fraction is transferred via a screw conveyor to a juice press which yields high purity sugar juice without addition of water. Pith moistures of 45% are typical. The billet half, now in the form of a rind strip, enters the epidermis removal station. The transporter and milling rolls are reversed, the transporter now grips the pith side surface while the milling roll removes the epidermis layer from the rind strip. The rind may be further processed as required. It may need to be de-sugared, depending on the application. Because the rind contains most of the impurities, this material is very low purity and would be considered either for addition to B molasses or to pass directly to final molasses.

The products from a conventional cane separator are: rind, pith, epidermis, high purity/high brix juice and low purity rind juice (if rind is extracted). Clearances can be set to provide the desired thickness of rind. Clearly, as billet diameters vary the amount of pith in the rind and/or rind in the pith will vary. Depending on the applications for the products, it may be desirable to have both rind free pith and pith free rind. CSR and AmCane are exploring the design of a separator with an additional separation station that would ensure pith free rind and rind free pith to produce an additional pith/rind product that could be conveyed to an existing milling train for processing.

Trials by CSR on a 1t h^{-1} cane separator unit yielded juice of 82% to 88% sugar. Considering the lower energy cost, the potential value of the epidermis, rind, and pith fractions, and the fact that mud and molasses losses will be reduced or eliminated; a recovery of 88% is very attractive. The modern machines, supplied by AmCane International Inc. have solved the mechanical problems of earlier versions. Replacement of the milling stations is a straightforward operation. Whilst AmCane has a design for a 100t h^{-1} unit, the largest unit in service (in India) is rated at 50t h^{-1}. Smaller units are in commercial operation in Jamaica, Mexico and South Africa. Current milling rates with existing technologies are up to 1000t h^{-1}.

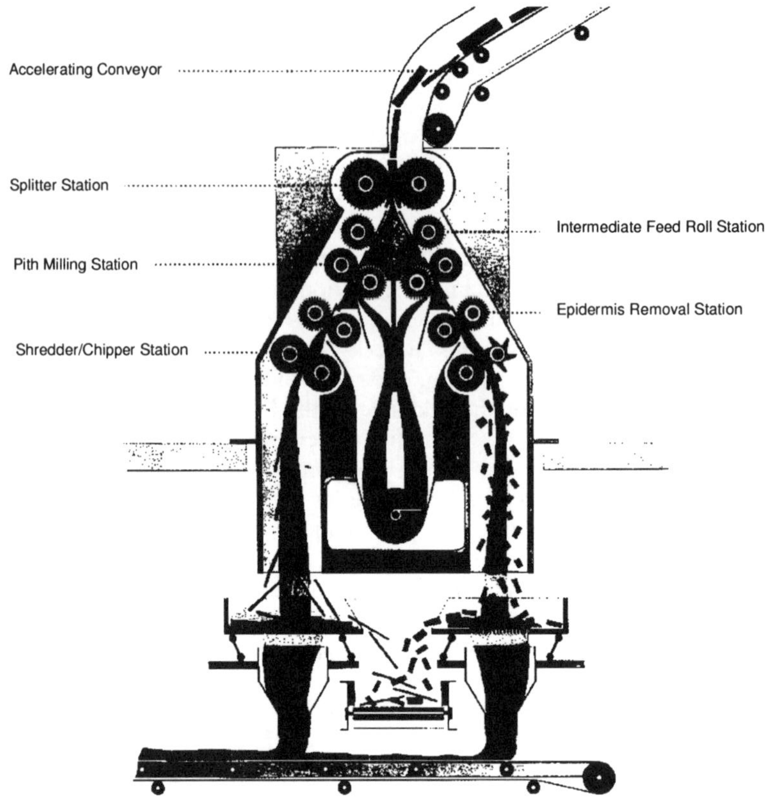

Fig. 15.2. Schematic of an AmCane separator. [Reprinted with the kind permission of AmCane International Inc.]

Liquid separation processes

Continuing developments in membrane, ion exchange and ion exclusion technologies will provide opportunities for these separation techniques to be successfully employed in sugar mills. These techniques can be utilised at the juice or syrup stages of the process to remove unwanted polysaccharides and colour forming components. Rodman (1996) reports that Puunene Factory in Hawaii is already using a 270t h^{-1} unit for juice ultrafiltration and softening to produce a very low colour sugar (super VLC). Flux rates of 2651 m^{-2} h^{-1} have been reported and a capital cost of US$4.5M for the unit makes it an attractive option to conventional decolourisation technologies. Saska (1995) provides a comprehensive review of the current membrane technologies applicable to the sugar industry and reports on significant improvements in crystal yield as a result of the removal of colloids and polysaccharides that contribute to high molasses and massecuite viscosities.

Coupled with chromatographic separation technology, it is possible to recover 90% of the initial sucrose content of molasses as a low-colour crystalline product and 95% of invert sugars as a relatively salt-free liquid product suitable for fermentation (Saska and Lancrenon, 1994). New, cost-effective processes to produce refined sugar at sugar mills are now viable. A leading company in this area, Applexion, has transferred membrane technology from the dairy, food and pharmaceutical industries to sugarcane processing. Pilot operations have shown that a combination of flocculation and ultrafiltration processes on raw sugar remelt remove 90% of turbidity and 50% of colour (Cartier *et al.*, 1996).

Ongoing research at James Cook University (JCU), in conjunction with Notre Dame University in the USA, is developing supported liquid membranes that may be able to separate valuable components from final molasses to upgrade its value. Pilot plant experiments in Australia have suggested that electrodialysis might be an appropriate technology to remove ash components, particularly potassium from factory syrups. The potassium has value as a fertiliser to be returned to the fields. The removal of potassium has a major impact on reducing sucrose solubility and increasing sugar recovery.

Impurity utilisation: waxes and sterols

Sugarcane wax is a generic term used to describe the waxy lipids and fatty lipids contained in sugarcane. The waxy lipids, or true cane wax, are located on the outside of the sugarcane stalk as a whitish, waxy coating. The wax is located mainly at the nodes and is thought to regulate evaporation from the surface of the plant. Wax is lost from the cane during burning and under prolonged exposure to ultra-violet light. The fatty lipids occur in varying amounts throughout the cane stalk. They are essential components of cell protoplasm.

There is a growing trend to replace synthetic materials with more natural ones and markets for natural waxes are showing substantial growth. There are many plant waxes available, however sugarcane wax has significant potential. CSR and JCU are developing new technology in this area and working closely with two Japanese companies towards new markets for sugarcane wax products. Most wax will be extracted from filter cake, but cane separation technology offers the opportunity to obtain very high quality wax from the epidermis without the damaging processing that it normally undergoes in the mill.

In conjunction with CSR, JCU researchers have developed a new technique for sugarcane wax refining and decolourisation. This process is currently under trial at Victoria Mill (in Queensland). Other valuable chemicals, such as sterols, triglycerides, and high molecular weight alcohols have also been identified in the crude wax. Future work will include optimisation of the process at the pilot scale, development of extraction process for other valuable components, and a study of the effect of process conditions and sugarcane plant variability on the quality and yield of sugarcane wax.

Chemical analysis of sugarcane wax is complicated by the fact that crude wax is a combination of both organic and inorganic compounds. A combination of various analytical techniques is being used to decipher the various chemical components of the wax. Analytical techniques, which include gas chromatography, nuclear magnetic resonance, Fourier transform-infra red, high performance liquid chromatography, carbon/hydrogen/nitrogen analyser and gas chromatography-mass spectrophotometry, are being utilised to investigate the organic components. Inorganic components are being studied using atomic absorption spectrophotometry, X-ray fluorescence and X-ray diffraction.

Impurity utilisation: other impurities

With improved understanding of the chemistry of some of the minor impurities in sugarcane, the possibilities for isolating small volume, high-value by-products will continue to increase. Some possibilities, already identified, are aconitic acid (animal feed supplement), polysaccharides (food thickening applications) and agents for health and cosmetic markets. Two such by-products extracted from sugarcane with increasing commercial potential are glycolic acid (an alpha hydroxy acid from the water soluble fraction) and octacosanol (a higher chain alcohol from the wax fraction). Glycolic acid has been demonstrated to produce a significant improvement on ichthyotic/xerotic human skin (Dinardo *et al.*, 1996) when applied topically, and is also used as a facial peel on sun damaged skin (Moy *et al.*, 1996b). It also has a photoprotective effect with an SPF of 2.4 and can accelerate resolution of erythema after UV exposure (Perricone and Dinardo, 1996). Its ability to decrease rough texture and fine wrinkling of skin (Newman *et al.*, 1996), possibly by its ability to stimulate collagen production (Moy *et al.*, 1996a), is becoming increasingly popular in cosmetics. In 1995, US$850M was spent on anti-ageing skin products in the US alone (Mestel, 1996). Octacosanol improves endurance training in mice, and lowers serum cholesterol (Shimura *et al.*, 1987). Octacosanol is marketed as a sports nutritional supplement in Australia, and as a patented cholesterol lowering agent and restorative for elderly people in Cuba (Anon, 1994).

Fibre utilisation

Cogeneration

A number of Australian sugar mills are moving into large-scale cogeneration of power. The returns from this type of fibre utilisation will be modest but guaranteed, so the risk factor is low. The implication for the mill of the future is that it will be necessary to conserve bagasse and maximise the amount of electricity exported. Optimum factory operations will have to be achieved to balance sugar recovered and energy exported.

Typically, sugar mills use 0.5 to 0.7 tonnes of steam per tonne of cane. In the beet sugar industry, where there is no free fuel, energy consumption is considerably lower (Fry, 1997). The 21st century mill will have to employ some of the energy efficient technologies of the beet industry, or develop its own. The application of techniques such as Pinch Technology, which have realised enormous energy savings in the process industries, will become more common.

Equipment will become continuous, stoppages and energy losses will be minimised, intermediate storage will be reduced, and new equipment will be used. New evaporator technology, efficient use of low grade vapour, and technologies such as vapour recompression will be employed. New technologies for storage and handling of large volumes of bagasse will need to be developed in order to minimise the risk of environmental damage from storm water run off. Bagasse gasification is an option for power generation that will come under increased scrutiny.

Paper pulp

Internationally, bagasse has been used for many years as a pulping feedstock for papers ranging from premium papers (for which it is usually mixed with a small proportion of long-fibre pulp to enhance drainage and strength characteristics) to newsprint. Annual international paper production is currently about 170Mt, of which more than 2Mt are produced from bagasse (Almazan, 1994). The largest bagasse paper producer is Mexico but it is also produced in many other countries. Several feasibility studies have been completed in Australia, although currently the only bagasse paper which can be purchased in Australia is imported.

Pulping characteristics and paper quality are directly related to the length of the fibres used. Bagasse consists of two principal types of fibre tissue; (i) the cellulose fibre of the rind and vascular tissue (vessels), and (ii) the pith (mainly parenchyma cells) of the cane core. On a dry weight basis these two components constitute approximately 55% of the bagasse residue. Bagasse cellulose fibres of the rind and vessels (at 1-1.5mm long) are longer than hardwood fibres (at 0.7-1.5mm) but shorter than premium softwood fibres (at 2.5-5mm), and are generally suitable for paper pulping. Conversely pith material, whilst chemically similar to cellulose fibre, consists of irregular cells which swell and absorb moisture. The presence of these parenchyma cells weakens the pulp, slows the drainage rate, decreases the pulping yield and increases the need for pulping chemicals during paper production (I.M. Wood, *pers. comm.*).

As pith is effectively a contaminant for the pulping process, and even small quantities of pith adversely affect paper strength and quality, it is desirable to remove as much pith material as possible from the bagasse prior to processing. The previously described cane separation process is being investigated in Japan to provide a pith-free rind fraction for very high quality paper.

In the absence of cane separation technologies, a new strategy for using bagasse to produce paper pulp has recently been evaluated by I.M. Wood and Associates and the CSIRO Division of Forest Products. The process involves the introduction of high performance cellulase enzymes to the fibre pulp immediately prior to bleaching. The objective is to increase the freeness or drainage rate of the pulp by digesting the parenchyma cell walls in a similar way to the solubilisation of xylan in wood pulp - an essential phase in the paper brightening process. Evidently the cellulase activity which removes pith material from the pulp will also cleave the desired longer cellulose fibres from the pith and vessels, reducing final pulp yield. The key to this new technology is the development and use of enzymes with controlled cellulase activity and reconciling yield loss against increased productivity through higher rates of drainage. Initial tests indicated a post treatment increase in drainage rate of up to 50% and a pulp yield loss of only 5% (I.M. Wood, unpublished data). Further work on the use of enzymes on non-wood fibre for paper should identify an improved combination of enzymes for the production of a quality pulp from fibre sources, such as bagasse.

Animal feed

The value of bagasse as a viable animal feed for ruminants and monogastrics is beginning to be realised in Australia. This has come about with the advent of the increasing live animal export trade to Asia from northern Australian ports. This, coupled with increased world human consumption of grains for protein, means that at some stage in the future there will be direct competition between humans and animals for grains. The shift from grain to fibre-based animal feeds in feedlots will make bagasse-based animal feed even more viable in the future.

Bagasse has a digestibility of c. 25% for ruminants. Research over many years has shown that chemical, thermo-mechanical and biological treatment of bagasse increases the digestibility to a minimum of c. 65% (Playne, 1984). This digestibility level makes the treated fibre suitable as a commercial stockfeed.

Alkali-treated bagasse, combined with molasses, protein and nutrients, produces a product that has a Metabolisable Energy (ME) content equivalent to that of a good quality lucerne hay. ME is a parameter of ruminant nutrition that measures the amount of energy in a product that can be utilised by the animal. Because of energy content, the bagasse-based animal feed has excellent commercial potential with diverse feed applications. Canefibre Products has established a pilot plant at CSR's Pioneer Mill in the Burdekin region south of Townsville. Pelletised bagasse-based products from this facility are already receiving wide acceptance in the feed industry.

Dietary fibre

The cane separator system opens up the possibility of accessing novel valuable markets. Milled pith can yield a fibre that is ideal for human consumption in such products as bread, breakfast cereals, snack bars, and juice drinks. A separation system like the cane separator can produce material that is clean enough for human consumption. Australia imports large amounts of dietary fibre, so a ready-made market exists for a competitive product.

Timber replacement

Depithed bagasse is widely used as a raw material for the production of medium density fibre board (MDF). A product as good as, or superior to, wood-based MDF can be produced. The lignins present in bagasse can be utilised as resins. To date, the distance from markets and the ready availability of plantation forests in Australia has prevented this potential being exploited.

The rind from a cane separator mill, however, offers the opportunity to develop new board products that can replace timber products such as plywood, particularly in the large Japanese market. The rind can be used in strips to produce an oriented strand board that is both decorative and structurally very strong. Alternatively, it can be chipped and used to produce a chip board. Tests, in Japan and Australia, have demonstrated that it is possible to meet the very strict Japanese standards for plywood.

Ethanol production from bagasse

Bagasse can be alkali-treated to extract the carbohydrates, cellulose and xylan, which can be further cleaved by an acid hydrolysis step to yield the monosaccharides, glucose and xylose, respectively. These sugars are highly fermentable by bacteria and yeast, and can thus be converted into useful by-products. Another useful end-product could be the bacteria or yeast biomass in its own right.

One particularly attractive fermentation by-product is fuel ethanol. Presently, when ethanol is produced conventionally from sugar or starch, the cost becomes uncompetitive when compared with fossil fuel. However, if lignocellulose (i.e. bagasse) is used as a feedstock, costs of production could decrease significantly. Such cost advantages would depend significantly on the market value of bagasse (which is anticipated to increase), introducing a further level of complexity to the feasibility of the options discussed below. Drawing on analogies from pilot wood chip to ethanol conversion plants, and assuming that the fermentation plant is at or nearby the sugar mill to avoid transportation charges, ethanol could be produced for A\$0.40 t^{-1}, with a potential yield of c. 200-400l t^{-1} of bagasse dry weight (Nguyen, 1993; Anon, 1996). At this price, ethanol starts to become cost effective as an alternative octane enhancer in lead-free petrol (Nguyen, 1993) but is still about 25% more expensive than the cost of refining petrol from crude oil (Anon, 1996).

The cost viability of this process could be further improved by the use of enzymes in place of acid to effect the conversion of bagasse to fermentable sugars. Once again, drawing on analogies from pilot wood chip plants, bagasse could be steam exploded and then saccharified by enzyme hydrolysis using powerful xylanases and cellulases produced on site. The soluble sugars so produced would then feed the fermentation reactor, the so-called separate hydrolysis and fermentation (SHF) process (Nguyen, 1993). An even more cost effective alternative would be to combine enzyme hydrolysis and fermentation in the one reactor vessel, the so-called simultaneous saccharification and fermentation (SSF) process This not only saves money on reactor hardware, but also alleviates the end-product inhibition of glucose and cellobiose on cellulase activity, as the sugars are fermented as soon as they are produced. It has been estimated by Nguyen (1993) that if a plant size of 1000t of wood per day is assumed, and a cost of wood material is US\$10 t^{-1}, the production costs of ethanol would be US25c l^{-1} using the SSF process.

Another significant improvement could be made by the genetic engineering of ethanol-producing yeasts which secrete their own package of cellulases and xylanases. These modified yeasts could then directly cleave and ferment the longer chain polymers of glucose and xylose produced from partial cleavage of bagasse during the first enzyme hydrolysis step, by a single step, the so-called direct microbial conversion (DMC) process.

Development of genetically engineered yeasts or bacteria which secrete cellulases and/or xylanases and are able to produce ethanol more efficiently is now feasible using present cloning techniques. Construction of such a recombinant microorganism would lower the cost of on-site enzyme production for the first hydrolysis step, which is considered to contribute 20% to the overall cost of production. The enzyme-dominant procedures have environmental appeal over the acid hydrolysis procedure for sugar production from lignocellulose, as the acid procedures can produce difficult-to-treat waste products. Enzyme produced on site could also be used in pulp bleaching for paper production, which may also help share the cost of enzyme production.

Another cost saving measure in the bagasse to ethanol conversion process may be to combine ethanol production with yeast manufacture for animal feed. Some strains of yeast have probiotic characteristics in ruminants and monogastrics, i.e., when added to feed they may improve animal health and production (Wallace and Newbold, 1993). It may also be possible to genetically engineer yeast to overproduce lysine, a limiting amino acid in animal feed and could itself be a valuable by-product. Lysine encased within the yeast cell allows the amino acid to avoid degradation in the rumen of cattle and sheep, and deliver direct benefits to the animal (Ohsumi *et al.*, 1994). Thus the lysine need not be extracted from the yeast cell. A lysine-producing yeast which also produces fuel ethanol could further lower the cost of the total fermentation.

Soluble short chain oligomers of xylose and glucose produced after partial enzyme hydrolysis of steam-exploded bagasse may also be useful by-products. Recently, similar oligomers have been demonstrated to have 'prebiotic' activity, stimulating the so-called beneficial intestinal microflora of monogastrics, and could be candidates as health-enhancing functional food ingredients (Gibson and Roberfroid, 1995).

Residual lignin 'waste' product after bagasse hydrolysis could be burnt on site for electricity cogeneration, as it has a higher calorific value than the fermentable components of bagasse. Alternatively, it could be transferred to the vanillin-resin stream, or possibly used as a component in slow release fertiliser (Garcia *et al.*, 1996).

Chemicals

Glucose, xylose and lignin derived from bagasse will offer the potential to develop low-volume, high-value chemical products either by chemical or biochemical processes. Activated carbon production is another possibility. Processes will need to be flexible and market driven as the possibilities for such products are explored.

Molasses utilisation

Where local economic or resource limitations make such use efficient, molasses is diluted and used for the production of ethanol and ethanol-derived products (Dwivedi, 1995) as well as production of yeasts for animal feed and human consumption (GEPLACEA, 1988). It is anticipated that fermentation of molasses to ethanol and subsequent modification to ethylene and then petro-chemical products will expand over the next decade. To date, molasses utilisation has been limited in Australia, however the production of high-value chemicals by biotechnology has major potential. A number of minor components, such as phytosterols and aconitic acid, might be economically extracted. The recovery of additional sucrose or the production of a pure glucose/fructose sweetener offer very attractive possibilities, and the extraction of glucose, fructose and sucrose from molasses may allow the production of an alternative invert syrup.

Novel products from sucrose

Sucrose as a sweetener

Despite the fact that the predominant dietary contributor of obesity in the Western diet is fat intake (Boltonsmith, 1996), there remains a public perception that controlling sugar intake equates with weight loss. To this end, there has been much research towards the development of low-joule, non-sucrose sweeteners. Some such sweeteners (saccharin and cyclamates) were found to be carcinogenic under some circumstances, whilst others (such as aspartame) exhibit temperature and low pH instability which limit their utility (Knight, 1994). A new sucrose-based low-joule sweetener is sucralose (1, 6-dichloro-1, 6-dideoxy-ß-D-fructofuranosyl-4-chloro-4-deoxy-α-D-galactopyranoside) (Hough and Khan, 1989). Sucralose is a tri-chloro substituted form of sucrose, which is produced from sucrose and is suitable for a wide range of food industry uses (Quinlan and Jenner, 1990). Despite the arguable need for low-joule sweeteners, it seems likely that the market will expand in the future, and there remains further scope for the development of new low-joule sweeteners as derivatives of sucrose.

The cane separator provides a clean, high-purity juice that has already been exploited in several places in the world (Jamaica, South Africa and Mexico). Opportunities exist to process this juice into fruit juices, table syrups, or total sugar products. The key marketing aspect is that these are natural products, produced without the addition of chemicals. It only requires a further step of developing organically grown sugarcane to be able to produce organic sugar.

Novel uses for sucrose

There are a number of valuable products which can be produced by chemical or microbial modification of sucrose (Abidi, 1995). Sucrose can be hydrolysed by acid, base or enzymes and used as a substrate for fermentation. Sucrose can also be used as a nutrient for the growth of micro-organisms which yield antibiotics (such as penicillin, aureomycin, terramycin and erythromycin), essential vitamins (such as riboflavin and cobamide) and polysaccharides (such as dextrans and gums). Sucrose is also being used as a feedstock for the production of a range of polyesters and polymers. There is considerable scope for the production of high value polymers from sucrose by microbial or chemical conversion. Valuable plastics such as polyhydroxybutyrate can be produced biologically, and there is exciting research into sucrose-based fixatives (Sachinvala *et al.*, 1995). A final area of development of novel sucrose-based products has been in the domain of organo-metallic complexes. There is potential for such complexes to be used as pesticides such as organostannyl sucrose pthalates (Poller and Perkin, 1977), and as anti-cancer agents such as sucrose-platinum complexes (Sachinvala *et al.*, 1993) although further development is necessary.

A Possible Future for Sugar Milling in Australia

As the Australian sugarcane industry moves into the 21st century, it is faced with a number of challenges. The opportunities for cost reduction will become limited and the real value of its current products, sugar and molasses, will continue to decline. Confronted by other political and economic necessities, the sugarcane agro-industries of some countries have already diversified profitably. Cuba for instance produces 28 products other than raw sugar, molasses and bagasse (refined sugar, five different types of alcohol, rum, *Saccharomyces* yeast, *Torula* yeast, paper, fluting, bagasse boards, syrups, dextran, confectionery, crude wax, filter cake oil and resins, high-protein molasses, molasses-urea mix, molasses-urea-pith mix, pre-digested pith, dehydrated filter cake, sugarcane harvest residues, glucose, fructose, sorbitol and furfural) (Almazan, 1994). The long term viability of the Australian industry will be enhanced by developing new products to supplement its income and reduce exposure to sugar price volatility. The sugarcane plant contains a wide range of valuable components which have enormous potential as new, high-value products. All sections of the industry will need to work together to maximise the returns available from the new technologies and exploit these exciting opportunities. We believe that with the adoption of new milling and processing technologies, the output diagram presented originally as Fig. 15.1 could be successfully and profitably amended as shown in Fig. 15.3. The future is exciting and will offer many challenges, frustrations and rewards in developing and expanding new markets for 'value-added' sugar products.

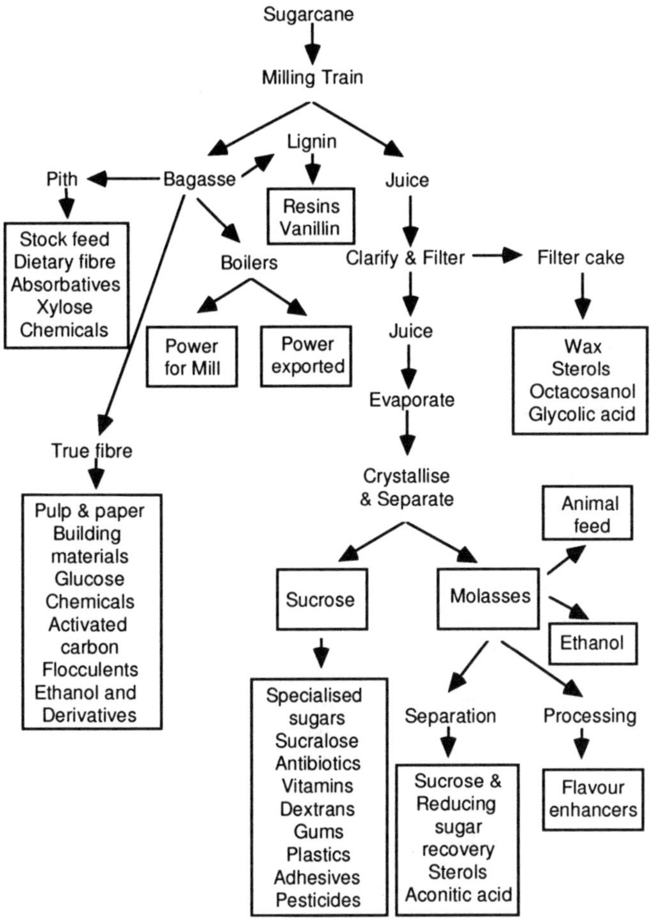

Fig. 15.3. Schematic diagram of Australian sugarcane processing as it might appear in the 21st century. Valuable end-products are shown in boxes.

Acknowledgments

The authors gratefully acknowledge the assistance of the following organisations in the preparation of this paper: CSR Sugar Mills Group, Qld; CSIRO Tropical Agriculture, St Lucia, Qld; CSIRO Division of Forestry and Forest Products, Clayton, Vic; I.M. Wood and Associates, Kenmore Hills, Qld; AmCane International Inc, St Paul, Minnesota, USA; James Cook University of North Queensland, Department of Chemistry & Chemical Engineering, Townsville; Sugar Research and Development Corporation; Australian Research Council; Department of Industry, Science & Technology; Chori Company Limited.

References

Abidi, A.B. (1995) Chemical and microbial products from sucrose: Industrial applications. In: Singh, G.B. and Solomon, S. (eds), *Sugarcane. Agro-industrial Alternatives.* Oxford & IBH, New Delhi, India, pp. 199-210.

Almazan, O. (1994) Past, present and future of sugarcane by-products in Cuba. *Cooperative Sugar* 26, 253-263.

Anon. (1994) A Mixture of Higher Primary Aliphatic Alcohols, its Obtention From Sugar Cane Wax and its Pharmaceutical Uses. *International Patent Application* WO 94/07830.

Anon. (1996) Ethanol comes of age. *Green and Gold* 3, 4-5.

Boltonsmith, C. (1996) Intake of sugars in relation to fatness and micronutrient adequacy. *International Journal of Obesity* 20, Suppl. 2, S31-S33.

CANEGROWERS (1996) *Information Handbook - 1996.* CANEGROWERS, Brisbane, Australia, pp. 4-12.

Cartier, S., Theoleyre, M.A., Lancrenon, X. and Decloux, M. (1996) Membrane Technology in the Sugar Industry. *Proceedings of the Workshop on Separation Processes in the Sugar Industry.* New Orleans, USA (In press).

Dinardo, J.C., Grove, G.L. and Moy, L.S. (1996) Clinical and histological effects of glycolic acid at different concentrations and pH levels. *Dermatologic Surgery* 22, 421-424.

Dwivedi, R.S. (1995) Development of molasses and alcohol based chemical industries. In: Singh, G.B. and Solomon, S. (eds), *Sugarcane. Agro-industrial Alternatives.* Oxford & IBH, New Delhi, India, pp. 173-198.

Fry, J. (1994) The world sugar outlook and opportunities. *Proceedings of the 1994 Australian Sugar Convention*, CANEGROWERS, Brisbane, Australia, pp. 145-159.

Fry, J. (1997) A global perspective of the sugar industry. In: Keating, B.A. and Wilson, J.R. (eds), *Intensive Sugarcane Production: Meeting the Challenges Beyond 2000.* CAB International, Wallingford, UK, pp. 1-16.

Garcia, M.C., Diez, J.A., Vallejo, A., Garcia, L. and Cartagena, M.C. (1996) Use of Kraft pine lignin in controlled-release fertiliser formulations. *Industrial and Engineering Chemistry Research* 35, 245-249.

GEPLACEA (1988) *Handbook of Sugar Cane Derivatives.* GEPLACEA, Mexico.

Gibson, G.R. and Roberfroid, M.B. (1995) Dietary modulation of the human colonic microbiota: introducing the concept of prebiotics. *Journal of Nutrition* 125, 1401-1412.

Gurgel, P.V., Mancilha, I.M., Pecanha, R.P. and Siquera, J.F.M. (1995) Xylitol recovery from fermented sugarcane bagasse hydrolysate. *Bioresource Technology* 52, 219-223.

Hough, L. and Khan, R. (1989) Enhancement of the sweetness of sucrose by conversion into chlorodeoxy derivatives. In: Grenby, T.H. (ed.), *Progress in Sweeteners.* Elsevier, London, UK, pp. 97-120

Knight, I. (1994) The development and applications of sucralose, a new high-intensity sweetener. *Canadian Journal of Physiology and Pharmacology* 72, 435-439.

Mestel, R. (1996) Wrinkle wars. *New Scientist* 150, 38-42.

Moy, L.S., Howe, K. and Moy, R.L. (1996a) Glycolic acid modulation of collagen production in human skin fibroblast cultures *in vitro. Dermatologic Surgery* 22, 439-441.

Moy, L.S., Peace, S. and Moy, R.L. (1996b) Comparison of the effect of various chemical peeling agents in a mini-pig model. *Dermatologic Surgery* 22, 429-432.

Newman, N., Newman, A., Moy, L.S., Babapour, R., Harris, A.G. and Moy, R.L. (1996) Clinical improvement of photoaged skin with 50-percent glycolic acid - a double-blind vehicle-controlled study. *Dermatologic Surgery* 22, 455-460.

Nguyen, Q.A. (1993) Economic analyses of integrating a biomass-to-ethanol plant in a pulp/saw mill. In: Saddler, J.N. (ed.), *Bioconversion of Forest and Agricultural Plant Residues*. CAB International, Wallingford, UK, pp. 321-340.

Ohsumi, T., Sato, H., Yoshihara, Y. and Ikeda, S. (1994) Selection and breeding of lysine-accumulating *Saccharomyces cerevisiae* as a stable source of lysine in the rumen. *Bioscience Biotechnology and Biochemistry* 58, 1302-1305.

Paturau, J.M. (1989) *By-products of the Cane Sugar Industry: An Introduction to Their Industrial Usage*. Third Edition. Elsevier Scientific Publishing, Amsterdam, The Netherlands.

Perricone, N.V. and Dinardo, J.C. (1996) Photoprotective and antiinflammatory effects of topical glycolic acid. *Dermatologic Surgery* 22, 435-437.

Playne, M.J. (1984) Increased digestibility of bagasse by pretreatment with alkalis and steam explosion. *Biotechnology and Bioengineering* 26, 426-433.

Poller, R.L. and Perkin, A.R. (1977) Organometallic derivatives of sucrose as pesticides. In: Hickson, J.L. (ed.), *Sucrochemistry*. American Chemical Society, Washington D.C., USA, pp. 145-152.

Quinlan, M.E. and Jenner, M.R. (1990) Analysis and stability of the sweetener sucralose in beverages. *Journal of Food Science* 55, 244-246.

Ramalinga Setty, T.K., Vishwanatha, G.K. and Panth MG (1995) Newsprint, paper and chemicals from sugarcane bagasse. In: Singh, G.B. and Solomon, S. (eds), *Sugarcane. Agro-industrial Alternatives*. Oxford & IBH, New Delhi, India, pp. 79-97.

Rodman, D. (1996) CSR/JCU Internal Report on Visit to Hawaii and USA.

Sachinvala, N.D., Chen, H., Niemczura, W.P., Furusawa, E., Cramer, R.E., Rupp, J.J. and Ganjian, I. (1993) Synthesis, characterization, and anticancer activities of the first platinum complexes from sucrose. *Journal of Medicinal Chemistry* 36, 1791-1795.

Sachinvala, N.D., Ju, R.F., Litt, M.H. and Niemczura, W.P. (1995) Preparation of polymers(methyl methacrylate) and copolymers having enhanced thermal stabilities using sucrose-based comonomers and additives. *Journal of Polymer Science* 33, 15-29.

Saska, M. (1995) Membrane based separations and countercurrent multicolumn systems for decolourisation, ion-exchange and ion exclusion: Emerging technologies for the sugar industry. In: Cock, J.H. and Brekelbaum, T. (eds), *International Society of Sugarcane Technologists Proc. XXII Congress*, 11-15 September, 1995, Cartagena, Colombia. Technicaña, Cali, Columbia, pp. 147-161.

Saska, M. and Lancrenon, X. (1994) Applications of continuous chromatographic separation in the sugar industry III. Desugarization of cane molasses. *International Sugar Journal* 96, 403-410.

Shimura, S., Hasegawa, T., Takano, S. and Suzuki, T. (1987) Studies on the effect of octacosanol on motor endurance in mice. *Nutrition Reports International* 36, 1029-1038.

Singh, G.B. and Solomon, S. (1995) (eds) *Sugarcane. Agro-industrial Alternatives*. Oxford & IBH, New Delhi, India.

Wallace, R.J. and Newbold, C.J. (1993) Rumen fermentation and its manipulation: the development of yeast cultures as feed additives. In: Lyons, T.P. (ed.)., *Biotechnology in the Feed Industry*. Alltech Technical Publications, Nicholasville, Kentucky, USA, pp. 173-192.

Chapter sixteen:

Opportunities for improving the use of limited water by the sugarcane crop

M.J. Robertson[1], N.G. Inman-Bamber[2] and R.C. Muchow[1]

1. *CSIRO Tropical Agriculture and CRC for Sustainable Sugar Production, 306 Carmody Road, St Lucia, Qld 4067, Australia*
2. *South African Sugar Association, Experiment Station, Mount Edgecombe 4300, South Africa*

Abstract

With increasing costs of water and declining availability, there is a requirement to make efficient use of water applied to the sugarcane crop. This paper considers options for improving the on-farm efficiency of use of irrigation water in sugarcane production. The major avenue for improving this efficiency is through increasing the proportion of the water applied that is stored in the root zone. Another avenue is varying the timing and amount of limited irrigation to maximise crop response. Choosing the best strategy is difficult not only because of the complex interaction between variable rainfall pattern, crop growth stage, soil type, and crop management; but the many outcomes that have to be considered by a decision maker, such as average returns and their variability from year-to-year, and impacts on soil and water quality. We present a case study to show the use of cropping systems models coupled with long term climate data and economic analysis in comparing the relative merits of two irrigation strategies in the contrasting rainfall regimes at Mackay in Australia and Pongola in South Africa. Finally, we evaluate the usefulness of cane water use efficiency, the ratio of cane yield to water use, for predicting response to irrigation, and calculating the irrigation requirement under full irrigation.

In: Keating, B.A. and Wilson, J.R. (eds), *Intensive Sugarcane Production: Meeting the Challenges Beyond 2000*, CAB International, Wallingford, UK, 1997.

Introduction

Sugarcane is grown in fully irrigated, supplementary irrigated and rainfed farming systems. Efficient use of rainfall and irrigation is central to profitable and sustainable sugarcane production. World-wide, irrigation water is becoming more scarce and costly (Jensen *et al.*, 1990). In districts with limited land for industry expansion, irrigation increases productivity on existing land (Chapman and Milford, 1997). There is also a need to manage irrigation efficiently to minimise impact on the resource base where irrigation supplies are plentiful and costs are low (Meyer, 1997).

Decisions by policy makers and farmers for the use of limited water are complex, as they involve the interaction of many factors, many of them uncertain, such as rainfall. Yield response to irrigation is highly variable across seasons, and it is difficult to identify one practice that is best for every season. Indeed, for a given practice there will be a proportion of seasons in which an alternative would have performed better than the long term optimum. Information is needed on the physical, biological, economic, and risk effects of a particular option. Judgements based on experience, the results of experimental trials or commercial production may be unduly influenced by only a few years' results. This poses particular problems where climatic (rainfall) variability is high, as is the case for the sugar industries of Australia and South Africa. Cropping systems models, which can specify flexible crop management rules, are powerful tools to analyse the consequences of different options using the historical climatic record at a location.

Systems modelling has been used in the past to evaluate various irrigation management strategies with limited water in soybean (Hearn and Constable, 1981) and sorghum (Harris and Mapp, 1986) production. Boggess and Ritchie (1988) demonstrated the use of models linked to economic and risk analysis in a humid cropping region. Wegener (1990) evaluated the economic response of sugarcane to irrigation at Mackay at two levels of irrigation and concluded that irrigation increased the riskiness of income, but that this was outweighed by the additional income.

There are several caveats to observe when using models. First, outputs from models do not examine all the tangible costs and benefits of irrigation (Ferguson and Hampson, 1975) and provide only part of the total information base required for decision-making. Second, evaluation of alternative strategies may not require precise biophysical output from models because the optimal net return strategy may be highly sensitive to factors such as commodity price. Similarly, there may be little need for refined estimates of crop yield response as there may be little difference in response to irrigation on different soil types. Third, one should evaluate management scenarios that are consistent with industry practices and the social context of decision making. For example, the inconvenience of operating an irrigation system at times of peak seasonal workload may count against some strategies. This is important in South Africa, where domestic chores limit options for small scale farmers. On the other hand, model analyses could identify potentially large increases in profit or efficiency that might be sufficiently valuable to warrant a change in current operational modes. Fourth, where there are significant yield-limiting constraints (e.g. salinity) that are not considered by the cropping systems model then a scaling factor on yields may be used to convert simulated yields to on-farm yields.

In the sugarcane literature, the concept of a cane water use efficiency (WUE), defined as the ratio of cane yield (= fresh weight of millable stalk) to evapotranspiration, has been widely used to predict and analyse crop response to irrigation. Cane WUE values have primarily been developed from experimental crops grown under full irrigation, and the notion of a constant value has developed, usually quoted as about 8-10t Ml^{-1} (Thompson, 1976; Kingston, 1994). Robertson and Muchow (1994) discussed the limitations of the cane WUE concept, and likely reasons for its lack of stability within and across production environments. Nonetheless, cane WUE is widely used as a rule-of-thumb in both the Australian and South African industries. An objective of this paper is to explore, by using a sugarcane cropping system model, variation in the cane WUE, particularly in rainfed and supplementary irrigated systems. Such an analysis may point to potential improvements in the use of cane WUE as an industry rule-of-thumb.

The purpose of this paper is to highlight the opportunities for improving the efficiency of use of limited water supplies in sugarcane production. To do this we demonstrate the usefulness of the crop production system simulation model, APSIM (McCown *et al.*, 1996), to define irrigation requirements, and evaluate opportunities for best use of limited water. APSIM-Sugarcane, a module of APSIM, simulates the growth of a uniform field of sugarcane in relation to climate, water and nitrogen inputs (Keating *et al.*, 1997). The key features of APSIM relevant to this paper are that irrigation management strategies can be dynamically simulated using conditional rules (McCown *et al.*, 1996). It can simulate aspects unique to the sugarcane production system such as harvesting, ratooning, and management of trash; and impacts of management strategies on sustainability indicators, such as drainage, can be evaluated. In this paper we also consider the opportunities for farm managers to improve use of limited water by greater irrigation application efficiency to stretch a given water supply further. Finally, we examine the usefulness of the cane WUE concept as an irrigation planning and management rule-of-thumb, and identify some limitations to its use in irrigation analyses.

Defining Irrigation Requirement

A need for best use of limited irrigation supplies is quantification of the irrigation requirement for crop production. In the Australian sugar industry, this information is needed for fully irrigated districts such as Kununurra, Mareeba and Ayr, and in districts where some farms have sufficient water supplies to practice full irrigation (e.g. Bundaberg, Mackay and Grafton).

Irrigation requirement is equal to the balance between crop water demand, determined by potential yield, the efficiency of water used to produce that potential yield, and the effective rainfall. Effective rainfall is defined as that portion of total rainfall that is stored in the crop root zone for use in evapotranspiration. In the past, irrigation requirements have been calculated with simple soil water budgets using rainfall and pan evaporation, as in the publication *Irrigation of Sugarcane* (Anon., 1991). Others have used crop models (e.g. Mason *et al.*, 1980) for soybeans in the Namoi Valley, Australia. The different approaches define irrigation requirement appropriate to the information needs of different end-users, such as policy makers, irrigation scheme engineers, or individual growers. In this section we use the cropping systems model, APSIM-Sugarcane, to examine the irrigation requirement in different sugar-growing districts. This approach can also consider within each region, where warranted, the impact of soil type, management regime, crop class and variety.

APSIM-Sugarcane was configured to simulate rainfed and fully irrigated production, and hence irrigation requirement, for a standard ratoon crop for Kununurra, Mareeba, Ayr, Mackay, Bundaberg, and Grafton, using the procedure outlined in Appendix 2 of Muchow *et al.* (1997) and soil inputs specified in Table 16.1. Parameters that did not vary with soil layer were bulk density (1.25g cm^{-3}), and drainage constant (0.50d^{-1}). Other soil parameters were: stage 1 soil evaporation coefficient (4mm), stage 2 soil evaporation coefficient (3mm), runoff curve number (70), maximum reduction in curve number due to cover (20), soil albedo (0.1), diffusivity constant for soil texture (40mm^2 d^{-1}), slope constant for diffusivity/soil water content (16).

Table 16.1. Inputs to the soil water balance model for each layer of the profile.

Soil layer (mm)	Air dry water content (mm mm^{-1})	Lower limit (mm mm^{-1})	Drained upper limit (mm mm^{-1})	Saturated water content (mm mm^{-1})	Root Extraction constant (d^{-1})
0-200	0.05	0.26	0.40	0.46	0.16
200-400	0.10	0.26	0.40	0.46	0.14
400-600	0.10	0.26	0.40	0.46	0.12
600-900	0.10	0.26	0.40	0.46	0.10
900-1200	0.10	0.30	0.40	0.46	0.05
1200-1500	0.10	0.38	0.40	0.46	0.01

Twelve month ratoon crops from October to September were simulated for a non-limiting nitrogen situation. Irrigation was applied whenever 30% of the available water in the root zone had been depleted, and no water was applied during 30 days preceding harvest to simulate a drying-off period. Crop water demand was defined as the evapotranspiration simulated from the fully irrigated crop.

The simulated mean annual crop water demand for 6 Australian locations and Pongola in South Africa is presented in Table 16.2. Mean crop water demand for the 12 month Q117 first ratoon crop varied almost three-fold from nearly 3000mm at Kununurra to 1021mm at Grafton. Demand at Ayr was simulated as 1502mm and this accords well with water use of 1460mm from a 12 month well-watered ratoon crop of Q80 growing in the Ayr lysimeter (Kingston and Ham, 1975). Hardy (1966) recorded similar water use of 1421mm by a 12 month ratoon crop in Mauritius (similar latitude to Ayr). While there is little published information on crop water use for southern locations in Australia to verify model predictions, Thompson (1976) quotes 12 month water use of 1100-1300mm from ratoon crops at locations between 27-30° S in South Africa, which are similar to those values simulated here for Grafton and Bundaberg. Crop water demand of 1300mm at Pongola agrees well with lysimeter measurements of water use by 12 month ratoon crops at this location (Thompson, 1976).

Rainfall at each location supplied effectively between 450 and 630mm of the crop water demand (Table 16.2), so that rainfall effectiveness declined as mean annual rainfall across the 6 locations increased. White (1970) found effective rainfall varied from 80% in years of high rainfall to almost 100% in a drought year. Many applications assume rainfall effectiveness to be 70% (e.g. Thompson, 1977), and others have calculated it based on the difference between monthly totals of rainfall and evaporation (Kingston, 1994). The model analysis here can provide a more refined estimate of effective rainfall for specific combinations of soil type and irrigation regime. Better estimates of effective rainfall are likely to improve, amongst other things, the accuracy of crop water use estimates from rainfall and irrigation inputs (Kingston, 1994).

Table 16.2. Determinants of sugarcane irrigation requirement for 6 locations in Australian and Pongola in South Africa. [Values are mean of all years simulated over the climatic record (see Muchow *et al.*, 1997) at each location. Simulations conducted for 12 month ratoon crop of Q117 (Australia) or NCo376 (South Africa), starting on 1st October]

Location	Rainfall (mm)	Crop demand (mm)	VPD (kPa)	Effective rainfall (mm)	Irrigation requirement (mm)	Cane WUE (t Ml^{-1})
Kununurra	765	2931	2.38	554	2377	7.2
Mareeba	915	1451	1.60	488	964	12.9
Ayr	1079	1502	1.48	448	1054	12.7
Mackay	1684	1220	0.96	630	570	15.4
Bundaberg	1128	1332	1.17	581	751	12.0
Grafton	981	1002	1.35	558	443	13.1
Pongola, SA	664	1388	1.41	502	886	10.1

VPD, vapour pressure deficit

VPD, vapour pressure deficit

Irrigation requirement, calculated as the deficit between crop water demand and effective rainfall, varied from about 24Ml ha^{-1} at Kununurra to about 4.5Ml ha^{-1} at Grafton. While these simulated irrigation requirements are similar to those quoted for Grafton, Bundaberg and Mackay at 2, 6.2 and 3.8, respectively (Anon., 1991), based on effective rainfall and estimated crop water use, they differ for the other locations by 2-8Ml ha^{-1}.

Improvement in Application Efficiency

An obvious way to increase efficient utilisation of limited water supplies is improvement in irrigation application efficiency. Application efficiency may be defined as the water stored in the root zone divided by the amount of water applied (Hillel, 1990). The difference between what is stored and what is applied represents losses due to deep percolation below the root zone within the field, and to runoff from the field. This difference also includes evaporative and seepage losses incurred in conveying water from its source to the field. However, these losses are usually beyond the control of the irrigator and are not often included in the definition of application efficiency.

Application efficiency varies widely within and between irrigation methods for sugarcane: flooding, furrow, sprinkler, drip (trickle) and sub-irrigation. In general, application efficiency increases from furrow to sprinkler to drip irrigation, and changing irrigation method has been one avenue used to stretch a limited water supply. For example, in Taiwan, drip irrigation has increased cane yields by more than 20% while reducing water use by almost 40% (Yang and Chang, 1976). In Hawaii, similar benefits have accrued with the widespread adoption of drip over furrow or sprinkler methods (Jones *et al.*, 1990). While the potential for high efficiency may be greater with some systems than with others, when the economics, including labour, capital costs and the cost of the water supply are considered, no one method has proven to be consistently the most practical. Lee *et al.* (1985) evaluated the economics of improving irrigation application efficiency with limited water supplies. They concluded that, under existing low prices for water, it is more economical to improve the application efficiency of existing systems (furrow or sprinkler) than to convert from one system to another. The benefit to the farmer from changing to drip or sprinkler systems often does not warrant the capital investment, increased training requirements, and increased operating and maintenance costs associated with a pressurised system.

Even the best designed, inherently efficient method will perform poorly if management of the system is inadequate. For example, in a survey of 60 farms using drip irrigation during the 1980s in California, the average application efficiency was 73% with individual systems ranging from a low of 24% to a high of 96%. The source of the problems was either system operation, environmental factors or system design and installation (Aslan *et al*., 1986). Large increases in application efficiency can be achieved in furrow irrigation, which is traditionally regarded as having poor efficiency, by matching furrow length, cross-sectional shape, water quality, discharge rate and cut-off time to soil type (Raine and Bakker, 1996) or by recovering and recycling tail waters. Raine and Bakker (1996) have shown that on-farm application efficiencies can vary from 14 to 90% for individual furrow irrigation events, or from 31 to 62% for seasonal application efficiencies. With sprinkler systems, application efficiency is effectively a function of application uniformity, which varies with climate, soil, terrain, and sprinkler spacing and design. Well-designed sprinkler systems can have application efficiencies of 85% (Shannon *et al*., 1996), but this is rarely attained (Kruse *et al*., 1990). Efficiencies above 75% can be attained, especially with mobile systems.

Clearly, with all irrigation methods there is scope for improving on-farm application efficiency by simple improvements in technical and management features. However, at least for the South African and Australian industries, there is a lack of on-farm data on application efficiency across production systems and locations, that could be used to benchmark the scope for improvement. There are also obvious advances to be made in improving grower knowledge of the biophysical system that they manage. Recent experience in Australia has shown that growers are often unaware of the quantities of water they are applying, resulting in waste (Shannon *et al*., 1996). Growers are being encouraged to use simple tools to monitor soil water depletion and measure amounts of irrigation applied, so as to tailor irrigation amount and avoid unnecessary wastage (Shannon *et al*., 1996).

Optimising Timing and Amount of Irrigation with Limited Water

While advances in application efficiency are an obvious means by which a limited water supply can be stretched, there are physical and economic limits to the improvement of application efficiency. For a fixed supply of water and application efficiency, the irrigator can concentrate irrigation on a limited area to maximise yields per hectare and minimise the investment of equipment, energy and labour, which are usually proportional to the area under irrigation. The other option is to spread the water so as to 'green-up' a greater area often producing more yield from the total farm, with additional outlays for water conveyance, labour and irrigation plant. The problem of determining the optimal size of the area to irrigate has no universal solution since the economic considerations are specific to each case. From the economic point of view, the criterion is to apply water to the irrigated area to the point where the marginal cost equals the marginal return from applying that water. A detailed treatment of this topic is beyond the scope of this paper. However, the reader is referred to Inman-Bamber *et al.* (1993) for an example of an approach to rationalising a limited water supply to a reduced number of fields in order to improve overall farm profitability.

Another dimension to the problem is the question of how to distribute the available supply in terms of timing and amount during the season. In this section, APSIM-Sugarcane is used to evaluate the yield and economic response to a range of seasonal irrigation allocations, under two management strategies for years 1895-1993 at Mackay (21.2° S) in tropical Australia and for 1967-1994 at Pongola (27.4° S) in South Africa. The analysis is intended to be illustrative, rather that definitive. While APSIM-Sugarcane has received limited validation for water-limiting conditions, confidence in the simulations will increase as the model is validated and improved to cope with a wider range of water deficit conditions.

Mackay has a highly variable rainfall with annual totals varying from 604 to 3359mm over the 99 years used in this analysis, with a CV of 39% (Fig. 16.1a). As a consequence irrigation responses (irrigated minus rainfed yield) are highly variable (see Wegener, 1990). About two-thirds of farms in the Mackay district have irrigation, ranging from small supplementary systems to those capable of supplying the total water requirements of the whole farm over prolonged periods. In contrast, Pongola is located in the northern region of the South African industry, where irrigation is necessary. Both mean annual rainfall (664mm) and year-to-year variability (417-1260mm, CV=26%) for the period 1967-1995 are much lower than at Mackay (Fig. 16.1c). At Pongola, water for irrigation can be limited in dry years if river flows are low.

The aim of the modelling analysis was to examine the relative merits of two irrigation strategies. The first represents a water-conserving strategy, where the crop is allowed to experience some water deficit before irrigating (irrigate 50mm every 130mm depletion), so as to stretch a limited supply of water through to the end of the season. This strategy may induce marginal crop stresses but may not affect yield enough to justify the cost of irrigation. The second represents a water-spending strategy, where the crop is irrigated more frequently (irrigate 50mm every 80mm depletion) to avoid most water stress, with the expectation that, in most years, rainfall later in the season would allow continued growth if irrigation supplies were exhausted. In addition, for both strategies 50mm was applied at ratooning if the surface 200mm of the profile contained less than 50% available water, since it is common practice to irrigate to establish crops if dry conditions prevail at ratooning. The model simulated water allocations of 0.5, 2.0, 4.0, and 6.0Ml ha^{-1} for the season, and also rainfed and fully irrigated production, which assumed an unlimited water supply. Inputs for the soil water balance are given in Table 16.1. The genetic coefficients for Q117 and NCo376 were used at Mackay and Pongola, respectively (see Appendix 2 in Muchow *et al.*, 1997). No scaling factor to convert simulated to commercial yields was used, so the results represent the maximum responses possible without other yield constraints. The difference in returns between the two strategies was evaluated for two cane prices (\$35 and \$25t^{-1}) for Mackay and two sucrose prices (R900 and R600t^{-1}) for Pongola. The different prices used for calculating income from sugarcane yield reflect the differences in payment systems in Australia and South Africa. Assumed costs of production were \$5 and R30t^{-1} for harvesting and hauling, and \$80 and R150Ml^{-1} for applied irrigation (water charges, pumping, etc.). No other cost differences between irrigated and rainfed strategies, such as initial capital investment costs were considered.

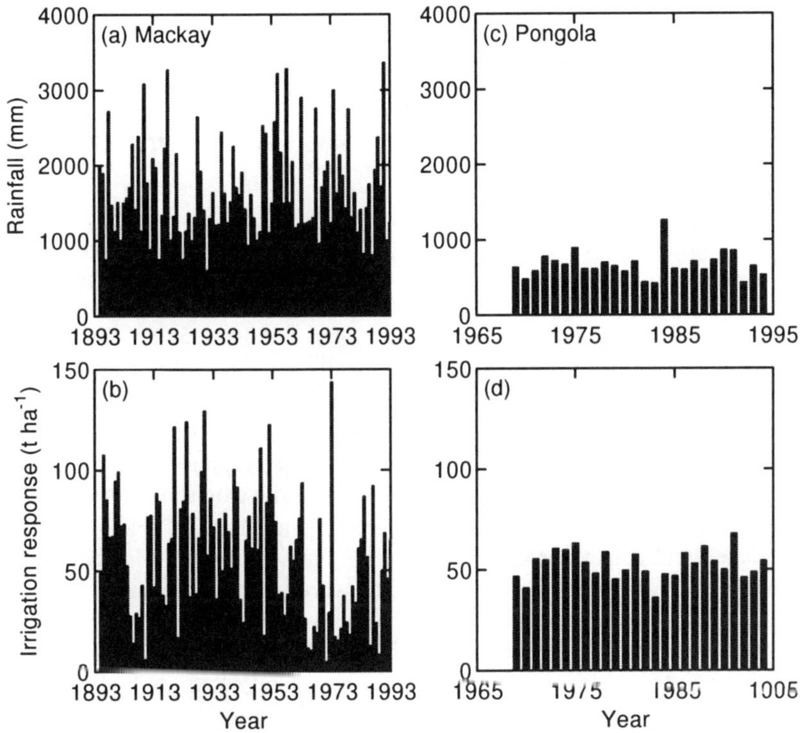

Fig. 16.1. Annual rainfall, and response to irrigation (t cane ha^{-1}) with water allocation of 4Ml ha^{-1} and water-conserving irrigation strategy at Mackay (a,b) and Pongola (c,d) for each year of simulation.

Rainfall variability from year-to-year was reflected in cane yields simulated for rainfed conditions. At Mackay mean cane yield was 114t ha^{-1}, but varied from 21 to 174t ha^{-1} over the 99 years. In contrast, at Pongola, cane yield was lower (34t ha^{-1}) and less variable (9-84t ha^{-1}) reflecting the consistently lower rainfall. For all allocations at Pongola (data not shown) and at 0.5 and 2.0Ml ha^{-1} allocations at Mackay (Table 16.3) all the irrigation water was utilised in all seasons. As allocation increased at Mackay, the frequency of years in which water was not fully used increased. At Mackay, the water-spending strategy ensured more thorough utilisation of the higher water allocations (Table 16.3).

Table 16.3. Mean values of output variates from irrigation case study for Mackay, simulated for 1894 to 1993.

	Rainfed	0.5Ml	2Ml	4Ml	6Ml	Full irrig.
Water-conserving strategy						
Irrigation applied (Ml ha^{-1})		0.50	1.94	3.06	3.39	
% seasons allocation used		100	96	42	13	
Sucrose yield (t ha^{-1})	14.9	17.4	21.0	24.5	25.7	
Seasonal deep drainage (mm)	477	488	519	529	528	
Effective rainfall (%)	58	60	58	57	56	
Water-spending strategy						
Irrigation applied (Ml ha^{-1})		0.50	2.00	3.71	4.50	5.70
% seasons allocation used		100	100	77	30	-
Sucrose yield (t ha^{-1})		17.4	20.0	23.9	26.2	26.3
Seasonal deep drainage (mm)		488	550	599	612	726
Effective rainfall (%)		60	56	52	51	42

An important measure of the response to irrigation is the extra cane yield produced above the rainfed yield. The cane yield response to irrigation at Mackay was highly variable from year to year, and the extent of this variability for a 4.0Ml ha^{-1} allocation, using the water-conserving strategy, is shown in Fig. 16.1b. Cane yield response varied over the 99 years from 5 to 144t ha^{-1} with a mean of 58t ha^{-1} and a CV of 54%. It is noteworthy that response to irrigation from the mid-1960s to the late-1970s, when much irrigation research was conducted at Mackay (Chapman and Chardon, 1979), was consistently lower than that in surrounding decades (Fig. 16.1b). This is reflected in higher annual rainfall (Fig. 16.1a) and rainfed cane yields for this period. In experiments at Mackay, for the years 1973-1984, the mean extra cane grown with irrigation compared to rainfed for ratoon crops was 13t ha^{-1} for a mean irrigation application of 200mm (assuming that the 3.2 irrigations of Chapman and Chardon (1979) corresponds approximately to 200mm ≡ 2Ml ha^{-1}). The model for the same period of years predicted a mean of 29t ha^{-1} extra cane grown for 2Ml ha^{-1}. Direct comparisons between modelled and experimental yields are difficult to make as the experimental data used ratoon crops of varying start time and duration, and applied different amounts of irrigation in each of the seasons. Also the model predicts the maximum response possible, without accounting for yield losses due to poor management, waterlogging or pest and diseases.

In contrast to Mackay, at Pongola with less variable rainfall (Fig. 16.1c), the extra cane grown with irrigation was more consistent year-to-year (Fig. 16.1d). At 4.0Ml ha^{-1} using a water-conserving strategy, yield response varied from 36 to 67t ha^{-1}, with a mean of 52t ha^{-1} and a CV of 14%.

At Mackay, for medium levels of water application (2 and 4Ml ha^{-1}) a water-conserving strategy out-yielded a water-spending strategy in most years (Fig. 16.2b,c), because water supply was metered out to last further into the season. Consequently, higher returns were generated with the water-conserving strategy at both \$35 and \$25 t^{-1} cane (Table 16.4). On the other hand, with plentiful water (6Ml ha^{-1}), the water-spending strategy was superior in 90% of years (Fig. 16.2d), but only marginally so in terms of average returns (Table 16.4). At the lower cane price of \$25 t^{-1}, there was less of an interaction between the type of strategy used and the amount of water available for irrigation (Table 16.3). At allocations of 0.5Ml ha^{-1} there was little advantage to either strategy.

At Pongola, where irrigation requirement is higher than Mackay (Table 16.2), season-to-season rainfall variability less, and irrigation responses more consistent, there was little interaction between the level of water allocation and the type of strategy used. For allocations of 2-6Ml ha^{-1}, the water-conserving strategy produced higher yields in most years (data not shown) and much higher average returns at both sucrose prices (Table 16.4) than the water-spending strategy. With 6Ml ha^{-1} allocation, in 50% of years, the advantage in using the water-conserving strategy was greater than 5t ha^{-1}, equivalent to R1500ha^{-1} at a sucrose price of R900t^{-1}.

Table 16.4. Advantage to the water-conserving over the water-spending strategy for four different water allocations in terms of gross margin per hectare [A\$ ha^{-1} for Mackay and R ha^{-1} for Pongola, based on two prices for crop yield. Values are the mean of all years simulated]

Allocation	0.5Ml ha^{-1}	2Ml ha^{-1}	4Ml ha^{-1}	6Ml ha^{-1}
Yield price		**Mackay**		
\$35t^{-1} cane	4	207	169	-8
\$25t^{-1} cane	3	139	130	24
		Pongola		
R900t^{-1}	-18	439	829	746
R600t^{-1}	-14	245	495	464

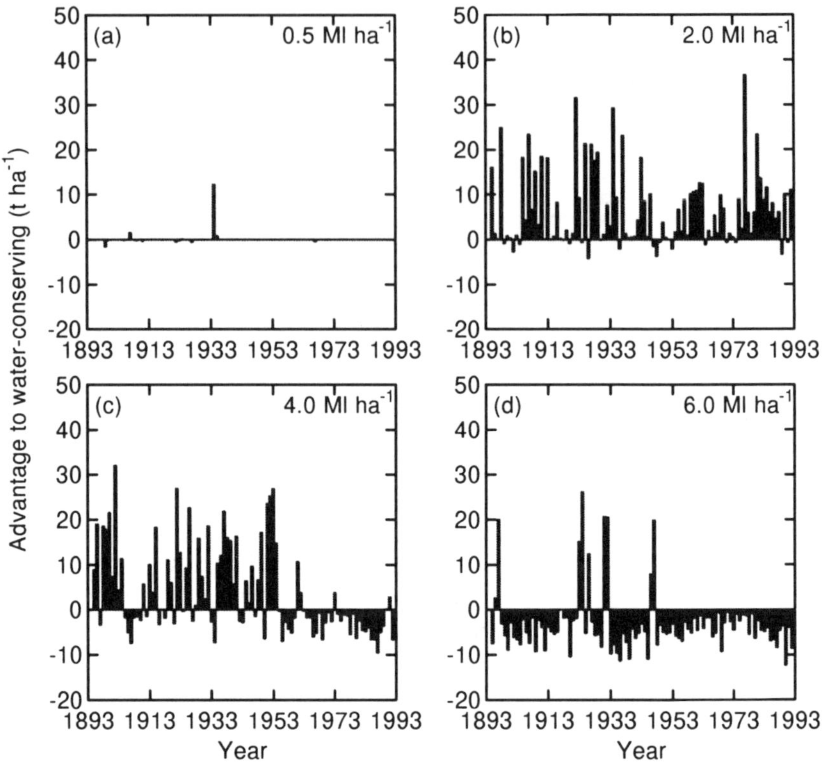

Fig. 16.2. Cane yield advantage of a water-conserving over a water-spending strategy for 0.5, 2, 4, and 6Ml ha^{-1} irrigation allocations for Mackay, for each year of the simulation.

The above case study highlights the following key points. No one strategy will consistently be the best in every season, and evaluation over the climatic record for a district allows the irrigator to assess the opportunity costs of not adopting one strategy. This is particularly so at locations with high rainfall variability, such as Mackay. At Pongola, where rainfall is low but more consistent, one strategy (viz. water-conserving) dominated in most seasons. Clearly, there is scope for improvement in within-season management of limited water with medium-sized water allocations. There will be limited scope for improvement in timing and amount for small allocations, whilst with large allocations, strategies will only differ in the driest of seasons where water is most limiting. For both locations, at lower product price the response of profit to irrigation applied was flatter than at the higher price (data not shown), so that, as the price falls, the profit-maximising level of irrigation also falls, and the maximum net return strategy calls for less water than the maximum yield strategy. The best-bet strategy for a given set of circumstances, will change as price changes. Clearly, price, and not only yield response, must be included as a variable when evaluating irrigation strategies.

Cropping systems models can also be used to indicate the impact of various management options on sustainability indicators and other non-production aspects. Effective rainfall % was lower under full irrigation than rainfed or supplementary irrigation. There was a trend for the water-conserving strategy to use rainfall more effectively because the greater deficit developed between irrigations allowed better storage of rainfall events. At both locations, seasonal drainage increased progressively with increasing water applications, particularly with the water-spending strategy (Table 16.3). It appears that for these examples, well-timed supplementary irrigation with the water-conserving strategy should not raise tables, nor increase leaching.

Analysis of Water Use Efficiency

In the sugar industry, cane WUE has been widely used to calculate the water requirement for a target level of production (Shih, 1985; Kingston, 1994), and to evaluate the unproductive use of water by comparing actual with maximum WUEs (Yates and Taylor, 1986). It has also been suggested that cane WUE can be used to calculate the likely yield (Thompson, 1977; Chapman and Chardon, 1979) and economic responses to irrigation (Ferguson and Hampson, 1975). This approach pre-supposes that WUE is relatively stable across a wide range of production systems and environments. However, cane WUE would be expected to vary in relation to saturation deficit, variety, amount of soil evaporation, the cane dry matter, and the proportion of cane in crop biomass, as detailed by Robertson and Muchow (1994).

Simulation output from the case studies above can be used to examine the robustness of the cane WUE benchmark, its potential variation with climate, crop class, soil type, level of water input, and most importantly whether it can be refined to predict irrigation requirement and response to irrigation.

At an inter-regional level, simulated cane WUE under full irrigation varied across the 6 locations studied (Table 16.2), it was lowest at Kununurra (7.2t Ml^{-1}) and ranged between 10 and 15t Ml^{-1} across the other locations. The cane WUE values predicted by the model (7-15t Ml^{-1}) are within the range of measured values reported in the literature (Robertson and Muchow, 1994). Values simulated for individual locations agree with experimental data, despite the problems inherent in comparing simulated results with field measurements (Robertson and Muchow, 1994). Kingston and Ham (1975) recorded cane WUE values of 9-12t Ml^{-1} at Bundaberg, compared here with a mean simulated value of 12.0t Ml^{-1}. Cane WUE values measured at Ayr between 1966 and 1976, and published in the annual BSES reports, vary from 8.2 to 12.5t Ml^{-1}, while the mean simulated value was 12.7t Ml^{-1}.

For each of the six Australian locations simulated there was little variability in WUE across years under full irrigation, and much more variation under rainfed conditions. Figure 16.3 highlights this point using Bundaberg as an example. Under full irrigation cane WUE varied from 11.1 to 13.0t Ml^{-1} over the 99 years of the simulated climate record, while under rainfed conditions it varied from 0.5 to 13.0t Ml^{-1}. It appears that constancy of cane WUE asserted by many may derive from the fact that it has nearly always been measured under well-watered conditions where, the simulations would indicate, there is little year-to-year variation.

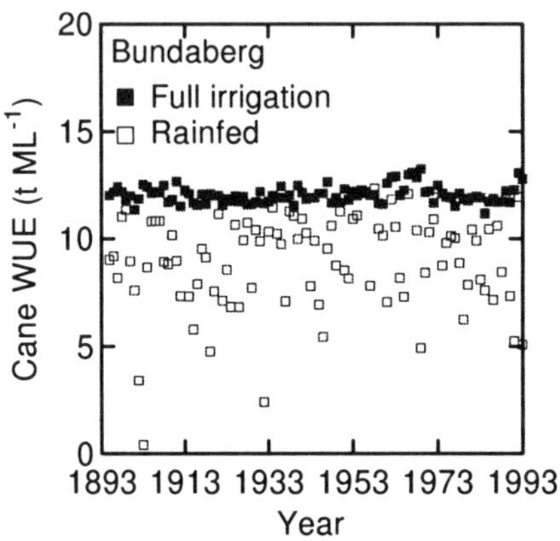

Fig. 16.3. Variation for each year of the simulation of cane water use efficiency (WUE = fw millable stalk per unit evapotranspiration) at Bundaberg under fully irrigated and rainfed conditions.

The case study of irrigation strategies at Mackay and Pongola indicate that cane WUE is likely to vary with irrigation input (Table 16.5). As seasonal applications of water increased from zero (rainfed) to fully irrigated production, mean cane WUE increased and the variation across years decreased (data not shown). Robertson and Muchow (1994, Table 16.1) review several datasets from irrigation experiments where cane WUE increases with irrigation input and yield level. The cane WUE of 10t Ml^{-1} simulated under full irrigation at Pongola is close to the 11.3t Ml^{-1} reported by Thompson (1976) for variety N14 in lysimeters at Pongola. Table 16.5 shows that at Pongola, the extra cane grown above the rainfed treatment per Ml of irrigation applied was similar at all irrigation levels, while irrigation response at either location was not well correlated with cane WUE. These results suggest that cane WUE is also a poor predictor of irrigation response, and that this indicator will not relate strongly to irrigation response because of the complex and variable interactions between rainfall patterns, irrigation and stage of crop growth.

Table 16.5. Mean simulated cane WUE (t Ml^{-1}) and irrigation response in cane yield (t Ml^{-1} applied) for four different irrigation applications using a water conserving strategy, and for rainfed and fully irrigated situations at Mackay and Pongola.

	Rainfed	0.5Ml	2Ml	4Ml	6Ml	Full irrigation
				Mackay		
Cane WUE	12.8	13.8	14.5	15.1	15.3	15.4
Irrigation response		21.6	13.4	12.9	13.0	8.0
				Pongola		
Cane WUE	5.3	5.8	7.2	8.6	9.5	10.1
Irrigation response		8.8	8.8	9.2	9.4	8.3

From a consideration of the components of cane WUE, we can examine underlying causes for its variation in relation to location and water supply.

$$\underset{(\mathrm{t\ Ml^{-1}})}{\frac{\text{Cane}}{\text{Water use}}} = \underset{(\mathrm{t\ t^{-1}})}{\frac{\text{cane yield}}{\text{biomass}}} \times \underset{(\mathrm{t\ Ml^{-1}})}{\frac{\text{biomass}}{\text{transpiration}}} \times \underset{(\mathrm{Ml\ Ml^{-1}})}{\frac{\text{transpiration}}{\text{water use}}}$$

Increased irrigation leads to better canopy development and maintenance which in turn leads to a greater proportion of evapotranspiration being used as transpiration. As sugarcane crops must reach a certain minimum level of biomass before cane stalk is produced, a greater proportion of biomass will be found as cane yield at higher yield levels. For these two reasons, crops with faster canopy closure and higher yield levels, due to longer crop durations or greater water inputs, will tend to have higher cane WUE.

Under fully irrigated conditions, where differences in yield due to location and rainfall are minimised, cane WUE will differ due to the influence of vapour pressure deficit (VPD). VPD is a chief determinant of how much water is transpired per unit of crop biomass produced. Generally, with higher VPD, more water needs to be transpired to achieve a given amount of biomass production (Tanner and Sinclair, 1983). VPD is between 0.9 and 1.6kPa for most Australian locations, except Kununurra where it is above 2.0kPa, indicating that less biomass will be produced per unit of water transpired at this location. Hence, while yields are high at Kununurra, WUE under full irrigation will be relatively low due to high VPD (Table 16.2).

The combination of these trends in the components of cane WUE suggests that cane WUE at a given location will be a relatively stable rule-of-thumb under fully irrigated conditions for a given crop class where yield levels are consistent from season-to-season. On the other hand, under rainfed conditions where yield levels and the amount of soil evaporation will vary across years according to rainfall patterns and the extent of canopy cover, cane WUE is much more likely to vary, with values increasing as water supply, and yield level, increases. As VPD affects the relationship between biomass and transpiration, at a given location there will be a weak correlation between WUE under fully irrigated and rainfed situations. This analysis suggests that cane WUE can be reliably applied in fully irrigated environments as a rule-of-thumb, but probably has limited applicability in rainfed or partially irrigated situations to estimate yield, water use, or response to irrigation.

Concluding Remarks

Sugar growing industries around the world must make better use of dwindling water resources in the face of rising costs and threats to the sustainability of water supplies. With limited water supplies, it appears that there is considerable scope at the farm level for growers to optimise the application of irrigation water, modify the timing and amount of irrigation applied, and adjust the size of the area of the farm to irrigate to achieve greater profits. Traditional experimentation will always be limited in accounting for climatic variability, and its interaction with soil type and crop management practices, in devising improved strategies. Well-tested cropping systems models have a role to play in developing these strategies, but should be coupled to economic analysis, risk assessment, and sustainability impacts. We also believe that the recent advances made in Australia in the area of short-term climate forecasting using the Southern Oscillation Index should be harnessed to improve irrigation management in a region with high climatic variability.

A key feature of past improvement in efficient water use in the sugar industry has been the development of useful, simple benchmarks to monitor changes in system performance. In the future, concepts such as cane WUE and effective rainfall will still have a role in benchmarking performance. However, the use of systems modelling in conjunction with field measurements has the potential to derive more useful system benchmarks by encompassing the wide range of factors affecting efficient water management.

References

Anon. (1991) *Irrigation of Sugarcane*. BSES/SRDC, Brisbane, Australia, 52 pp.

Aslan, S., Ackley, D. and Willey, P. (1986) *A Summary of the Drip Irrigation Problems of the Coachella Valley.* USDA Soil Conservation Service, Coachella Valley Resource Conservation District, Escondido, CA, USA, 53 pp.

Boggess, W.G. and Ritchie, J.T. (1988) Economic and risk analysis of irrigation decisions in humid regions. *Production Agriculture* 1, 116-122.

Chapman, L.S. and Chardon, C.W. (1979) An evaluation of irrigation at Mackay. *Proceedings of the Australian Society of Sugar Cane Technologists* 1, 117-123.

Chapman, R.N. and Milford, B.J. (1997) Economic constraints and opportunities in the Australian sugar industry. In: Keating, B.A. and Wilson, J.R. (eds), *Intensive Sugarcane Production: Meeting the Challenges Beyond 2000.* CAB International, Wallingford, UK, pp. 55-70.

Ferguson, G.A. and Hampson, R.J. (1975) Some factors influencing the decision to irrigate sugar cane. *Proceedings of the Queensland Society of Sugar Cane Technologists 42nd Conference*, 99-103.

Hardy, M. (1966) Water consumption of the cane plant. *Annual Report Mauritius Sugar Industry Research Institute*, Mauritius, pp. 104-106.

Harris, T.R. and Mapp, H.P. (1986) A stochastic dominance comparison of water-conserving irrigation strategies. *American Journal of Agricultural Economics* 68, pp. 298-305.

Hearn, A.B. and Constable, G.A. (1981) Irrigation for crops in a sub-humid environment. V Stressday analysis for soybeans and an economic evaluation of strategies. *Irrigation Science* 3, 1-15.

Hillel, D. (1990) Role of irrigation in agricultural systems. In: Stewart, B.A. and Neilson, D.R (eds), *Irrigation of Agricultural Crops.* Agronomy Monograph No. 30, American Society of Agronomy, American Society of Crop Science, American Society of Soil Science, Madison, WI, USA, pp. 5-30.

Inman-Bamber, N.G., Culverwell, T.L., and McGlinchey, M.G. (1993) Predicting yield responses to irrigation of sugarcane from a growth model and field records. *Proceedings of the South African Sugar Technologists Association* 67, 66-72.

Jensen, M.E., Rangeley, W.R., and Dieleman, P.J. (1990) Irrigation trends in world agriculture. In: Stewart, B.A. and Neilson, D.R (eds), *Irrigation of Agricultural Crops.* Agronomy Monograph No. 30, American Society of Agronomy, American Society of Crop Science, American Society of Soil Science, Madison, USA, pp. 31-67.

Jones, C.A., Santo, L.T., Kingston, G.K. and Gascho, G.J. (1990) Sugarcane. In: Stewart, B.A. and Neilson, D.R (eds), *Irrigation of Agricultural Crops.* Agronomy Monograph No. 30, American Society of Agronomy, American Society of Crop Science, American Society of Soil Science, Madison, WI, USA, pp. 835-858.

Keating, B.A., Robertson, M.J., Muchow, R.C., and Huth, N.I. (1997) Modelling sugarcane production systems. I. Description and validation of the APSIM sugarcane module. (in preparation).

Kingston, G. (1994) Benchmarking yield of sugarcane from estimates of crop water use. *Proceedings of the Australian Society of Sugar Cane Technologists 16,* 201-209.

Kingston, G. and Ham, G.J. (1975) Water requirements and irrigation scheduling of sugar cane in Queensland. *Proceedings of the Queensland Society of Sugar Cane Technologists 42nd Conference*, 57-65.

Kruse, E.G., Bucks, D.A., and von Bernuth, R.D (1990) Comparison of irrigation systems. In: Stewart, B.A. and Neilson, D.R (eds), *Irrigation of Agricultural Crops.* Agronomy Monograph No. 30, American Society of Agronomy, American Society of Crop Science, American Society of Soil Science, Madison, WI, USA, pp. 475-508.

Lee, J.C., Ellis, J.R. and Lacewell, R.D. (1985) Valuation of improved irrigation efficiency from an exhaustible groundwater source. *American Water Resources Association Water Resources Bulletin* 21, pp. 441-447.

Mason, W.K., Constable, G.A. and Smith, R.C.G. (1980) Irrigation for crops in a sub-humid environment. II Water requirements of soybeans. *Irrigation Science* 2,13-22.

McCown, R.L., Hammer, G.L., Hargreaves, J.N.G., Holzworth, D.P. and Freebairn, D.M. (1996) APSIM: a novel software system for model development, model testing and simulation in agricultural systems research. *Agricultural Systems* 50, 255-271.

Meyer, W.S. (1997) The irrigation experience in Australia: lessons for the sugar industry. In: Keating, B.A. and Wilson, J.R. (eds), *Intensive Sugarcane Production: Meeting the Challenges Beyond 2000.* CAB International, Wallingford, UK, pp. 437-454.

Muchow, R.C. Robertson, M.J. and Keating, B.A. (1997) Limits to the Australian sugar industry: climatic and biological factors. In: Keating, B.A. and Wilson, J.R. (eds), *Intensive Sugarcane Production: Meeting the Challenges Beyond 2000.* CAB International, Wallingford, UK, pp. 37-54.

Raine, S.R. and Bakker, D. (1996) Increased furrow irrigation efficiency through better design and management of cane fields. *Proceedings of the Australian Society of Sugar Cane Technologists* 18, 119-124.

Robertson, M.J. and Muchow, R.C. (1994) Future research challenges for efficient crop water use in sugarcane production *Proceedings of the Australian Society of Sugar Cane Technologists* 16, 193-200.

Shannon, E., McDougall, A., Kelsey, K. and Hussey, B. (1996) Watercheck - a co-ordinated extension program for improving irrigation efficiency on Australian cane farms. *Proceedings of the Australian Society of Sugar Cane Technologists* 18, 113-118.

Shih, S.F. (1985) Use of sugar cane yield and evapotranspiration relationships to improve water management. *Sugar Cane* 1, pp. 3-7

Tanner, C.B. and Sinclair, T.R. (1983) Efficient water use in crop production: research or re-search? In: Taylor, H.M., Jordan, W.R., and Sinclair, T.R. (eds), *Limitations to Efficient Water use in Crop Production.* ASA, Madision, Wisconsin, USA, pp.1-27.

Thompson, G.D. (1976) Water use by sugarcane. *South African Sugar Journal*, 60, 593-600, 627-635.

Thompson, G.D. (1977) Irrigation of sugarcane. *South African Sugar Journal* 61, 126-131, 161-174.

Thompson, G.D. and Harding, R.L. (1986) A computerised model for evaluating irrigation schemes for sugarcane. *Proceedings of the South African Sugar Technologists Association* 60, 177-182.

Wegener, M.K. (1990) Analysis of risk in irrigated sugarcane at Mackay. *Proceedings of the Australian Society of Sugar Cane Technologists* 12, 45-51.

White, B.J. (1970) Climate, sugar yields and irrigation response in Queensland. *Proceedings of the Queensland Society of Sugar Cane Technologists 37th Conference*, 57-64.

Yang, S.J. and Chang, Y.T. (1976) Forced ripening by properly controlled irrigation. *Taiwan Sugar Research Institute Annual Report 1977-78*, pp. 21-22.

Yates, R.A and Taylor, R.D. (1986) Water use efficiencies in relation to sugarcane yields. *Soil Use and Management* 2, 70-76.

Chapter seventeen:

Opportunities for Amelioration of Soil Physical and Chemical Constraints under Intensive Cropping

M.E. Sumner

Department of Crop and Soil Science, University of Georgia, Athens, GA 30602, USA

Abstract

The nature of the processes which enhance chemical and physical degradation of soil once it is brought into intensive cultivation for sugarcane and other major crops are discussed. These processes are organic matter depletion, crust and seal formation, acidification, and salinisation and sodification. Evaluation of these processes is a necessary step in developing an understanding of the phenomena involved in soil degradation before proceeding to assess the consequences and potential impacts of any degradation on productivity. Available strategies for amelioration and some case studies with successful outcomes are discussed.

Introduction

Immediately a soil is cultivated, a train of events is set in motion which leads to a greater or lesser degree of soil degradation. As a result, various chemical and physical constraints to productivity which may already be present become amplified. The objective of this paper is to discuss the nature of these constraints in terms of their impact on productivity and to suggest potential strategies for amelioration.

In: Keating, B.A. and Wilson, J.R. (eds) *Intensive Sugarcane Production: Meeting the Challenges Beyond 2000*. CAB International, Wallingford, UK, 1997.

Organic Matter Depletion

Processes

When virgin soil is brought into cultivation, the organic matter content decreases due to increased oxidation and crop removal. This loss weakens interparticle bonding and causes structural degradation (Nelson and Oades, 1997). Being rich in nutrients, particularly N and S, organic matter when mineralised releases these nutrients, a portion of which may be lost due to leaching. Gifford *et al.* (1982) estimated that 39% of the organic C in Australian surface soils was lost between 1880 and 1980 by mineralisation and erosion.

Consequences

As a consequence of organic matter depletion, the clay becomes more readily dispersible in water particularly under the influence of impacting raindrops. This mobile clay blocks the pore space forming seals or hardsetting topsoils (Sumner, 1993). At a given rainfall intensity, infiltration is reduced (Fig. 17.1) and runoff and erosion accelerated as organic matter decreases. In addition, soils become more readily compacted under load and soil fertility is degraded in terms of the quantities of nutrients (N and S) lost after mineralisation. In variable charge and sandy soils, loss of cation exchange capacity (CEC) impairs the ability of the soil to retain basic cations (Ca, Mg, K).

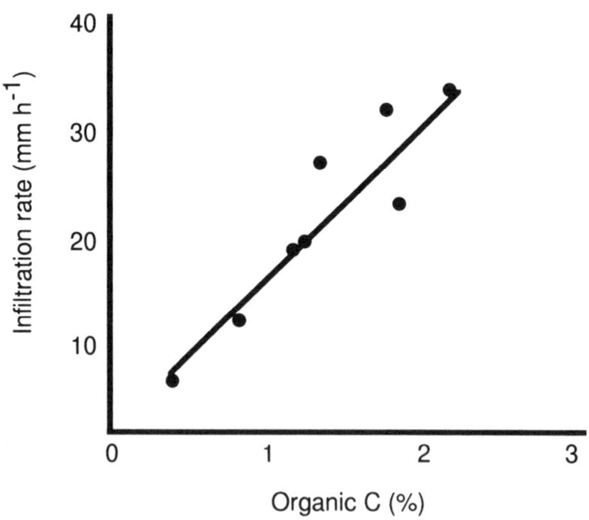

Fig. 17.1. Effect of organic carbon content on the infiltration rate of Australian Red-Brown Earths. [Grierson *et al.*, 1972]

Impacts on productivity

Other than in terms of nutrients, the impacts of organic matter loss on productivity are not direct but arise from the resulting degraded structure. An example of the effect of organic matter loss on soybean production is reflected in Table 17.1.

Potential remedies

The negative train of events that is induced by cultivation can be partially stemmed by using reduced or no-till systems of production (Table 17.1) or by planting permanent pastures or forests (Donal and Williams, 1954).

Table 17.1. Effect of organic matter loss induced by tillage on nutrients, physical condition and yield of soybeans on a Typic Hapludult. [Langdale *et al.*, 1992]

Tillage system	Organic C	Organic N	Water stable aggregates	Rainfall retention	Soybean yield
	(%)				($t\ ha^{-1}$)
Conventional	0.70	0.05	61	60	1.5
Conservation	2.50	0.21	91	80	3.0

Seal Formation

Causes

Seal formation is the process of aggregate breakdown followed by clay dispersion occurring at the soil surface forming a thin seal that is less permeable than the underlying soil. On drying, a hard crust often forms. Clay dispersion is promoted by elevated levels of Na and low electrolyte concentration (salt) in the soil solution. Thus in sodic (high exchangeable Na) soils with low salt contents, clay tends to disperse spontaneously sealing the pore space at the surface (Shainberg *et al.*, 1989). In addition, there are many soils which under optimal conditions do not seal, but after prolonged periods of cultivation lose substantial proportions of their organic matter content, rendering the clay particles prone to dispersion under the influence of applied energy usually in the form of impacting raindrops. As a result of the particle sorting which takes place during a rainfall event, a 'washed-in' layer in which the pore space has become clogged by the mobile clay forming the seal and a 'washed-out' layer, often consisting of clean sand grains on the surface, are formed (McIntyre, 1958). Such soils, which make up a large proportion of the world's row crop cultivated area, can form seals even at very low levels of exchangeable Na provided that the salt content is very low (Sumner and Miller, 1992). In addition, many soils in Australia and Africa are hardsetting which is caused essentially by the same processes except that the plough layer is adversely affected upon drying (Mullins *et al.*, 1990).

Consequences

Infiltration, runoff and erosion

Because of the reduced pore space in the seal, the rate of water acceptance (infiltration rate) at the soil surface is greatly reduced as illustrated in Table 17.2. The mulch treatment which intercepts the impacting raindrops reduces their energy to zero, successfully preventing seal formation. Because a seal readily forms on a bare surface, runoff increases which causes rills to form thus promoting soil loss. Miller and Radcliffe (1992) demonstrated that dispersible clay was inversely and directly related to infiltration rate and soil loss, respectively. Sediment transported in the runoff water severely impacts adjacent water bodies particularly with nutrients and pesticides. Not only does seal formation lead to runoff and erosion but also to loss of available water as that which is lost in runoff can never be regained for the crop. Consequently, under cultivated conditions which promote structural decline, many soils exhibit the propensity to form seals with disastrous consequences for infiltration, runoff and erosion.

Table 17.2. Effect of crust formation on infiltration rate, runoff and erosion from a Rhodic Haplorthox in Brazil. [Roth *et al.*, 1988]

Mulch ($t\ ha^{-1}$)	Crust	Soil cover (%)	Infiltration rate ($mm\ h^{-1}$)	Runoff (mm)	Soil loss ($g\ m^{-2}$)
0	Yes	1	18.5	56	71
6	No	96	56.1	23	0

Seedling Emergence

Once a seal has formed at the surface, drying leads to hardening of the crust or hardsetting of the topsoil which can take a heavy toll on seedling emergence (Miller *et al.*, 1988). With increasing content of mechanically dispersible clay, which promotes crust formation, resistance to penetration (measured with a needle penetrometer, which simulates a root) increases markedly, resulting in decreased seedling emergence. Crops vary widely in their ability to penetrate a crust at the soil surface.

Impacts on productivity

The impact of poor physical conditions at the soil surface can be evaluated based on the improved production of sugarcane treated with surface applied and dissolved gypsum under irrigation in the Burdekin (Table 17.3). Plant cane yields were poor because of prolonged wet weather but the substantial improvements in the ratoon crop due to the application of gypsum indicate that seal formation and associated problems can take a heavy toll on yield. This is confirmed by the work of So and Onus (1984) and Cochrane and Alymore (1991) who showed that wheat yields in New South Wales and Western Australia, respectively, were severely reduced on soils whose surface layers were readily dispersible.

Table 17.3. Effect of surface applied and dissolved gypsum at 10t ha^{-1} on sugarcane yields and commercial cane sugar (CCS) on a soil with poor physical condition in the Burdekin. [Ham *et al.*, 1995]

Treatment	Plant crop		First ratoon	
	Yield (t ha^{-1})	CCS	Yield (t ha^{-1})	CCS
Control	50.4	15.01	92.2	14.12
Dissolved gypsum	42.1	14.95	114.9	14.74
Surface applied gypsum	45.2	14.24	109.3	14.49
$LSD_{0.05}$	ns	ns	10.96	ns

Erosion can have a profound effect on crop productivity (Bruce *et al.*, 1995) as a result of nutrient and organic matter loss, and degraded rooting conditions. Often topsoils are sandier than subsoils which on eroded sites can pose additional problems in the cultivation of the soil.

Strategies for amelioration

Because the mechanisms involved in crust formation are known, strategies for amelioration can readily be implemented. Basically three approaches are available, the first approaching the problem by reducing the energy of impacting raindrops, the second by enhancing the electrolyte concentration at the soil surface and the third by forming water stable aggregates at the surface by raising the organic matter content or using synthetic polymers.

Reducing raindrop energy

Maintenance of cover on the soil surface by vegetation or mulching is the most effective and usually the most economic way of countering the dispersive effect of rain. This can be achieved by applying mulch on the soil surface (Table 17.2) or using a crop such as sugarcane or pasture which protects the soil from impacting raindrops for most of the year (Table 17.4), or by practising minimum or no-till systems of crop production (Table 17.5). Keeping the soil surface protected increases infiltration and drastically reduces runoff and soil loss to low levels.

Enhancing electrolyte concentration

Clay dispersion at the soil surface can be reduced by the application of solid by-product gypsum on the soil surface (Miller and Radcliffe, 1992) under rainfed conditions and by dissolving gypsum in irrigation water (Ham, 1988). In the former case, mined gypsum has been found to be much less effective because it dissolves more slowly (Keren and Shainberg, 1981). For this strategy to be effective, the gypsum must dissolve instantly in the impacting raindrops to subtend an electrolyte concentration sufficient to prevent clay dispersion. Both strategies can lead to improved infiltration as illustrated in Fig. 17.2 for surface applied phosphogypsum (PG). The beneficial effects of gypsum in promoting clay flocculation and hence reduced soil loss under conventional tillage can be seen in Fig. 17.2. However, the effect is not as pronounced as that of no-tillage nor is it as long-lasting because gypsum being slowly soluble is leached fairly rapidly from the surface layer (Shainberg *et al.*, 1989). Supporting data for sugarcane in South Africa are presented in Table 17.4 where PG aided in preventing surface structural deterioration. Surface application of PG also reduced crust strength improving stand establishment for maize and soybeans in Queensland (Ham *et al., 1995).* Ford *et al.* (1995) demonstrated that gypsum was highly effective in reducing the penetration resistance of red duplex soils in Victoria.

Table 17.4. Effect of soil cover (scattered tops [ST] and trash blanket) and surface applied phosphogypsum (PG) on runoff and soil loss from a Longlands soil in South Africa. [Dewey and Meyer, 1989]

Treatment	Runoff rate[1] (l min^{-1})		Runoff (% of rainfall)		Soil loss (t ha^{-1})	
	Storm 1	Storm 2	Storm 1	Storm 2	Storm 1	Storm 2
Control	17.5	22.0	42	90	2.4	5.3
ST	7.5	14.5	24	61	0.8	1.6
ST + 2.5 PG[2]	7.6	14.5	20	55	0.5	1.6
ST + 5 PG	3.5	11.5	10	43	0.4	1.3
Trash blanket	--	--	5	21	0.3	0.3

[1] At 60 min. of rain at 63 mm h^{-1} (EC = 20 mS m^{-1}; SAR = 0.03)
[2] 2.5 PG = 2.5t ha^{-1} of phosphogypsum

Table 17.5. Effect of minimum tillage[1] on runoff and soil loss under sugarcane in South Africa (Haywood and Mitchell, 1987), data averaged over four soil types.

Treatment	Soil loss ($t\ ha^{-1}$)		Runoff (%)	
	Storm 1	Storm 2	Storm 1	Storm 2
Minimum till	0.4	1.4	19.9	36.0
Conventional till	5.2	12.0	48.2	85.0

[1] Ratoon cane sprayed with Roundup herbicide and plant cane established in interrow

Synthetic polymers

Organic polymers such as polyacrylamide (PAM) and polysaccharides can be used as surrogates for organic matter in bonding soil particles together into water-stable aggregates thus improving structure and water acceptance at the soil surface (Shainberg, 1992). Because only the soil at the surface requires amelioration, relatively small amounts (5-20kg ha^{-1}) of polymer are effective making the technology attractive economically. The improved infiltration rate with PAM (Fig. 17.2) is similar to that obtained with 5t ha^{-1} PG. A combination of both gave a further improvement in infiltration rate, a result confirmed by Smith *et al.* (1990) and Warrington *et al.* (1991).

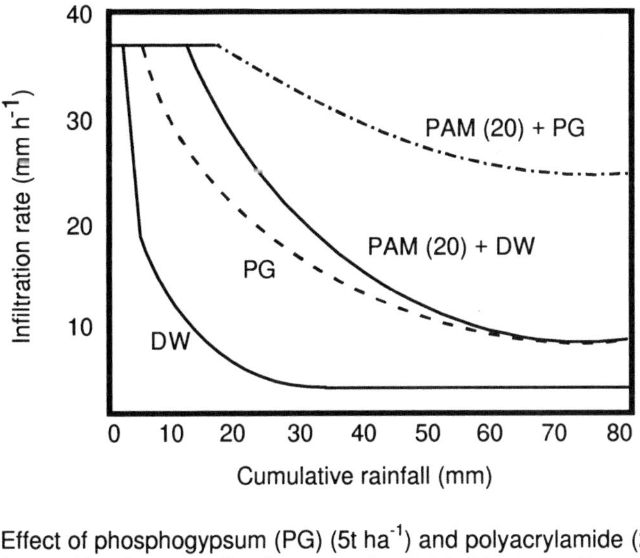

Fig. 17.2. Effect of phosphogypsum (PG) (5t ha^{-1}) and polyacrylamide (20kg ha^{-1}) alone and in combination on infiltration rate of a Calcic Hapluxeralf under distilled water (DW) rainfall (Shainberg *et al.*, 1990).

Compaction

Causes

In intensive agriculture, soil compaction manifest as an increase in soil bulk density (BD) and penetration resistance is usually caused by in-field wheeled traffic which is particularly heavy in mechanised sugarcane production where harvesters and transport vehicles make frequent passes often when thc soil is wet (Swinford and Boevey, 1984). Tillage implements can also cause compaction manifest as tillage pans. Compaction increases with moisture content to a maximum and then decreases under very wet conditions.

Consequences

Because bulk density increases when a soil is compacted, the marked decrease in the large pores in the soil disturbs the balance between water and air space reducing plant available water (van Antwerpen and Meyer, 1996). This together with the increased penetration resistance results in poor root proliferation, and consequently reduced yields (Swinford and Boevey, 1984).

Impacts on productivity

Although many studies have been conducted on the effects of compaction on crop productivity (Ellington, 1986; Owen, 1988), very few have dealt with sugarcane. Early work in South Africa indicated that compaction may be a problem in the sugar industry, but the effects were not significant (Johnston and Wood, 1971). Subsequent work (Swinford and Meyer, 1985) has shown that increased compaction due to wheeled traffic has a profound effect on sugar production as illustrated in Fig. 3.

Strategies for amelioration

Compacted soil can be ameliorated physically by deep ripping to break-up compacted layers, but the effects are often short-lived (Mead and Chan, 1988), and chemically by the use of gypsum and deep-rooted perennial crops (Radcliffe *et al.*, 1986). In the latter case improvement is slow and arises from improved clay flocculation and aggregation.

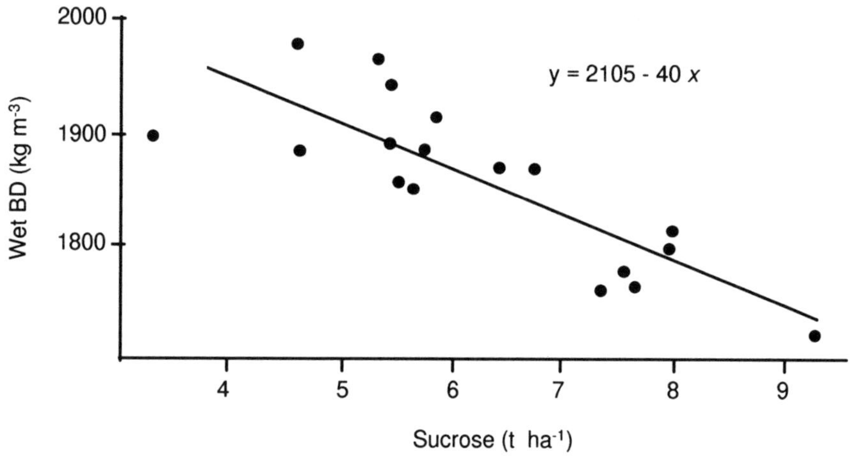

Fig. 17.3. Effect of wheeled traffic compaction (bulk density, BD) on sucrose yield. [Swinford and Meyer, 1985]

Acidification

Causes

The major causes of soil acidification are N fertilisers containing or producing NH_4^+, organic matter mineralisation, and legumes (Sumner, 1996a). Legumes play a minor role in sugarcane production systems and will not be considered here. All ammoniacal N fertilisers develop acid when NH_4^+ is oxidised to nitrate (NO_3) by nitrifying microorganisms (*Nitrosomonas* and *Nitrobacter*). The quantity of acidity produced depends on: (i) how much ammonia (NH_3) is volatilised prior to nitrification, (ii) how much NO_3^- is denitrified which consumes H^+, (iii) how much NO_3^- is taken up by roots in exchange for OH^- or HCO_3^-, and (iv) how much NO_3^- is leached with basic cations. Mineralisation of organic matter can contribute to soil acidity by the oxidation of N and S to HNO_3 and H_2SO_4 in a similar manner. In addition, the CO_2 liberated during decomposition dissolves in water to form a weak acid (H_2CO_3) which only really plays a role above pH 5.

Consequences

As soils become more acid, basic cations on exchange sites are progressively replaced by H^+ and then by Al^{3+} (and Mn^{2+}) below pH 5.5; variable charge soils lose cation (CEC) and gain anion exchange capacity (AEC) as illustrated in Table 17.6. In addition, biological N fixation by legumes can be impaired under acid conditions. The Al^{3+} together with the loss of basic cations create potentially toxic conditions for plants with the severity depending on the type of crop. Generally, many legumes are sensitive to Al^{3+} toxicity whereas grasses are often tolerant. On soils with limited CEC, the availability of cationic nutrients (Ca, Mg, K) may become limiting, particularly on variable charge soils.

Table 17.6. Effect of continued application of ammoniacal N fertilisers for 15 years on soil chemical properties and yield of sorghum. [Sumner, 1991]

Soil pH	CEC at field pH	Extractable Al	Extractable H	Sum of basic cations	Sorghum yield
		(cmol$_c$ kg^{-1})			(t ha^{-1})
4.7	2.72	1.88	0.16	0.61	0.16
5.1	2.89	1.07	0.09	1.49	2.00
5.7	3.10	0.35	0.145	2.64	3.60
6.2	4.62	0.00	0.08	4.66	4.20

Impacts on productivity

The effects of acidification on exchange properties and the resulting loss in yield are illustrated in Table 17.6. Soil acidification has been reported in many sugar producing countries. In South Africa, Schroeder *et al.* (1994) found average pH declines of 0.1 pH unit per year with corresponding decreases in basic cations and increases in exchangeable Al^{3+}. However, recent experiments have shown only small to modest sugarcane responses to lime on moderately acid soils in South Africa (Turner *et al.*, 1992) and Brazil (Vitti *et al.*, 1992). Subsequently, Schroeder *et al.* (1994) demonstrated that this lack of response was partially due to the high levels of SO_4^{2-} in many South African sugar soils as a result of the prolonged use of single superphosphate and $(NH_4)_2SO_4$, and mineralisation of the humic soils. These high levels of SO_4^{2-} contribute to the 'self liming effect' (Reeve and Sumner, 1972) and ion pair formation (Pavan *et al.*, 1982). In Australia, Quinan and Wood (1989) and Ridge *et al.* (1980) have obtained substantial yield responses to lime in the Hambledon (Table 17.7) and Mossman mill areas as a result of marked decreases in exchangeable Al and increases in Ca. Quinan and Wood (1989) showed that the yield responses to 5 and 10t ha^{-1} of lime were highly economic highlighting the negative impact that soil acidification can have on productivity. On the other hand, responses to lime were not obtained in the Ingham mill area (Haysom *et al.*, 1986).

Table 17.7. Effect of lime (t ha^{-1}) on sugarcane yield and soil acidity parameters at three sites in Australia. [Quinan and Wood, 1989]

Attribute	Lime rate			
	0	5	10	15
Red Schist Soil				
Cane yield (t ha^{-1})	73	81	85	84
pH (cmol$_c$ kg^{-1})	4.4	4.9	5.7	6.5
Ca (cmol$_c$ kg^{-1})	0.1	0.9	3.1	5.1
Al (cmol$_c$ kg^{-1})	1.6	1.0	0.1	0.1
White Schist Soil				
Cane yield (t ha^{-1})	83	97	99	102
pH (cmol$_c$ kg^{-1})	4.4	4.9	5.2	5.4
Ca (cmol$_c$ kg^{-1})	0.2	0.7	1.5	1.6
Al (cmol$_c$ kg^{-1})	1.3	0.8	0.3	0.2
Red Basaltic Soil				
Cane yield (t ha^{-1})	82	93	99	103
pH (cmol$_c$ kg^{-1})	4.4	4.8	5.2	6.1
Ca (cmol$_c$ kg^{-1})	0.4	1.5	2.5	5.2
Al (cmol$_c$ kg^{-1})	1.2	0.3	0.1	0.0

Strategies for amelioration

Topsoils

The most economic approach to ameliorating topsoil acidity is the use of lime which has proved to be highly effective on a wide range of crops as a result of the precipitation of Al^{3+} and Mn^{2+} and enhanced levels of Ca^{2+} and Mg^{2+} (Sumner, 1996a). The quantity of lime required for optimal crop production (lime requirement), which can be readily determined by simple chemical tests (van Lierop, 1990; Sumner, 1996b), depends mainly on the acid sensitivity of the crop and the buffer capacity of the soil. To be effective, lime should be thoroughly incorporated into the volume of soil to be ameliorated because lime does not readily move in soil (Sumner, 1994).

Subsoils

When acidification becomes so severe that acidity is transferred to the subsoil, amelioration becomes much more difficult because lime applied to the topsoil does not readily move downward (Sumner, 1994). Deep incorporation of lime is expensive and the cost can rarely be borne by crop production. However, one economic strategy is to adequately lime the topsoil and apply acidifying N fertilisers. The ensuing reaction produces soluble and mobile $Ca(NO_3)_2$ which in conjunction with plant roots which take up more N than Ca (thereby excreting an excess of OH^- over H^+) allows alkalinity to be transferred to the subsoil. That these reactions are effective over time is illustrated in Fig. 4. When lime was not applied, N fertilisation severely acidified the entire profile. When lime and N inputs $[(NH_4)_2SO_4]$ were made together, the excess uptake of NO_3^- over Ca^{2+} by the grass roots in the subsoil raised the pH above the original down the whole profile. Without roots, this strategy does not work (Pleysier and Juo, 1981).

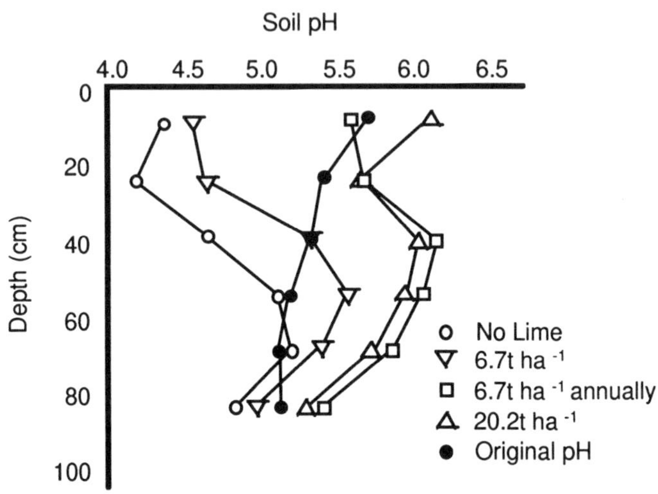

Fig. 17.4. Effect of lime on profile acidity after 4 years under bermuda grass sod fertilised with NH_4NO_3 at an annual N rate of 900kg ha^{-1}. Single lime applications were made at the beginning of the experiment except for the treatment labelled '6.7 t ha^{-1} annually'. [Adams *et al.*, 1967]

Surface applied gypsum has also proved to be an effective ameliorant of subsoil acidity in naturally acid highly weathered soils with substantial yield responses in a wide range of crops being recorded (Sumner, 1994). The soil amelioration has been attributed to the formation of a non-toxic ion pair ($AlSO_4^+$), and to the 'self liming effect' (Reeve and Sumner, 1972). Improved conditions for root growth are created by the reduced and increased levels of Al^{3+} and Ca^{2+}, respectively, which permit roots to extract water previously beyond their reach. Under water limiting conditions, this improved water regime translates into increased yields. In Brazil, yield responses in sugarcane to this strategy have been modest ranging from 7 to 17% (Dematte, 1992; Vitti *et al.*, 1992), but have been substantial in other more acid sensitive crops such as lucerne (100%) (Sumner, 1993) and corn (82%), coffee (59%), leucaena (81%), wheat (72%) and soybeans (40%) (Sumner, 1994).

Case study: subsoil acidity amelioration under sugarcane

In Brazil, on Oxisols with very low levels of Ca in the subsoil, increased yields of cane have been obtained to treatments in which both lime and gypsum have been applied (Table 17.8). This effect has been largely the result of subsoil amelioration and has resulted in increases in subsoil Ca levels (Table 17.9) and reductions in exchangeable Al (Vitti *et al.*, 1992). The changes in both these parameters create an improved rooting environment as has been shown by Sumner (1993a) and illustrated in Fig. 5. As a result, the crop is able to extract more water from the profile which under drought conditions results in improved productivity.

Table 17.8. Effect of phosphogypsum in combination with lime on sugarcane yield in t ha^{-1} (average of plant and 2 ratoons) on a Dark Red Latosol (Oxisol) in Brazil. [Vitti *et al.*, 1992]

Lime rate (t ha^{-1})	Phosphogypsum rate (t ha^{-1})			
	0	2	4	6
0	99.1	106.3	111.2	112.0
2	110.3	114.0	117.3	114.4
4	112.7	121.1	117.7	118.3
6	110.7	117.4	113.7	118.4

Table 17.9. Main effect of phosphogypsum on profile distribution of exchangeable Ca ($cmol_c\,kg^{-1}$) in a Dark Red Latosol in Brazil. [Vitti *et al.*, 1992]

Depth (mm)	Phosphogypsum rate (t ha^{-1})			
	0	1.5	3	6
0-200	1.4	1.6	1.8	2.2
200-400	0.3	0.6	0.7	0.9
400-600	0.3	0.4	0.5	0.7

Salinisation and Sodification

Causes

Salt-affected soils occur both naturally and as a result of anthropogenic influences, mainly irrigation and ecosystem disturbance. In the context of this paper, only the effects of irrigation on salinity and sodicity will be considered.

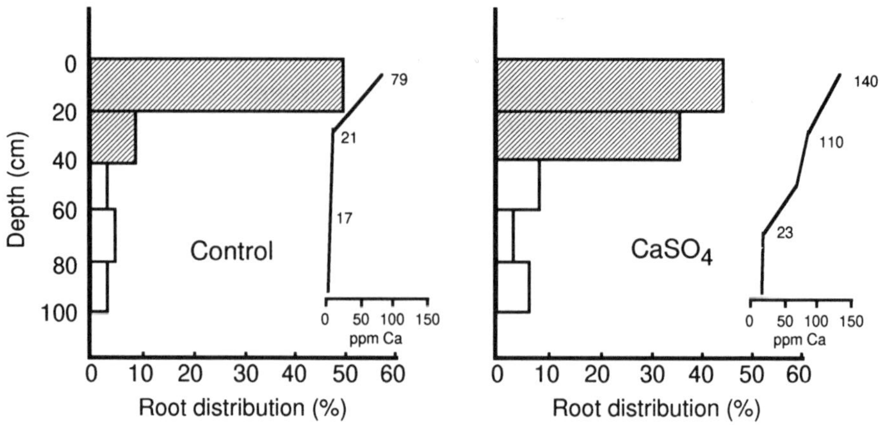

Fig. 17.5. Effect of gypsum application on sugarcane root distribution in a Red Yellow Podzolic soil in Brazil. [Vitti *et al.*, 1992)]

Salinity

When soils weather, the salts which are liberated are either transported to water bodies under conditions of adequate rainfall or may accumulate in the soil when water is limited. Those salts transferred to surface or groundwaters can reappear in the soil as a result of irrigation or rising watertables. As all natural waters contain salts and plants transpire pure water, any irrigation will contribute salt load to the soil. Whether or not salts accumulate creating saline soils depends on the fraction of applied water (irrigation and rain) which leaches through the soil and is removed in drainage. Globally, irrigated and salinised areas are closely related. The salts in the drainage water degrade its quality in many cases rendering the water unsuitable for further reuse for agricultural, municipal or industrial purposes. If there is insufficient leaching of salts, salinity will develop, the extent of which is determined by the salt concentration of the water, the length of time it is applied, and the evaporative demand of the environment. Alteration of ecosystems such as the clearing of forests can result in the transfer of 'fossil' salts stored in upland positions to lower points in the landscape where crop production is often impaired (Sharma and Williamson, 1984). The accumulation of salts in soil represents severe degradation as the osmotic potential developed impairs the ability of plants to extract water. Currently, a saline soil is defined as one having an electrical conductivity in a saturated paste extract (EC_e) > 4dS m^{-1} above which the productivity of most crop plants decreases (USSL Staff, 1954).

Sodicity

Although many soils are naturally sodic, sodicity, which is the accumulation of Na^+ relative to Ca^{2+} and Mg^{2+}, can develop as a result of the leaching of a Na-rich saline soil not containing weatherable minerals when irrigation is introduced or when the irrigation water is relatively rich in Na^+. The exact level of Na^+ in a soil which makes it sodic is being debated, but lies between an exchangeable Na percentage (ESP) of 3 and 15 (Sumner, 1993b; Sumner and Naidu, 1997) where ESP is (100*Exch. Na)/CEC. Many soils exhibit sodic behaviour at ESP values well below this level (Sumner, 1993b). What is important from a soil degradation point of view is whether or not a soil exhibits sodic behaviour characterised by clay dispersion and swelling. This is determined by the relative balance between ESP and EC of the soil solution. As ESP increases, more and more salt is required to keep the clay in a flocculated condition to prevent declines in hydraulic conductivity.

Consequences

Salinity

Salts in the soil promote flocculation and good physical condition but restrict water availability to plants. Thus, the presence of a little salt is usually beneficial both from the points of view of the soil and the crop, but too much results in drought stress and reduced yield (Oster *et al.*, 1996) or poor germination (Meiri, 1984).

Sodicity

With increasing sodicity at any given salt level, soil physical properties deteriorate with such problems as crusting at the surface and impaired internal drainage becoming more acute. The effects of Na addition on infiltration rate and soil loss are illustrated in Table 17.10. A relatively small addition of Na (160kg ha^{-1}) drastically reduced infiltration and increased soil loss. Hydraulic conductivity is also adversely affected by increasing ESP of the soil but the presence of a little salt in the percolating solution (5mmol$_c$ l^{-1}) largely offsets the decline (Chiang *et al.*, 1987).

Table 17.10. Effect of $NaNO_3$ addition on average infiltration rate, sediment concentration and soil loss from a Greenville soil during a 50mm h^{-1} rainfall event on a 9% slope under a rainfall simulator. [Miller and Radcliffe, 1992]

Treatment	Infiltration rate (mm h^{-1})	Dispersed clay (%)	Soil loss (t ha^{-1})
Control	36	18	200
$NaNO_3$[1]	9	55	3560
$NaNO_3$ + gypsum[2]	47	0	31
Gypsum after $NaNO_3$[3]	35	5	470

[1] 0.6t ha^{-1} including 100kg N ha^{-1}
[2] 5t ha^{-1} as phosphogypsum
[3] After $NaNO_3$ and wetting, soil was dried, sieved and repacked before rain applied

Impacts on productivity

Salinity

Salts osmotically induce drought stress in crops with yields of many crops being adversely affected at average root zone EC_e > 4dS m^{-1} (Oster *et al.*, 1996).

Sodicity

In addition to unfavorable physical condition, Na-induced Ca deficiency is often observed on sodic soils due to competition for uptake. Because of a limit to water entry and passage through the soil, crops growing on sodic soils can be stressed by alternating periods of drought and waterlogging. Some indication of the impact of sodicity on sugarcane production is presented in Table 17.11. While the yields on the gypsum ameliorated soil were much better, this profile had not yet been completely reclaimed as revealed by the high ESP values in the subsoil.

Table 17.11. Comparison of sugarcane production and profile exchangeable Na percentage (ESP) on control and ameliorated (11.25t ha^{-1} phosphogypsum added) sodic soil in the Burdekin. [Ham *et al.*, 1995]

Crop	Cane yield (t ha^{-1})		Depth	ESP	
class	Control	Gypsum	(m)	Control	Gypsum
Plant	105	143	0.0-0.1	12	2
First ratoon	68	83	0.2-0.3	18	14
Second ratoon	70	80	0.5-0.6	25	19
Third ratoon	34	50	0.9-1.0	27	26
Average	69	89			

Strategies for amelioration

Salinity

When salts accumulate to the extent that crop production is impaired, leaching is required to reestablish an appropriate condition. Under irrigation, salt accumulation can be prevented by ensuring that adequate water to remove the salts passes through the profile either to groundwater or via drainage to surface water. In soils where salts have been allowed to accumulate, amelioration by leaching is the only remedy. As a general rule, 60% or more of the accumulated salts will be leached under continuously ponded conditions by a volume of water equal to the bulk volume of soil to be leached. Use of intermittent ponding or sprinkler application of water greatly increases the efficiency of leaching (Oster *et al.*, 1996), but care must be taken to prevent the development of sodic behaviour when most of the salts have been removed. This can be achieved by leaching with water containing low levels of salt or by applying gypsum to the soil.

Sodicity

For soils which exhibit sodic behaviour, application of gypsum fulfils two roles, namely, (i) supplying electrolyte which prevents clay dispersion and crust formation, and (ii) supplying Ca^{2+} ions which can replace Na^+ on exchange sites thus lowering the ESP and increasing hydraulic conductivity. The former function has been illustrated in Table 17.10 where application of gypsum negates the effect of Na^+ on crust formation. The second role of gypsum in reducing ESP takes considerably longer to achieve and is discussed in the case study below.

Case study: reclamation of a sodic soil under sugarcane in Swaziland

On an 80ha field in which cane production had declined severely due to poor physical condition, perforated plastic drains at 20m spacings 1.2m deep were installed with a trenchless drain layer. Gypsum was then applied to part of the field at a rate required to reduce the ESP of the top 1m to 5; the remainder of the field served as an untreated control. Paddy bunds were built and the area was ponded (150mm deep) and leached for 9 months prior to being planted to two crops of rice. After draining, the field was levelled with land planes and sugarcane was established under a furrow irrigation system and a plant and 6 ratoon crops were harvested. Reclamation at the end of the trial was complete with the entire gypsum-treated profile having an acceptable ESP and EC_e consistent with stable physical properties (Table 17.12). Redevelopment of the field (levelling, drains and gypsum) which cost US$1225 ha^{-1}, resulted in an average increase of 27t cane ha^{-1} (Table 17.13). At the existing price of US$25 t^{-1} cane, the redevelopment in Swaziland would have been paid for by the increased yield in < 2 years. Similar successes have been reported in South Africa (Johnston, 1977) and Australia (Ham *et al.*, 1995).

Table 17.12. Effect of gypsum application at a rate required to lower initial exchangeable Na percentage (ESP) to 5 on subsequent soil ESP and EC_e at the end of the trial after two rice and seven sugarcane crops.

Treatment	Depth (m)	ESP (%)	EC_e (ds m^{-1})	EC_e required for stability (ds m^{-1})
Control	0-0.3	8.5	0.5	0.6
	0.9-1.2	16.2	0.7	1.5
Gypsum	0-0.3	2.0	0.8	0.2
	0.9-1.2	7.0	1.2	0.4

Table 17.13. Effect of reclamation using phosphogypsum (at rate applied to lower exch. Na % to 5) and drainage on whole cycle yield of sugarcane in Swaziland.

Year	Crop class	Cane yield (t ha^{-1})	
		Control	Gypsum
1974	Plant	120	126
1975	First ratoon	84	102
1976	Second ratoon	95	97
1977	Third ratoon	96	132
1978	Fourth ratoon	94	122
1979	Fifth ratoon	72	123
1980	Sixth ratoon	77	124
Average for cycle		91	118

Soil Degradation and Sustainability: Implications

Severe degradation such as soil loss by erosion definitely leads to a situation which is not sustainable in the long term as the rate of soil formation is, in most cases, much slower than that of erosion. The best strategy to adopt here is prevention as seldom are attempts at amelioration successful from a productivity point of view. On the other hand, acidification is largely a reversible process, particularly in the topsoil where liming readily restores productivity. However, if subsoils are allowed to acidify, amelioration is much more difficult, expensive and time consuming. Consequently, in most cases, sustainability will be compromised. Development or the presence of saline conditions drastically impair productivity, but in many cases, solutions involving adequate drainage and leaching are available which will permit sustainability. However, this is not true in terms of dryland salinity development resulting from the clearing of natural vegetation where sufficient water is seldom available for leaching. Sodification always has negative impacts on sustainablity, and therefore, the major thrust should be aimed at prevention rather than cure. Although the productivity of sodic soils can be restored, the very high cost requires cropping with highly remunerative crops to cover the costs.

References

Adams, F., White, A.W. and Dawson, R.N. (1967) Influence of lime sources and rates on Coastal bermudagrass production, soil profile reaction, exchangeable Ca and Mg. *Agronomy Journal* 59, 147-149.

Bruce, R.R., Langdale, G.W., West, L.T. and Miller, W.P. (1995) Surface soil degradation and soil productivity restoration and maintenance. *Soil Science Society of America Journal* 59, 654-660.

Chiang, S.C., Radcliffe, D.E., Miller, W.P. and Newman, K.D. (1987) Hydraulic conductivity of three southeastern soils as affected by sodium, electrolyte concentration, and pH. *Soil Science Society of America Journal* 51, 1293-1299.

Cochrane, H.R. and Aylmore, L.A.G. (1991) Assessing management-induced changes in the structural stability of hardsetting soils. *Soil and Tillage Research* 20, 123-132.

Dematte, J.L.I. (1992) Aptidao agricola de solos e uso do gesso. In: Anonymous (ed.), *II Seminário Sobre o Uso do Gesso na Agricultura.* IBRAFOS, Uberaba, Brazil, pp. 307-324.

Dewey, F.J. and Meyer, J.H. (1989) The use of phosphogypsum as an ameliorant for poorly structured crust-forming soils. *Proceedings of the South African Sugar Technologists Association* 63, 147-152.

Donal, C.M. and Williams, C.H. (1945) Fertility and productivity of a podsolic soil as influenced by subterranean clover (*Trifolium subterraneum* L.) and superphosphate. *Australian Journal Agricultural Research* 5, 664-687.

Ellington, A. (1986) Effects of deep-ripping, direct drilling, gypsum and lime on soils, wheat growth and yield. *Soil and Tillage Research* 8, 29-49.

Ford, G.W., Martin, J.J., Rengasamy, P., Boucher, S.C. and Ellington, A. (1995) Soil sodicity in Victoria. In: Naidu, R., Sumner, M.E. and Rengasamy, P. (eds), *Australian Sodic Soils: Distribution, Properties and Management.* CSIRO Publications, East Melbourne, Australia, pp. 277-298.

Gifford, R.M., Cheney, N.P., Noble, J.C., Russell, J.S., Wellington, A.B. and Zammit, C. (1992) Australian land use, primary production of vegetation and carbon pools in relation to atmospheric carbon dioxide concentration. In: Gifford, R.M. and Barson, M.M. (eds), *Australia's Renewable Resources: Sustainability and Global Change.* Bureau of Rural Resources Proceedings 14, Canberra, Australia, pp. 151-187.

Grierson, I.T., Jijne, J.W. and Greenland, D.J. (1972) Changes in some physical properties of red-brown earths of different texture associated with increasing content of organic matter. *Experimental Record* 6, 16-22.

Ham, G.J. (1988) Dissolvenator: A result of Burdekin research on water infiltration. *Bureau of Sugar Experiment Stations Bulletin* 23, 12-13.

Ham, G.J., McMahon, G.G., Elliot, P.J. and Smettem, K.R.J. (1995) Cropping sodic soils in the Burdekin River irrigation area. In: Naidu, R., Sumner, M.E. and Rengasamy, P. (eds), *Australian Sodic Soils: Distribution, Properties and Management.* CSIRO Publications, East Melbourne, Australia, pp. 139-146.

Haysom, M.B., Hurney, A.P. and Nielsen, P.J. (1986) The influence of lime on the acid soils of Ingham. *Proceedings of the Australian Society of Sugar Cane Technologists* 8, 55-61.

Haywood, R.W. and Mitchell, A.G. (1987) A comparison of soil and water losses from conventional and minimum tillage replanting methods using a rainfall simulator. *Proceedings of the South African Sugar Technologists Association* 61, 146-149.

Johnston, M.A. (1977) Reclamation of a saline sodic soil in the Nkwaleni Valley. *Proceedings of the South African Sugar Technologists Association* 51, 42-46.

Johnston, M.A. and Wood, R.A. (1971) Soil compaction at Pongola. *Proceedings of the South African Sugar Cane Technologists* 45, 261-269.

Keren, R. and Shainberg, I. (1981) Effect of dissolution rate on the efficiency of industrial and mined gypsum in improving infiltration of a sodic soil. *Soil Science Society of America Journal* 45, 103-107.

Langdale, G.W., Mills, W.C. and Thomas, A.W. (1992) Use of conservation tillage to retard erosive effects of large storms. *Journal of Soil and Water Conservation* 47, 257-260.

McIntyre, D.S. (1958) Permeability measurements on soil crusts formed by raindrop impact. *Soil Science* 85, 185-189.

Mead, J.A. and Chan, K.Y. (1988) Effect of deep tillage and seedbed perforation on the growth and yield of wheat on a hardsetting soil. *Australian Journal of Experimental Agriculture* 28, 491-498.

Meiri, A. (1984) Plant response to salinity: Experimental methodology and application to the field. In: Shainberg, I. and Shalhevet, J. (eds), *Soil Salinity under Irrigation.* Springer-Verlag, Berlin, Germany, pp. 284-297.

Miller, W.P. and Radcliffe, D.E. (1992) Soil crusting in the southeastern United States. In: Sumner, M.E. and Stewart, B.A. (eds). *Soil Crusting: Chemical and Physical Processes.* Lewis Publishers, Boca Raton, FL, USA, pp. 233-266.

Miller, W.P., Truman, C.C. and Langdale, G.W. (1988) Influence of previous erosion on crusting behavior of Cecil soils. *Journal of Soil and Water Conservation* 43, 338-341.

Mullins, C.E., MacLeod, D.A., Northcote, K.H., Tisdall, J.M. and Young, I.M. (1990) Hardsetting soils: Behavior, occurrence and management. *Advances in Soil Science* 11, 38-108.

Nelson, P.N. and Oades, J.M. (1997) Organic matter, sodicity, and soil structure. In: Sumner, M.E. and Naidu, R. (eds), *Sodic Soils.* Oxford University Press, New York, NY, USA.

Oster, J.D., Shainberg, I. and Abrol, I.P. (1996) Reclamation of salt-affected soil. In: Agassi, M. (eds), *Soil Erosion, Conservation, and Rehabilitation.* Marcel Decker, Inc., New York, NY, USA, pp. 315-352.

Owen, G.T. (1988) Soil disturbance associated with deep subsoiling in compact soils. *Canadian Agricultural Engineering* 30, 33-37.

Pavan, M.A., Bingham, F.T. and Pratt, P.F. (1982) Toxicity of aluminium to coffee in Ultisols and Oxisols amended with $CaCO_3$, $MgCO_3$, and $CaSO_4.2H_2O$. *Soil Science Society of America Journal* 46, 1201-1207.

Pleysier, J.L. and Juo, A.S.R. (1981) Leaching of fertilizer ions in a ultisol from the high rainfall tropics: Leaching through undisturbed soil columns. *Soil Science Society of America Journal* 45, 754-760.

Quinan, D.J. and Wood, A.W. (1989) The effects of lime applications on acid soils in the Hambledon mill area. *Proceedings of the Australian Society of Sugar Cane Technologists* 11, 70-75.

Radcliffe, D.E., Clark, R.L. and Sumner, M.E. (1986) Effect of gypsum and deep-rooting perennials on subsoil mechanical impedance. *Soil Science Society of America* 50, 1566-1570.

Reeve, N.G. and Sumner, M.E. (1972) Amelioration of subsoil acidity in Natal Oxisols by leaching of surface-applied amendments. *Agrochemophysica* 4, 1-6.

Ridge, D.R., Hurney, A.P. and Haysom, M.B.C. (1980) Some aspects of calcium and magnesium nutrition in North Queensland. *Proceedings of the Australian Society of Sugar Cane Technologists* 2, 55-61.

Roth, C.H., Frede, H.G. and Derpsch, R. (1988) Effect of mulch rates and tillage systems on infiltrability and other soil physical properties of an Oxisol in Parana, Brazil. *Soil and Tillage Research* 11, 81-91.

Schroeder, B.L., Robinson, J.B., Wallace, M. and Turner, P.E.T. (1994) Soil acidification: Occurrence and effects in the South African sugar industry. *Proceedings of the South African Sugar Technologists Association* 68, 70-74.

Shainberg, I. (1992) Chemical and mineralogical components of crusting. In: Sumner, M.E. and Stewart, B.A. (eds), *Soil Crusting: Chemical and Physical Processes.* Lewis Publishers, Boca Raton, FL, USA, pp. 33-53.

Shainberg, I., Sumner, M.E., Miller, W.P., Farina, M.P.W., Pavan, M.A. and Fey, M.V. (1989) Use of gypsum on soils: A review. *Advances in Soil Science* 9, 1-111.

Shainberg, I., Warrington, D.N. and Rengasamy, P. (1990) Water quality and PAM interactions in reducing surface sealing. *Soil Science* 149, 301-307.

Sharma, M.L. and Williamson, D.R. (1984) Secondary salinization of water resources in Southern Australia. In: French, R.H. (ed.), *Salinity in Watercourses and Reservoirs.* Butterworth Publishers, Boston, MA, USA, pp. 571-582.

Smith, H.J.C., Levy, G.J. and Shainberg, I. (1990) Water-droplet energy and soil amendments: Effect on infiltration and erosion. *Soil Science Society of America Journal* 54, 1084-1087.

So, H.B. and Onus, G.C. (1984) The extent of unstable gray and brown clays in the Lower Gwydir Valley New South Wales and its relationship to wheat yields. In: McGarity, J.H., Hoult, E.H. and So, H.B. (eds), *The Properties and Utilization of Cracking Clay Soils.* Reviews in Rural Science, University of New England, Armidale, NSW, Australia, pp. 132-135.

Sumner, M.E. (1991) Soil acidity control under the impact of industrial society. In: Bolt, G.H. (ed.), *Interactions at the Soil Colloid - Soil Solution Interface.* Kluwer Academic, Netherlands, pp. 517-541.

Sumner, M.E. (1993a) Gypsum and acid soils: The world scene. *Advances in Agronomy* 51, 1-32.

Sumner, M.E. (1993b) Sodic soils: New perspectives. *Australian Journal of Soil Research* 31, 683-750.

Sumner, M.E. (1994) Amelioration of subsoil acidity with minimum disturbance. In: Jayawardane, N.S. and Stewart, B.A. (eds), *Subsoil Management Techniques.* Lewis Publishers, Boca Raton, FL, USA, pp. 147-185.

Sumner, M.E. (1997) Acidification. In: Lal, R., Blum, W.H. and Valentin, C. (eds), *Methodology for Assessment of Soil Degradation.* Lewis Publishers, Boca Raton, FL, USA.

Sumner, M.E. (1997) Methodology for assessing lime requirement of soils: A critical review. In: Schaffler, R. (ed.), *Plant Soil Interactions at Low pH.* Kluwer Academic Publishers, Dordrecht, The Netherlands.

Sumner, M.E. and Miller, W.P. (1992) Soil crusting in relation to global soil degradation. *American Journal of Alternative Agriculture* 7, 56-62.

Sumner, M.E. and Naidu, R. (1997) *Sodic Soils.* Oxford University Press, New York, NY, USA.

Swinford, J.M. and Boevey, T.M.C. (1984) The effects of soil compaction due to infield transport on ratoon cane yields and soil physical characteristics. *Proceedings of the South African Sugar Technologists Association* 58, 198-203.

Swinford, J.M. and Meyer, J.H. (1985) An evaluation of a nuclear density gauge for measuring in field compaction in soils of the South African sugar industry.

Proceedings of the South African Sugar Technologists Association 59, 218-224.

Turner, P.E.T., Meyer, J.H. and Wood, R.A. (1992) Some results of field experiments to test lime and phosphogypsum for aluminium amelioration. *Proceedings of the South African Sugar Technologists Association* 66, 44-49.

USSL Staff (1954) *Diagnosis and Improvement of Saline and Alkali Soils.* United States Department of Agriculture, Washington D.C., USA.

van Antwerpen, R. and Meyer, J.H. (1996) Evaluation of soil degradation under sugarcane cultivation in northern Kwazulz-Natal. In: Wilson, J.R., Hogarth, D.M., Campbell, J.A. and Garside, A.L. (eds), *Sugarcane: Research Towards Efficient and Sustainable Production.* CSIRO Division of Tropical Crops and Pastures, Brisbane, Australia, pp. 244-247.

van Lierop, W. (1990) Soil pH and lime requirement determination. In: Westerman, R. (ed.), *Soil Testing and Plant Analysis.* American Society of Agronomy, Madison, WI, USA, pp. 73-126.

Vitti, G.C., Mazza, J.A., Pereira, H.S. and Dematte, J.L.I. (1992) Resultados erperimentais do uso de gesso na agricultura. In: Anonymous (ed.), *II Seminário Sobre o Uso do Gesso na Agricultura.* IBRAFOS, Uberaba, Brazil, pp. 191-224.

Warrington, D., Shainberg, I., Levy, G.J. and Bar-On, Y. (1991) Polysaccharide and salt effects on infiltration and soil erosion. A rainfall simulation study. *Soil Technology* 4, 79-91.

Chapter eighteen:

Raising Profitability and Productivity by Improving Harvesting Procedures

L.E. Brennan[1], D.R. Ridge[2] and B.J. Milford[3]

1. *Department of Agriculture, The University of Queensland, Qld 4072, Australia*
2. *Bureau of Sugar Experiment Stations, PO Box 651, Bundaberg, Qld 4670, Australia*
3. *CANEGROWERS, GPO Box 1032, Brisbane, Qld 4001, Australia*

Abstract

The cane harvesting and transport process is a high cost activity for the Australian sugar industry. In this paper, opportunities for increasing industry profitability are explored through the technical and socio-economic dimensions of harvesting. Although the focus is on in-field harvesting processes, the importance of cooperation and closer coordination of the in-field and milling sectors to achieve maximum industry efficiency is highlighted. A socio-economic profile of the harvesting sector was developed and used in conjunction with a harvesting-cost model to examine in-field harvesting systems. Changing the structure and operational basis of harvesting groups, particularly through increasing group size, appears to be the best option for increasing profitability. The group enlargement process has been slow in most regions and factors which may reduce the incentive for the industry to increase group size include: the risk of wet weather, avoidance of transaction costs, inappropriate or individualistic accounting and lifestyle considerations. The high ground speeds necessary to achieve large throughputs with single shift harvesting that lead to high levels of soil and extraneous matter in cane are also a factor. Despite the large historical impact of technical innovations on harvesting productivity, they do not currently offer significant opportunities to improve efficiency. Recent technological improvements have been of an incremental nature only and there appear to be diminishing returns to research efforts concentrating on the current design.

In: Keating, B.A. and Wilson, J.R. (eds) *Intensive Sugarcane Production: Meeting the Challenges Beyond 2000*. CAB International, Wallingford, UK, 1997.

The Development of Cane Harvesting in Australia: 1956 to Present

The mechanisation of harvesting operations is recognised as one of the most significant technical advances in the Australian sugar industry. Prior to its introduction in the 1960s, sugarcane was manually harvested using seasonal labour. Labour was always scarce; in the last century Kanakas from the Pacific Islands were imported. After Federation, migrants, predominantly from southern Europe, were the main source of new recruits. The shortage of labour and its high cost provided the stimulus for the invention and adoption of mechanical cane harvesters. Although some experimental work in mechanical harvesting began when the Pacific Islanders were deported early this century, it was largely the increasing difficulty in obtaining sufficient suitable labourers after the Second World War which stimulated the development of mechanical harvesting (Kerr and Blyth, 1993). Commercial production of the Massey Ferguson 515 harvester commenced in 1956, marking the transition between experimental and commercial development. In 1960, only 1% of the Queensland crop was cut by chopper harvesters. Rapid adoption and technical development followed for the next two decades (Fig. 18.1).

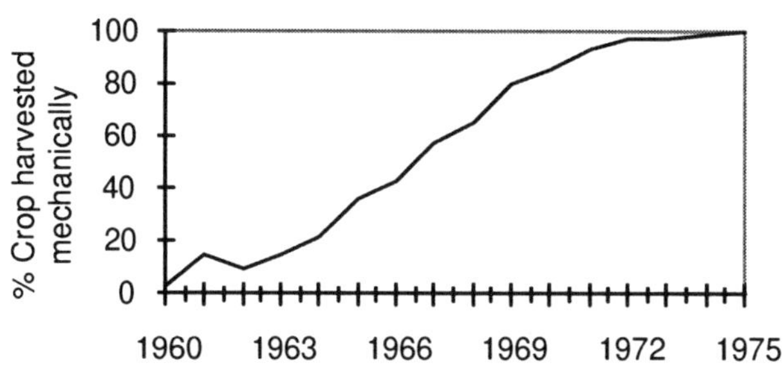

Fig. 18.1. Mechanical harvesting of sugarcane in Australia (as % of total crop) from 1960 to 1975.

By 1971, 96% of cane was harvested mechanically (Churchward and Belcher, 1972) and manual harvesting ceased altogether in 1978. Accompanying these technical changes was a significant structural change in the industry, characterised by intensified capital investment, revised transport arrangements and significant labour displacement. By international standards, advanced harvesting technology is used in the Australian sugar industry. The rate of technical change, however, has slowed considerably in recent years, reflecting the maturation of mechanical harvesting technology in the industry.

Technical Advances

Harvester innovations

Harvester capacity

Harvester development has favoured the production of higher capacity machines designed to cut in a wide range of crop conditions. Current model harvesters can achieve pour rates (a measure of the machine's capacity in tonnes cut per hour) of 140t h^{-1} in burnt cane. In contrast, pour rates were only 35-50t h^{-1} in the early 1970s. Harvester capacity has steadily increased. The rate of increase, however, has slowed since the introduction of the Toft 7000 models around 1985. It is expected that the rate of increase in capacity will slow further over the next few years.

Green cane capacity

The common practice for many years in mechanical harvesting in Australia was to burn before harvest. This reduced pest and disease control problems and increased harvesting efficiency. Disruptions to the harvest through wet weather in north Queensland prompted the first introduction of commercial green cane harvesters in 1976. Adoption of green cane harvesting was initially slow due to limited green cane capacity of most harvesters (Fuelling *et al.*, 1978). The expansion of green cane harvesting was made possible by the release of the Austoft 7000 series harvester and by successful modification of older model harvesters to cut green (Ridge and Dick, 1989). Rapid adoption followed in north Queensland, with almost the entire crop harvested green in Ingham by 1994. Overall, adoption has been steadily increasing in Queensland, although at a slower rate (Table 18.1).

Table 18.1. Trends in green cane harvesting for selected Queensland regions from 1986-1994.

Region	% Green cane harvesting				
	1986	1988	1990	1992	1994
Ingham	47.7	87.0	96.4	97.3	98.0
North Queensland	33.6	60.0	74.8	80.1	83.0
South Queensland	n.a.	20.0	17.5	18.9	23.9
Queensland	11.6	30.0	33.6	35.9	41.2

Improvements to the feeding characteristics and cleaning capacity in modern commercial harvesters have increased throughput in both green and burnt cane, but green cane pour rates still range from 50 to 80% of burnt cane rates, depending on crop size. The remaining technical challenge in green cane harvesting is cutting high-yielding, lodged crops such as two year cane where feed rates are uneven and very low and operators have difficulty in locating the cane row.

Cleaning efficiency

Cleaning of sugarcane prior to processing is carried out solely by the cane harvester and the mills require as clean a product as is commercially feasible. Extraneous matter is removed by topping the cane where possible and pneumatic cleaning after the cane is cut into billets. With this type of cleaning system there is a trade off between cleaning efficiency and cane loss. Cane varietal characteristics have a strong influence on both cane loss during cleaning and final extraneous matter levels in harvested cane. Trials conducted in green cane with major industry varieties showed initial extraneous matter levels in the field ranging from 17-30%, final levels of 4-13% and cane losses through harvester cleaning extractors of 4-16% (Ridge and Dick, 1989). A more recent industry-wide survey showed average cane losses of approximately 7t ha^{-1} for green cane and 2t ha^{-1} for burnt cane (Linedale *et al.*, 1993).

There has been considerable research aimed at reducing cane losses during green cane harvesting while maintaining acceptable cleaning efficiency. Recent model harvesters are fitted with vertical drive extractors to improve cleaning efficiency and variable speed primary extractors so that cleaning intensity can be adjusted for different crop conditions. A cane loss monitor has also been developed so that losses can be monitored during harvesting and settings adjusted accordingly. The ability to tune harvester cleaning systems provides an acceptable compromise between cane loss and extraneous matter in the cane supply. Industry wide this has had a significant impact on losses in green cane with a 1995 survey showing average losses of about 4t ha^{-1}. Despite these advances there is industry concern about higher leaf and trash levels when harvesting green due to their impact on transport system capacity and mill processing rates. With increased harvester throughput capacity in green cane, and the trend to larger contracts, it is unlikely that this cleaning problem will be overcome with current technology. It is timely for the industry to re-evaluate the system for cleaning cane and consider options such as partial cleaning at the mill and use of extraneous material as a value-adding fuel source.

Soil in cane

Chopper harvesters have increased soil levels in the cane supply (Ridge and Dick, 1992), and this has had an important impact on maintenance, operational costs and efficiency of filtration and clarification in sugar mills. It is estimated that soil in the cane supply increases mill maintenance and operational costs by approximately \$1 t^{-1} cane. Harvester maintenance costs are also increased by the abrasive action of soil on components. The higher soil levels are attributed to the ability to cut lodged cane, the pushing of erect cane forward prior to basecutting so that cane is fed butt first into the harvester, and the ability of modern harvesting equipment to operate in wet conditions.

Research into methods of reducing soil levels in chopper-harvested cane has highlighted a number of areas where improvements can be made. These include improved cultural practices so that cane can be cut at ground level without intake of soil; matching of basecutter angle to the row profile; use of dome shaped basecutter discs to increase rejection of soil and design of feed train rollers to improve rejection of soil. The effectiveness of these measures has been limited by several factors including the lack of a suitable industry-wide method for measuring soil in cane, the active encouragement of harvesting in wet conditions to maintain cane supply to the mill and high harvester throughput, the wish to maximise cane yields which leads to more lodged crops, insufficient monetary incentive for growers and harvesting contractors to adopt measures for reducing soil levels, and the pressure for high harvesting rates that leads to more soil intake and less soil rejection by the harvester. Despite these difficulties, current extension activity to highlight approaches to minimising soil in the cane supply is expected to have a significant impact on industry soil levels.

Farming practices

The mechanical harvesting process has a detrimental effect on cane yields unless appropriate land preparation is undertaken. Compaction of the rows by harvesting equipment, as a result of narrow row spacing, has caused average yield reductions of 10% (M. Braunack, *pers. comm.*). Cultural measures and operational procedures that have been adopted to improve harvester efficiency have been aimed at reducing stool damage, cane loss and extraneous matter, all of which affect crop harvestability.

Cultural practices identified as playing a key role in harvesting efficiency include: block grading and complete filling in of the row in plant cane, wider row spacings to accommodate harvesters, cultivation to round the shoulders of hills in ratoon rows, and creating wide headlands to minimise compaction damage. Operational procedures, such as maintenance of cutting blades, reducing cutting speeds where appropriate, and taking care to avoid crossing rows while turning, can minimise damage to stools and billets, and reduce cane loss.

Systems

The transition to chopper harvesters was a major departure from whole stalk harvesting. The realisation that deterioration of cane was much more rapid and severe with chopped, burnt cane came slowly (Young and Vallance, 1959). This potential for deterioration necessitated greater coordination between mills and growers. Mill transport systems evolved to keep pace with harvester development

(Ridge, 1987). The rail network expanded in most mill areas, bins were designed for chopped cane, and portable line disappeared from cane blocks to be replaced by roll-on/roll-off trailers or motor transport.

The rapid investment in mechanical harvesting equipment changed the industry from a labour-intensive to a capital-intensive system. By the early 1970s, a seasonal workforce of 8000-10,000 cane cutters had been displaced. In 1970, however, there were 1920 harvesters operating, which would have directly employed around 5000 workers, thereby taking up some of the slack. At the same time, a commercial sector developed to service the maintenance and repair needs of the expanding harvesting sectors.

The diversity of technical innovations and the manner in which they have evolved to form a complex system have been key features in the development of mechanical harvesting. The coordination of interrelated innovations has enabled greater efficiencies to be gained from technical advances, highlighting the importance of adopting a systems approach when viewing components of the harvesting sector.

A Socio-Economic Profile of the Harvesting Sector

Group size

The harvesting sector has experienced structural adjustment characterised by the amalgamation of harvesting groups, resulting in fewer but larger groups in the industry (Table 18.2). A 'group' in this context is a single harvesting contract: the farms (or farm) harvested by a single harvester-operator.

A major stimulus for structural change has been the rising cost of purchasing and maintaining harvesting equipment. The total cost of a harvester and two infield transporters is at least $500,000. Formation of larger groups can reduce costs by enabling more efficient use of resources, particularly capital. Also, larger groups are in a stronger position to take advantage of innovative technology, which often involves expenditure on large capital items that are indivisible, making their purchase difficult to justify economically for smaller enterprises.

Recognition that groups should enlarge to counteract rising machinery costs began as far back as the early 1970s (Churchward and Belcher, 1972) and several cost studies examining the group size issue followed in the 1980s (Petersen *et al.*, 1984; Page *et al.*, 1985; Ridge and Dick, 1985; Connell and Borrell, 1987; Ferguson, 1987; McWhinney *et al.*, 1988). The consensus emerging from these studies was that increasing throughput reduced group harvesting costs. Despite this information, group sizes have increased only slowly, with large variations in size amongst regions, and most are well below the operating potential of modern harvesters.

Table 18.2 shows group sizes have gradually increased, with all regions following a similar trend characterised by rapid growth over the periods 1970 to 1975 and 1985 to 1995, except the Southern region over the latter period. Coinciding with depressed sugar prices, there was minimal or no growth in between these periods. Average group sizes in the Central and Southern regions are considerably smaller than in the Burdekin, Northern and NSW regions. The large increase in group size for the Central region between 1990 and 1995 was largely attributable to rapid change in the Proserpine mill area.

Table 18.2. Average group size (tonnes per harvester) on a district basis and the percent increase from the previous five years.

Year	NSW	%	South	%	Central	%	Burdekin	%	North	%
1970	n/a		6,522		6,137		12,582		11,873	
1975	18,535	n/a	9,353	43	9,882	61	23,149	83	14,780	24
1980	31,465	70	10,649	13	9,814	0	25,176	8	15,742	7
1985	31,338	0	12,112	14	8,923	0	26,808	6	16,679	6
1990	37,359	19	13,895	15	12,774	43	35,827	33	23,566	41
1995	65,454	75	19,478	40	20,820	63	49,896	40	39,589	67

Group structure

The structure of harvesting groups has developed along highly individual lines. Consequently, there is considerable variation in the levels of investment in harvesting equipment, the type of ownership of harvesting equipment, as well as harvesting group sizes. Group ownership of harvesting takes on several forms. These are normally classified into five categories: (A) growers cutting their own cane, (B) growers cutting own cane and others, (C) group of growers cutting own cane (cooperative ownership), (D) group of growers cutting own cane and others, (E) independent contractors with no farm involvement.

In the amalgamation process over the last 25 years, contract harvesting (category E) and farmer-contractor harvesting (category B) have emerged as the most common forms of harvesting (Table 18.3). Rising equipment costs, together with the potential to achieve economies of size, have encouraged this type of harvesting. For farmer-contractors, contracting out provides an economic procedure to enlarge their business and so become large enough to justify large investments in harvesting equipment. While still accounting for a relatively small proportion of the harvest, cooperative harvesting where group members share or pool equipment has become more popular in the past 5 to 10 years, especially in NSW where almost 100% of the crop is harvested in this manner.

In contrast to the other regions, a large proportion of groups (34%) in the Central region are growers cutting their own cane (Category A), although collectively, they harvest only 14% of the region's crop. This helps explain the high proportion of small groups in the region. Cane in the Northern region is harvested mainly by farmer-contractors while the Burdekin crop is predominantly harvested by independent contractors. The variation in type of ownership appears well entrenched in each region. Ownership patterns for the 1995 season matched those reported in Petersen *et al.* (1984) for the 1982 and 1978 seasons.

Table 18.3. Harvester ownership categories (A-E, see text) for North, Burdekin, Central and NSW cane regions.

Region	A		B		C		D		E	
	%[1]	%[2]	%[1]	%[2]	%[1]	%[2]	%[1]	%[2]	%[1]	%[2]
North	13	6	52	60	3	2	8	2	24	29
Burd'n	3	<1	26	26	0	0	0	0	71	74
Central	33	14	36	33	10	12	2	12	20	31
NSW	0	0	0	0	100	100	0	0	0	0

[1] Per cent of total groups in region
[2] Per cent of total crop harvested

Although there are significant numbers of old harvesters in all Queensland regions, > 60% of harvesters in the Central and Southern regions are more than ten years old. In contrast, most machines in the Burdekin and NSW regions are less than three years old. This suggests that group size and structure is a key factor in the diffusion of new harvesting technology in the industry.

On the surface, it appears that increasing group size would reduce industry costs. There are currently 1225 harvesters in Queensland. If all groups were 60,000 tonnes, which is a realistically achievable target in all regions, then the number of harvesters required would fall by over half to 588; the number of cane haulout units would also be reduced. The largest reductions would occur in the Central and Southern regions (Table 18.4). Major adjustments would not be necessary to accommodate this scenario because there is sufficient capacity amongst the existing new harvesters in most regions to absorb the additional tonnage. Impediments to increasing group size are discussed later in the paper.

Table 18.4. The effect of increasing group size on the number of harvesters requirements for the four Queensland cane regions.

Region	No. Harvesters Current	No. harvesters if 60,000t groups	Reduction in numbers
North	311	205	106
Burdekin	149	123	26
Central	486	169	317
South	279	91	188

Prospects for Raising Profitability

Technical innovation

The rate of technical innovation in harvesting and transport systems has slowed in recent years. By the early 1980s the extent of mechanisation in the industry was such that there was little scope for further increases in labour productivity (Committee of Inquiry into Technological Change in Australia, 1980). Technical innovation in the last decade has been of an incremental nature and research has concentrated on refinements within the established design. The improvements have been in individual components, but the underlying core design concepts and the links between them have remained the same. The question which arises is whether it is time to move to radical innovation with a new dominant design of harvesting technology and a new set of design concepts.

Possible developments include the use of double or multi-row harvesters in conjunction with upgraded infield transport and supply of mill transport capacity; and, in conjunction with use of extraneous matter for fuel, a simpler, high capacity harvester with reduced cleaning of cane at harvest and greater cleaning at the mill. Such innovative approaches could offer potential benefits from reduced labour input and lower capital investment per unit of output.

Field layout

In line with suggestions for technical innovation in harvesting, radical changes to agronomic practices could offer opportunities for raising profitability. Current research on narrow row spacings for increasing crop production could give benefits in harvester productivity through the ability to harvest multiple rows. If high yields can be obtained without crop lodging this has benefits for green cane harvesting and limiting extraneous matter levels in harvested cane.

Within the limits of the current technology, harvesting costs can be reduced by manipulating field conditions to maximise harvester throughput. Table 18.5 considers the example of row length, which is an important field factor affecting harvester throughput. On an equal block area basis, increasing the length of rows reduces the number of rows; this in turn results in less total time spent turning harvesting equipment at the end of rows and increased cutting rates. In this example, a saving of A\$0.37 t^{-1} is achievable by increasing row length from 100m to 600m. Longer rows reduce total operating time, and savings are largely attributable to the improved efficiency of labour utilisation.

While technical improvements are an ongoing means of improving harvesting systems, there appears to be scope to achieve harvesting efficiency in the socio-economic dimension of harvesting systems.

Table 18.5. Effect of row length on operating time and cost (A$).

Row length (m)	Time lost in turns (min)	Cutting rate ($t\ h^{-1}$)	Operating time (h)	Cost* ($\$\ t^{-1}$)
100	72.0	72	3.07	5.28
200	36.9	90	2.46	5.06
400	18.4	102	2.15	4.95
600	12.3	108	2.05	4.91

* Calculations for 22,000t group, 2 haulouts, 5.4t x 2 double bin haulouts, 0.5 km haul distance, haulout speed 40km h^{-1}, crop density 100t ha^{-1}, 120t h^{-1} pour rate, no harvester waiting time. [Using model outlined in Ridge *et al.,* 1996]

Group size

Increasing the size of harvesting groups appears to be the best option for increasing profitability. There is no single optimum group size for the industry. The minimum average costs achievable by efficiently managed harvesting businesses will differ according to the inherent capacity of the machines and operating conditions. The in-field system described below does not provide a benchmark for determining optimum group size but serves only as an illustration of the behaviour of harvesting costs.

The throughput-cost model developed by Ridge *et al.* (1996) is used in this paper to evaluate the impact of group size on harvesting cost. Figure 18.2 shows the fixed, variable (fuel, repairs and labour) and total cost curves for a harvesting business operating one new 1995 harvester and three new cane transport units of the elevating-tipper type.

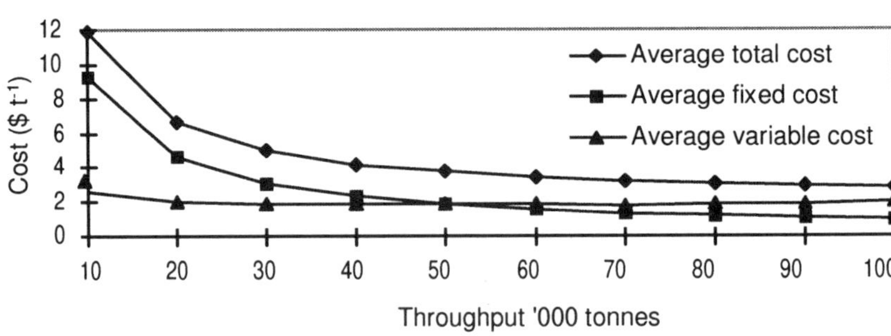

Fig. 18.2. Costs per tonne (A$) vs seasonal throughput of cane per harvester.

Fixed costs per tonne fall as seasonal throughput increases because these costs are being spread over larger tonnages. The average variable cost curve is relatively flat, reflecting constant $ t^{-1} costs for most operating inputs, such as fuel usage and materials costs for maintenance. The curve falls slightly at low tonnages; labour economies arise because there is a basic level of maintenance which must be carried out regardless of the tonnes cut by the equipment. At higher tonnages, this cost curve shifts slightly upwards due to the extra costs associated with employing extra labour to take the harvesting operation into a second shift and the increased pre-season labour required to harvest large seasonal throughputs. Variable costs do not increase to the extent of turning the average total cost curve upwards. Consequently, for this example, the economic optimum is obtained by operating the equipment to the harvester's seasonal physical capacity.

This example could justly represent a harvesting group in any mill area and it suggests that most groups in all four regions do not have sufficient throughput to operate at the economic optimum using new equipment. Many groups could economically justify their current groupings only with the use of much older equipment. Small groups can reduce their capital costs by investing with older machines. A number of inefficiencies, however, are associated with this practice. The efficiency of inputs is reduced with old equipment. Although the capital investment costs are reduced, some of the variable costs of old machines, such as maintenance, are higher than for new machines. Slower pour rates also decrease the labour productivity per tonne because of the longer time needed to harvest the allotment. The industry forgoes the benefits of technical change which were discussed earlier, particularly the ability to cut green cane in addition to the other cane quality aspects related to the harvesting process. Small groups, in large numbers, impose costs on other parts of the harvest-transport system. These cost transfers are discussed later in this paper.

Impediments to Change

Given the apparent economic advantages of increasing group size, the question that arises is how can the industry move to a higher level of profitability in this way? For this to occur, cognisance has to be taken of the factors that exist to reduce the incentive for group size increases. The main issues are discussed below.

Risk of wet weather

Farmers may prefer to operate in a small group, particularly in an area that suffers frequent disruption due to wet weather during the crushing season. In a smaller group there is better ability to 'catch up' and to ensure that all cane is removed in a timely manner. The risks associated with losing large areas of burnt cane to wet weather are possibly a major deterrent to increasing the size of groups, particularly in the Southern and Central regions where large areas are still burnt for harvesting. Big groups burn large areas at a time which presents smaller producers with a large risk when a significant proportion of their crop could be standing burnt prior to harvest.

Farmers also worry about the loss in sugar content which commonly occurs in wet weather. One response to this has been a move from 'weekly relative' schemes to 'daily relative'. Relative payment schemes are designed so that farmers are indifferent to the time of the season in which they harvest, despite the fact that sugar content changes throughout the season.

Measures available to minimise the risk of wet weather include harvesting cane green where there are suitable conditions, and using harvesting equipment which has wet weather capacity.

Lifestyle considerations and avoidance of transaction cost

Harvester operators do not base their investment decisions solely on economic incentives and there are various non-cash benefits that growers and farmer-contractors claim to obtain from their group structure. Many place a high value on their lifestyle and their work becomes not only a productive activity but a lifestyle preference and leisure activity. For many growers cutting their own cane, or small-sized farmer contractors and independent contractors, the reduction in the capital-output ratio achievable by increasing group size is no compensation for the loss of lifestyle that could be involved.

Farmers may also value highly the effort involved in negotiating with a group and in fitting in with other farmers during the rotation. Cutting one's own cane or belonging to a smaller group are ways of reducing this cost. Independence is sought for a number of reasons. Firstly, it avoids personal disputes which may arise from negotiating harvesting arrangements with other growers in the absence of contractual agreements. Maintaining independence may also enable the farmer to utilise or tailor the work required on a farm of a particular size to the labour available. This may provide a reasonable return on a smaller farm if, for reasons of risk avoidance or labour availability, a farmer chooses not to expand onto a larger farm. Thirdly, some growers harvest independently because they feel there is no other alternative. This may be the case where the farm is isolated, or where poor farm layout such as many small blocks or stony, hilly terrain discourage contractors or cooperatives from harvesting the farm, or contractors charge a higher than average price to harvest.

Inappropriate or individualistic accounting

Some harvesting costs may be regarded as part of the farming cost which would tend to reduce the perceived harvesting cost. Many growers harvesting small allotments with old machinery do so because they see it as a low-cost harvesting method, but may fail to account accurately for expenditure such as labour, fuel and maintenance costs which are recorded as farm costs. If these were costed into the harvesting process, the end cost of the operation per tonne would be much higher.

Information enabling growers and contractors to make more informed decisions about optimal grouping arrangements is not widely available. Current industry group sizes arguably reflect the lack of a reliable information source providing details of the costs associated with operating various in-field harvesting systems. The process of group enlargement could be facilitated by the provision of cost information which could be easily used by individual groups. One way is to release costing models, such as Ridge *et al.* (1996), which would allow groups to make improvements tailored to their specific set of harvesting conditions.

Also, better understanding of what drives harvesting costs could allow for more customised price setting, with prices set to reflect field conditions on a block by block basis, rather than apply the current pricing system which does not vary significantly amongst individual farms. Harvesting costs are relatively constant across a district with variations of only 0.3-0.5 \$ t^{-1} in each district. Thus prices set by different harvesting groups tend to be similar regardless of the harvesting group's cost curve. In theory, however, harvesting prices charged to farmers should be bid down so that marginal costs of harvesting equal marginal returns. This indicates the perception that harvesting groups do not see themselves in a competitive market with other groups. There is therefore the likelihood that, because of the nature of the investment in harvesting and the differing motivations of harvesting group operators, prices charged for harvesting are not being completely set in a free market. Some element of economic surplus may accrue to harvester operators to the extent that this reduces the impetus to move from one group to another, and this may hamper an overall move to large groups.

If there were clearer harvesting price signals, individual growers may take measures to improve field conditions in order to lower their harvesting costs. Also, the practice of cross-subsidisation which involves growers with good field conditions subsidising the harvesting cost of farms with poor harvesting conditions could be substantially reduced.

Physical limits

There are physical limits to group size. These relate to the hours of harvest, the availability of bins, the pour rate and ground speed of the machine and the level of loss which is acceptable. In order to achieve throughputs of more than around 80,000 tonnes during normal harvesting hours, ground speeds of up to 8-9km h^{-1} are necessary. It has been shown that at ground speeds of over 7km h^{-1} the losses from inadequate cane pick-up and spillage increase. Another problem is damage to the cane stool at high operating speeds. Concern about these losses is one of the factors which make growers less inclined to form larger groups.

A Systems Approach

There is a mutually dependent relationship between growers and millers, and at the physical and financial interface between the in-field and milling sectors is the harvest-transport system. Harvesting systems cannot be viewed in isolation but rather as a single process commencing at the point of cutting in the field through to the arrival of cane at the mill gate. To achieve major reductions in harvesting and transport costs for the Australian sugar industry, it must be recognised that the harvest-transport system is an entity and both sectors must be considered jointly when addressing future directions.

Harvesting costs, from the growers' point of view, are affected by the number of points of delivery which determines the average haul distance from the farm to the mill, transport schedules which determine the availability of containers and the ability for groups to cut large tonnages without interruption, and the hours of harvesting. Mill costs, in addition to the factors above, are affected by the number of harvesters operating, the movement of harvesters from one line to another thereby changing transport schedules, the reliability of harvesters, and amount of extraneous matter which raises the mills' costs by increasing transport costs and causing wear and quality problems in the factory.

Without changing the number or size of groups in the system, savings can be made by spreading the harvesting hours so that some groups start early in the day and others start in the afternoon. Benefits arising from this practice include the reduced requirement for mill rolling-stock and a reduction in delay between harvesting and milling. The latter is reflected in higher CCS values. Some districts are looking towards harvesting 24 hours a day, allowing for large savings in mill operating costs. Extended-hours harvesting does not require a major restructure of the harvesting sector. Many harvester operators and growers, however, view extended hours harvesting into the night as a significant threat to their quality of life because of the 'anti-social' working hours, noise problems, and occupational health and safety issues. The challenge, however, remains to implement these changes in a way which adequately rewards both parties.

Increasing group sizes is a more difficult process to coordinate because of the considerable adjustments required by the in-field sector. Larger group sizes result in fewer groups in the system and can provide savings to the mills as well as growers provided the necessary mill-transport infrastructure is in place to ensure large groups are able to supply large daily allotments without delays. For mills, it is really the total number of harvesting groups in the system, rather than the size of individual groups, which affects the efficiency of mill transport (McWhinney, 1995). In a benchmark study, Page *et al.* (1985) examined the relationship between in-field and mill costs at the interface. Minimum-cost group sizes were calculated for a range of harvesting hours and the impact of these grouping arrangements on mill transport costs was examined. They proposed that a system utilising large groups with a greater spread of harvesting hours brought about savings to the in-field sector, largely by improved utilisation of capital and, to a lesser extent, improved CCS resulting from lower cane age. Importantly, savings also flowed to mill transport with fewer locomotive shifts and bins required, and from improved CCS. If the system has fewer groups, fewer active sidings are needed on any one day, allowing for more efficient transport scheduling. Savings stemming from this include reduced requirement for capital stock, and lower labour and operating costs, all resulting from fewer locomotive shifts, and from reduced maintenance on mill delivery points made possible by rationalising their numbers.

Concluding Remarks

The technical and socio-economic dimensions of in-field harvesting systems are discussed in this paper. Technical innovations have historically had large impacts on harvesting productivity, although recent improvements have been of an incremental nature only, and there appear to be diminishing returns to research efforts concentrated on improving the current design. Increasing group size appears to be the best option for increasing profitability. The challenge is to successfully coordinate the in-field adjustment process with the other components of the harvest-transport system, particularly the milling sector, to enable maximum savings to be effected in the sugar industry.

Acknowledgments

The authors acknowledge Joseph Evans and Heidi Hopkins at CANEGROWERS for their assistance in preparing data for this paper.

References

Churchward, E.H. and Belcher, R.M. (1972) Some economic aspects of mechanical mechanical cane harvesting in Queensland. *Proceedings of the Queensland Society of Sugarcane Technologists 1972 Conf.*, pp. 31-38.

Committee of Inquiry into Technological Change in Australia (1980) *Technological Change in Australia, Volume 2: Report of the Committee of Inquiry into Technological Change in Australia.* Australian Government Public Service, Canberra, Australia, 448pp.

Connell, P. and Borrell, B. (1987) *Costs and Regulations of Cane Harvesting Practices.* Occasional paper 101, Bureau of Agricultural Economics, Australian Government Public Service, Canberra, Australia, 42pp.

Ferguson, G.A. (1987) A model to measure harvest and haulout costs. *Proceedings of the Australian Society of Sugarcane Technologists* 9, 81-86.

Fuelling, T.G., Henkel, C.R., Leverington, K.C. and Wegener, M.K. (1978) Sugarcane harvester performance. *Proceedings of the Queensland Society of SugarCane Technologists 1978 Conf.*, pp. 209-216.

Kerr, J. and Blyth, K. (1993) *They're All Half Crazy: 100 Years of Mechanical Cane Harvesting,* CANEGROWERS, Brisbane, Australia, 117pp.

Linedale, A.I., Ridge, D.R. and Chapman, F.L. (1993) A coordinated extension approach for maximising returns from mechanical harvesting. *Proceedings of the Australian Society of Sugarcane Technologists* 15, 45-51.

McWhinney, W. (1995) Harvest control and cane transport - the vital link from field to factory. *Proceedings of the Australian Society of Sugarcane Technologists* 17, 46-54.

McWhinney, W., Crawford, D.M., Ridge, D.R., Cameron, J.M. and Absolon, J.E. (1988) The economics of cane from field to factory. *Proceedings of the Australian Society of Sugarcane Technologists* 10, 7-18.

Page, W.A., Couchman, J.A. and Bathgate, R.R. (1985) A method for determining the effect on industry economics of a change to present harvesting systems. *Proceedings of the Australian Society of Sugarcane Technologists* 7, 1-11.

Petersen, A.J., Mackson, J.V., Crawford, D.M. and Robinson, T.O. (1984) Capacity and trends in mechanical harvesting machines in Australia. *Proceedings of the Australian Society of Sugarcane Technologists* 6, 115-120.

Ridge, D.R. (1987) Efficiency through mechanisation - 1957-87. *BSES Bulletin*, 19, 4-8.

Ridge, D.R., Brennan, L.E. and Powell, J. (1996) Factors influencing industry harvest-transport performance. *Proceedings of the Australian Society of Sugarcane Technologists* 18, 6-13.

Ridge D.R. and Dick R.G. (1985) Factors affecting cane harvesting and transport costs. *Proceedings of the Australian Society of Sugarcane Technologists* 7, 139-145.

Ridge, D.R. and Dick, R.G. (1989) The performance of modified harvesters in green cane. *Proceedings of the Australian Society of Sugarcane Technologists* 11, 65-69.

Ridge, D.R. and Dick, R.G. (1992) Soil intake with cane during chopper harvesting - minimising the problem. *Proceedings of the Australian Society of Sugarcane Technologists* 14, 24-30.

Young H.E and Vallance L.G.(1959) Deterioration of cut-up cane with reference to mechanical harvesting. *Proceedings of the Queensland Society of Sugarcane Technologists 1959 Conf.*, pp. 9-13.

Chapter nineteen:

Opportunities to Improve Economic Performance on Sugarcane Farms

M.K. Wegener

Department of Agriculture and CRC for Sustainable Sugar Production, The University of Queensland, Qld 4072, Australia

Abstract

Profitability is a key aspect of the sustainability of farm businesses and no business can operate in the long term without making sufficient profits to cover all costs and provide an adequate return on the capital invested in it. By world standards, Australian raw sugar production is a low-cost, efficient industry but the majority of the industry's output is sold on the world market which is one of the most volatile commodity markets in the world.

Management techniques such as comparative analysis or benchmarking are being developed as effective ways to improve the economic performance of canefarming businesses, and an understanding of the relationship between the various cost components and output can be used to advantage to increase the margin between total farm returns and production costs.

Research and the application of technology to increase productivity, substitution between production inputs and increases in farm size to reduce costs can provide useful opportunities to increase profitability in sugarcane farming. However, the best opportunity to increase profits appears to lie in changes to sugar marketing. Increased flexibility in marketing, to provide growers with the opportunity to exercise control over the most important aspect of their business operation, is recommended as an obvious strategy to increase profitability. Until growers can exercise that option, cane production will remain subject to much unnecessary uncertainty and the risk-averse nature of decision-makers will allow sub-optimal use of resources to continue.

In: Keating, B.A. and Wilson, J.R. (eds) *Intensive Sugarcane Production: Meeting the Challenges Beyond 2000*. CAB International, Wallingford, UK, 1997.

Introduction

Profitability is a key aspect of sustainability of farm businesses. No business can be viable in the long term without making sufficient profits in the short term to cover all costs and provide a return to the capital invested in it.

The Australian sugar industry is technically efficient and low cost by world standards but it is highly regulated in comparison with other rural industries in Australia. Regulations still control the payment for cane, although the quantity of sugar produced is being determined by market conditions. There is scope for continued improvement in economic performance through increased competition among production units.

The world sugar market is one of the most volatile of agricultural commodity markets and the Australian industry is exposed to world market fluctuations, but individual producers are unable to make decisions on where and how to market their output. The industry has recently experienced a period of relatively stable prices but it is inevitable that lower and more fluctuating prices will return, hence individual growers and millers need flexible management policies to take advantage of changing market conditions.

This paper describes economic changes that have occurred in the Australian sugar industry and outlines management strategies to ensure the sustained profitability of sugarcane farming. Changes in the Australian sugar industry should be noted by other countries because of the potential relevance to the development of their own industries.

Alternative Methods for Assessing Profitability on Farms

Profitability in farming is seldom defined adequately and four measures are commonly used to measure it: cash in the farmer's bank account; taxable income of the farm operator; true or managerial profit; and the annual change in net worth of the farm operator.

Although there are linkages between them, these four indicators of profit do not measure the same thing. A profitable farmer may have little cash in the bank if periodic payments are deducted for machinery or land purchases. Alternatively, money may be transferred into a superannuation fund for retirement. These both represent legitimate uses of cash for a well-managed farm but, if interpreted incorrectly, can create the impression of poor economic performance.

Taxable income has been widely used as a measure of economic performance by Australian farmers but special accounting standards apply to the preparation of income tax returns and agricultural enterprises are taxed differently to other businesses. Accounting statements prepared for taxation purposes must be modified to serve as reliable indicators of economic performance.

Profit, as measured by a managerial profit and loss statement, is equivalent to the annual change in net worth calculated from the statement of assets and liabilities or balance sheet. Its calculation is based on correct accounting assumptions, and all costs related to the production of farm output are taken into account.

Management Techniques to Improve Profitability

Comparative analysis

The comparison of farm performance among groups of similar farms has been one of the principal methods used to help producers improve profitability. Now often known as 'benchmarking', it is a simple, though powerful technique for farm managers to use to improve their farm performance.

A number of recent studies have highlighted the differences between the most profitable and the least profitable farms in Australia. In a 1994 survey, the National Australia Bank divided its farmer clients into five categories: real achievers, established clients, average clients, battlers, and part-time farmers. Real achievers attained at least 8% return on their capital investment in the 1992-94 period, a time of poor seasonal conditions and low prices for major agricultural and pastoral commodities. Established clients achieved at least 5% return on capital but all other groups, which represented over 80% of the bank's farm clients, had negative returns to capital. The high-achieving farmers took a business-like approach to farm management. They planned ahead and used cashflow budgets to manage their businesses. Most owned a computer and used it for record-keeping and business analysis. Many of these farmers were involved in producer groups to compare their performance and to improve the efficiency of their production systems. Half of the top achieving group employed consultants, showing that they sought management help and information beyond their own resources. Most of the high-achieving farmers intended to buy more land, an expression of their confidence in the future of agriculture, and 70% of the group had prepared estate plans.

Graham (1995) reported the economic performance of 83 Mackay canegrowers. While the difference in costs between the top 20% and bottom 20% of producers was only A\$30 ha^{-1}, the difference in profit was A\$1164 ha^{-1}. Cane yields on the most profitable farms were significantly higher.

Bureau of Sugar Experiment Stations (BSES) extension officers and the RCS Group have collected three years' financial data from 150 canegrowers in the Bundaberg, Maryborough, Burdekin, and Herbert River districts to identify the best farm management practices (Hanlon, 1996). The study showed that productivity gains and opportunities to increase profits are available to canegrowers. As in the case reported by Graham (1995), most growers spent almost the same amount on inputs to grow the crop but the surplus after costs was highly variable. Differences in profit per hectare were as much as 60% between individual farms.

As a result of this project, some interesting observations have been made about the allocation of expenditure among inputs. Hanlon (1996) noted that canegrowers, like other farmers, invested much of their capital in land which often reduced the funds available for farm development and the creation of profits. In spite of their strong desire to own their own farmland rather than lease or rent it, Australian farmers will ultimately have to adopt the same ownership rules for assets as other businesses. The return on capital invested in land and fixed improvements is comparatively low and long term, and most businesses need to allocate scarce capital resources to growth of the business rather than to ownership of fixed assets.

Hanlon (1996) also observed that canegrowers tended to invest in too much machinery and paid too little attention to their own personal development. Largely because the rural industries measure performance in physical terms, then items such as high yielding crops, best seasonal CCS, well-kept machinery, and well-fed animals are seen as measures of success and less emphasis is placed on the ability to run a profitable farm or manage cash resources. Many farmers unfortunately believe that physical indicators of performance are surrogates for profit but output does not equate with profit and many physical measures of performance are not linked to profit at all.

Alternative Ways to Increase Farm Profits

Comparative analysis can be used to compare the performance of individual farms with others in the same industry and to highlight reasons for poor profitability. Such comparisons provide an incentive for farmers identified as being in the less profitable groups to improve performance and those in the best group are encouraged to stay ahead. By comparing farm performance data expressed in a common form, growers are able to question why their costs or returns are different from others in the group. If there are no obvious reasons for the difference in the level of farm output, or total costs, it is necessary to examine the performance of the business systematically, like unpacking a series of suitcases (Errington, 1991).

Net farm income, or management and investment income, is available for family living expenses, loan repayments, new investment, and the payment of taxation. It is the focus of all farm financial management and needs to be as large as possible. In principle, there are four routes by which a farm manager can conduct a systematic search to make an existing farm more profitable: by increasing profitability of existing enterprises; by substituting more profitable enterprises for low-profit ones; by cutting back on existing fixed costs; or by adding a totally new element to the business base (Giles, 1992).

Normally, farmers do not find it necessary to be involved on a regular basis in elaborate whole-farm re-planning exercises to bring about a major change in farm structure and potential profitability but they are involved in making numerous decisions about the farm's operations each year: whether to plant a different variety, to extend the crop cycle, to apply soil ameliorants, or to vary the amount of irrigation. Essentially, these are fine-tuning exercises within the existing business plan. Occasionally, there will be a need for a major re-organisation of the farm business such as retirement of the business principal, purchase of another farm, introduction of new technology such as green cane harvesting or trickle irrigation, or the need to meet changing standards in environmental management. Such major upheavals to the farm business occur infrequently, perhaps two or three times in the average farmer's lifetime. In contrast, farmers' planning activities are dominated by the need to contemplate and make continuous, marginal adjustments to the farm plan to counteract the continuous deterioration in their terms of trade.

Systematic Farm Business Analysis

A systematic framework by which to undertake a farm business analysis with the aim of increasing profitability is outlined in Table 19.1. It shows how the various types of expenditure contribute to farm profitability.

A brief discussion of some of the factors included in Table 19.1 highlights their relevance to improving the economic performance of canefarms.

Table 19.1. Effect of farm expenditure on farm financial management.

Value of output	-	Costs	=	Net Farm Income
Influencing factors		Type of costs		Farm financial management
Quantity of output (i) size of business (ii) productivity of business		Fixed costs (occur irrespective of level of output)		Family living expenses Income tax
Value of output (i) quality (ii) marketing		Variable costs (i) necessary costs (ii) income stimulating costs		Loan repayments New investment

Quantity

Quantity of output is a function of farm size and productivity of the resources used. Size can be measured by the number of productive units (land area, number of labour units, etc.) or by the amount of capital invested.

Productivity is measured in standard units, usually yield per assigned hectare. In sugarcane production, the simplest measure of productivity (t cane ha^{-1} harvested) is not the most useful. There is a trade-off between productivity and profitability because of the extended cropping cycle. The most productive crops (highest yields per harvested hectare) are not always the most profitable. The plant cane crop takes longer to grow than a ratoon crop and the cost of crop establishment is considerably higher than the cost of ratooning. Queensland canegrowers have expressed their interest in profits as well as productivity and the number of ratoon crops has increased over time. A recent survey of canegrowers in four mill areas (Tully, Pioneer, Proserpine, and Maryborough) showed that they were practising the most profitable crop cycle and that there was little difference in profitability from a crop cycle one ratoon longer, or one ratoon shorter, than the optimal length (McLeod, 1995).

Price

Price is normally affected by marketing alternatives and quality. There is less flexibility in marketing cane than most other agricultural commodities. There are no realistic, profitable outlets for cane apart from sending it to the nearest mill for processing. Sales proceeds from raw sugar are pooled and shared between cane suppliers and processors by an empirical formula in which an estimate of the commercial sugar to be made from the cane is the main variable. In other agricultural industries, the farmer can ask whether average or better price was received for farm produce but various schemes have been introduced in the sugar industry to modify the impact of sucrose content on the price received for millable cane and to achieve price equity among growers. Thus the price paid for cane is not a true reflection of all its quality attributes. The pricing system was developed to deal with the seasonal pattern of sucrose in cane and the equity question is compounded by the fact that all growers cannot deliver cane to the local mill at its maximum sugar content. Many growers could now benefit from a more flexible approach to cane harvesting, delivery, and sugar marketing.

Production costs

Annual farm production costs can be separated into fixed and variable components. Fixed costs do not vary with the level of output and less than 50% of the total costs of operating a farm business should be spent on fixed costs. The cost of providing the capacity for production should not exceed the amount spent on productive inputs. Where fixed costs exceed this proportion, output needs to be increased.

Variable costs change in proportion to output and should exceed 50% of total costs since they are, in large part, income stimulating costs. The optimum level of expenditure on these costs can be determined by marginal analysis and use of these inputs should be decided by the contribution to income from the last unit to be used.

Responding to Changes in the Economic Environment

Australian farmers' terms of trade, expressed as a ratio of an index of prices received by farmers to an index of prices paid by farmers for inputs, have declined steadily over the past 25 years (Fig. 19.1). These changes are part of a long-term trend for agricultural commodities, influenced by their underlying characteristics. While real prices for agricultural commodities in Australia have risen steadily over the past 25 years, the rate of increase has not been as rapid as the increase in costs that farmers pay for inputs.

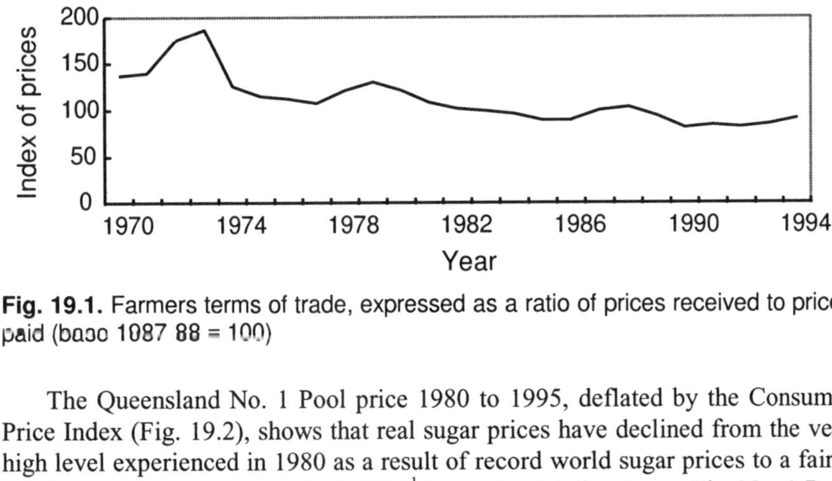

Fig. 19.1. Farmers terms of trade, expressed as a ratio of prices received to prices paid (base 1987 88 = 100)

The Queensland No. 1 Pool price 1980 to 1995, deflated by the Consumer Price Index (Fig. 19.2), shows that real sugar prices have declined from the very high level experienced in 1980 as a result of record world sugar prices to a fairly consistent price range around A$300 t^{-1} in constant dollar terms. The No. 1 Pool price represents a convenient benchmark against which to measure trends in sugar prices, although the final return to growers is actually based on a mix of No. 1 Pool and No. 2 Pool or Excess sugar sales.

Many primary producers in Australia have benefited from a slow upward trend in product prices, but sugarcane growers have not experienced much change in the real sugar price over the past 15 years. This has occurred in spite of both cyclical variation and reasonable increases in the nominal pool price in recent years.

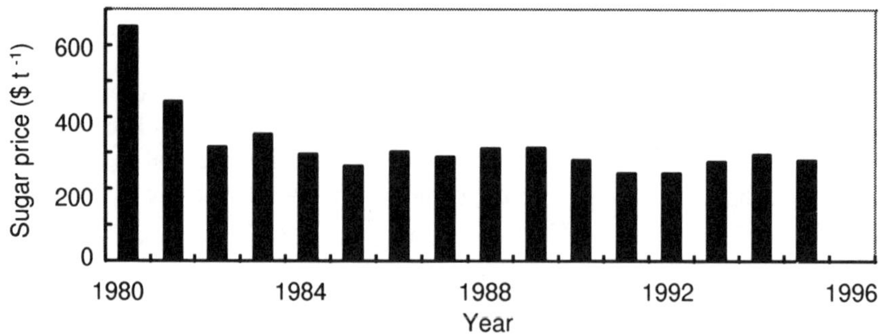

Fig. 19.2. Queensland No. 1 Pool price, 1980-1995, deflated by CPI index (base 1987-88 = 100).

Changing Prices, Changing Technology

Cane farm production can be described in simple terms by a single-input production function, with all productive inputs conveniently regarded as a single, bundled package. For a farm of fixed size, the Law of Diminishing Marginal Returns usually applies and output declines marginally as more inputs are used. An optimal or profit maximising level of inputs can be defined where the value of the marginal product is equal to the cost of the marginal unit of inputs. Blocks of the same variety in the same area may produce quite different amounts of sugar (Fig. 19.3). The diagram encapsulates the effect of holding one input, land, constant and combining it with increasingly higher levels other inputs to generate output.

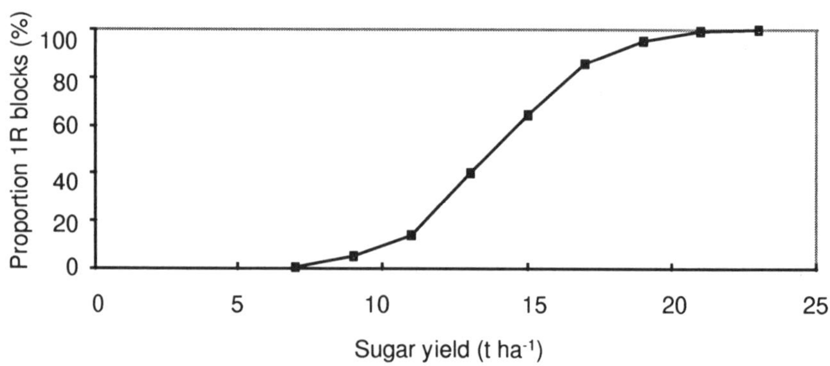

Fig. 19.3. Cumulative distribution of sugar yield for first ratoon (1R) blocks of Q120 cane at Babinda. [Data from Muchow 1996, *pers. comm.*]

The exact nature of the relationship between inputs and output is not known, but it is obvious that a relationship exists and in 1992 the majority of blocks at Babinda were concentrated in the range producing 13 to 19t sugar ha^{-1}. The number of blocks in each category declined sharply above that level because producers faced the effects of the Law of Diminishing Marginal Returns. Producers supplying inputs above those required to yield about 20t sugar ha^{-1} from first ratoon cane at Babinda in 1992 were probably involved in uneconomic production. There were nearly 5% of blocks in that category.

If the price of sugar or the cost of inputs changes, growers need to make appropriate adjustments in their farm management to maintain production at its most profitable level. The real price of sugar has been slowly dropping over the past 15 years while the price of farm inputs has continued to rise and some canegrowers have rationally reduced inputs. The alternative approach is to move onto a higher production function. This occurred, for example, with the adoption of new varieties like Q124 which replaced NCo310, CP44-101, and other less productive varieties in southern and central districts after the mid-1980s. The intensification of production through irrigation also illustrates the move on to a higher production function.

Cane output can also be described as a function of a major input variable, water, assuming that the level of all other inputs remains constant. The example from Willcox (1994) showed how responsive cane can be to important input variables like water (Fig. 19.4). The diagram also shows the enormous variation in response to inputs achieved by growers. In 1994, some canegrowers in the Fairymead Mill area recorded a yield of 120t cane ha^{-1} from an input of 6Ml of irrigation water while others produced only 60t cane ha^{-1}. Such differences are a major source of inefficiency in the industry and a potential source of increased productivity and profitability. The average production function for this group of growers indicated that 10Ml water was optimal and would maximise profits but many growers needed to improve the technical efficiency of water application before trying to fine-tune the amount applied.

Production functions have traditionally been defined as the maximal output attainable with a given set of inputs and this definition precludes the use of traditional, mean-based regression methods. To many growers, the location of the 'frontier' production function is more relevant than the 'average' function defined in the diagram below. An innovation in estimation techniques developed by Bloesch and Beard (1995) represents a new statistically based method that can be used to determine the location of the frontier production function for any level of technical efficiency by defining a function based on limiting rather than central values (Beard *et al.*, 1995). This method of determining the production function could lead to more precise, and more relevant, recommendations to growers about the most profitable quantity of irrigation water to apply to cane.

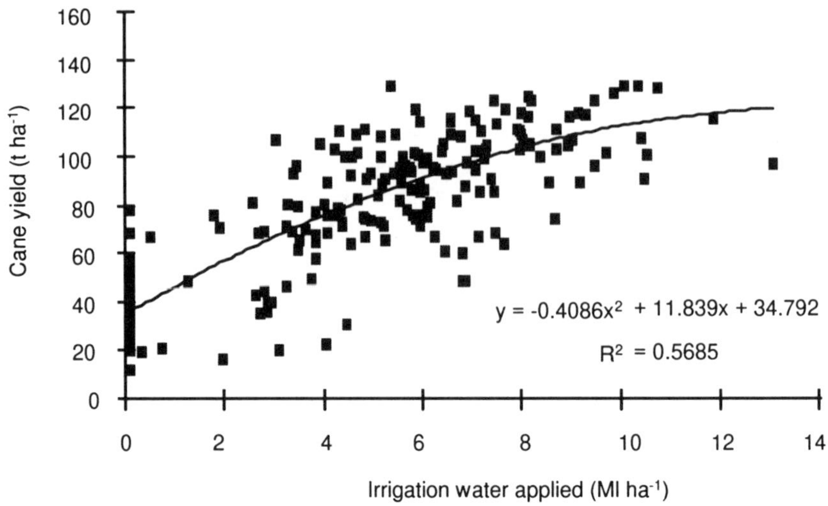

Fig. 19.4. Response of cane yield to irrigation water, Fairymead Mill area, Queensland, 1992-93. [After Willcox, 1994]

Changing Relativity Between Cost of Inputs

Cane production involves the use of a large number of production inputs and the relative cost of them changes constantly. Individual farmers adjust the mix of inputs accordingly, using both economic principles and commonsense so they can organise their production system to produce their chosen level of output with more of the cheaper inputs and less of the expensive ones.

Tilley (1984) noted that the price relativity between fuel, machinery, and herbicide costs had changed since Australian canegrowers substituted weedicides for mechanical cultivation. He showed how the purchasing power of canegrowers had changed over the 10-year period 1974 to 1983 in terms of diesel fuel, machinery, and labour. Since then, some inputs have risen in real price more slowly than others and a few, e.g. glyphosate which is used only minimally on cane farms, have actually dropped in real cost (Fig. 19.5).

Marketing Options

Deaton and Laroque (1992) noted that sugar prices are extremely volatile and Fig. 19.6 shows an index of the average annual price of sugar from 1900 to 1987 in US dollars, deflated by the US consumer price index since 1990.

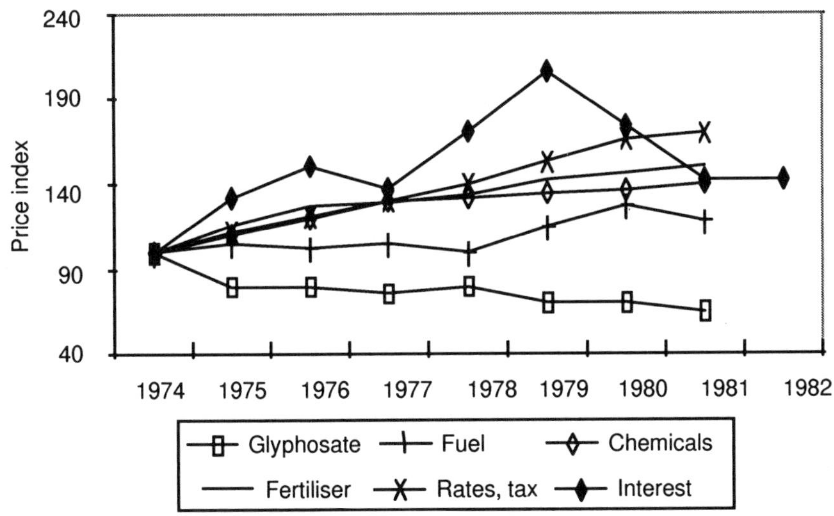

Fig. 19.5. Changing relativity in cost of various farm inputs. [ABARE and Monsanto, unpublished data]

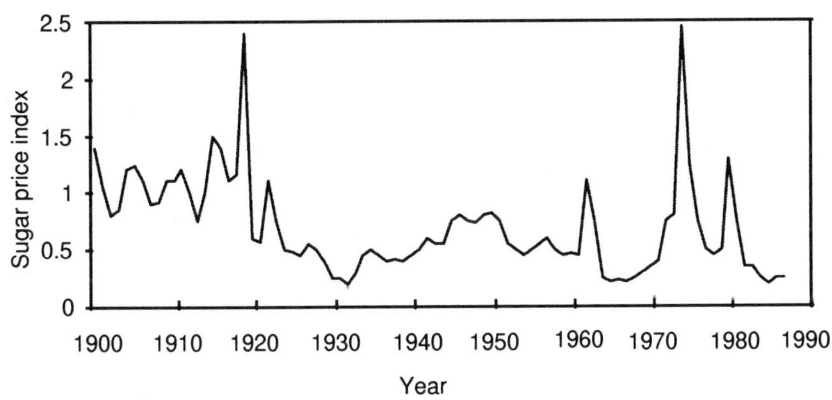

Fig. 19.6. Index of annual sugar prices, 1900-1987 deflated by US CPI. [Adapted from Deaton and Laroque, 1992]

The real price of sugar has followed a steady downward trend during this century although there were several sharp peaks, most notably in 1920, in the mid-1970s, and in 1980. Some of these price rises were very sharp indeed. The world sugar price in 1975 was nearly ten times its value less than a decade earlier.

For countries like Australia, whose export earnings and gross domestic product are partly dependent on sugar sales, such volatility poses major problems for macro-economic policy setting and for farm management. An understanding of the stochastic nature of these price movements is essential for management of the macro economy, for pricing policies, and for the design of risk management mechanisms for farmers, resource owners, and the government.

Since 1987, sugar prices have continued their erratic course (Fig. 19.7) although the two graphs are not comparable since the price in the latter one has not been indexed. In the past ten years, the daily spot price for the New York No. 11 contract has ranged from 5 USc lb^{-1} in early January 1986 to a peak of over 16 US cents in the first half of 1990. Recent price movements were more subdued but in 1996 the price still moved through the range from 9.5 to nearly 12.0 USc lb^{-1}.

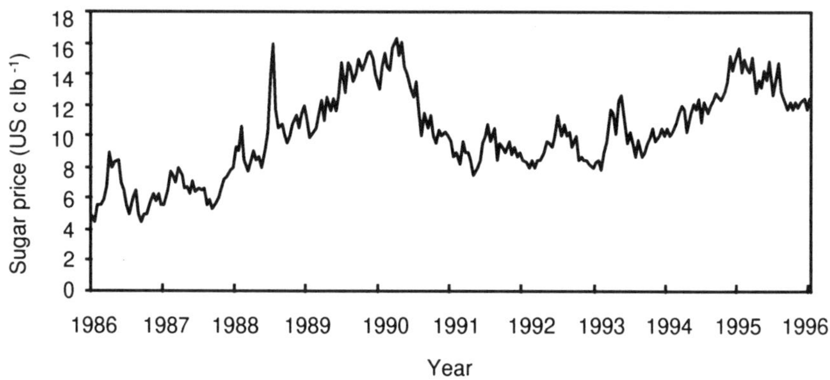

Fig. 19.7. Daily movements for New York No. 11 contract spot price, 1986-1995. [Czarnikow Sugar Review 1986-1996]

Past marketing strategies for many agricultural industries were designed to reduce the impact of such variability on individual farmers, and the sugar industry in Australia has a long history of government intervention in producing and marketing its product. Bartley and Connell (1991) outlined how economic performance in the sugar industry had been affected by this intervention. There have been many changes in sugar industry administration since the mid-1980s which have achieved a much less regulated environment. The most important changes allowed the sale and leasing of assignment and farm peak within a mill area in 1986 and introduced more flexibility in the location of canegrowing areas in the same year, effectively increasing the area of cane that growers could harvest each year. The import embargo on sugar and sugar products was replaced by a system of tariffs in 1989. Continuous crushing has gradually been introduced into the industry since 1990 and a policy of small, regular increases in the productive capacity of the industry was introduced in place of large, irregularly spaced expansions. This policy has allowed sugar production to increase by about 5% per year since 1990.

In September 1995, the Sugar Industry Review Working Party was established with representatives from government and the sugar industry to review both the tariff on raw and refined sugar and the current Queensland legislation that controls the industry.

The regulatory changes introduced in the past 10 years have benefited the industry enormously. Fortunately, their introduction coincided with a period of relatively stable sugar prices and considerable industry restructuring has occurred free of the economic disruption that normally accompanies structural adjustment in primary industries. However, the process of adjustment must continue to allow farms, contract harvesting businesses, raw sugar mills, refineries and marketing agencies, to respond to the industry's changing commercial environment.

Producer Pricing for Raw Sugar

The recent agricultural economics literature has included many articles in which the future of statutory marketing of agricultural commodities has been examined. Trebeck (1995) questioned whether faster moving, more flexible, more astute, private sector agencies could achieve better returns for farmers in the current economic environment than statutory marketing authorities. He suggested that farmers themselves, acting individually or collectively, could outperform the statutory marketing bodies in some instances.

Rutledge (1994) proposed the option of producer pricing for the sugar industry, noting that raw sugar was one of the few significant primary commodities in Australia where individual producers were not able to exercise direct pricing discretion over their product. He argued that pooling and single-desk marketing arrangements for sugar had generated significant benefits for industry stakeholders and recognised that individual producers cannot influence the market price but they should have the discretion whether or not to sell and so accept the prevailing price on a particular day. In exercising this discretion, the individual producer can consider factors such as cost of production, his or her view of the market and its subsequent movements, personal cash flow situation, and risk preference. From experience in other industries, risk is an important consideration and the satisfaction of individual producers might be enhanced, even if their monetary returns were not improved, if they had control of the final step in the production and disposal of their product.

Rutledge (1994) outlined a possible form of individual pricing that could be developed within the structure of acquisition and single-desk selling. While single-desk selling is not universally acclaimed, the scheme he outlined had a number of attractive features for canegrowers while still having a number of operational difficulties. In 1994, Rutledge was optimistic that a trial system of producer pricing could be implemented in the 1995 season. It should be revived and promoted as a valuable addition to individual canegrower's management options.

Such a change to individual grower pricing for sugar would mean that growers absorbed the market risks themselves rather than have the pool bear them on their behalf. Compared to pooled pricing, individual growers might then benefit from their own decision-making in times of rising prices but they could suffer greater losses from poor decision-making when prices fell. Clearly, some growers could benefit while others, who might not have the same marketing skills, could be disadvantaged. Pooling should be retained as an option in the marketing system, rather than as the exclusive method, so that its advantages over other marketing methods can become apparent over time or growers could adopt other marketing methods if its performance was poor.

Some of the potential attractions of producer pricing are evident from other industries, particularly cotton, and more recently wheat. Individual cotton growers have access to a wide range of marketing options to suit their own situations including selling on the futures market, fixed price contracts with marketing companies, or pooling and they can compare the returns from the alternatives over time to decide which is the most attractive.

The cotton industry in Australia has grown very rapidly since its deregulation and cotton growers can now typically sell cotton 18 months to two years in advance of planting their crop. In late April 1996, for example, individual Australian cotton growers sold much of their crop to be harvested and delivered in May 1997 and May 1998 respectively. The price was fixed (in excess of A$500 per bale) and with production costs about A$300 per bale, growers were assured of a significant profit.

Over the past 20 years, cotton growers in Australia have learned to cope with a cotton price that typically moves around 20% during a season without any price support from governments, marketing boards, or a single-desk seller. An active group of merchants, buyers and sellers/exporters of Australian raw cotton, has emerged to meet the growers' needs for a range of marketing services. The cotton industry has totally deregulated its marketing system and the sugar industry should consider doing the same thing.

The wheat industry is also experimenting with alternative marketing methods. The Sydney Futures Exchange launched a new domestic wheat futures contract in April 1996. The contract unit is 50 tonnes of wheat for delivery in the months of January, March, May, June, July, September, or November up to eighteen months ahead. More than 300 contracts, representing 15,000 tonnes of wheat worth over A$30 million, were traded in the first three days after their introduction and an opportunity was created to sell wheat at prices significantly higher than the existing price range. Prices for forward delivery wheat rose A$20 to A$30 t^{-1} following the start of futures trading, largely because merchants were able to reduce their exposure to price risk. Prior to the introduction of these futures contracts, the Australian Wheat Board pool price was not translating the high world price at that time into an equivalent price for local farmers.

After its disastrous experience with the Reserve Price Scheme, the Australian wool industry is trying to re-build industry self-reliance with changes to its marketing system. The number of wool contracts traded on the Sydney Futures Exchange is slowly increasing and, although woolgrowers could only sell about three months forward in 1996, further opportunities for forward selling may develop in time. In 1996, woolgrowers received very low prices for wool because there was no opportunity a year earlier to sell wool forward when prices were considerably higher. The marketing opportunities that are developing in other Australian primary industries should be available to canegrowers.

Economic Advantages from Producer Pricing

A comparison between spot prices for the New York No. 11 contract and the average pool price paid by the Queensland Sugar Corporation over past seasons is presented in Fig. 19.8. This comparison has involved the conversion of daily spot market prices (Czarnikow Sugar Review 1996), expressed in USc lb^{-1} to A\$ t^{-1} using appropriate assumptions about the US\$ / A\$ exchange rate.

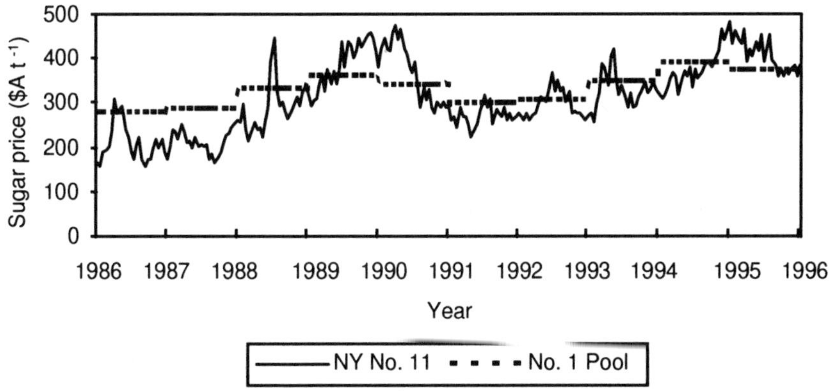

Fig. 19.8. Queensland No. 1 Pool price compared with New York No. 11 spot price

The effect of pooled pricing has been to 'even-out' the volatility in world market prices and provide a much more stable return to growers. While pooled pricing did provide some advantages over the spot market in the early 1980s when some high-priced contracts with Japan were still in force, such advantages were not so apparent in later years. Of concern is the fact that pooling prevents individual growers taking advantage of short periods of high prices. While advocates of statutory marketing claim that these higher prices are captured by the pool and therefore benefit all growers, it is hard to accept that the aggregate benefits are as great as they could be if individual growers were marketing their own sugar.

However, growers marketing their own sugar on the international market would need to be aware that they could be swapping a higher price subject to greater fluctuation for a lower but more certain pool price as well as subjecting their income to further risk because of exchange rate changes. Appropriate strategies for hedging against changes in the exchange rate would be required where, at present, these risks are managed by the marketing authority.

If growers were responsible for selling their own sugar, they would almost certainly sell on the futures market rather than take a spot market price. Fig. 19.9 shows the extent of variation in futures prices over time and the price advantages that are possible through judicious selling. Futures prices for four contracts (March, May, July, and October 1996) are shown over the period 18 December 1995 to 1 July 1996. A price between 10.5 and 12.5 USc lb^{-1} could have been obtained on a May 1996 delivery contract, depending on the date when the price was fixed. This was equivalent to returns between A$300 and A$350 t^{-1}. Similar prices were available for the other three contracts for 1996 delivery.

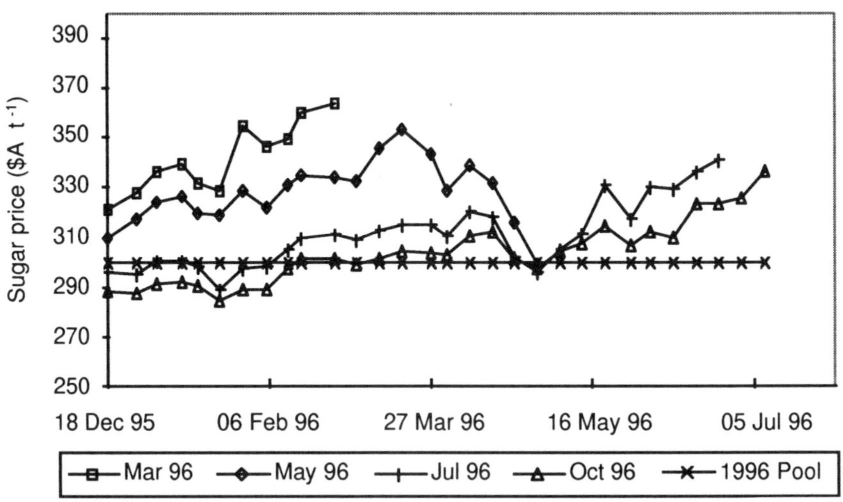

Fig. 19.9. Price for sugar on New York No. 11 futures contracts for 1996 delivery between December 1995 and July 1996 compared to the Queensland No. 1 Pool price in 1996. [Data from Czarnikow Sugar Review (1996) and ABARE (1996)]

For 1997 contracts, prices were generally lower and there was not nearly as much variability although all four contracts (March, May, July, and October 1997) rose in value, from 9.5-10.0 USc lb^{-1} in December 1995, to 10.5-11.0 USc lb^{-1} by July. Two contracts for 1998 delivery opened during this period with a lead time of approximately two years. Prices for the March and May 1998 contracts have been quoted since April and June 1996 respectively in the range 10.0 to 10.5 USc lb^{-1}. These futures contract prices can also be compared with the New York spot price and with the expected pool price. Only when producers have access to all of this information will they be in a position to make satisfactory judgments about the appropriate method of disposing of their produce, both in respect of timing as well as method.

Marketing costs for individual growers through these alternatives may be lower than for organised marketing. Farmers, usually, do not count the opportunity cost of their time spent in farming activities and it is presumed the same situation would apply for marketing. Also, farmers are independently minded individuals and many of them might feel more comfortable handling their own marketing rather than entrusting it to someone else. This may be illogical reasoning if the professional marketers perform better than enthusiastic amateurs in this game but there is a strong argument that those who profit from farming, or other activities, should be the ones who bear the risk.

While there is no guarantee that the advantages noted by wheat and cotton growers could be repeated in the sugar industry, if there was an opportunity for individual canegrowers to trade in futures and price their crop forward instead of receiving pooled returns, the concept is sufficiently attractive for it to be investigated thoroughly.

References

ABARE (1996) *Australian Commodities: Forecasts and Issues*, Australian Government Publishing Service, Canberra, Australia, 330pp.

Bartley, S.W. and Connell, P.J. (1991) *Impacts of Regulatory Changes on Sugar Cane Growers*. ABARE Discussion Paper 91.10, Australian Bureau of Agricultural Economics, Canberra, Australia.

Beard, R., Bloesch, A. and Willcox, T.G. (1995) The impact of irrigation and soil type on sugarcane production: A whole production set approach. *Australian Agricultural and Resource Economics Society Annual Conference*, Perth, Australia, 1995. Australian Agricultural and Resource Economics Society, Melbourne, Australia (microfiche).

Bloesch, A. and Beard, R. (1995) *Estimation of Frontier Production and Other Related Functions by Density Methods*. Agricultural Economics Discussion Paper, Department of Agriculture, The University of Queensland, Brisbane, Australia.

Czarnikow Sugar (1996) *The Czarnikow Sugar Review*. C. Czarnikow, London, UK.

Deaton, A. and Laroque, G. (1992) On the behavior of commodity prices. *Review of Economic Studies* 59, 1-23.

Errington, A. (1991) How well am I doing? A simple system for assessing the performance of your farm business. In: Giles, A. and Errington, A. (eds), *Farm Business Data*. Department of Agricultural Economics and Management, University of Reading, Reading, UK, pp. 19-20.

Giles, A. (1992) Looking for more profit? A systematic search. In: Giles, A. and Errington, A. (eds), *Farm Business Data*. Department of Agricultural Economics and Management, University of Reading, Reading, UK, pp. 46-48.

Graham, D. (1995) Survey of growers costs. *Australian Canegrower*, 12 June 1995, 22.

Hanlon, D. (1996) Management is key to making more money . *Australian Canegrower*, 24 June 1996, 25.

McLeod, R.S. (1995) An Integrated Approach to the Evaluation of Research Options in the Sugar Industry. PhD Thesis, The University of Queensland, Brisbane, Australia.

Rutledge, D. (1994) Producer pricing - Is it an option. *Proceedings Australian Sugar Convention*, Brisbane, Australia, 17-19 May 1994, pp. 139-144.

Tilley, L.G.W. (1984) 'Herbicides - better value than cultivation'. *BSES Bulletin*, October 1984, 4-5.

Trebeck, D. (1996) Farming's future and statutory marketing. *Australian Farm Journal*, April 1996, 8.

Willcox, T.G. (1994) Irrigation: benchmarks for efficiency. *Australian Canegrower*, 21 March 1994, 21-22.

Chapter twenty:

Sharing the Land - the Sugar Industry as Part of the Wider Landscape

A.K.L. Johnson[1], G.T. McDonald[2], D.A. Shrubsole[3] and D.H. Walker[1]

1. *CSIRO Tropical Agriculture and CRC for Sustainable Sugar Production, PMB PO Aitkenvale, Qld 4814, Australia*
2. *Department of Geographical Sciences and Planning, The University of Queensland, Qld 4072, Australia*
3. *Department of Geography, University of Western Ontario, London, Ontario NCA 5C2, Canada*

Abstract

Competition for natural resources among conflicting uses is a world-wide phenomenon but is an issue that has received little public attention for the Australian sugar industry. This industry is largely confined to the narrow, high rainfall coastal zone from near Grafton in northern New South Wales in the south, to Mossman in north Queensland in the north. In addition to satisfying the edaphic and climatic requirements of sugarcane production, land in this zone is also attractive to a wide range of alternative agricultural and non-agricultural uses. Competition with tourist and recreational uses, urban encroachment, public utilities and hobby farms has increased significantly in recent years. This increase has occurred in the context of an ever-increasing community expectation for the preservation of the natural environment. Given the finite nature of the resource base, the range of possible uses, and the growth in demand arising from growth in population and incomes, conflict over the use of these scarce resources is likely to grow. Environmental concerns, community goals, and a range of political, institutional and social factors, are also shown to be important. We argue that planning, in concert with a range of policy and economic instruments, can play an important role in the resolution of such problems. Options for efficient and equitable allocation of resources between the sugar industry and its neighbours at the policy level and on-farm are discussed and a vision for the future presented.

In: Keating, B.A. and Wilson, J.R. (eds) *Intensive Sugarcane Production: Meeting the Challenges Beyond 2000.* CAB International, Wallingford, UK, 1997.

Introduction

The sugar industry is one of the nation's largest rural industries. Sugarcane is predominantly grown along 2100km of coastline from Grafton in northern New South Wales to Mossman in north Queensland. About 507,000ha of land are currently under cultivation and the industry is a major user of natural resources in the coastal zone of eastern Australia north of latitude 30°S. For most of its history, the sugar industry has been free to decide how to grow cane, how much to grow and how to transport, mill and sell it. This situation is changing. In addition to satisfying the edaphic and climatic requirements of sugarcane production, land along the Queensland coast has, over the last 30 years, attracted economic and population growth unmatched by any other Australian region (Holmes, 1994). The finite nature of the natural resource base and the range of alternative uses has led to conflict between existing and potential users. It has done so in the context of an ever-increasing community expectation for preservation of the natural environment, involvement in decision making and effective conflict resolution.

This paper addresses the theme 'sharing the land', where 'land' refers to all 'natural resources', viz. land, water, air and biota. It examines questions about the relationship between the sugar industry and the land. Why does the industry have to share? What are the key forces driving the need for sharing the land? With whom does the industry have to share? How can governments influence the sugar industry to share the land and what opportunities exist for the industry to contribute to government policy? The paper is organised into four main sections. The first identifies and discusses the changing context (socio-political, economic and environmental) in which the sugar industry operates. The second overviews how markets and governments can influence the behaviour of individuals and highlights the role of planning in guiding and addressing change. The policy environment which currently guides sugarcane management is outlined in brief and the environmental impacts on the industry are discussed in the third section. The fourth section presents conclusions and a vision for the future. Throughout the paper, it is argued that improved industry planning can lead to effective, equitable and efficient approaches to sharing the land.

Why Share? The Changing Socio-Political, Economic and Environmental Context

Socio-political context

An increasing need to share natural resources with other users has irreversibly changed the socio-political context in which the sugar industry operates. While sugar is still the dominant economic force in many regions such as the Burdekin and

Herbert, in other regions, e.g. south-east Queensland and Cairns, economic diversification has diminished the predominance of the industry. In all regions, the industry is more frequently being confronted by other interests with different needs, values and aspirations. These interests range from environmental conservation to fisheries productivity, recreation, tourism and urban uses. The list of stakeholders concerned about natural resource use in cane growing regions has expanded greatly in recent years and the political complexity has increased accordingly.

The changing face of cane growing regions has been paralleled by a shift in society's attitudes to the use of natural resources. The environment is an important public policy area, and has assumed both 'mainstream' and 'bi-partisan' status as witnessed during the 1996 Australian federal election. Since the late 1980s, the environment has been consistently ranked by public opinion polls as one of the most important issues facing the nation (Cocks, 1992). This finding is consistent among age and income groups (Sinden, 1988). This high level of public and political support cannot be ignored as a short-term aberration.

Economic context

Australia approaches the 21st century with the state of its natural resources reflecting over 200 years of development, exploitation and experimentation. The rewards from this experience have undoubtedly been large, but 'success' in the context of the Australian sugar industry has been achieved with substantial cost to the natural environment.

Since federation in Australia, the ability of the rural sector to influence policy has gradually decreased. The 1980s in Australia saw a fundamental shift in the role of resource-based industries and a reshaping of the general economic policy framework (Musgrave and Pigram, 1994). Accordingly, governments responded with macro- and micro-economic reforms aimed at reducing the economic burden on resource-based industries. Micro-economic reform also introduced the requirement that many natural resources (particularly water) be priced according to the full social cost of their use (Young, 1993) and investment in publicly owned resources (e.g. dams and irrigation schemes) was expected to generate satisfactory rates of return for tax-payers' dollars (DPIE, 1989). Hence, current resource management activity in Australia is occurring in the context of reform within the greater national economy.

Competition for sugarcane land

Urban encroachment, public utilities, hobby farms, tourist and recreational developments have been responsible for re-allocation of valuable land away from the sugar industry (Anstey *et al.*, 1996). Between 1980 and 1995, 29,350ha of assigned sugarcane land was lost to other uses (≡ c. 6.4% of the area assigned in 1994), and while this may appear to be a relatively small figure for the whole industry, regional losses of land in some mill areas (e.g. Mourilyan, Bingera,

Mulgrave, Maryborough and Babinda) are more pronounced.

The concern for the loss of cane growing land in some parts of Queensland illustrates the need for well-informed planning. 'Planning' activities in the sugar industry have traditionally been directed at examining mill capacity and identifying suitable new land for expansion. In areas where expansion in land area is severely limited or no longer possible, production increases must rely on technological improvements. In such cases, mills are not as well buffered to losses of land or other fluctuations in supply. Accordingly, planning activities in these areas have shifted from the identification of new land areas to the identification and protection of the most important existing areas.

Sugar industry expansion

Most agricultural industries in Australia have seen a progressive historical increase in farm enterprise size. Significant expansion in both growing area and milling capacity has also occurred in the sugar industry, especially over the past 10 years. Between 1991 and 1995, increases of 15,000ha have occurred in the northern region, 44,700ha in the Herbert and Burdekin regions, 14,000ha in the Mackay region and 6,600ha in the Southern region (Queensland Sugar Corporation, 1995). In 1992 and 1993, c. 20,000ha *per annum* were allocated across the state. Increases over time for average cane production, average sugar production and average farm size per grower are indicative of the changes in both industry efficiency and exposure.

In general, recent sugarcane expansions have been onto more marginal land (e.g. Wilson, 1985) or onto land with a high conservation/habitat significance (Anon., 1993; Queensland Department of Environment and Heritage, 1995). Land tenure, mill zoning, ownership, and industry policy in regard to new growers have aggravated the problem (Dawson *et al.,* 1983). However, biophysical and economic data on potential areas at scales relevant to the level of decision making have been scarce. Given that much of the new land is of a relatively reduced quality, the need to allocate land and water resources carefully is fundamental. The provision of improved information will be important in future decision making. However, while much is known about the biophysical characteristics of many sugarcane growing regions, few districts have fully used these data in planning mill area expansions (Johnson *et al.*, 1994).

Environmental aspects

Since the mid-1980s, coastal Queensland and north-eastern New South Wales have experienced strong economic growth, underpinned by their agricultural and tourist sectors (ABS, 1995). While the long-term performance of these sectors will reflect the quality of its supporting natural resource base, current trends suggest a decline in terrestrial and riverine systems, and the adjacent marine environment (Anonymous, 1993; Drummond and Associates, 1993).

Erosion and sedimentation

Severe soil erosion is well recognised as a threat to long-term productivity of sloping land growing sugarcane (Prove *et al.*, 1995). Soils losses > 200t ha^{-1} were recorded at Mackay (Sallaway, 1979) and > 500t ha^{-1} on the wet tropical coast (Prove and Hicks, 1991; Prove *et al.*, 1995). Adoption of green cane harvesting-minimum tillage practices has decreased the potential sediment supply from sugarcane lands. Agronomic, infra-structure and socio-economic impediments have limited the acceptance of green cane harvesting-minimum tillage (S. Norrish, *pers. comm.*) to c. 40% of the Queensland crop harvested in 1995.

Although in many areas soil erosion remains a problem, the precise rate and impact of sediment delivery to estuarine and marine environments remains poorly understood (Pailles *et al.*, 1993; Pailles and Moody, 1996). Whilst it can be argued that there has been some increase in sediment accession to riverine and marine environments since European settlement, there is little evidence to support claims that sediment deposition resulting from sugarcane cultivation has had a major impact on the characteristics of the rivers of sugar catchments over the last 50-100 years (Drummond and Associates, 1993). Only in the Johnstone catchment, where drainage lines continue to contain fine sediments eroded and accumulated over the last 50 years and where this sediment has been shown to be flushed out during major floods (Pailles and Moody, 1996), does solid evidence exist; although similar effects have been reported in marine areas adjacent to the Herbert and Burdekin rivers (G. Brunskill, *pers. comm.*).

Maintenance and establishment of riparian vegetation

Riparian vegetation plays an important role in maintaining the biological health of the aquatic and terrestrial riverine ecosystems (Raisin, 1995). Large areas of this have been removed from stream banks in sugarcane-growing regions. In the Herbert River catchment, for example, about 20 and 24km of the right and left bank vegetation, respectively, has been cleared since 1990 (Perry, 1995). Mangrove areas have largely remained intact, although some clearing occurred in the 1960s and 1970s. Large areas of riverine rainforest with patches of eucalypts and the riverine rainforest, have also been cleared. Most of the major tributaries still have riparian vegetation, albeit narrow in most places. Most of the cleared stream banks have been invaded by weeds, including para grass, bamboo, thunbergia and African tulip trees, which tend to suppress natural revegetation. This situation is common in other catchments in which sugarcane is the dominant agricultural land use (Anon., 1993).

Preservation of coastal wetlands

Wetlands, swamps, billabongs and waterholes of coastal areas are important wildlife habitat areas and form an integral part of the hydrological regime. They are particularly important as nursery areas for several species of commercially important fish such as barramundi (McLeod, 1996). Clearing for sugarcane has significantly reduced the area of this habitat type in most cane-growing regions. For example, in the Johnstone Rivers system, there has been a 60% reduction in the area of coastal wetlands between 1951 and 1992 (Russell and Hales, 1996).

Diffuse source pollution

Given the size of the sugar industry, large potential exists for diffuse source pollution to generate water quality problems in both ground and surface water quality. Major pollution sources have increased in recent years, predominantly associated with expansion of sugarcane areas. Recently, Bramley and Johnson (1996) and Mitchell *et al.* (1996) have shown that nutrient concentrations in streams in selected sugar catchments are generally below the guidelines detailed by ANZECC (1992) for freshwater. Exceptions were mostly associated with peak flow events. In north Queensland, nutrient concentrations in streams associated with sugarcane production were generally greater than in those associated with other land uses, but differences between these were only seen during wet season peak flow events.

Further work by Furnas *et al.* (1995) in the marine environment adjacent to the Wet Tropical Coast (WTC) has shown that direct external inputs of nitrogen (N) and phosphorus (P) through point sources are currently small relative to natural nutrient inputs. Riverine inputs of N and P comprise a large proportion of external inputs, but are still small relative to internal recycling fluxes. Although nutrient levels are currently low in marine waters adjacent to sugar-producing regions and external inputs are small relative to natural fluxes and stocks, understanding of ecosystem behaviour is not sufficiently developed to allow the assimilative capacity of these waters to be predicted with certainty.

Information on accession, persistence and behaviour of pesticides in cane-growing and adjacent environments is sparse. While the use of pesticides such as dieldrin and DDT has ceased, little is known of their persistence in the environment. Many chemicals in this group are known to accumulate in produce suitable for human consumption. Analyses from the Mackay region (A. Garside, *pers. comm.*) have shown dieldrin levels in peanuts harvested from selected cane farms to be 25 times the acceptable level for human consumption.

Understanding of the accession of nutrients and pesticide into groundwaters remains limited. Evidence from the Bundaberg area (Keating *et al.*, 1996) indicates that extensive nitrate accessions to the area's aquifers have occurred, although no unequivocal data are available to assess whether nitrate concentrations are continuing to rise. Similar results have been reported in the Burdekin delta (J. Bauld, unpublished data). Similarly, evidence to support pesticide contamination has also been scarce, although again, low levels of atrazine and heptachlor have been detected in the Burdekin delta aquifer system (Brodie *et al.*, 1984; Keating *et al.*, 1996).

Current knowledge suggests that downstream effects of agricultural production are largely governed by the periodicity, volume and intensity of rainfall. Within catchments, significant climatic, spatial and temporal variability occur. Nutrient and pesticide loss from land under intensive agriculture is likely to exceed that from other land uses irrespective of the type of intensive production (Bramley and Johnson, 1996). Hence, it is the high inputs intensive agricultural land use involves, rather than the specific crop, which results in increased export of pollutants relative to lower input land uses, such as low intensity grazing or forestry. Minimisation of diffuse source pollution is therefore likely to depend on carefully matching actual land use to land use suitability.

Heavy metals

Little is known of the contribution of the sugar industry to heavy metal contamination. Zinc and copper are applied only in very small quantities, but with the high use of phosphatic fertilisers, large areas of cane-growing land could have elevated levels of cadmium. This is potentially a serious problem because cadmium is extremely toxic to aquatic and marine animals (Hart, 1982). Also, cadmium is known to accumulate in produce for human consumption, therefore much of the current cane-growing area may be unsuitable for alternative agricultural land uses. Analyses (A. Garside, *pers. comm.*) of peanuts from the Mackay and Tully areas have shown cadmium levels 3 times the acceptable level.

The New Context: Ecologically Sustainable Development

The changing role of resource-based industries and general economic policy in the 1980s and early 1990s has been paralleled by an increased interest in the environment and in the demand for environmental goods and services. This change is illustrated in the 1992 Intergovernmental Agreement on the Environment signed by all levels of Government who agreed to pursue the following seven principles of ecological sustainable development (ESD) (Australia, 1992):

1. Decision-making processes should effectively integrate both long and short-term economic, environmental, social and equity considerations.
2. Where there are threats of serious or irreversible environmental damage, lack of full scientific certainty should not be used as a reason for postponing measures to prevent environmental degradation.
3. The global dimension of environmental impacts of actions and policies should be recognised and considered.
4. The need to develop a strong, growing and diversified economy which can enhance the capacity for environmental protection should be recognised.
5. The need to maintain and enhance international competitiveness in an

environmentally sound manner should be recognised.

6. Cost-effective and flexible policy instruments should be adopted, such as improved valuation, pricing and incentive mechanisms.

7. Decisions and actions should provide for broad community involvement on issues which affect them.

From an environmental perspective, the principles of ESD are the most important influence on all planning activities in Australia at the present time. Many recent federal initiatives, such as the National Soil Conservation, National Tree and National Landcare Program, the establishment of nine working groups to consider the implementation of the Commonwealth's Ecologically Sustainable Development Discussion Paper (Australia, 1992) and the Prime Minister's 1996 *Our Land* Statement, support ESD principles. At the state level, Whole Catchment Management (Syme *et al.*, 1994) and Property Management Planning (Queensland Landcare Council, 1991) have been prominent and will, in time, change the management of natural resources by the sugar industry. Commitment to the 'precautionary principle' will have ramifications on the acceptability or otherwise of impacts, placing the onus of proof on the proponent to show that minimal risks of environmental degradation are involved rather than on the planning agency or other stakeholders to show that the impacts are unacceptable.

How to Share the Land: Implementing Natural Resource Policy

While urban Australians accept government planning and management as a form of everyday life, it remains an anathema to many rural residents - a threat to their individual liberties to do what they want with land. The extent of intervention is the subject of much debate between planners, economists and politicians and vested interests in land. The level of success in this respect has been somewhat varied, with public policies having altered sound markets unnecessarily or unintentionally and made poorly performing markets even less efficient. What most authors agree on, however, is the need for some form of planning in the development of natural resources. Planning can be achieved by a combination of education, economic incentives, regulations, or public sector management.

Resource planning initiatives affecting the sugar industry

The Australian sugar industry is highly regulated and the policy environment in which it operates is complex. Government and industry regulations in Queensland effectively control the number of cane growers, the areas of cane planted, the location of plantings and the quantity of sugar produced.

The current resource management legislative and administrative framework within which the sugar industry operates incorporates a maze of government agencies with often conflicting statutory responsibilities (Table 20.1). These complex arrangements combined with high expectations from the 'public' for effective involvement in decision-making suggest that the sugar industry will have to respond to, and indeed lead, the implementation of ESD in cane areas.

ESD principles, to some extent, have been incorporated in some of the industry's operations. The environmental audit of the industry sponsored by the CANEGROWERS organisation demonstrates a concern for the environment. At a local level, many cane assignment boards have established a number of conditions on new assignments. While this trend is encouraging, effective establishment and implementation of assignment conditions relies on adequate information about the natural environment (e.g. hydrology, land suitability, flora and fauna). Without this information, local boards are unable to rigorously determine the desirability and feasibility of assigning new cane land. Current arrangements do not allow adequate monitoring of either adherence to assignment conditions or the impact of conditions on the natural environment.

On-farm, ESD principles have been recognised with the use of Property Management Plans (PMPs). Voluntary development and implementation of PMPs can play a key role in minimising environmental degradation (Queensland Landcare Council, 1991). Supported by Landcare and ICM groups, PMPs can be developed with a high level of landholder involvement. They provide another facility by which cane growers can more actively participate in farm management planning with the assistance of government officials. While PMPs are achieving acceptance for new cane lands, the industry must foster a similar approach on existing cane lands. Its future is dependent not only on an increased acceptance by existing growers but also improvements in the quantity and quality of natural resource and other information to guide effective decision making.

Opportunities for Improvement

In detailing the opportunities for improvement in natural resource management in the sugar industry, it is instructive to first provide a broader framework for future development. This can be done by drawing heavily from the landmark work of Young *et al.* (1996) and placing it in a sugar industry context. We believe, in terms of a broad policy, it is incumbent on the sugar industry, government and the broader community to work towards ensuring that the following objectives are fulfilled:

1. Natural resource management in the sugar industry is undertaken as a fundamental part of, and necessary precondition for, ecologically sustainable development and the implementation of the precautionary principle.
2. Responsibility for natural resource management is shared between government, community and the sugar industry in a transparent manner.

Table 20.1. Important legislative and administrative initiatives affecting natural resource management in the Qld. sugar industry.

Initiative	Lead Agency	Resources Effected	Objectives	Processes	Outputs
Local					
Planning Environment and Development Assessment (Bill)	Dept. of Housing Local Government and Planning	All private sector development applications	Integrated planning, development, assessment and approval	Local Development Management, community involvement, regional forums, state concurrence and referral	Statutory plans, Regional plans. State planning policies, integrated development assessment, approval and appeals
State					
Nature Conservation Act	Dept. of Environment	State conservation reserves Habitat conservation	Flora and fauna conservation Endangered species, biodiversity conservation	Land acquisition Cooperative agreements	Regulations, management plans Voluntary agreements Guidelines
Environmental Protection Act	Dept. of Environment	All air, land and water	Air, noise & water quality protection and waste management	Standards Guidelines	Environmental Protection Policies and Plans
Lands Act	Dept. of Natural Resources	Crown Lands	Use and development of land	Leasehold conditions	
State Public Works and Development Organisation	Premier's Department	Prescribed (large) projects	Integrated development	Coordination	Infrastructure coordination and impact assessment
Natural Resources Management (Discussion paper)	Dept. of Natural Resources	State owned or controlled water and forests, privately controlled natural resources			NRM plans for water, forests, fisheries, private lands (critical areas), licences, codes of practice

Table 20.1. (cont.)

Integrated Catchment Management	Dept. of Natural Resources	Defined river basins	Integrated natural resources management	CCCs, community involvement	Catchment management strategies
Coastal Management (proposed)	Dept. of Environment	State controlled tidal and coastal lands	Coastal protection, habitat management		Coastal Plans
National					
National Landcare Program	Dept. of Primary Industries and Energy	All land and water in Australia	Community based land conservation	Facilitation community groups	Property management plans; Integrated resources management
Australian Heritage Commission	Dept. of Environment, Sports and Tourism	All significant natural, historical and cultural heritage	Identification and protection of heritage	Inventory which must be taken into account	Australian Heritage Register; Heritage Guidelines
World Heritage Properties Conservation Act	Dept. of Environment, Sports and Tourism	Australia's eleven World Heritage Areas	Conservation, rehabilitation and presentation of WHAs	Management Authorities, including joint management	Strategic Plans, licenses and infrastructure
Commonwealth Environmental Protection Authority	Dept. of Environment, Sports and Tourism	All of Australia (potentially)	Environmental quality protection	Environmental standards setting Assessment	Environmental standards; Impact Assessment
Rural Adjustment Scheme	Dept. of Primary Industries and Energy	All of Australia's agricultural and grazing areas	Integrated economic, social & environmental restructuring	Regional programs	Structural adjustment funds

Source: Adapted from McDonald (1995)

3. Appropriate incentives are put in place to encourage the protection of natural resources and to encourage their use only in ways which are ecologically sustainable.
4. Appropriate mixes of incentives are developed and appropriate weighting given to motivational, voluntary, property-right, price-based, and regulatory instruments in ways which vary according to the issue and local, regional and social characteristics.
5. The Australian community as a whole - as well as users and beneficiaries of natural resources - contribute towards the provision of incentives to cane growers, whose primary responsibility is for the protection of the environment.

To implement this framework, it is essential that governments at all levels develop the necessary supporting institutional capacity. They must develop structures that empower local communities and industries to protect and manage natural resources and provide access to instruments necessary for their operation. Conversely, governments must address problems where local initiatives fail or are incomplete and importantly, must give consistent signals to resource users and consumers about the importance of ESD. This issue is particularly relevant in the sugar industry where government is currently advocating environmental conservation on the one hand and major infrastructure developments with its attendant environmental problems on the other (e.g. in the Herbert and Burdekin).

In order to overcome some of these problems, we suggest that the appropriate boundaries for government decision-making on natural resource management are bioregional. However, local governments and the sugar industry, whose boundaries are rarely consistent with bioregional boundaries, have close and identifiable ties with the community, and provide an appropriate base for implementation. Furthermore, cane growers and other stakeholders must be involved in the establishment (where they do not already exist), and operation, of relevant decision-making and advisory bodies (e.g. ICM).

Stakeholders, particularly cane growers, need to ensure that their own internal processes allow real opportunities for the involvement of others at a 'grassroots level'. Greater involvement of the community with the sugar industry in key decision-making represents a fundamental change. In the sugar industry, there is potential for self-interest and local needs to dominate important resource management issues. Conflict within decision-making is a major potential threat to improved practice. Visible monitoring and accountability mechanisms are essential pre-requisites for increased community and sugar industry involvement, as are the development of effective conflict resolution and facilitation processes. This is especially the case for the operation of the cane assignment system. We believe it is more efficient to build on and adapt existing administrative and institutional structures where possible to address management of natural resources in the sugar industry, than to create new structures.

Motivational and voluntary incentives

Incentives aimed to increase levels of knowledge and understanding, or encourage voluntary action, are premised on the belief that environmentally responsible behaviour is more likely to result when people have a basic knowledge of the issues at stake and/or a commitment to the principles of ESD (Young *et al.,* 1996). Motivational and voluntary incentives encourage all stakeholders to share information and contribute to improved management. Motivational and voluntary incentives have already been used in cane-growing regions, and include activities such as ICM and the Herbert Resource Information Centre (Johnson and Walker, 1996). Motivational incentives must be a core mechanism for improved natural resource management in the sugar industry. For incentives to succeed, it is crucial that both the content of the information presented and the method of transferring information are appropriate and well targeted. This issue is addressed in the context of R&D in a later section.

Property right incentives

Two types of property right incentives: (i) covenants or management agreements, which cannot be separated from a specific resource, and (ii) those like licences to use water, which can be moved from one location to another (Young *et al.,* 1996), have applicability to the sugar industry. Management agreements can be used to reimburse people for the costs of resource management which are not able to be recovered through normal market mechanisms. An example would be to encourage canegrowers to undertake specified management activity on private land adjacent to public conservation reserves.

Conservation covenants are particularly effective in protecting remnant vegetation, and can be used to underpin approaches such as management agreements. This use can be linked to other incentives, e.g. the local government rating systems, to recognise the value of uncleared land. Licence and permit systems are used where there is a need to link natural resource management with an economic activity, e.g. pollution permits and land clearing. In the design of licence systems for the sugar industry, there is a need to place emphasis on the dependability of the licence to protect natural resources in an efficient and equitable manner. Central to this must be the establishment of effective monitoring strategies that assess not only specific activities such as cane assignment but also general environmental performance

Tradeable rights systems are frequently advocated by economists, but generally have a poor track record (Nelson, 1977). The benefit of transferable property rights are more apparent in areas where equivalent units (e.g. water) can be traded than where units traded may not necessarily be equivalent (e.g. wetland mitigation schemes). Hence they may be more useful in areas such as Bundaberg and the Burdekin than in other cane growing regions. Where natural resources are used commercially, tradeable rights have a role but should never be seen as the sole solution. In most cases, they need a support network of regulatory, financial and institutional measures (Young *et al.,* 1996).

Regulatory incentives

Despite advocating motivational, social and market-based incentives, there remains an important role for regulation, because acting alone, non-regulatory instruments are not always effective (Napier, 1994). Regulations provide precautionary standards and an essential safety net to protect against the recalcitrant few not persuaded by other incentives. Each level of government needs to set such standards and no one level should undermine the standard set by another or by the community or industry itself. It is crucial that the sugar industry and community be given the opportunity to contribute to the development of standards. As regulatory incentives provide protection against those who do not respond to other measures, they are particularly important when threats to natural resources are likely to become irreversible, e.g. land clearance on a coastal floodplain.

Financial considerations

Fundamental to developing and implementing any policy or instrument mix are the private and public cost implications. The cost of controlling and preventing damaging processes in cane growing areas should be borne primarily by those who cause the problems, the sugar industry. However, the community should accept financial responsibility for protecting the natural environment when the costs cannot be recovered using market mechanisms. All who benefit from non-market dimensions of protecting the environment, whether directly and indirectly, should contribute to the cost of its maintenance (Young *et al.,* 1996).

The tax system is an important area for reform, particularly when incentives are recognised as the most cost effective means of encouraging altruistic investments in resource conservation by the private sector (Davenport, 1995). Another method of ensuring that users pay is the imposition of levies and charges. These have the additional benefit of making users of natural resources aware of the cost of their activities (Young *et al.*, 1996). Acceptance of such levies is likely to be improved if the funds raised are clearly directed towards resource management and protection, preferably by being placed in resource management funds administered by the local community or the sugar industry.

Implications for the Australian Sugar Industry

The policy environment

The policy environment for natural resource management in the sugar industry will continue to respond to changing political, economic, social and environmental considerations. Management procedures will vary but public expectations for, and support of, effective environmental management policies will remain. The principles for ESD will likely form a general basis for the development of new, and hopefully improved initiatives. The general trend of shifting management responsibilities to lower levels of government is likely to continue, although the extent will be guided by overriding State and Federal government policies. There will be a continued emphasis on regional planning to establish a sound foundation for policy implementation. The conflicts between sugarcane and other uses are likely to become more intense.

Threats to the achievement of ESD

Adequacy of effort to meet the challenge

Devolution of Federal and State government responsibilities to lower levels of government is potentially problematic given the limited resources of local government to implement resource management policies and plans. Evidence from the United States, where voluntary compliance programs and subsidies have operated since the 1930s (Napier, 1994), clearly illustrates the ramifications of a failure to change institutional structures for implementing conservation programs. The positive behaviour of U.S. farmers involved in programs to control soil loss and improve water quality seldom continued once subsidies or compliance programs were removed. There is no reason to suggest that cane farmers in Australia will be any different unless institutional changes are made. Hence, State and Federal governments must provide sufficient financial and human resources for local government and the community to conduct natural resource management activities in an objective and effective manner.

Complexity and conflict - a role for science and information

Managing resources for ESD is more complex than managing them solely for maximum productivity or conservation. More facets of the resource base need to be monitored and managed, and a more holistic view of the range of potential impacts - particularly cumulative impacts - must be considered. To address the issues in ecologically sustainable natural resource use, managers need the best current scientific understanding of the physical impacts of alternative decisions or actions and effective means of collating, interpreting and using this information. Policy and planning cannot achieve societal goals without adequate data relating to biophysical, economic and social factors that impinge on resource use (Cocklin *et al.*, 1987).

Research is required to better understand the physical and socio-economic systems in which cane production occurs, and further research is needed to organise and integrate that knowledge using information technologies such as

GIS, expert systems, other knowledge-based systems and simulation modelling into decision support systems to help address ESD issues (Fedra, 1995). The role and management of such innovations demands attention and the use of information technology tools in some capacity appears vital for sound decisions at regional or local government level.

The challenge for research is to address the needs of decision-makers in three key areas: (i) problem identification; (ii) data collection, collation and analysis; and (iii) rational use of the outputs of that analysis (Walker and Johnson, 1996). A shift towards a broader view incorporating natural resource management as well as productivity issues suggests a role for increased collaboration in data capture, storage and distribution. Projects such as the Herbert Mapping Project and the Herbert Resource Information Centre (Johnson and Walker, 1996), where stakeholders from government, industry and community have invested corporately in acquiring key resource data and GIS, provide an example of what the future may hold.

Length of time to complete plans

As illustrated by the ICM experience in the Johnstone catchment, the time required to complete a planning exercise is often in the order of 5 years. While this and other local and state resource planning efforts are in progress, cane assignment committees, mills and other participants in the sugar industry must and will make resource-related decisions. It would be prudent for the sugar industry to adequately consult with others during this formative period to ensure the compatibility of its decisions with the interests of other resource users. Failure to do so could lead to increased levels of conflict.

It is important that the purpose of all planning activities is clear, and that they are coordinated with other relevant initiatives. In its absence, there is little chance that a whole government approach can be effectively realised. The feasibility of a whole government approach is also uncertain. While there is a tendency to want to streamline public agencies' reviews, there is a demand that all constituents have effective access to the decision-making process. This demand could increase the time required to implement decisions.

A Vision for the Future

The challenges facing the Australian sugar industry in its management of natural resources are substantial. It must embrace rather than resist the inevitable trend toward greater public intervention in its affairs. It cannot do this alone, hence the challenge facing government is to provide a stable environment in which locally relevant decision-making can occur. Government must support this with appropriate and viable monitoring, cost-sharing and regulatory arrangements. The sugar industry, unlike many other industries, has the rare opportunity to have a significant say in its future direction. As such, it needs to develop its own policies and activities to guide its future developments within a wider context. It must recognise that external pressures will evolve with time, hence planning must be a continuous activity rather than a singular exercise. Further, the industry must develop its own strategic view on where and how cane will be grown. For this to succeed, the sugar industry must be proactive in involving itself in non-sugar industry programs related to natural resource management and planning.

In terms of a vision for the future, the Australian community of coming generations will benefit from development processes in the sugar industry that are consistent with the principles of ESD. Community and industry participation will occur at all levels of government decision-making and will become a central tenet of operational management of the natural environment. Incentives for sustainable resource management will be embedded in economic policy at the national and state levels and use of natural resources in cane-growing regions will occur only within limits necessary to preserve future options. Planning and decision-making will be based on scientific information and will occur at a regional, catchment and local level. Where information is not available, actions with irreversible consequences will be avoided and barriers to more individual and community action will be diminished by the distribution of essential information. Rather than natural resource management being an issue of conflict, the sugar industry will work together with others who also have a stake in the coastal lands, to develop common goals and strategies. The approach taken will be strategic and adaptive, and partnership arrangements between the sugar industry and other stakeholders in sugar catchments and the adjacent marine environment will be developed in response to specific natural resource management needs and the aspirations of local communities.

Protection of the natural environment will be the priority consideration in future planning for economic development in the sugar industry. A precautionary approach will underpin all developments which in itself takes into consideration precautionary standards for natural resource management. Non-interventionist and financially attractive incentives embedded in a set of precautionary standards designed to prevent irreversible decline in the natural resource base will form the basis of government and sugar industry policy and practice. The complexity of natural resource values and the range of objectives and options for their use and protection means that no single policy opportunity is likely to offer a general solution.

In achieving this vision, the Australian sugar industry is seen both nationally and internationally as a leader in the way in which it manages Australia's precious natural resources. In so doing the Australian sugar industry will clearly demonstrate that environment-economy linkages can lead to positive economic, social and environmental benefits for all Australians and that the endowment of natural resources provided to future generations of Australians will be equal to or better than that afforded to the current generation.

Acknowledgments

The authors would like to thank Dr Phil Price and two anonymous referees for constructive comments on early drafts of this manuscript.

References

ABS (1995) *Queensland Statistical Yearbook - 1995.* ABS Brisbane.

Anon. (1993) *The Condition of River Catchments in Queensland - a Broad Overview of Catchment Management Issues.* Queensland Department of Primary Industries, Brisbane, Australia, pp. 85.

Anstey, G., McDonald, G.T. and Wegener, M.K. (1996). Conversion of caneland to hobby farms: is planning intervention warranted? *Proceedings of the Australian Society of Sugarcane Technologists* 18, 50-57.

ANZECC (1992) *Australian Water Quality Guidelines for Fresh and Marine Waters.* Australian and New Zealand Environment and Conservation Council. Canberra, Australia.

Australia (1992) *National Strategy for Ecological Sustainable Development.* Australian Government Publishing Service, Canberra, Australia.

Bramley, R.G.V. and Johnson, A.K.L (1996) Land use impacts on nutrient loading in the Herbert River. In: Hunter, H.H., Eyles, A.G. and Rayment, G.E. (eds), *Downstream Impacts of Land Use.* Department of Natural Resources, Brisbane, Australia, pp. 93-96.

Brodie, J.E., Hicks, W.S., Richards, G.N. and Thomas, F.G. (1984) Residues related to agricultural chemicals in the groundwaters of the Burdekin River Delta, North Queensland. *Environmental Pollution* 8, 187-215.

Cocklin, C., Smit, B. and Johnson, T. (1987) Rural land-use analysis and planning: an overview. In: Cocklin, C., Smit, B. and Johnston, T. (eds), *Demands on Rural Lands - Planning for Resource Use.* Westview Press, London, UK, pp. 3-12.

Cocks, K.D. (1992) *Use With Care - Managing Australia's Natural Resources in the Twenty First Century.* New South Wales University Press, Sydney, Australia.

Davenport, S. (1995) The role of income taxation in natural resources management. *Review of Marketing and Agricultural Economics* 63(1), 200-208.

Dawson, N.M., Berndt, R.D. and Venz, B. (1983) Land use planning - Queensland canelands. *Proceedings of the Australian Society of Sugarcane Technologists* 5, 43-52.

DPIE (1989) *The National Landcare Program: Sustainable Natural Resources Management - Integration of Primary Industries and Energy Programs.* Department of Primary Industries and Energy, Canberra, Australia.

Drummond and Associates (1993) *Stream Management Plan Herbert River District.* Ian Drummond and Associates, Wangaratta, Australia.

Fedra, K. (1995) Decision support for natural resources management: Models, GIS and expert systems. *AI Applications* 9(3), 3-19.

Furnas, M., Mitchell, A.W. and Skuza, M. (1995) *Nitrogen and Phosphorus Budgets for the Central Great Barrier Reef Shelf.* Research Publication No. 36, Great Barrier Reef Marine Park Authority, Townsville, Australia.

Hart, B.T. (1982) *Australian Water Quality Criteria for Heavy Metals.* Australian Water

Resources Council Technical Paper No. 77. Australian Government Public Service, Canberra, Australia.

Holmes, J. (1994) Coast versus inland: two different Queenslands? *Australian Geographical Studies* 32(2), 167-182.

Johnson, A.K.L., Cramb, R.A. and McAlpine, J.R. (1994) Integrated land evaluation as an aid to land-use planning in northern Australia. *Journal of Environmental Management* 40, 101-116.

Johnson, A.K.L. and Walker, D.H. (1996) Utilising GIS in the sugar industry - development of the Herbert Resource Information Centre. *Proceedings of the Australian Society of Sugarcane Technologists* 18, 32-40.

Keating, B.A., Bauld, J., Hillier, J., Ellis, R., Weier, K.L., Sunners, F. and Connell, D. (1996) Leaching of nutrients and pesticides to Queensland groundwaters. In: Hunter, H.H., Eyles, A.G. and Rayment, G.E. (eds), *Downstream Impacts of Land Use.* Department of Natural Resources, Brisbane, Australia, pp. 151-164.

McDonald, G.T. (1995). Recent developments in natural resource management. *Developments in Government Policy and Their Impacts on Land and Water Resource Management in North Queensland.* Proceedings Australian Institute of Agricultural Science North Queensland Branch 1995 Symposium, pp. 2-11.

McLeod, J. (1996) Commercial fishing industry's perspective in assessing/adopting scientific/technical information. In: Hunter, H.H., Eyles, A.G. and Rayment, G.E. (eds), *Downstream Impacts of Land Use.* Department of Natural Resources, Brisbane, Australia, pp. 427-428.

Mitchell, A.W., Reghenzani, J.R., Hunter, H.M. and Bramley, R.G.V. (1996) Water quality and nutrient fluxes of river systems draining to the Great Barrier Reef Marine Park. In: Hunter, H.H., Eyles, A.G. and Rayment, G.E. (eds), *Downstream Impacts of Land Use.* Department of Natural Resources, Brisbane, Australia, pp. 23-33.

Musgrave, W. and Pigram, J. (1994) Social, economic and institutional considerations affecting soil and water conservation in Australia. In: Napier, T.L. Camboni, S.M. and El-Swaify, S.A. (eds), *Adopting Conservation on the Farm: an International Perspective on the Socioeconomics of Soils and Water Conservation.* Soil and Water Conservation Society of America, Ankeny, IA, USA, pp 475-498.

Napier, T.L. (1994) Regulatory Approaches for Soil and Water Conservation. In: Swanson, L.E. and Clearfield, F.B. (eds), *Agricultural Policy and the Environment: Iron Fist or Open Hand.* Soil and Water Conservation Society, Ankeny, IA, USA, pp. 189-202.

Nelson, R.H. (1977) *Zoning and Property Rights*. MIT Press. Cambridge, UK.

Pailles, C., McConchie, D., Arakel, A. and Saenger, P. (1993) The distribution of phosphate in the sediments and soils of the Johnstone Rivers catchment-estuary system, north Queensland, Australia. *Sedimentary Geology* 85, 253-69.

Pailles, C. and Moody, P.W. (1996) Phosphorus dynamics of sediments from the Johnstone Rivers. In: Hunter, H.H., Eyles, A.G. and Rayment, G.E. (eds), *Downstream Impacts of Land Use.* Department of Natural Resources, Brisbane, Australia, pp. 359-364.

Perry, T. (1995) *Vegetation Patterns of the Herbert River Floodplain*. Queensland Department of Primary Industries, Townsville, Australia.

Prove, B.G., Doogan, V.J. and Truong, P.N. (1995) Nature and magnitude of soil erosion in sugarcane land on the wet tropical coast of north-eastern Queensland. *Australian Journal of Experimental Agriculture* 35, 641-649.

Prove, B.G. and Hicks, W.S. (1991) Soil and nutrient movements from rural lands of north Queensland. In: Yellowlees, D. (ed.), *Land Use Patterns and Nutrient Loading of the Great Barrier Reef Region.* James Cook University of North Queensland, Townsville, Australia, pp. 67-76.

Queensland Department of Environment and Heritage (1995) *Conservation of the Mahogany Glider Petaurus gracilis.* Conservation Plan Series S95/0359. QDEH, Brisbane, Australia.

Queensland Landcare Council (1991) *A Guide to Property Management Planning in Queensland.* Queensland Landcare Council, Brisbane, Australia.

Queensland Sugar Corporation (1995) *Annual Report.* Queensland Sugar Corporation, Brisbane, Australia.

Raisin, G. (1995) The use of small wetlands in catchment management for the control of diffuse agricultural pollution. *Proceedings, National Conference on Wetlands for Water Quality Control,* James Cook University of North Queensland, Townsville, Australia, pp. 245-254.

Russell, D.J. and Hales, P.W. (1996) Major fisheries resource and habitat issues in the Johnstone River catchment. In: Hunter, H.H., Eyles, A.G. and Rayment, G.E. (eds), *Downstream Impacts of Land Use.* Department of Natural Resources, Brisbane, Australia, pp. 353-358.

Sallaway, M.M. (1979) Soil erosion studies in the Mackay district. *Proceedings of the Australian Society of Sugarcane Technologists* 1, 125-132.

Sinden, J.A. (1988) Community support for soil conservation. *Search* 18(4), 188-194.

Syme, G.J., Butterworth, B.E. and Nancarrow, B.E. (1994) *National Whole Catchment Management: A Review and Analysis of Processes.* Report to Land and Water Resources Research and Development Corporation by Australian Research Centre for Water in Society ARCWIS, Perth, Australia.

Walker, D.H. and Johnson, A.K.L. (1996). Delivering flexible decision support for environmental management: a case study in integrated catchment management. *Australian Journal of Environmental Management* 3(3), 174-188.

Wilson, P.R. (1985) Land use planning in the Ingham area using the relationship between soil attributes, water table heights and sugarcane yields. *Proceedings of the Australian Society Sugar Technologists* 6, 65-72.

Young, M.D. (1993) *For Our Children's Children: Some Practical Implications of Inter-Generational Equity and the Precautionary Principle.* Report to the Department of Prime Minister and Cabinet, Canberra, Australia.

Young, M.D., Gunningham, N., Elix, J., Lambert, J., Howard, B., Grabosky, P., and McCrone, E. (1996) Reimbursing the future: an evaluation of motivational, voluntary, price-based, property-right and regulatory incentives for the conservation of biodiversity. *Biodiversity Series, Paper No 9,* Biodiversity Unit, Department of Environment Sport and Territories, Canberra, Australia.

Chapter twenty-one:

The Development and Application of Environmental Policy: the Everglades and Florida Sugar Industry[3]

D.L. Anderson[1] and P. Rosendahl[2]

1. *Everglades Research & Education Center, University of Florida, Belle Glade, FL 33430-8003, USA*
2. *Environmental Relations, Flo-Sun Inc., Palm Beach, FL 33480, USA*
3. *Contribution from: University of Florida, Institute of Food and Agricultural Sciences. Fla. Exp. Sta., J. Series No. R-05080*

Abstract

Historically, US Federal and State government policies have encouraged extensive drainage and hydraulic modifications within southern Florida to foster agricultural development. By the early 1900s, these efforts had allowed the rapid development of agriculture in the northern Everglades adjacent to Lake Okeechobee, which subsequently supported the evolution of the Florida sugar industry. During the current decade, environmental restoration and conservation efforts have changed water/wetland management policies. Currently, the impacts of agricultural land use on the environmental and ecological health of adjacent wetlands of the Everglades are being assessed. Policies that once favoured development are now being reversed by policies and regulating efforts to restore natural ecosystems. The Florida sugar industry and other agricultural interests are adjusting to more stringent environmental standards and policies. This paper outlines the historical and current changes in Florida's physical and political landscape and how the sugar industry is adapting to these changes.

In: Keating, B.A. and Wilson, J.R. (eds) *Intensive Sugarcane Production: Meeting the Challenges Beyond 2000.* CAB International, Wallingford, UK, 1997.

System Description - Landscapes and Boundaries

The historic Everglades region comprises 19,400 km^2 of land encircling Lake Okeechobee and extending to the southern end of the Florida Peninsula and including the Kissimmee River Basin (Davis, 1943; Jones, 1948; SFWMD, 1990). Discussions regarding contemporary Everglades restoration efforts often refer to the 1856 Military Map (US Congress, 1911) that portrays information of the region prior to large-scale development. The region can be divided into a number of landscapes (Fig. 21.1): (i) the northern watersheds of Lake Okeechobee including the Kissimmee River basin; (ii) the Everglades Agricultural Area (EAA); (iii) the Everglades Protection Area, consisting of water conservation areas (WCAs) and the Everglades National Park (ENP); (iv) the Atlantic coastal ridge on the eastern coast with sandy flatlands, (v) the interior uplands of the Big Cypress Swamp; (vi) the southeastern urbanised regions including coastal estuaries; (vii) Whitewater Bay and southwest Florida estuaries; and (viii) Florida Bay. Extending southward of Florida Bay are the Florida Keys and coral reef formations.

South Florida is characterised by poorly drained soils with many marshes and sloughs. Water moves through a network of canals and water control structures managed by the South Florida Water Management District (SFWMD) and the U.S. Army Corps of Engineers. Surface water from Lake Okeechobee moves east through the St Lucie Canal to the Atlantic Ocean, west through the Caloosahatchee River to the Gulf of Mexico, and south and southeast through a number of lesser man-made canals (West Palm Beach, Hillsboro, North New River, and Miami Canals) to the Everglades Protection Area and Atlantic Ocean. Lake Okeechobee is used for the storage of freshwater and flood control. The lake is surrounded by a dike to prevent storm surge flooding resulting from hurricanes and heavy rains during the wet season. Inflows, except for the Fisheating Creek, are controlled structures, such as locks, gates, and pumps.

South Florida watersheds receive an average of 1320mm rainfall per year, with regional variations ranging from 965 to 1651mm. Late May through August is the hot rainy season, and November through March is cooler and drier. Large areas are inundated with water for periods ranging from 1-12 months each year. The soils in the northern watersheds of Lake Okeechobee are principally Spodosols with alluvial and organic soils in the Kissimmee River basin, where cattle, dairy, citrus, vegetables, and small grains crops are the principal agricultural commodities. Soils in the EAA are predominantly Histosols formed from the decaying remains of sawgrass (*Cladium jamaciense*) and other marsh plants accumulated under historically flooded conditions (Davis, 1946; Anderson, 1990; Izuno and Bottcher, 1994). Land in the EAA (210,229ha; SFWMD, 1996) is primarily used for sugarcane, vegetables, horticultural sod, and rice, supplying about 25% of the sugar and 40% of the winter vegetable demands of the United States. Sugar production contributes over US$1B annually to the south Florida economy (Polopolus and Alvarez, 1991).

Fig. 21.1. General map of south Florida and the Everglades today.

The Everglades Protection Area consists of marsh areas typically inundated with water 9 to 12 months each year, and soils range from peats and sands to rocky soils. Mangrove swamps are found along fresh and salt water boundaries of Biscayne Bay, the ENP, Florida Bay, and the southwest coastline. Urbanised areas such as Miami have grown tremendously during the last decade, taking over unprotected wetlands and estuaries bordering the ENP. Rapid urban expansion has also occurred in the western and southwestern coastal areas. Demographers forecast that the current (1996) urban population of 7M will increase to over 20M by 2020. Florida Bay has been affected from diminishing supplies of freshwater and urban pollutants along the Florida Keys (NOAA, 1995). Prior to the construction of canals and water control structures, 80% of south Florida's surface freshwater moved slowly from the Kissimmee River basin into Lake Okeechobee and eventually into coastal estuaries adjacent to Whitewater Bay and Florida Bay (Davis, 1943). Today, 80% of the surface freshwater is diverted to the Atlantic and Gulf of Mexico by conveyance canals and rivers (Larsen, 1994, 1995). Extensive roads and urbanised developments also restrict overland flow of water. Thus, the historical Everglades has dramatically changed with respect to the quantity and timing of freshwater it receives (McIvor *et al.*, 1994).

The Everglades ecosystem has been negatively impacted by freshwater shortages, pollutants from urban and agricultural areas, and various recreational uses. The Everglades is a source of water for agriculture, urban uses, industry, and recreation. Thus, the development of integrated water management plans must consider the balance between a viable Everglades ecosystem and a sustainable south Florida for agricultural, urban, and industrial sectors.

Hydrologic Changes

The earliest and most comprehensive scientific investigation of Everglades hydrology was by the U.S. Geological Survey (Parker *et al.*, 1955), which described drainage conditions and hydraulic changes through the 1940s. Later publications by Parker (1974), Tebeau (1974), and Light and Dineen (1994), summarise many of the pervasive hydrologic changes in south Florida and the Everglades. A more quantitative study of hydrologic changes was reported by Larsen (1994, 1995) using results of the South Florida Water Management District's Natural Systems Mathematical Model. It documents current annual discharges exceeding 2.5 billion m^3 of freshwater to the Atlantic Ocean. This is over and above the historical (1900 to current) easterly discharge of 0.9 billion m^3. Correspondingly, freshwater flows to the Everglades National Park were reduced from 1.5 billion m^3 to only 0.6 billion m^3. These hydrologic changes have negative ecological consequences to the Everglades and downstream estuaries.

Environmental Concerns

Intense development within the historic Everglades ecosystem during the past century has had major environmental implications. This includes development of the Florida sugar industry since 1928 (Fig. 21.2). Channelisation of the Kissimmee River, construction of roads and canals, and the agronomic and urban development in south Florida have greatly reduced wetlands (SFWMD, 1990).

In reclamation efforts spanning more than 100 years, wetland drainage projects were executed for flood protection and agricultural endeavors. Improved flood protection allowed for more intensive cattle and dairy production in the northern Kissimmee River Basin region with consequent increased inputs of phosphorus to Lake Okeechobee (Ritter and Allen, 1982; SFWMD, 1992 a,b; SFWMD, 1993b). Environmental concerns led to early development of 'Best Management Practices' (BMPs) for the cattle and dairy industry north of Lake Okeechobee and agronomic operations in the Everglades (Anderson and Flaig, 1995; Bottcher and Izuno, 1995). In recent years, reduction of wetland area has been slowed due to acquisition of environmentally sensitive lands (Cox *et al.*, 1994). This is a major policy change from development to restoration.

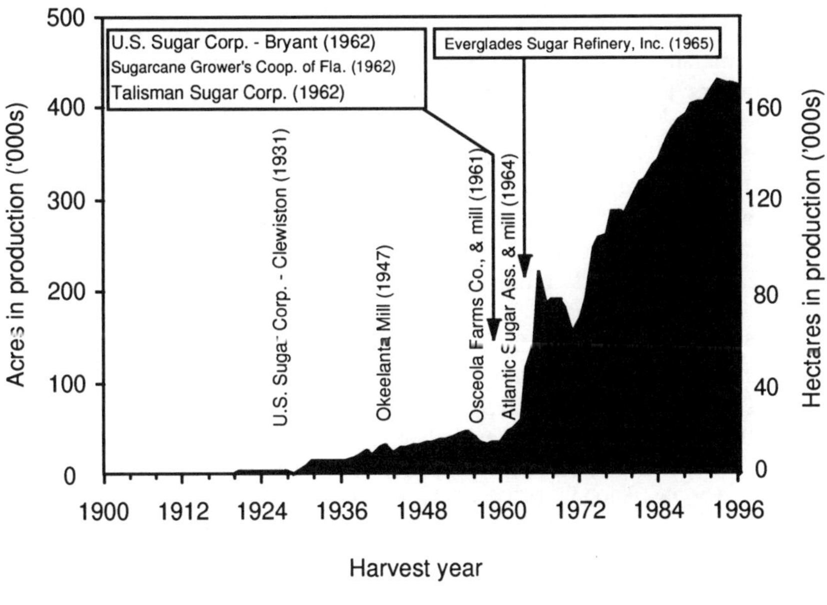

Fig. 21.2. Growth of the Florida sugar industry (data Florida Sugar Cane League, Clewiston, FL) with major events indicated.

As a result of hurricanes in 1926 and 1928 that led to the destruction of much property and loss of life, Lake Okeechobee was encircled by a muck levee to protect surrounding urban centres and agricultural endeavors. This levee was later improved with crushed rock. During times of drought the Lake is a source of freshwater for areas south of it. Today, sports and commercial fishing industries also benefit from the Lake. Management of lake levels is essential for a number of water users. However, there are conflicts between lake management for providing adequate storage to meet downstream user needs (lowering water levels) versus desires to also maintain marshlands within the lake (maintaining higher water levels). These issues have important economic and environmental consequences and are currently being reviewed by resource managers, policy makers, environmentalists, and private landowners.

Within the EAA, sugar and vegetable operations depend on a hydraulic farm infrastructure consisting of canals, pumps and levees which were constructed and are operated by farmers. Water is eventually pumped from private land holdings to a publicly operated regional system for ultimate conveyance to the Everglades Protection Area and ENP. Concerns surrounding the drainage of waters with elevated P concentrations into oligotrophic marshes of the Everglades Protection Area and ENP prompted the Florida legislature to pass the Everglades Forever Act of 1994. The Act set time lines and funding mechanisms for the construction of 16,187ha of stormwater treatment marshes (STAs). The STAs will be designed to reduce P discharge concentrations to 0.05ppm prior to entering the Everglades Protection Area and ENP (Nolte and Associates, 1992). Research is underway to determine P concentrations in water which will not cause an imbalance to the fauna and flora of the Everglades. If this determination is scientifically defensible, this concentration will become the new water quality P standard. If a determination is not made by the year 2002, then 0.01ppm total phosphorus (TP) will be adopted as the standard. Interestingly, this is well below the 1.00ppm TP standard set for waste water treatment effluent and non-point runoff in water basins adjoining the Great Lakes; 0.05 to 1.00ppm TP for combined primary and secondary wastewater treatment systems; 0.65ppm nationally for highway and road runoff; 0.18ppm for improved pastures north of Lake Okeechobee; 0.09 to 0.11 ppm for Lake Okeechobee; and 0.048 to 0.09ppm in regional rainfalls (Water Pollution Control Federation, 1990, 1991; Sedlak, 1991; SFWMD, 1992b, 1992c, 1993b). Farm drainage waters in the EAA are expected to meet or exceed the concentrations listed above, including rainfall. The future TP standard set for the EAA will be the lowest water quality standard to be set for any industry in the world. At present, it is highly questionable whether STAs alone will be able to reduce TP concentrations sufficiently (Water Pollution Control Federation, 1990, p.228; Kadlec and Knight, 1995) without supplemental technologies such as chemical treatment.

The Everglades has been reduced to half its original area and lies adjacent to expanding urban centres and downstream of the EAA farms. The Everglades wetland ecosystem was reduced and altered by federal/state drainage and flood control projects. Wading bird populations have been reduced 75% to 80% (from 1934 to 1976), as an indicator of the declining health of the Everglades (National Audubon Society, 1990; Frederick, 1992; Ogden, 1994). A major effort is underway by federal, state, and private interests to restore the Everglades through water quality improvements, hydrological projects, and sensitive lands acquisition (SOR, 1996). The reestablishment of 'sheet-flow' is planned for the upper Everglades, and the need for increased 'storage' is recognised. Detailed hydrological studies have concluded that tremendous volumes of freshwater are essentially wasted through large discharge events into the Atlantic Ocean and Gulf of Mexico. Along the heavily populated lower southeast 'gold coast' of South Florida, urban stormwater runoff (2.5-3.7 billion m^3) currently diverted to the ocean could go to the lower Everglades (Larsen, 1994, 1995). This has limited the productivity of coastal mangrove areas where wading birds once nested in great numbers (Ogden, 1995). Increased freshwater supply for the Everglades is being sought. There have been proposals to capture water west of the urban centres of the lower southeast coast and to capture flood discharge waters from Lake Okeechobee within storage areas located on farms in the southern EAA.

Bays and estuaries which eventually flow to the Gulf of Mexico are located downstream of the Everglades. Between the Florida Keys and mainland Florida lies Florida Bay, a troubled ecosystem which in recent years has experienced declining sport and commercial fishing and massive algal blooms (NOAA, 1995). Hypersaline conditions have resulted from diversion of freshwater south of Miami and poor exchange rates with the adjacent Atlantic Ocean. A correlation between greater than normal rainfall and improved estuarine bay conditions is additional evidence that suggests the need for increased freshwater flow (NOAA, 1995). Lower freshwater levels, breaching barrier islands, and degraded water quality have negatively affected estuarine systems along the entire fringes of the lower east coast of Florida (SFWMD, 1993a). The Florida Bay, Florida Keys, and coral reefs have been affected by: excessive boating, fishing, extended land use, recreation, and poor water quality related to urban and boating pollutants (NOAA, 1995). Essential issues for the management of these regions that are being addressed are: growth management, marinas and boat discharges, water use and re-use, dredge and fill, domestic wastewater, uncontrolled stormwater, canals, and zoning (NOAA, 1995). Because the EAA is distant and not hydraulically connected to these regions, EAA farm nutrient pollutants do not affect ecosystem health of Florida Bay, Florida Keys, and barrier islands.

Soil Subsidence Concerns

In Florida, sugarcane is largely produced within the EAA on organic soils (Anderson, 1990). These soils are characterised by an organic epidedon overlying limestone cap rock or sand. Subsidence is a loss of soil volume and mass resulting from shrinkage, compaction, and biological oxidation of organic matter. Loss of surface elevation occurs when these soils are drained. Since 1900, many areas of the EAA have lost from 2.7-4m of surface elevation. Substantial losses of soil have occurred as a result of fires and erosion (wind and water). Subsidence rates of 25-38mm y^{-1} have been recorded (Stephens and Johnson, 1951; Stephens, 1969; Tate, 1976). Volk (1972) reported that 58 to 73% of the subsidence was due to microbiological oxidation caused by heterotrophic bacteria and fungi. Subsidence of organic soils has a somewhat predictable environmental consequences for N and P release into surface discharge waters and the disappearance of soil sufficient for future crop production. Both issues are important in discussion of the sustainability of the Florida sugar industry.

Various methods of stopping or minimising soil subsidence have been widely discussed. Both short-term and long-term flooding reduce or eliminate subsidence, but with less than desirable effects on sugarcane production. The use of sugarcane varieties tolerant to high water tables and flooding has been studied to a limited extent (Porter *et al.*, 1991; Glaz, 1995). Additions of copper and other compounds to the soil may also inhibit or slow microbial activity (Mathur, 1983). Farming on soils less than 100mm deep over crushed rock has also been considered. Unfortunately, the economic and environmental practicalities of these methods have not been determined.

Continued soil subsidence in the EAA will have negative consequences for the sugar industry. Although Stephens and Johnson (1951) predicted that subsidence would reduce agricultural production 45% by the year 2000, no decline in production has been recorded. To date, the largest losses of agricultural lands have been due to the acquisition of over 16,187ha for stormwater treatment areas (STAs). The industry has coped with subsidence by farming on higher pH soils, dealing with poorer drainage by efficient water control, fertilising to overcome micronutrient deficiencies, selecting sugarcane varieties resistant to root diseases associated with wetter soils, and by developing BMPs to reduce nutrient release into farm discharge waters. Some areas in the EAA have subsided with soils less than 150-300mm deep, and growers have developed specialised water control systems. Also, some of these areas have been acquisitioned for conversion to STAs.

The recent observation that subsidence rates have diminished may be a result of the higher water tables used by growers, but they are still not constant. Rice and other rotational crops have been used, in which 10 to 20% of the area is flooded for 2 to 4 months during the summer rainy season. Flooding inhibits aerobic microbial oxidation of organic soils as well as redefining the population of heterotrophic bacteria and fungi in the soil after flooding. However, unless the region is flooded for durations of 6 to 12 months, soil subsidence may not be completely inhibited. If the EAA is flooded for extended periods, private ownership of agricultural lands in the EAA may not be economical.

Phosphorus Water Quality Concerns

Phosphorus is the principal water quality concern in surface waters of south Florida and the Everglades ecosystem. Nitrogen concentrations are not considered as an important criterion for surface waters because of denitrification and biological cycling in an open system. Phosphorus watershed inputs into Lake Okeechobee originate from the Kissimmee River and various sloughs. During the late 1980s, Fluck *et al.* (1992) and Boggess *et al.* (1995) found that 3447Mt P y^{-1} were annually imported into the watershed as animal feed, fertiliser, and human consumables. An additional 300Mt P y^{-1} originated from rainfall. Phosphorus exports from the watershed (milk, animals, and crops) accounted for 900Mt P y^{-1}. The total P imports to the Lake were 300Mt P y^{-1}, resulting in a net watershed retention of 90% of the net P imports.

Phosphorus loads to the EAA originate from Lake Okeechobee waters flowing into the EAA (irrigation), rainfall, fertilisers, and soil mineralisation. The amount in agricultural discharges into the Everglades Protection Area (220Mt P y^{-1}) has been a source of concern, despite the fact that water quality concentrations are considerably lower than urban stormwaters or non-point watershed discharges from farming regions in other parts of the US and the world. Phosphorus originates from suspended/resuspended sediments, mineralised organic matter, soluble inorganic/organic P (*i.e.* fertilisers, 5800Mt P y^{-1} and rainfall, 108Mt P y^{-1}) (Anderson and Howell, 1993; Anderson *et al.*, 1994). Mineralisation rates for the organic soils in the EAA range from 1.4-40kg P ha^{-1} y^{-1} (Anderson *et al.*, 1994) and account for 18,000Mt P yr^{-1} (SFWMD, 1993b). Also, nutrient exports from the EAA account for 1996Mt P yr^{-1} (Fluck *et al.*, 1992). With this taken into consideration, the EAA retains > 99% of TP mineralised or imported into the region.

Routine monitoring of surface waters leaving the EAA has shown that > 56% of TP discharged from the EAA is in particulate form (Table 21.1). Total particulate P (TPP) is present as suspended and bottom sediments in organic and inorganically bound forms (Anderson *et al.*, 1994). Biological debris such as algae and phytoplankton contribute to this sediment load, although this source may be of minimum importance compared to the contribution from eroded soils. Notably, variability in P concentration is very high (Table 21.1). In farm stormwater-monitoring studies reported by Anderson and Ceric (1992), TPP in farm canals ranged from 35% to 85% of TP, with an average of 66%. Seasonal variation of TPP in canal waters was also reported by Coale *et al.* (1994a,b) and Izuno *et al.* (1995), with the largest variation in particulate loads during rainfall events in August (ranging from 20% to 52% as TPP). No differences in discharge P water quality have been observed between farms cropped in sugarcane or in fallow (Coale *et al.*, 1994b).

Phosphorus is also released into surface waters from physical disturbances of fire, high water, and erosion. Temperatures may fall below 0 °C for several hours or longer from December through February each year. This results in damaged

Table 21.1. Summary of phosphorus in different samples of EAA basin waters discharged to the Everglades Protection Areas from 1973-1994. [Analysis of grab or composite samples when both TP and SRP were determined on the same sample, after Anderson and Flaig (1995)]

EAA basin	TP[1] min.	max. (mg l^{-1})	mean	SRP[2] min.	max. (mg l^{-1})	mean	TPP[3] (%)
S5A	0.026	0.581	0.167	0.002	0.465	0.095	55.5
S6	0.011	0.872	0.110	0.002	0.847	0.067	61.1
S7	0.006	1.030	0.085	0.002	0.922	0.049	57.8
S8	0.008	0.933	0.117	0.001	0.437	0.051	43.4
Mean			0.120			0.065	56.4

[1]TP, total P concentrations; [2]SRP, soluble reactive P concentrations; [3]TPP, total particulate P (%)

vegetation and potential fire hazard. Although freezing temperatures are not certain, the likelihood of frost damage to vegetation in many of the public lands in south Florida is high. Also, during low rainfall periods, water control in natural areas is difficult or impossible. Commonly, under these conditions, dry vegetation is ignited by fires or lightning (Tebeau, 1968; Taylor, 1981). Each year, fires become a significant management problem in all public lands, although fire is an important ecosystem management tool (Bancroft, 1976; Carlson *et al.*, 1993). Phosphorus is subsequently released when rains and water tables increase during May through September. As much as 0.8 to 1.6Mha of public lands come under threat of fire and widely variable water tables each year. These natural events also contribute to the overall increase of P loading to the Everglades (Bayley and Odum, 1976).

Agricultural Best Management Practices (BMPs)

State and regional agency rules mandate use of on-farm BMPs (Rule 40E-63). A regulatory program was developed to reduce P loads in non-point agricultural runoff from the entire EAA by 25% (SFWMD, 1992b, p. 63-64; State of Florida, 1992). The EAA regulatory program was developed, in cooperation with the land owners, as a comprehensive approach to establish and monitor BMPs (Table 21.2). A provision allowed for BMP credits as an incentive to farmers to establish BMPs quickly and immediately reduce P exports to the Everglades. This incentive is working (Fig. 21.3).

Table 21.2. Agricultural best management practices (BMPs) in the Everglades Agricultural Area. [After Anderson and Flaig, 1995]

BASIN-WIDE BMPs	
Algal-turf for phosphorus removal Chemical treatment of basin discharge waters Deep well injection Limit back-pumping into Lake	Regional sedimentation basins Sediment removal from primary canals Stormwater Treatment Areas (STAs)
CROP BMPs	
Aquatic cover crop in rotation with vegetables	Coordinate from high fertility to low fertility crops
FERTILIZER BMPs	
Band and split fertiliser under appropriate conditions Calibrated soil test for fertility recommendations	Improve fertiliser application practices Prevent fertiliser spills into surface waters Utilise slow-release fertilisers where appropriate
FARM MANAGEMENT BMPs	
Develop surface water management control plan: • coordinate/educate water control personnel for implementing water-control plan	• develop water table control policy/plan under temporal conditions (seasonal) • establish farm-specific discharge rules
HYDRAULIC OR WATER MANAGEMENT BMPs	
Hydraulic linkage of farm units/cropping areas Farm interior water control structures (i.e. area pump stations, culvert riser controls, etc.) Flash pumping to avoid discharge of soil pore waters On-farm water retention from aquatic cover crops and fallow-flood crop rotations Parallel canal water seepage treatment Percolation (seepage) ponding of water • Stormwater retention areas during high frequency rainfall events (seepage)	Water management rules established: • avoid pumping during rainfall events • pump canal storage before rainfall events • pump water only to the 'level of risk' (i.e. to the acceptable high water table level) • use all on-farm water storage before discharging • use all on-farm stored water before irrigating Water table control: • temporal management (seasonal) • optimise levels • minimise fluctuations • account for ET to assist drainage
SEDIMENT/EROSION CONTROL BMPs	
Aquatic weed control (physical removal) Canal bottom barriers to intercept sediment load Canal bottom sediment load sump traps Canal cleaning program for sediment removal Chemical treatment of discharge/flood waters Cover crops/flooding to reduce erosion Culvert placement to reduce flow/sediment transport Culvert sumps to minimise soil erosion Direct filtration Ditch bank berming/resloping	Ditch bank stabilisation: • polyacralamides (PAM) erosion inhibitors • vegetative cover Expanded canal construction (sedimentation) Laser-level production fields to reduce erosion No-till cultivation to reduce soil erosion Stormwater retention/seepage areas during high frequency rainfall events Stormwater Treatment Areas (STAs) Rock pit drainage diversions Upstream sumps from pump intake

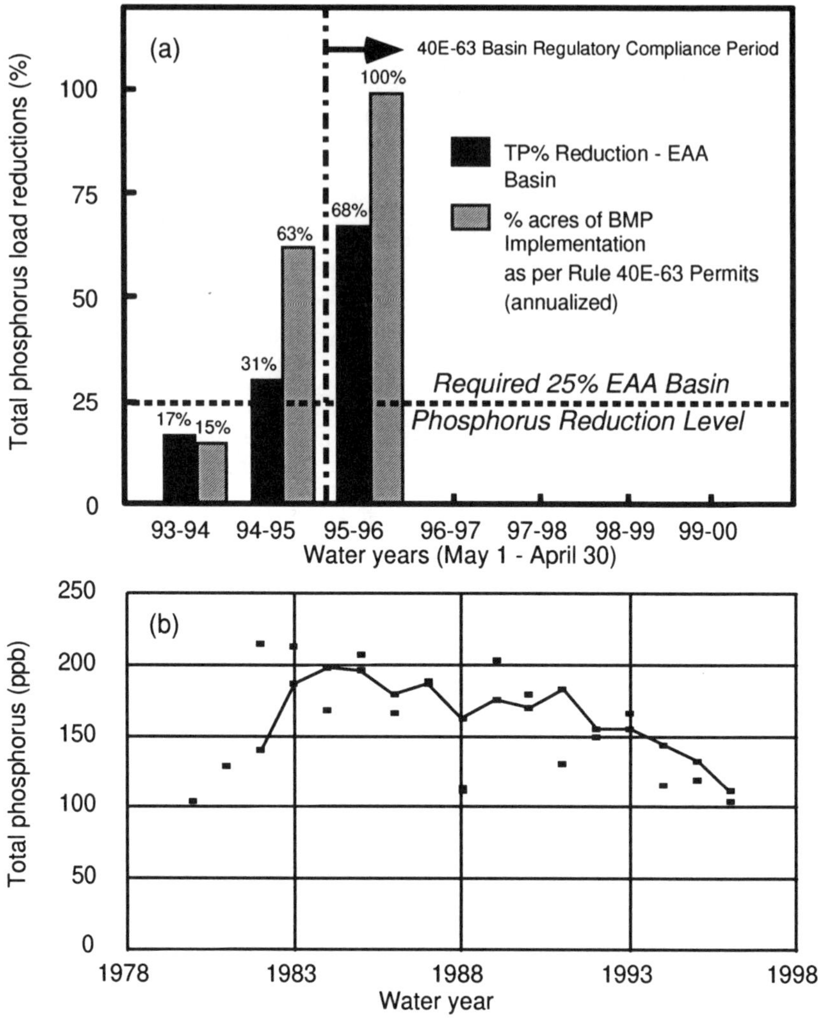

Fig. 21.3. (a) Timetable for implementation of total phosphorus load reductions in the EAA and (b) total phosphorus concentrations discharged from EAA basin structures during 1979 through 1996. [SFWMD, 1996; Whalen and Whalen, 1996]

The EAA regulatory program was designed to be comprehensive, requiring landowners to develop nutrient and water management plans, BMPs, and monitor water quality from each farm. Landowner development plans were required to include rationales for selected BMPs and time for implementation. The monitoring plans included self-monitoring of discharge and P loads and quarterly reports of water quality data, as well as BMP implementation.

To increase the success of the program, EAA farm managers and field technicians are required to be certified by the SFWMD through University of Florida Cooperative Extension Service education and training programs for development and implementation of BMPs, and operation of BMPs and monitoring equipment (Whalen and Whalen, 1996). A procedural guide to BMPs in the EAA was developed by Bottcher and Izuno (1995). Table 21.2 outlines typical BMPs for the EAA (Anderson and Flaig, 1995). Hydraulic or water management BMPs were shown to have a greater potential for reducing P loads than were fertility practices (Coale *et al.*, 1994a; Izuno and Capone, 1995; Izuno *et al.*, 1995). Percentage P load reductions since the early 1980s have increased to over 68% in 1996, and likewise, EAA basin water discharge P concentrations have decreased since 1984 (Fig. 21.3; SFWMD, 1996). This success was a result of improved land management following involvement of the land owners in development of the regulatory program (Whalen and Whalen, 1994, 1996). Best management practices requiring simple management changes have been quickly accepted and utilised in the EAA. Total P loads are highly variable, because they are greatly affected by rainfall frequency, intensity, and distribution, independent of BMPs. Compliance with regulation has been shown to require management flexibility to engage specific BMPs on a case-by-case basis. Although this program in the EAA appears to be motivated from a regulatory point-of-view, many BMP, monitoring and science developments were financed by the industry (Izuno *et al.*, 1991; Anderson and Ceric, 1992; Capone *et al.*, 1995; Izuno and Capone, 1995).

Principles of Balanced Water Resource Use

In concept, there are three dimensions of water management making up the 'water cube' as discussed by van Rooy *et al.* (1993). In south Florida, surface and ground water issues associated with restoration of the Everglades are affecting all uses of water. Water demands arise from urban and recreation uses, drinking water, industry, agriculture, and nature. Developing and implementing policies that balance water resource use will be difficult.

Historically, management of ground water quantity has been the business of national soil and water authorities. Regional or provincial authorities have been in charge of surface water quality. Many aspects of water management fall within the domain of district water boards, while local authorities are responsible for collection and treatment of drainage water and waste within urban areas. Because water systems are managed by separate management authorities, elements within the whole system are seldom treated equally. Economic and demographic developments have a direct influence on one or more elements within the water system. The entire water system will be affected by the interaction between all elements.

Water resources have specific value which should be protected for long-term utility. Although user demands add value to the water system, each function has its impact. For instance, recreational use of water (e.g. swimming, fishing, sailing) requires water of high quality as well as quantity which may directly be affected by utilisation by industrial, urban, or agricultural users. However, not all impacts are negative. Ground and surface waters are manipulated and managed in various manners: flood/drought protection, hydro-period stabilisation for nature, storage for multiple uses and interests, or salt water intrusion protection for drinking water. When several functions or users compete for the same water resource, often one use may be managed while the other is not.

In reality, the number of interests and functions which compete for the same resource is likely to increase. Good policies and management are absolutely essential. Without these, the water system would be overwhelmed. Policy and management are directed toward water control, urban and rural planning, and environmental protection. Water authorities' greatest responsibility is to prioritise water use, although it may be impossible to reconcile all demands for the same resources within a limited area and time. Within the framework of integrated water management, prioritisation is the responsibility of both policy makers and managers. Unequivocal rankings are assigned to each interest after weighing one demand against the other. However, rational water use decisions have far-reaching effects, especially when past policies for established users change irreconcilably. Water management from the 1850s to the 1970s was directed under policies structured for flood and drought protection. The environmental movement has changed past policies toward restoration of the Everglades, which has thus changed how water must be managed for all users. In these cases, policy takes precedence over management until problems are solved through research, development, and application.

Resource User Changes

Prioritising the use of water and setting acceptable impacts will be difficult to reconcile. In 11 counties of south Florida, the public lands set aside for preservation, conservation, or wildlife restoration purposes approximate 50 to 60% of the total area. Acquisition of public lands has increased 15% over the last five years (3% per year) in some counties, and is expected to continue (SOR, 1996). At the same time, agricultural water demands in the EAA will decrease. Over 16,187ha of agricultural land in the EAA will be used for STAs, and during 1996, another 12,950ha from Talisman Sugar (of an estimated 21,044ha total) could be acquisitioned for this purpose. Interestingly, agricultural conversion to STAs could actually increase water use in the EAA.

Urban development of other land is also rapidly increasing in south Florida. Population is expected to increase from 7M to over 20M by the year 2020. For example, in Lee County between 1978 and 1985, population increased 6% per year, by 4% for years between 1985-1989, and by 3% for years between 1989 and 1995. If annual population increase of 4% per year is anticipated, then 20M people can be expected to live on diminishing land and water resources by the year 2020 (Fig. 21.4). Population demands for water (e.g. recreation, drinking, industrial) will increase.

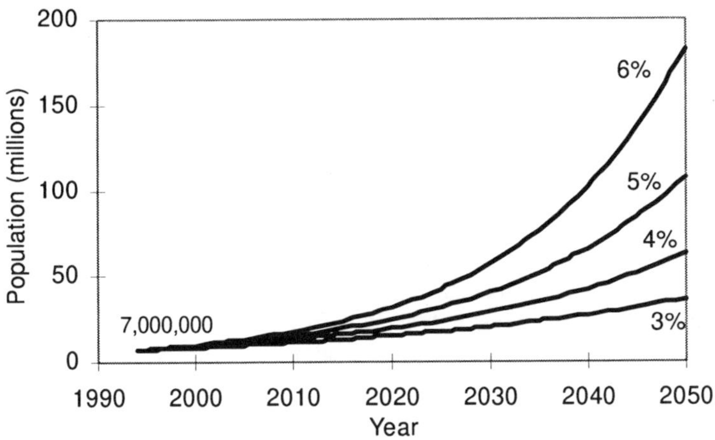

Fig. 21.4. South Florida population growth projections for 3 to 6% annual increases from 1995 to 2050.

Because public land acquisition (primarily for conservation and environmental protection) and population have both increased at similar rates during the last five years, and because lands already in public possession are 50-60% south Florida, prioritising water resources for all users will be increasingly more difficult. Sugarcane production area in 1996 has expanded to 179,640ha (Florida Sugar Cane League, unpublished data). However, some lands currently in production will be converted to STAs, and the total production area is expected to decrease. Politically, farmers have not been strong enough to fight land acquisition, farming privilege taxes (Everglades Restoration), nor various regulations focusing on impacts on neighbouring environmentally-sensitive and protection areas. Should 'zero impact' conditions be set along boundaries between lands set aside for natural protection and other land uses, all other resource users may be deemed incompatible with ecosystem preservation and protection. Development of rational policies and sustainable management systems that will protect and guarantee that all users have access to water resources in south Florida will be an intensely debated issue well past the year 2000.

Environmental Politics and Policy

Many lasting physical changes have been made to the southern Florida landscape. Part of the vast Everglades wetland ecosystem was transformed to one of the highest urban growth regions in the US. The accompanying sugar industry located between Lake Okeechobee and the Everglades has also flourished. More than 100 years of Federal and State policies encouraged successful drainage of wetlands for urban and agricultural development within the historical Everglades. Today, the Florida sugar industry is one of the most sophisticated producers of sugarcane in the world. In the last five years, major sustainable and value-added projects have been also completed which provide renewable energy co-generation power plants and state-of-the-art distribution, packaging, and warehouse centres. Renewable biomass energy power plants are located adjacent to two sugar mills co-generating steam and enough electricity for 76,000 households. These power plants burn both urban woody wastes and bagasse.

Environmental concerns starting in late 1960s and early 1970s focused on ecosystem protection and restoration. Initially oriented to protect the downstream Everglades National Park, the environmental movement now focuses on restoring the entire remnant Everglades, including the designated water conservation areas. The Everglades Forever Act of 1994 allowed Federal government agreements to be codified into Florida law, providing for water quality and water distribution solutions. Stormwaters from Lake Okeechobee, urban areas, and farming regions are to be cleaned using macrophyte treatment marshes known as STAs and delivered to the adjacent Everglades under 'sheet-flow' conditions that simulate historical natural conditions.

A great number of environmental groups have claimed that the Everglades Forever Act was too conservative and their own proposals typically include appreciably larger and more costly solutions which threaten the sustainability of the sugar industry. Another proposal involves the removal of 60,700ha from production in addition to approximately 16,187ha already ear-marked for STA construction. Such a proposal would destroy three of the seven sugar mills operating within the region. Presidential election politics have also surfaced in this debate, with one candidate calling for the removal of 40,468ha of productive farmland and taxation of the remaining farmers to purchase this land. In the press release by Vice President Al Gore (1996), the 'penny-a-pound' tax of sugar was proposed which would financially cripple farmers. These initiatives are proposed at the Federal and State level.

Farmers in south Florida have striven to be responsible environmental stewards. Improvements in water quality and reduction of P discharged from the region is being realised (Fig. 21.3). Farming adjacent to the Everglades is both politically sensitive and challenging. New technologies are being investigated to reduce P to the part-per-billion level. Sugar farmers have embraced concepts such as BMPs and 'sustainable agriculture' and are making significant progress towards natural resource conservation and balancing their farming livelihood with the ecological requirements for a healthy Everglades. The future of farming in south Florida depends on the recognition of the economic importance of the industry and its positive efforts to protect the environment.

Lessons Learned

The first lesson is adapting to changing water management policy. At the turn of the century, the principle thrust was to drain swamps for urban and agricultural development. Today, Federal and State policies lean toward wetland ecosystem preservation and restoration. The Florida sugar industry is caught in the middle of this transition.

Secondly, the public desire for 'green space' adjacent to rapidly expanding urban centres will continue. Expansion is focused on converting productive farmlands to STA wetlands by reflooding. The future threat of taking additional farmlands for water storage or treatment pumping has created an economic and political crisis for agriculture and local government. In hindsight, if Federal and State planners had foreseen today's crisis, ecosystem design modifications could have been accomplished during the early design phase (1948-1960s). Unfortunately, farmers are being held accountable for past design flaws which they had no part in developing.

The third lesson refers to the drafting of rational legislation that is scientifically acceptable. One consequence of shifting public policies is inevitable litigation that results from changing environmental law and regulation. The Central and South Florida Flood Control Project was authorised in 1948, well before the National Environmental Policy Act of 1972 which set into motion water quality criteria. In addition to successfully growing high-yielding sugarcane in the organic soils of the EAA, this region has also yielded significant legal debate. For Class I, II, and III waters, Florida State law (State of Florida, 1996) specifies that '*in no case shall nutrient concentrations of a body of water be altered so as to cause an imbalance in natural populations of aquatic flora or fauna*'. In the strictest interpretation, this can be considered an intolerance law and has prompted extensive debates between private landowners, scientists, and the State and Federal governments. Incorporation of water quality concerns within the scope of the original project design of 1948 would have averted costly land acquisition, redesign and reconstruction, and legal implications currently underway.

The fourth lesson has been learned by the Florida cane growers in the last five years. Implementation of BMP programmes have greatly improved farm water quality discharges. Greater on-farm retention and discharge control of stormwaters have had greater and more favourable effects on water quality than other BMPs (Table 21.2). These programmes have been successful in achieving regulatory objectives and improving public opinion. For sugar industries that are adjacent to environmentally sensitive regions (i.e. the Australian sugar industry adjacent to the Great Barrier Reef), the experiences of the Florida sugar industry should be noted.

If, after spending hundreds of millions of dollars and engaging hundreds of scientists and engineers, a balance between agriculture and the Everglades is not obtained, then it may be concluded that there are no solutions. A sustainable system must have the continuity and balance between the dimensions of environment, users, and management policy. The development and management of south Florida, the Everglades, and the Florida sugar industry is an experiment of world-wide interest, with world-wide consequences.

References

Anderson, D.L. (1990) A review: Soils, nutrition, and fertility practices of the Florida sugarcane industry. *Soil Crop Science Society Florida Proceedings* 49, 78-87.

Anderson, D.L. and Ceric, A. (1992) Reduction of phosphorus concentrations in agricultural drainage of the EAA by precipitation, coagulation, and sedimentation. Update Report to the Florida Sugar Cane League (Rep. 92-11). University of Florida /IFAS, Everglades Research & Education Center, Belle Glade, FL, USA, p. 74 .

Anderson, D.L. and Flaig, E.G. (1995) Comprehensive water management in south Florida. Agricultural best management practices (BMPs) and surface water improvement and management (SWIM). *Journal of Water Science Technology* 30, 109-121.

Anderson, D.L., Hanlon, E.A., Bottcher, A.B. and Ceric, A. (1994) Phosphorus in drainage waters from organic soils (Histosols) of Florida. University of Florida/IFAS, SSLAH-1, USA, 3 pp.

Anderson, D.L. and Howell, M.D. (1993) Implementation of the Everglades Surface Water Improvement and Management Plan in the EAA. *Sugar Journal* 56, 5-10.

Bancroft, L. (1976) Fire management in Everglades National Park. *Fire Control Notes, U.S. Forest Service* 37(1), 18-21.

Bayley, S. and Odum, H.T. (1976) Simulation of interrelations of the Everglades' marsh, peat, fire, water, and phosphorus. *Ecological Modelling* 2(3), 169-188.

Boggess, C.F., Flaig, E.G. and Fluck, R.C. (1995) Phosphorus budgets-basin relationships for the Lake Okeechobee tributary basins. *Ecological Engineering* 5,143-162.

Bottcher, D. and Izuno, F.T (1995) Procedural guide for the development of farm-level best management practices plans for phosphorus control in the Everglades Agricultural Area. Ver. 1.1, IFAS, Florida Cooperation Extension Service University of Florida, Gainesville, USA, 47 pp.

Capone, L.T., Izuno, F.T., Bottcher, A.B., Sanchez, C.A., Coale, F.J. and Jones, D.B. (1995) Nitrogen concentrations in agricultural drainage water in south Florida. *American Society Agricultural Engineering* 38, 1089-1098.

Carlson, P.C., G.W. Tanner, J.M. Wood, and S.R. Humphrey. (1993) Fire in Key deer habitat improves browse, prevents succession, and preserves endemic herbs. *Journal of Wildlife Management* 57, 914-928.

Coale, F.J., Izuno, F.T. and Bottcher, A.B. (1994a) Phosphorus in drainage water from sugarcane in the Everglades Agricultural Area as affected by drainage rate. *Journal Environment Quality* 23, 121-126.

Coale, F.J., Izuno, F.T. and Bottcher, A.B. (1994b) Sugarcane production impact on nitrogen and phosphorus in drainage water from an Everglades Histosol. *Journal Environment Quality* 23, 116-120.

Cox, J., Kautz, R., MacLaughlin, M. and Gilbert, T. (1994) *Closing the Gaps in Florida's Wildlife Habitat Conservation System.* Office of Environmental Services Florida Game and Fresh Water Fish Commission, Talahassee, FL, USA, 239 pp.

Davis, J.H. (1943) *The Natural Features of Southern Florida, Especially the Vegetation, and the Everglades.* Geological Bulletin No. 25. State of Florida, Department of Conservation, Florida Geological Survey, Tallahassee, USA, 311 pp.

Davis, J.H. (1946) *The Peat Deposits of Florida. Their Occurrence, Development, and Uses.* Geological Bulletin No. 30. State of Florida, Department of Conservation, Florida Geological Survey, Tallahassee, USA, 247 pp.

Fluck, R.C., Fonyo, C.A. and Flaig, E.G. (1992) Land-use-based phosphorus balances for Lake Okeechobee, Florida, drainage basins. *Journal Applied Engineering* 8, 813-820.

Frederick, P.C. (1992) Wading bird nesting success studies in the Water Conservation Areas of the Everglades, 1992. Department Wildlife and Range Sciences, University of Florida, Gainesville, FL. Contract C3138, South Florida Water Management District, West Palm Beach, FL, USA.

Glaz, B. (1995) Research seeking agricultural and ecological benefits in the Everglades. *Journal Soil Water Conservation* 50, 609-612.

Gore, A. (1996) Statement by Vice President Al Gore announcing a new plan to restore the Florida Everglades. The Clinton/Gore administration's Everglades Restoration Plan. Principles and Elements. Everglades National Park, 19 February. Press Release, The White House, Office of the Vice President, Washington, DC, USA. 5 pp.

Izuno, F.T. and Bottcher, A.B. (1994) The history of water management in south Florida. In: Bottcher, D. and Izuno, F.T. (eds), *The Everglades Agricultural Area: Water, Soil, Crop, and Environmental Management.* University Press of Florida, Gainesville, FL, USA, pp. 13-26.

Izuno, F.T., Bottcher, A.B., Coale, F.J., Sanchez, C.A. and Jones, D.B. (1995) Agricultural BMPs for phosphorus reduction in south Florida. *American Society Agricultural Engineering* 38, 735-744.

Izuno, F.T. and Capone, L.T. (1995) Strategies for protecting Florida's Everglades: The Best Management Practice approach. *Journal Water Science Technology* 31, 123-131.

Izuno, F.T., Sanchez, C.A., Coale, F.J., Bottcher, A.B. and Jones, D.B. (1991) Phosphorus concentrations in drainage water in the Everglades Agricultural Area. *Journal Environmental Quality* 20, 608-619.

Jones, L.A. (1948) *Soils, Geology, and Water Control in the Everglades Region.* University of Florida, Bulletin 442, USA, 168 pp.

Kadlec, R.H. and Knight, R.L. (1995) *Treatment Wetlands.* Lewis Publishers, Boca Raton, FL, USA.

Larsen, P. (1994) Everglades water budget presentation - Principles and overview. Governor's Commission for a Sustainable South Florida. Larsen & Associates, First Union Financial Center, Suite 4910, Miami, FL, USA.

Larsen, P. (1995) Everglades water budget presentation. Technical Advisory Committee to the Governor's Commission for a Sustainable South Florida. Larsen & Associates, First Union Financial Center, Suite 4910, Miami, FL, USA.

Light, S.S. and Dineen, J.W. (1994) Water Control in the Everglades: A Historical

Perspective. In: S.M Davis and J.C. Ogden (eds), *Everglades. The Ecosystem and Its Restoration.* St, Lucie Press, Delray Beach, FL, USA, pp. 47-84.

Mathur, S.P. (1983) A lack of bactericidal effect of subsidence-mitigating copper in organic soils. *Canadian Journal Soil Science* 63(3), 645-649.

McIvor, C.C., Ley, J.A. and Bjork, R.D. (1994) Changes in freshwater inflow from the Everglades to Florida Bay including effects on biota and biotic processes: A review. In: S.M. Davis and J.C. Ogden (eds), *Everglades. The Ecosystem and Its Restoration.* St. Lucie Press, Delray Beach, FL, USA, pp. 117-146.

National Audubon Society (1990) Wading bird populations and distributions in the Water Conservation Areas of the Everglades: 1985-1988. W. Hoffman, G.T. Bancroft, and R.J. Sawicki, National Audubon Society, Tavernier, FL. Contract 396-M87-0345, South Florida Water Management District, West Palm Beach, FL, USA.

NOAA (1995) Florida Keys National Marine Sanctuary. Draft Management Plan/Environmental Impact Statement. Volume I: Management Plan, 323 pp. Volume II, Development of the management plan: environmental impact statement, 245 pp. Volume III, Appendices, 203 pp.

Nolte and Associates (1992) Technical advisory panel review of the Everglades Protection Project conceptual design of stormwater treatment areas. Nolte and Associates, Sacramento, CA Report to the South Florida Water Management District, West Palm Beach, FL, USA.

Ogden, J.C. (1994) A comparison of wading bird nesting colony dynamics (1931-1946 and 1974-1989) as an indication of ecosystem conditions in the southern Everglades. In: Davis, S.M. and Ogden, J.C. (eds), *Everglades. The Ecosystem and its Restoration.* St. Lucie Press, Delray Beach, FL, USA, pp. 533-570.

Parker, G.G. (1974) Hydrology of the pre-drainage system of the Everglades in Southern Florida. In: P.J. Gleason (ed.), *Environments of South Florida: Present and Past.* Memoir 2, Miami Geological Society, Miami, FL, USA, pp. 18-27.

Parker, G.G., Ferguson, G.E. and Love, S.K. (1955) Water resources of southeastern Florida. U.S. Geological Survey Water-Supply Paper No. 1255, Miami, USA.

Polopolus, L.C. and Alvarez, J. (1991) *Marketing Sugar and Other Sweeteners.* Elsevier Science, New York, USA, 361pp.

Porter, P.S., Snyder, G.H. and Deren, C.W. (1991) Flood-tolerant crops for low input sustainable agriculture in the Everglades Agricultural Area. *Journal Sustainable Agriculture* 2, 77-101.

Ritter, G.J. and Allen, L.H. Jr. (1982) Taylor Creek Headwaters Project. Phase I: Water quality. Technical Publication 82-8. South Florida Water Management District, West Palm Beach, FL, USA.

Sedlak, R.I, (ed.) (1991) *Phosphorus and Nitrogen Removal from Municipal Wastewater: Principles and Practice.* 2nd ed. Lewis Publishers. Soap and Detergent Assoc., NY, USA.

SFWMD (1990) Kissimmee River restoration: Alternative plan, evaluation, and preliminary design report. South Florida Water Management District, West Palm Beach, FL, USA. June.

SFWMD (1992a) Everglades Agricultural Area Regulatory Program. Chapter 40E-63, Florida Annotated Code, South Florida Water Management District, West Palm Beach, FL, USA. (amended 7/7/92).

SFWMD (1992b) Surface Water Improvement and Management Plan for the Everglades. Supporting Information Document. South Florida Water Management District. West Palm Beach, FL, USA. 6th Draft, 2 January.

SFWMD (1992c) Surface Water Improvement and Management Plan for the Everglades. Three volumes: Supporting Information Documents, Planning Document, and Appendices. South Florida Water Management District. West Palm Beach, FL, USA. 13 March.

SFWMD (1993a) Working document in support of the Lower East Coast regional water supply plan. Appendices and Technical Information. March. South Florida Water Management District, West Palm Beach, FL, USA.

SFWMD (1993b) Surface Water Improvement and Management (SWIM) Plan. Update for Lake Okeechobee. Vol. 1: Planning Document. Vol. 2: Appendices. 14 January. South Florida Water Management District, West Palm Beach, FL, USA.

SFWMD (1996) Everglades Regulation Program: EAA Basin. Delivery under Contract no. C-6543 'Everglades regulation technical support'. W.W. Walker, Jr. South Florida Water Management District, West Palm Beach, FL, USA 15 June.

SOR (1996) *Save Our Rivers 1996 Five Year Plan.* South Florida Water Management District, FL, USA, 96 pp.

State of Florida (1992) Surface Water Quality Standards. Department of Environmental Regulations, Chapter 62-302, Surface Water Quality Criteria,530,48b, Florida Administrative Code. Tallahassee, FL, USA.

Stephens, J.C. (1969) Peat and muck drainage problems. *Journal Irrigation Drainage Division Proceedings American Society Civil Engineering* 95, 285-305.

Stephens, J.C. and Johnson, L. (1951) Subsidence of organic soils in the upper Everglades region of Florida. *Soil Science Society Florida Proceedings* 11, 191-237.

Tate, R.L. (1976) Nitrification in Everglades Histosols: A potential role in soil subsidence. *International Association Hydrological Science Proceedings Anaheim Symposium* 121, 657-663.

Taylor, D.L. (1981) Fire history and fire records for Everglades National Park, 1948-1979. South Florida Research Center. Report T-619. Homestead, FL, USA, 121 pp.

Tebeau, C.W. (1968) *Man in the Everglades: 2,000 Years of Human History in the Everglades National Park.* Copeland Studies in Florida History, No. 3. University Miami Press, Coral Gables, FL, USA, 192 pp.

Tebeau, C.W. (1974) Exploration and early descriptions of the Everglades, Lake Okeechobee, and the Kissimmee River. In: Gleason, P.J. (ed.), *Environments of South Florida: Present and Past.* Memoir 2, Miami Geological Society, Miami, FL, USA, pp. 1-7.

U.S. Congress. (1911) *Everglades of Florida.* 62nd Congress. Doc. 89. Government Printing Office. Washington, DC, USA. 208 pp.

van Rooy, P.T.J.C., Anderson, D.L. and Verstraelen, P.J.T. (1993) Dimensions of integrated water management in the Netherlands and the United States of America. *HyData* 12, 1-3.

Volk, B.G. (1972) Everglades Histosol subsidence. 1. CO_2 evolution as affected by soil type, temperature, and moisture. *Soil and Crop Science Society Florida Proceedings* 32, 132-135.

Whalen, B.M. and Whalen, P.J. (1994) Nonpoint source regulatory program for the Everglades Agricultural Area. Annual Meeting, Florida ASAE, Orlando, FL. *ASAE Paper* FL94-101, St. Joseph, MI, USA.

Whalen, P.J. and Whalen, B.M. (1996) Nonpoint source best management practices program for the Everglades Agricultural Area. Annual Meeting, Florida ASAE, Phoenix, AR. *ASAE* Paper no. 962071, St. Joseph, MI, USA.

Water Pollution Control Federation (1990) *Natural Systems for Wastewater Treatment.* Manual of Practice FD-16. Water Pollution Control Federation, Alexandria, VA, USA.

Water Pollution Control Federation (1991) *Nutrient Control. Manual of Practice FD-7 Facilities Design.* Water Pollution Control Federation, Alexandria, VA, USA.

Chapter twenty-two:

Potential Impact of Sugarcane Production on Riparian and Freshwater Environments

A.H. Arthington[1], J.C. Marshall[1], G.E. Rayment[2], H.M. Hunter[2] and S.E. Bunn[1]

1. *Centre for Catchment and In-Stream Research, Griffith University, Nathan, Qld 4111, Australia*
2. *Resource Sciences Centre, Department of Natural Resources, 80 Meiers Road, Indooroopilly, Qld 4068, Australia*

Abstract

The catchments of many lowland streams in tropical and subtropical Queensland, and northern New South Wales have been extensively cleared for cultivation of sugarcane. This narrow coastal strip is also under considerable pressure from a wide range of non-agricultural land uses. Economic and social benefits from agricultural, urban and tourism development in the coastal zone have been achieved at a high cost to the natural environment. This paper demonstrates how the cultivation of sugarcane may interfere with and modify the functional linkages between vegetated riparian zones, the surrounding land, streams, riverine floodplains and the adjacent marine environment. Three consequences of sugarcane production have been particularly detrimental in coastal catchments: (i) extensive vegetation clearing in the riparian zones of rivers and floodplain wetlands, (ii) soil erosion and stream sedimentation, and (iii) contamination of waterbodies with nutrients, pesticides and other discharges from diffuse sources. Hydrological impacts related to irrigation water use and regulation of river flows are also of concern. Large knowledge gaps suggest the need for biological monitoring and ecological research to enhance environmental management and ensure ecological, as well as economic sustainability of the sugar industry.

In: Keating, B.A. and Wilson, J.R. (eds), *Intensive Sugarcane Production: Meeting the Challenges Beyond 2000*. CAB International, Wallingford, UK, 1997.

Introduction

The cultivation of sugarcane has necessitated extensive clearing of the catchments of many lowland streams in tropical and subtropical Queensland and northern New South Wales. This narrow coastal strip is also under considerable pressure from a wide range of non-agricultural land uses. Economic and social benefits from agricultural, urban and tourism development in the coastal zone have been achieved at a high cost to the natural environment.

Major environmental concerns for the sugar industry are: (i) extensive vegetation clearing extending into the vegetated riparian strip along watercourses and floodplain wetlands, (ii) soil erosion and stream sedimentation, and (iii) effects on water quality and biota related to diffuse sources of contaminants. Rising watertables and increasing salinity, saline intrusions and acid sulfate soils pose additional water quality hazards. Hydrological impacts related to irrigation water use, ground and surface water pumping and the regulation of natural river flows are also important.

The following review is principally concerned with the Australian sugar industry, largely for want of sound environmental data for other sugar producing countries. Even the Australian situation suffers from a limited understanding of biological systems and ecological processes in catchments affected by sugarcane production. This paper demonstrates how the various processes associated with the cultivation of sugarcane may interfere with and modify the functional linkages between vegetated riparian zones, the surrounding land, streams, riverine floodplains and the adjacent marine environment.

Riparian and Floodplain Vegetation

Effects of clearing

The catchments of many lowland streams in tropical and subtropical Queensland have been extensively cleared for the cultivation of sugarcane (Russell and Hales, 1993; Sattler, 1993). *Melaleuca* (paper-bark) wetlands once covered large areas of coastal floodplain which are now used to grow sugarcane (Tait, 1994). Before clearing these wetlands would have provided extensive buffer strips protecting coastal river systems, estuaries and shorelines. Clearing, notably for sugarcane cultivation, has left only remnants of this ecosystem (Russell *et al.,* 1996a). The situation is similar in northern New South Wales where sugarcane is often grown right to the banks of streams leaving no natural riparian vegetation (A. Raine, NSW Department of Land and Water Conservation, *pers. comm.*).

Riparian zones in sugarcane areas are frequently narrow and sparsely vegetated and have been invaded by exotic weeds, particularly para grass (*Brachiaria mutica*), bamboo, thunbergia and African tulip trees, which tend to suppress the process of natural revegetation (Johnson *et al.,* 1997). It is likely that riparian zones in such poor condition have suffered a corresponding degradation of their intrinsic ecological values. To appreciate the significance of riparian vegetation clearing and decline, it is necessary to understand the functions of a healthy riparian zone. These include effects on water quality, stream morphology and habitat complexity, provision of shade and energy to the stream ecosystem, and the role of vegetated riparian zones as habitat for wildlife (Bunn, 1993; Campbell, 1993; Catterall, 1993; Cummins, 1993).

Water quality effects

The riparian zone forms a buffer between streams and agricultural systems, performing a filtering function which reduces concentrations of nutrients, pesticides and suspended sediment in runoff water en route to streams. The mechanisms of this filtering function are both physical and biological including deposition, vegetative uptake and conversion by micro-biological processes such as denitrification (Lowrance *et al.,* 1984; Hubbard and Lowrance, 1994). In a similar manner, the riparian zone can act as a sink for stream nutrients carried in solution over the floodplain during floods. Clearing and draining riparian zones destroys their capacity to filter the water entering stream systems.

Effects on stream morphology and habitat complexity

The roots of riparian plants bind stream bank materials and help to reduce slumping. This in effect constrains the width of stream channels and may allow banks to become under-cut without collapsing, thus creating habitat for aquatic organisms (Bunn, 1993). Clearing of riparian vegetation increases the chances of stream channels migrating (Barling and Moore, 1994).

Riparian vegetation contributes woody debris which has an integral and often profound influence on stream morphology and ecological processes in most natural streams (O'Connor, 1986). The surface of the wood itself provides habitat for invertebrates (O'Connor, 1991; Sweeney, 1993) and this is particularly significant in large silty and sandy rivers where woody debris can be the major invertebrate habitat and is consequently also very important as feeding habitat and shelter for fish (Benke *et al.,* 1984).

Effects on riparian shading

Riparian vegetation provides shade and reduces both light and heat reaching the stream. Riparian shading also alters the wavelength of light reaching streams, filtering out much of the ultraviolet spectrum and favouring green in the visible spectrum. It is likely that the altered wavelength of incident light favours algae with certain pigment classes (Sweeney, 1993). Streams cleared of riparian vegetation in agricultural areas are often choked by the expansive growth of aquatic and semi-aquatic plants (Knight and Bottorff, 1981). In sugarcane areas of north Queensland, the exotic semi-aquatic ponded pasture grasses *Brachiaria mutica* and *Hymenachne amplexicaulis* have become major aquatic weeds (Sattler, 1993; Tait, 1994; Bunn *et al.,* 1997).

Water temperature increases in unshaded streams (Knight and Bottorff, 1981; Pearson and Penridge, 1992), and this is frequently associated with a reduction in dissolved oxygen levels in the water and the progressive death of sensitive aquatic organisms. This can be particularly significant in north Queensland during the dry season when natural biological oxygen demand can be high. An increase in water temperature at this time may eliminate some taxa from a stream (Pearson and Penridge, 1992). In unshaded streams with dense growth of aquatic plants, trophic status may fluctuate between production and respiration depending on cloud cover; several overcast days in succession may result in anoxia in such streams (Murphy *et al.*, 1981; Bunn *et al.*, 1997).

Shade also renders fish less visible to predators such as birds and thus provides them with some protection (Koehn, 1992). This may be especially important in streams with little structural in-stream cover.

Effects on energy input

Leaves falling and washing into streams from the riparian zone may contribute as much as 99% of the energy input into forested headwater streams (Knight and Bottorff, 1981). Clearing the riparian zone decreases inputs of leaf litter and woody debris and reduces populations of aquatic invertebrates (Knight and Bottorff, 1981; Delong and Brusven, 1994). Other inputs from the riparian zone into streams may be important food sources for aquatic animals. Terrestrial insects falling into the water of Australian streams form a major component of the diets of surface living insects such as whirly-gig beetles (Gyrinidae), freshwater turtles, freshwater crocodiles and many species of fish (Pusey and Kennard, 1995).

A poorly understood yet important management issue is the consequence of converting whole catchments from forest to sugarcane on the flux of carbon to streams and rivers. Recent studies using stable isotope tracing have identified that carbon from C4 plants (e.g. grasses like *Brachiaria mutica* and sugarcane) is not taken up into aquatic food webs (Hamilton *et al.*, 1992; Bunn and Boon 1993, Forsberg *et al.*, 1993; Bunn *et al.*, 1997). The reason is unknown, but neither aquatic herbivores nor detritivores consume C4 carbon directly, as living or dead plant matter, nor indirectly via a microbial loop. This is despite much of the organic matter in the streams in canelands being of C4 origin (Bunn *et al.*, 1997).

This raises a very important issue for the protection and management of streams draining canelands. What is the fate of carbon (dissolved organic matter, fine- and coarse-particulate organic matter) entering streams from canelands? Some of it accumulates in the channel (in pools and trapped sediments) and contributes to the observed high microbial respiration (Bunn *et al.*, 1997). The latter leads to deterioration in water quality through low dissolved oxygen and potential release of nutrients and other contaminants from anoxic sediments. The microbial breakdown product of this unconsumed organic matter is likely to include methane, which also raises questions about greenhouse gas emissions. The fate of this unconsumed source of organic matter in streams is likely to be as, if not more, important a management issue as nutrient exports from canelands into streams, rivers and receiving estuaries.

Effects on wildlife habitat

Riparian ecosystems are often characterised by higher biological diversity and productivity than adjacent ecosystems (Decamps *et al.,* 1990; Catterall, 1993). Variation in climatic factors such as temperature and humidity are often less extreme here than in adjacent ecosystems (Catterall, 1993). The adults of many insects with aquatic larvae complete metamorphosis, rest, feed, disperse and mate in the riparian zone. Others lay their eggs on vegetation overhanging the stream from where the larvae hatch and drop into the water (Knight and Bottorff, 1981). Vegetation loss may greatly increase the mortality and reduce the reproductive output and dispersal potential of the adults of insects with aquatic larvae (Sweeney, 1993).

Riparian zones may also act as critical refuges for the maintenance and replenishment of wildlife populations during times of drought or after fire (Catterall, 1993). Riparian habitats provide important corridors for wildlife movement and dispersal, especially in catchments cleared for human use where zones of riparian vegetation link bushland remnants (Noss, 1983; Catterall, 1993).

Remnant wetland habitats in the coastal province land system of the Tully-Murray catchments, such as bulkuru sedge swamps, are important as nursery areas for silver barramundi (*Lates calcarifer*), and waterfowl are also dependent upon these wetlands. The Mahogany Glider (*Petaurus gracilis*), once believed extinct, was rediscovered in the lower Murray area in 1989 and now occupies a very restricted range (Tait, 1994). Clearing on the Murray floodplain since 1992 has resulted in the loss of significant areas of Cassowary (*Casuarius casuarius*) habitat (Tait, 1994). Further clearing of lowland riparian wetlands and floodplain vegetation must be questioned considering the generally low agricultural suitability of these areas and their important functional and habitat roles. It has been suggested that they be conserved and incorporated into farm drainage systems (Tait, 1994).

Agricultural benefits of riparian vegetation

As well as the many intrinsic ecological advantages reviewed above, there are practical agricultural benefits to be gained from conserving and restoring riparian vegetation adjacent to sugarcane farms. Riparian zones, particularly riparian wetlands, act as sponges, soaking up water and minimising flood peaks following heavy rainfall. They help reduce soil loss from bank erosion (Moss *et al.,* 1996) and from runoff. The presence of well vegetated riparian zones along natural waterways increases the abundance of beneficial insects in adjacent crops (Lowrance *et al.,* 1985). In north Queensland, the canefield rat (*Rattus sordidus*) and Burton's climbing rat (*Melomys burtoni*) cause severe damage to sugarcane by chewing through cane stalks, leaving access points for bacterial infection. Rats can be controlled within two years by establishing closed canopy vegetation in harbourage areas (riparian zones, drainage lines and hillsides). Shade trees impede grass and weed growth, providing a long-term solution to rat control, and less effective and costly methods (slashing, grazing and burning, and spraying with herbicides) can be replaced (Tucker and Brodie, 1996).

Riparian zone management

There is opportunity in Queensland for better protection of riparian ecosystems under several acts such as the Land Amendment Act 1992 and the Nature Conservation Act 1992. In some circumstances the planning provisions of local authorities may also offer protection. For example, the Johnstone Shire Council protects vegetation within 30m of minor water courses and 40m of major watercourses (Tait, 1994). Recent examples of the sugar industry acknowledging the importance of riparian ecosystems are found in the preliminary water management scheme for the Riversdale-Murray valley (Anon., 1995) and the water management projects proposed for the Russel and Mulgrave catchments under the Sugar Industry Infrastructure Package (Russell *et al.*, 1994).

Soil Erosion and Sedimentation

Steep slopes, erodible soils and high rainfall contribute to very high erosion rates in some areas of sugarcane production along the wet tropical coast of north Queensland (Prove *et al.*, 1986). Annual soil losses exceeding 380t ha^{-1} have been recorded in catchments of this region (Prove *et al.*, 1996) compared to erosion rates from natural rainforest areas of only 4t ha^{-1} (Capelin and Prove, 1983). Despite the importance of soil erosion, the sources, quantities and precise fate of terrestrially derived sediment within the major storage compartments of coastal catchments are poorly understood (Pailles and Moody, 1995).

Eroded soil transported through a river basin may be deposited on cultivated lands, within local drainage lines, within stream and river channels, in the estuarine and nearshore zone or further offshore on outer reefs (Arakel *et al.*, 1989). This paper considers the fate and ecological effects of eroded materials within the river network while Crossland *et al.* (1997) consider the effects of sediments deposited offshore and in the Great Barrier Reef lagoon.

In the lower South Johnstone River, Arakel *et al.*, (1989) estimated an average aggradation rate of 300,000t y^{-1}, equivalent to caneland erosion rates of 150t ha^{-1} y^{-1}. The effect of this has been the formation of large sand spits and islands in the river mouth, with minimal sedimentation in the estuary. Aggradation in river channels has caused elevated flood levels in both the upper and lower catchment areas. Soil erosion, sedimentation of stream channels and flooding have also been identified as major problems in the lower catchment of the Russell-Mulgrave river system (Russell *et al.*, 1996a).

Rates of sediment accumulation in streams are influenced by processes interacting to determine the balance of erosion, deposition, resuspension and redeposition. The presence of vegetation may influence this balance by acting as a trap for suspended particles and encouraging aggradation, followed by further colonisation and plant growth. In Bamboo Creek near Innisfail, where 63% of the upper catchment has been cleared for sugarcane, loss of direct riparian shading, high insolation and increased nutrient inputs have led to prolific growths of introduced para grass along banks and into stream channels. A dense matrix of living vegetation and detritus acts as a trap for sediments, causing accumulation to channel depths exceeding 2.5m and marked loss of channel capacity (Bunn *et al.*,1997). Caneland streams probably represent only a temporary store for sediments as high discharge events inevitably lead to scouring of the channel and downstream transport of sediment and associated plant matter, shifting the problem from the freshwater to the marine environment.

Fluvial transport, deposition and redeposition of organic and inorganic materials have a variety of environmental effects. Suspended sediments often lower water temperatures by increasing reflection and this may affect temperature-sensitive species (Ryan, 1991). The dissolved oxygen content of water may decline where organic materials are transported together with sediment particles (Guy and Ferguson, 1970). Suspended solids interfere with light penetration and light dependent processes such as photosynthesis and fish foraging (Bruton, 1985). Suspended matter clogs the fine respiratory membranes of fish gills, and the delicate filaments of invertebrate gills and may cause cellular damage and interference with normal gaseous exchange (Ryan, 1991).

The fate of settled solids is also important. Stream sedimentation is considered a major national problem in Australia (Campbell and Doeg, 1989). Sediment fills in pools, degrades the normal pool-riffle sequence of streams and fills the spaces between particles in the stream bed which are utilised by benthic invertebrates and fish (Ryan, 1991; Koehn, 1992). This may reduce the density of benthic invertebrates and alter community structure (Ryan, 1991). Indiscriminate filter feeders are killed as their guts are filled with indigestible silt (Bruton, 1985). Suspended silt reduces the hatching success of eggs and the survival of larvae of some fish species (Auld and Schubel, 1978) and sensitive fish and invertebrate species may be killed by silt clogging their gills (Berkman and Rabeni, 1987; Doeg *et al.*, 1987; Ryan, 1991).

The impacts of sedimentation and turbidity on ecological processes in streams flowing through sugarcane lands remain to be investigated systematically. The present emphasis on the effects of eroded terrestrial sediments and nutrients on the Great Barrier Reef may have led to oversight of the damage caused in the freshwater environment.

Water Quality Issues Related to Sugarcane Production

Sources of stream pollution

Sugar mill effluent has long been recognised as a source of stream pollution, with reports from the 1960s of low dissolved oxygen levels, increased temperature, odour problems and fish kills occurring downstream of sugar mill discharges (Henry, 1966). Improved effluent management practices have since helped minimise point sources of stream pollution from sugar mills (and other point sources). Some problems have persisted however, such as the depletion of dissolved oxygen levels downstream of discharge from some sugar mills, particularly during periods of low stream flow (Moss and Bennett, 1991).

A recent study in the Burdekin River Irrigation Area (BRIA) has shown that irrigation drainage from older established canelands has higher nutrient concentrations than elsewhere in the BRIA (Congdon and Lukacs, 1995). A study is being undertaken in the BRIA to evaluate the ability of artificial wetlands to improve the quality of irrigation drainage, prior to its discharge into downstream watercourses. Preliminary results suggest that the wetlands may be useful for reducing levels of suspended solids and nutrients (H. Hunter, unpubl. data). Artificial wetlands are also being assessed as a tertiary treatment option for sugar mill effluent, again with encouraging results (Dawson *et al.,* 1995).

Diffuse sources of pollution such as storm runoff are more difficult to quantify and control. In the cane industry, as in agriculture generally, soil erosion and the use of fertilisers and pesticides are key issues (Hunter, 1992).

Nutrients

Nutrients such as nitrogen (N) and phosphorus (P) occur naturally at low concentrations in stream waters (Meybeck, 1982) and form an essential part of aquatic food chains. Elevated concentrations of one or more nutrients, however, can upset the ecological balance of aquatic systems and cause problems of eutrophication, such as algal blooms and excessive growth of aquatic macrophytes (Wood, 1975).

Major expansion of the cane industry has contributed to the increased use of fertilisers in many Queensland coastal catchments. Quantities of fertilisers used annually for sugarcane production in Queensland have increased from negligible amounts in 1910 to around 63,870 tonnes of N and 7670 tonnes of P in 1994, with the industry in 1990 accounting for 71% of N and 55% of P fertilisers used in Queensland coastal catchments from the Mary River north (Pulsford, 1993)

Rainfall in most sugar producing areas, particularly in north Queensland, occurs predominantly in summer and often in high intensity, sometimes cyclonic falls. Catchment runoff generated by these events markedly increases stream concentrations of suspended sediment and nutrients during the resulting high stream flows (Hunter *et al.,* 1996; Mitchell *et al.,* 1996). Moreover, most nutrients transported in flood waters are attached to the suspended sediment load (Hunter, 1996; Mitchell *et al.,* 1996). Sources of sediment in flood waters can include erosion from farm lands, roadsides, urban developments and stream banks, as well as remobilisation of bottom sediments. Clearly, land management strategies that help minimise soil erosion in catchments can contribute significantly to reducing stream contamination by sediment and nutrients (and other sediment bound contaminants such as pesticides). The widespread adoption of greencane trash blanketing practices in some areas of north Queensland has been a very positive step towards reducing soil erosion.

Pesticides and pesticide residues

Sugarcane production, in common with most facets of modern agriculture, relies on a wide range of pesticides for the control of insects, diseases, weeds and other pests. Government departments, local authorities and urban communities also use pesticides for similar reasons. One negative consequence has been the movement and at times persistence of pesticide residues in biotic and abiotic compartments of riparian areas and the aquatic environment.

Hamilton and Haydon (1996) reviewed the environmental fate of pesticides used in the Queensland sugar industry, and for each catchment recommended monitoring for 'priority' pesticides (including degradation products) in ground waters, rivers and estuaries. The order of annual usage of active ingredients (a.i.) was herbicides > insecticides > fungicides, with usage of individual pesticides sometimes differing between catchments. The herbicide atrazine was the most used pesticide (annual usage of 331 tonnes a.i.), while on an a.i. basis, the ratio of herbicides to insecticides exceeds 11:1.

Monitoring of surface waters in the BRIA (Rayment, 1991; G. Rayment and H. Hunter, unpubl. data) indicates the occasional presence of pesticide residues, including traces of superseded organochlorine insecticides and the herbicides 2,4,D, 2,4,5-T (in sediments) and atrazine. Moreover, Bauld *et al.* (1996) have confirmed the occurrence of low concentrations of atrazine in some BRIA ground waters. A survey of 39 stream and ground water sites in the Johnstone River catchment in late 1995 detected few pesticide residues, the main exceptions being the presence of low concentrations of atrazine at seven sites and 2,4-D at 20 sites (Hunter, 1996b; H. Hunter, unpubl. data). In southern Queensland, samples from the Mary River under high-flow conditions showed only trace concentrations of 2,4-D and triclopyr, of twelve herbicide residues tested (Moss, 1996). None of these findings can be linked directly to sugarcane production, although it is an important industry in all three catchments.

Between 1990 and 1993, Russell *et al.* (1996b) detected low concentrations of 2,4-D in some samples of aquatic biota taken from both the Johnstone and Daintree River systems, as well as low levels of 2,4,5-T and atrazine in a few samples from the Johnstone River. Only low concentrations of insecticide residues (DDT and dieldrin) were detected during the study, in 23% of samples from both rivers, in contrast to a much higher detection frequency (69%, predominantly DDT plus metabolites and dieldrin) found in similar samples from the 1970s (Russell *et al.,* 1996b). The encouraging downward trend in positive detections (despite lower levels of detection in recent years) can be attributed partly to changing pesticide formulations, including the withdrawal from use of most organochlorine insecticides. It may also reflect improved pesticide application practices over the last twenty years. The trend needs to continue, as the long term consequences of trace concentrations of pesticides in aquatic ecosystems remain unclear, particularly concerning chronic, sublethal effects and the sensitivity of native flora and fauna. Integrated pest management practices that minimise the need for pesticides warrant strong support.

Acid sulfate soils

Acid sulfate soils (ASS) form when soils with high concentrations of pyritic minerals are drained or otherwise oxidised. Prior to oxidation, these soils are termed potential ASS (PASS) and are of no adverse environmental consequence. Following oxidation, soil pH levels typically drop to < 3.8 (Ford and Calvert, 1970; Dent, 1986), with mobilisation of soluble forms of aluminium and iron.

Both PASS and ASS have an extensive distribution along the Australian coastline (Kawalec, 1973; White *et al.*, 1996), often in association with mangroves and coastal wetlands (Melville *et al.*, 1993). Topographic maps are available showing likely areas of ASS in coastal New South Wales (NSW), while areas of NSW canelands know to be underlain by ASS have been recorded (Nielsen, 1993). Mapping areas of ASS risk has commenced in southern Queensland, where it is suggested around 21% of canelands from Bundaberg to the NSW border could be affected. Despite these estimates, only small areas (mostly < 0.5ha) have hitherto been linked directly to ASS problems in Queensland canelands (Reghenzani, 1995). However, continued clearing of coastal wetlands may expose new areas of PASS to oxidation, a concern raised by Russell and Hales (1993a) in the Johnstone River catchment, where extensive reclamation of freshwater wetlands has occurred over the last 40-50 years.

Off-site impacts from ASS arise when strongly acidic runoff and leachate drain into streams, lowering water pH to < 4.0, with values as low as 2.6-2.9 reported in northern NSW (Nielsen, 1993). The acidity persists until neutralised naturally by mixing with estuarine waters or by the addition of limestone. Drainage and flood mitigation measures, including the installation of flood-gates, can significantly accelerate drainage, while one-way flood-gates also allow the build-up of acidic waters, particularly in dry times (White *et al.*, 1996).

In northern NSW, acidified drain waters have been linked to the prevalence of epizootic ulcerative syndrome (red-spot disease) in estuarine fish species, particularly when rainfall follows extended dry periods (Sammut *et al.*, 1993). Major fish kills in the Tweed River catchment were recorded as early as 1880 (Nielsen, 1993), while a major event in 1987 affected about 23km of the Tweed River estuary, causing the death of fish, crustaceans and annelid worms (Easton, 1989). Fortunately, mobile aquatic fauna actively escape the acidic water when not impeded by barrages and floodgates (Sammut *et al.*, 1993). In contrast, the empire gudgeon (*Hypseleotris compressa*) and a possible new species of gudgeon (*Hypseleotris* sp.) search purposefully for acidic drainage and, with acclimatisation, can exist for extended periods at pH levels as low as 4 (J. Sammut, unpublished data). The existence of such species points to the longevity of acid drainage (White *et al.*, 1996). Nevertheless, land management practices designed to minimise the oxidation of ASS to levels that prevent major fish kills is a desirable environmental goal.

Heavy metals

Heavy metals are important environmental pollutants when present in sufficient quantities to cause recognisable toxic effects or a diminution of amenity and quality of life (Livett, 1988). At lower concentrations, some heavy metals are essential plant or animal nutrients. Little is known of baseline levels of heavy metals in Australian canelands, although early indications are that their concentrations are generally within normal limits (G.E. Rayment, unpublished data). The main exception is cadmium (Cd) in surface soils from sites with a long history of cane production, where concentrations are up to seven-fold higher than corresponding unfarmed sites. The cause of high Cd levels is probably its presence as an impurity in phosphatic fertilisers, particularly those used prior to 1992. Estimated inputs of Cd from this source vary from around 340g ha^{-1} for the Barron-Daintree and Murray-Mulgrave catchments to 38g ha^{-1} for the Burdekin-Haughton (Rayment, 1995). Nevertheless, all sites tested to date contain < 1mg Cd kg^{-1}, a level considered representative of background concentrations in surface soils of rural Australia (Tiller, 1992).

Heavy metals can be mobilised in the environment by human activity (Rayment and Barry, 1993) and it is suspected, but not yet confirmed, that cane fires could dissipate Cd offsite, while trash blanketing may, over time, concentrate heavy metals such as Cd at the soil surface, making it more vulnerable to erosion. However, agricultural activities are generally considered to have little influence on heavy metal loadings in streams, compared with other sources, such as mining and industrial effluent (Garman and Sutherland, 1983). Weathering of natural minerals may also be a source of heavy metals in streams, as suggested, for example, by results of Russell and Hales (1993b), who found violative levels of heavy metals in occasional samples of aquatic biota from both the Johnstone and Daintree Rivers in north Queensland.

Water quality monitoring

There is presently no fully coordinated ambient water quality monitoring program for canelands in Australia. New South Wales has several monitoring sites located in the Tweed, Richmond and Clarence River catchments at which electrical conductivity, turbidity and total P are monitored (Daly *et al.,* 1995), but all present sites are located upstream of cane growing areas in these catchments. Queensland now has monitoring sites in all major sugarcane catchments, with at least one site per catchment located within a caneland area (W. Poplawski, *pers. comm.*). Monitoring in these catchments includes nutrients, turbidity, dissolved oxygen, pH and major ionic constituents, with plans to also include pesticide residues. Most records date from 1992 or later, too recent for detection of any water quality trends.

Irrigation water use, flow regulation and environmental flows

Sugarcane is the most irrigated crop in Queensland, accounting for > 40% of all irrigation (Anon., 1996). Irrigation is essential for crop growth in the Burdekin district and is necessary for consistent and profitable cane growing in many other areas. Construction of major irrigation schemes has facilitated recent expansion of the industry into newer and more marginal areas (Anon., 1996).

Infrastructure development and the allocation of large quantities of river flow for irrigation purposes have serious environmental implications (Petts, 1989). These are most apparent in basins such as the Murray-Darling, where irrigation water use is widely regarded as a key factor in the development of blue-green algal blooms, proliferation of pest species (e.g. introduced carp), decline of rare indigenous species, shrinking wetland systems and a widescale loss of biological diversity and ecosystem function. In addition, accessions to groundwaters from low efficiency water delivery and application have caused water tables to rise, predisposing large areas to increasingly wet and saline soils and declining productivity (Meyer, 1997).

Single crop industries such as cotton tend to be the target of blame associated with irrigation water use and the deteriorating condition of rivers and wetlands. The sugar industry as a major user of irrigation water in Queensland is therefore vulnerable in this aspect of its operations. The extent to which the sugar industry has impacted upon river flow regimes could be assessed relatively easily from stream flow records, whereas impacts on aquatic ecosystems are more difficult, although not impossible, to evaluate.

The allocation of river and ground water for environmental purposes is guided by principles set down by the Council of Australian Governments (COAG). These principles imply that further allocation of water for consumptive use will be severely constrained in many areas of Australia, as is now occurring in the Murray-Darling Basin. The sugar industry must be prepared to operate in this new policy environment.

By-Products of Sugarcane Production

Processing of sugarcane results in various by-products: bagasse, ash and filter mud, ethanol and molasses. Virtually all by-products from most Queensland sugar mills are recycled, adding to the efficiency of milling and benefiting the natural environment (Anon., 1996). Bagasse provides most of the fuel required for steam and electricity generation at sugar mills plus about 20 megawatts of surplus power supplied to the Queensland electricity grid. From 1998, five Queensland mills will provide an additional 49 MW of power per annum to the grid under a $70M 10-year agreement with the Queensland Electricity Commission. This will slightly reduce the quantity of coal burnt and reduce greenhouse gases (Anon., 1996). Bagasse is also being trialed in stock feed, and in some areas is used as garden mulch; its use in paper production is under consideration. Ash and filter mud are used as soil conditioners on cane farms and gardens.

A new proprietary distillation process 'Biostil' results in a concentrated form of dunder used as fertiliser. Clayton and Pearson (1992) investigated the effects of dunder applications on water quality in the Sarina/Mackay region. Stream sites potentially affected by dunder application were significantly more productive and had higher nitrogen and phosphorus loadings than other stream areas. Some invertebrate taxa were more abundant and others less abundant at these sites but the differences were rapidly ameliorated with time and could not be related to the timing and location of dunder applications (Clayton and Pearson 1992).

A Forward View of Environmental Issues for the Sugar Industry

The geographic extent of sugarcane cultivation in Australia is regarded as a key factor in the success of the industry (Anon., 1996). However, from an environmental perspective, this geographic extent and its concentration in a narrow coastal strip usually within 50km of the sea, increase and complicate the demands on the industry as it strives to become ecologically sustainable as well as economically efficient. The ecosystems of this coastal strip are highly modified, fragmented, generally degraded and not particularly well protected by conservation estates. Existing sugarcane assignments and areas of expansion in the future will be forced to compete with many other forms of land use in the coastal zone, all having an impact on the natural environment. When many industries and sectors contribute to environmental degradation it is difficult to assign responsibility, or to understand the relationship between cause and effect. The 'tragedy of the commons' will tend to prevail unless every industry accepts some of the responsibility for environmental degradation.

The sugar industry is expanding, often into marginal lands where the cost of preventing or repairing environmental damage may be high and commercial returns relatively low. Increasing demands for irrigation will add to the potential for environmental side-effects in marginal lands; salinity problems, waterlogging and saline intrusions are more likely to emerge in such areas.

Our greatest difficulty in preparing this review has been the dearth of quantitative information about biological systems and ecological processes in environments relevant to the sugar industry. This suggests that insufficient attention has been paid in the past to monitoring and environmental research in sugarcane lands. To date, monitoring of the downstream effects of the sugar industry has focused on water quality issues and improvements in the containment and management of pollution will undoubtedly result from this work. In the future, however, the use of sensitive, robust indicators of riparian and aquatic ecosystem condition will be equally if not more important. Biological monitoring will be especially valuable in the assessment of cumulative stress (e.g. Wright *et al.,* 1993). Indicators which measure aspects of ecosystem function, such as the production to respiration ratio, may also be appropriate in caneland streams with highly modified riparian zones and food web structure (Bunn, 1995). All indicators have their limitations and a suite of different measures and indicator systems will be more powerful and informative than the few commonly applied in water quality assessments (Bunn, 1995). The whole issue of monitoring stream condition in canelands will require a substantive review before a suitable program can be recommended.

Australia is one of 120 countries producing sugar, with China and India being two of the largest sugar producing countries. The latter countries do not have the capacity, incentive or wealth to address environmental issues to the extent that is possible in Australia. The Australian sugar industry is in a strong position to set an example for other sugar producing nations by establishing environmental monitoring, rehabilitation and protection programs, as well as a relevant research agenda. Sugar R&D expenditure in Australia in 1995 was over $32M (Anon., 1996) and the program is comprehensive. Nevertheless, environmental research will need to receive a larger share of this commitment if we are to ensure real improvements in environmental management. A wider appreciation and scientific understanding of how the sugar industry affects riparian and aquatic ecosystems will help to ensure rapid progress towards the industry's mission of environmental responsibility, as well as economic sustainability, as enunciated in Vision 2000.

Acknowledgments

Angela wishes to thank Dr Brian Keating for his oversight of this contribution to Sugar 2000. We also thank the following colleagues for providing information and other forms of assistance: Dr Peter Chudleigh, Dr Paul Clayton, Mr Alf Hogan, Dr Anne Kemp, Ms Romeny Lynch, Mr Glen Moller, Ms Julie Murdoch, Dr Ngaire Phillips, Dr Brad Pusey, Mr Allen Raine, Dr David Russell, Ms Donna Smith, Mrs Pam Soper, Dr Jim Tait, Mr Stephen Tapsall and Mr Damon Telfer. Funding for this review was provided by CSIRO Tropical Agriculture, and the Centre for Catchment and In-Stream Research, Griffith University, Brisbane.

References

Anon. (1995) *Draft report of the project steering committee on the formation of the Riversdale-Murray valley water management board incorporating the preliminary design report for the Riversdale-Murray valley water management scheme*. Sugar Industry Infrastructure Package, Riversdale-Murray valley infrastructure project, Draft for Consultation.

Anon. (1996) *Information Handbook - 1996*. CANEGROWERS, Brisbane, Australia.

Arakel, A.V., Hill, C.M., Piorewicz, J. and Connor, T.B. (1989) Hydro-sedimentology of the Johnstone River estuary. In: Sly, P.G. and Hart, B.T. (eds), *Sediment Water Interaction. Hydrobiologia* 176/177, 51-60.

Auld, A. and Schubel, J.R. (1978) Effects of suspended solids on fish eggs and larvae: A laboratory assessment. *Estuarine Coastal and Shelf Science* 6, 153-164.

Barling, R.D. and Moore, I.D. (1994) Role of buffer strips in management of waterway pollution - a review. *Environmental Management* 18, 543-558.

Bauld, J., Leach. L.L. and Sandstrom, M.W. (1996) Impact of land use on groundwater quality in the Burdekin River Irrigation Area. In: Hunter, H.M., Eyles, A.G. and Rayment, G.E. (eds), *Downstream Effects of Land Use.* Department of Natural Resources, Brisbane, Australia, pp. 195-106.

Benke, A.C., Van Arsdall, T.C., Gillespie, D.M. and Parrish, F.K. (1984) Invertebrate productivity in a subtropical blackwater river: The importance of habitat and life history. *Ecological Monographs* 54, 25-63.

Berkman, H.E. and Rabeni, C.F. (1987) Effects of siltation on fish communities. *Environmental Biology of Fishes* 18, 285-294.

Bruton, M.N. (1985) The effects of suspensoids on fish. *Hydrobiologia* 125, 221-241.

Bunn, S.E. (1993) Riparian-stream linkages: Research needs for the protection of in-stream values. *Australian Biologist* 6, 46-51.

Bunn, S.E. (1995) Biological monitoring of water quality in Australia: Workshop summary and future directions. *Australian Journal of Ecology* 20, 220-227.

Bunn, S.E. and Boon, P.I. (1993) What sources of organic carbon drive food webs in billabongs? A study based on stable isotope analysis. *Oecologia* 96, 85-94.

Bunn, S.E., Davies, P.M. and Kellaway, D.M. (1997) Contributions of sugar cane and invasive pasture grass to the aquatic food web of a tropical lowland stream. *Marine and Freshwater Research* 48: 173-179.

Campbell, I.C. (1993) Riparian stream linkages: an Australian perspective on in-stream issues. In: Bunn, S.E., Pusey, B.J., Price, P. (eds), *Ecology and Management of Riparian Zones in Australia.* Land and Water Resources Research and Development Corporation Occasional Paper Series No: 05/93, Canberra, Australia, pp. 21-30.

Campbell, I.C. and Doeg, T.J. (1989) Impact of timber harvesting and production on streams: a review. *Australian Journal of Marine and Freshwater Research* 40, 519-39.

Capelin, M.A. and Prove, B. (1983) Soil conservation problems of the Humid Coastal Tropics of North Queensland, *Proceedings of Australian Society of Sugarcane Technologists* 5, 125-130.

Catterall, C.P. (1993) The importance of riparian zones to terrestrial wildlife. In: Bunn, S.E., Pusey, B.J. and Price, P. (eds), *Ecology and Management of Riparian Zones in Australia.* Land and Water Resources Research and Development Corporation Occasional Paper Series No: 05/93, Canberra, Australia, pp. 41-52.

Clayton, P.D. and Pearson, R.G. (1992) *Instream Effects of Dunder Application to Canefields in the Sarina Region, Queensland.* Australian Centre for Tropical Freshwater Research, James Cook University, Townsville, Australia.

Congdon, R.A. and Lukacs, G.P. (1995) *Limnology and Classification of Tropical Floodplain Wetlands, with Particular Reference to the Effects of Irrigation Drainage. Part 1, Effects of Irrigation Drainage.* James Cook University, Australian Centre for Tropical Freshwater Research Report No. 95/12, Townsville, Australia.

Crossland, C.J., Done, T.J. and Brunskill, G.T. (1997) Potential impact of sugarcane production on the marine environment. In: Keating, B.A. and Wilson, J.R. (eds), *Intensive Sugarcane Production: Meeting the Challenges Beyond 2000.* CAB International, Wallingford, UK, pp. 423-436.

Cummins, K.W. (1993) Riparian stream linkages: In-stream issues. In: Bunn S.E., Pusey B.J., and Price P. (eds), *Ecology and Management of Riparian Zones in Australia.*

Land and Water Resources Research and Development Corporation Occasional Paper Series No: 05/93, Canberra, Australia, pp. 5-20.

Daly, H., Martin, V. and Bales, M. (1995) *Window on Water, the State of Water in NSW 1993/1994*. Department of Land and Water Conservation (Water Resources), Water Resources Division, Parramatta, NSW, Australia.

Dawson, M.W., Lukacs, G.P. and Sockhill, D.G. (1995) Establishment of a constructed wetland at Victoria Mill. *Proceedings of the Australian Society of Sugarcane Technologists* 17, 270-278.

Decamps, H., Fornier, F., Naiman, R.J. and Peterson, R.C. (1990) An international research effort on land/inland water ecotones in landscape management and restoration 1990-96. *Ambio* 19, 175-176.

Delong, M.D. and Brusven, M.A. (1994) Allochthonous input of organic matter from different riparian habitats of an agriculturally impacted stream. *Environmental Management* 18, 59-71.

Dent, D. (1986) *Acid Sulfate Soils: a Baseline for Research and Development*. Publication No. 39. International Institute for Land Reclamation and Improvement, Wageningen, The Netherlands.

Doeg, T.J., Davey, G.W. and Blyth, J.D. (1987) Response of the aquatic macroinvertebrate communities to dam construction on the Thomson River, Southeastern Australia. *Regulated Rivers: Research and Management* 1, 195-209.

Easton, C. (1989) The trouble with the Tweed. *Fishing World*, March 1989, 58-59.

Ford, H.W. and Calvert, D.V. (1970) A method for estimating the acid sulfate potential of Florida soils. *Soil and Crop Science Society Florida Proceedings* 30, 304-7.

Forsberg, B.R., Araujo-Lima, C.A.R.M., Martinelli, L.A., Victoria, R.L. and Bonassi, J.A. (1993) Autotrophic carbon sources for the fish of the central Amazon. *Ecology* 74, 643-52.

Garman, D.E.J. and Sutherland, P.D. (1983) Pesticides, fertilisers, heavy metals and salinity - an assessment of the non-point source problems in Australia. In: *Workshop on Non-point Sources of Pollution in Australia Proceedings*. Australian Water Resources Council Conference Series No. 9. AGPS, Canberra, Australia, pp. 27-47.

Guy, H.P. and Ferguson, G.E. (1970) Stream sediment: An environmental problem. *Journal of Soil and Water Conservation* 25, 217-221.

Hamilton, D. and Haydon, G. (1996) *Pesticides and Fertilisers in the Queensland Sugar Industry - Estimates of Usage and Likely Environmental Fate. Vol. 1. Pesticides.* Department of Primary Industries, Brisbane, Australia.

Hamilton, S.K., Lewis, W.M. and Sippel, S.J. (1992) Energy sources for aquatic animals in the Orinoco River floodplain: evidence from stable isotopes. *Oecologia* 89, 324-30.

Henry, L. (1966) *Pollution of Coastal Streams in Queensland*. Unpublished Report, Queensland Department of Local Government, Brisbane, Australia

Hubbard, R.K. and Lowrance, R.R. (1994) Riparian forest buffer system research at the Coastal Plain Experiment Station, Tifton, G.A. *Water, Air and Soil Pollution* 77, 409-432.

Hunter, H.M. (1992) *Agricultural Contaminants in Aquatic Environments - A Review*. Queensland Department of Primary Industries Bulletin QB92002, Brisbane, Australia.

Hunter, H.M. (1996a) Nutrients and suspended sediment discharged from the Johnstone River catchment during cyclone Sadie. *Proceedings of Workshop on Cyclone Sadie, Australian Institute of Marine Science, Townsville, Nov. 1994*. GBRMPA Workshop Series.

Hunter, H.M. (1996b) *Water quality in the Johnstone River Catchment.* Unpublished Catchment Care Leaflet, Department of Natural Resources, Brisbane, Australia.

Hunter, H.M., Walton, R.S. and Russell, D.J. (1996) Contemporary water quality in the Johnstone River catchment. In: Hunter, H.M., Eyles, A.G. and Rayment, G.E. (eds), *Downstream Effects of Land Use.* Department of Natural Resources, Brisbane, Australia, pp. 339-345.

Johnson, A.K.L., McDonald, G.T., Shrubsole, D.A. and Walker, D.H. (1997) Sharing the land - the sugar industry as part of the wider landscape. In: Keating, B.A. and Wilson, J.R. (eds), *Intensive Sugarcane Production: Meeting the Challenges Beyond 2000.* CAB International, Wallingford, UK, pp. 361-380.

Kawalec, A. (1973) World distribution of acid sulfate soils, references and map. In: Dost, H. (ed.), *Acid Sulphate Soils, I. Introductory Papers and Bibliography.* Proceedings of the International Symposium, Wageningen, 13-20 August, 1972 Wageningen. International Institute for Land Reclamation and Improvement, Wageningen, The Netherlands, pp. 292-295.

Knight, A.W. and Bottorff, R.L. (1981) The importance of riparian vegetation to stream ecosystems. In: Warner, R.E., Hendrix, L.M. (eds), *California Riparian Systems Ecology.* University of California Press, Davis, California, USA, pp. 160-167.

Koehn, J. (1992) Habitat: the essential component of fisheries management. *Freshwater Fisheries in Victoria: Today and Tomorrow Symposium.* 11 Oct 1992, Melbourne. Victorian Department of Conservation and Natural Resources, Melbourne, Australiam, pp. 5-20.

Livett, E.A. (1988) Geochemical monitoring of atmospheric heavy metal pollution: theory and applications. *Advances in Ecological Research* 18, 65-177.

Lowrance, R.R., Leonard, R. and Sheridan, J. (1985) Managing riparian ecosystems to control nonpoint pollution. *Journal of Soil and Water Conservation* 40, 87-91.

Lowrance, R.R., Todd, R., Fail, J., Hendrickson, O., Leonard, R. and Amussen, L.E. (1984) Riparian forests as a nutrient filter in agricultural watersheds. *Bioscience* 34, 374-377.

Melville, M.D., White, I. and Lin, C. (1993) The origins of acid sulfate soils. In: Bush, R. (ed.), *Proceedings of National Conference on Acid Sulfate Soils.* Coolangatta, 24-25 June, 1993, NSW Agricultural Research Institute, Wollongbar, Australia, pp. 19-25.

Meybeck, M. (1982) Carbon, nitrogen and phosphorus transport by world rivers. *American Journal of Science* 282, 401-450.

Meyer, W.S. (1997) The irrigation industry in Australia - lessons for the sugar industry. In: Keating, B.A. and Wilson, J.R. (eds), *Intensive Sugarcane Production: Meeting the Challenges Beyond 2000.* CAB International, Wallingford, UK, pp. 437-454.

Mitchell, A.W., Reghenzani, J.R., Hunter, H.M. and Bramley, R.G.V. (1996) Water quality and nutrient fluxes from river systems draining to the Great Barrier Reef Marine Park. In: Hunter, H.M., Eyles, A.G. and Rayment, G.E. (eds), *Downstream Effects of Land Use.* Department of Natural Resources, Brisbane, Australia, pp. 23-33.

Moss, A. J. (1996) *Flood Related Pesticide Monitoring in the Mary River and Hervey Bay.* Interim Unpublished Report, Department of Environment, Brisbane, Australia.

Moss, A. J. and Bennett, J. (1991) Focus on some north Queensland water quality issues. *Water*, December issue, 16-19.

Moss, A.J., Bennett, J., Poplawski, W., Shar, R. and Moller, G. (1996) Land use factors affecting the condition of rivers, estuaries and bays in southern Queensland. In: Hunter, H.M., Eyles, A.G. and Rayment, G.E. (eds), *Downstream Effects of Land Use.* Department of Natural Resources, Brisbane, Australia, pp. 35-44.

Murphy, M.L., Hawkins, C.P. and Anderson, N.H. (1981) Effects of canopy modification and accumulated sediment on stream communities. *Transactions of the American Fisheries Society* 110, 469-478.

Nielsen, P.J. (1993) Acid sulphate soils - a sugar industry viewpoint. In: Bush, R. (ed.), *Proceedings of National Conference on Acid Sulfate Soils.* Coolangatta, 24-25 June, 1993, NSW Agricultural Research Institute, Wollongbar, Australia, pp. 57-61.

Noss, R.F. (1983) A regional landscape approach to maintain diversity. *Bioscience* 33, 700-706.

O'Connor, N.A. (1986) Ecological role of wood in streams. *Australian Society for Limnology Congress,* 25th, 9-12 May 1986, Lorne, Abstracts.

O'Connor, N.A. (1991) The effects of habitat complexity on the macro-invertebrates colonising wood substrates in a lowland stream. *Oecologia* 85, 504-512.

Pailles, C. and Moody, P.W. (1996) Phosphorus dynamics of sediments from the Johnstone Rivers. In: Hunter, H.M., Eyles, A.G. and Rayment, G.E. (eds), *Downstream Effects of Land Use.* Department of Natural Resources, Brisbane, Australia, pp. 359-364.

Pearson, R.G. and Penridge, L.K. (1992) *An Ecological Survey of Selected Rivers in Queensland with Particular Reference to the Effects of Sugar Mill Effluents.* Report No. 92/02, Australian Centre for Tropical Freshwater Research, James Cook University of North Queensland, Townsville, Australia.

Petts, G.A. (1989) Perspectives for ecological management of regulated rivers. In: Gore, J.A. and Petts, G.E. (eds), *Alternatives in Regulated River Management.* CRC Press, Boca Raton, Florida, USA, pp. 3-24.

Prove, B.G., Doolan, V.J. and Truong, P.N. (1995) Nature and magnitude of soil erosion in sugarcane land on the wet tropical coast of north-eastern Queensland. *Australian Journal of Experimental Agriculture* 35, 641-649.

Prove, B.G., Truong, P.N. and Evans, D.S. (1986) Strategies for controlling caneland erosion in the wet tropical coast of Queensland. *Proceedings of the Australian Society of Sugarcane Technologists* 8, 77-84.

Pulsford, J.S. (1993) *Historical Nutrient Usage in Coastal Queensland River Catchments Adjacent to the Great Barrier Reef Marine Park.* Research Publication No. 40. Great Barrier Reef Marine Park Authority, Townsville, Australia.

Pusey, B.J. and Kennard, M.J. (1995) *The Direct Utilisation of Allochthonous Plant and Animal Matter by Australian Freshwater Vertebrates.* Report for the Land and Water Resources Research and Development Corporation, Canberra, Australia.

Rayment, G.E. (1991) Monitoring downstream effects of agricultural practices in North Queensland. Australian Marine Science Association Annual Scientific Conference, Brisbane, July 1991, Abstracts No 102.

Rayment, G.E. (1995) Sources of cadmium in agricultural products. *National Cadmium Workshop,* May 1995., DPIE/NFA, Canberra, Australia, (in press).

Rayment, G.E. and Barry, G.A. (1993) Heavy metals: toxicity, sources, chemistry and loading rates. In: Fergus, I.F and Coughlan, K.J. (eds), *Environmental Soil Science,* Australian Society of Soil Science Incorporated, Qld Branch, Brisbane, Australia, pp. 211-235.

Reghenzani, J.R. (1995) *Acid Sulfate Soils in Canegrowing Regions of Northern New South Wales,* Technical Report TE95001, Bureau of Sugar Experiment Stations, Brisbane, Australia.

Russell, D.J. and Hales, P.W. (1993a) *Stream Habitat and Fisheries Resources of the Johnstone River Catchment.* Northern Fisheries Centre, Department of Primary Industries, Cairns, Australia.

Russell, D.J. and Hales, P.W. (1993b) Downstream effects of agricultural chemicals on aquatic biota - the Johnstone River, a case study. Poster Paper, Vol 4. *Proceedings of 15th Federal Convention, Australian Water and Wastewater Association, Gold Coast, 18-23 April, 1993.* Australian Water and Wastewater Association, pp. 834-840.

Russell, D.J., Hales, P.W. and Helmke, S.A. (1994*) Environmental Aspects of the Sugar Industry Infrastructure Package: Russell and Mulgrave River Catchment Projects.* Queensland Department of Primary Industries, Brisbane, Australia.

Russell, D.J., Hales, P.W. and Helmke, S.A. (1996a) *Stream Habitat and Fish Resources in the Russell and Mulgrave Rivers Catchment.* Queensland Department of Primary Industries, Northern Fisheries Centre Cairns. Information Series QI96008, Australia.

Russell, D.J., Hales, P.W. and Moss, A. (1996b) Pesticide residues in aquatic biota from north-east Queensland coastal streams. *Proceedings Royal Society of Queensland* 106, 23-30.

Ryan, P.A. (1991) Environmental effects of sediment on New Zealand streams: a review. *New Zealand Journal of Marine and Freshwater Research* 25, 207-221.

Sammut, J., Callinan, R.B. and Fraser, G.C. (1993) The impact of acidified water on freshwater and estuarine fish populations in acid sulfate environments. In: Bush, R. (ed.), *Proceedings of National Conference on Acid Sulfate Soils.* Coolangatta, 24-25 June, 1993, NSW Agricultural Research Institute, Wollongbar, Australia, pp. 26-40.

Sattler, P.S. (1993) Riparian zone management in Queensland and the Northern Territory: policy and practice. In: Bunn, S.E., Pusey, B.J. and Price, P. (eds), *Ecology and Management of Riparian Zones in Australia.* Land and Water Resources Research and Development Corporation Occasional Paper Series No: 05/93, Canberra, Australia, pp. 157-171.

Sweeney, B.W. (1993) Effects of streamside vegetation on macroinvertebrate communities of White Clay Creek in eastern North America. *Proceedings of the Academy of Natural Sciences of Philadelphia* 144, 291-340.

Tait, J. (1994) *Lowland Habitat Mapping and Management Recommendations: Tully-Murray catchments. Final Report.* An ICM Initiative of the Queensland Department of Primary Industries. Cardwell Shire Catchment Coordinating Committee, Brisbane, Australia.

Tiller, K.G. (1992) Urban soil contamination in Australia. *Australian Journal of Soil Research* 30, 937-957.

Tucker, N. and Brodie, A. (1996) Rats have an expensive sweet tooth. *RIP-RAP* 4, 2-3.

White, I., Melville, M.D., Sammut, J., Wilson, B.P. and G.M. Bowman, G.M. (1996) Downstream impacts from acid sulfate soils. In: Hunter, H.M., Eyles, A.G. and Rayment, G.E. (eds), *Downstream Effects of Land Use.* Department of Natural Resources, Brisbane, Australia, pp. 165-172.

Wood, G. (1975) *An Assessment of Eutrophication in Australian Inland Waters.* Aust. Water Resources Council Tech. Paper No. 15., Australian Government Publishing Service, Canberra, Australia.

Wright, J.F., Furse, M.T. and Armitage, P.D. (1993): RIVPACS - a technique for evaluating the biological quality of rivers in the UK. *European Water Pollution Control* 3(4), 15-25.

Chapter twenty-three:

Potential Impacts of Sugarcane Production on the Marine Environment

C. J. Crossland[1], T. J. Done[2] and G. J. Brunskill[2]

1. *CRC Reef Research Centre, Townsville, Qld 4811, Australia*
2. *Australian Institute of Marine Science, Townsville, Qld 4810, Australia*

Abstract

Human settlements and land use practices in catchments impact on coastal wetlands, estuaries and coastal environments. Activities which influence coastal habitats include: urbanisation and sewage, dredging and infrastructure, habitat alienation (e.g. mariculture in wetlands), agriculture (cropping and grazing), and forestry. Intensive agricultural practices contribute significantly to pollution of coastal waters, particularly through increased sediment erosion and the application of fertilisers and agrochemicals.

Sugarcane is a mainly tropical crop grown on islands and coastal catchments. Tropical rainfall patterns and cane production practices are conducive to soil erosion. The agricultural practices, which include application of fertilisers (N, P), pesticides and herbicides, and trace metals often associated with fertilisers, affect the quality of both groundwater and surface runoff to estuarine and coastal environments. Australia's sugarcane is grown on coastal plains along 2100km of the eastern coastline; 95% in Queensland and 5% in northern New South Wales. We address our current knowledge of sugarcane production and its effects on the marine environment, using Queensland and the Great Barrier Reef (GBR) area as a case study.

In: Keating, B.A. and Wilson, J.R. (eds), *Intensive Sugarcane Production: Meeting the Challenges Beyond 2000*. CAB International, Wallingford, UK, 1997.

Introduction

Concern about agricultural practices, and sugar production in particular, in catchments (Brown, 1986; Hatcher *et al.*, 1989) reflects perception of impacts on the Great Barrier Reef World Heritage Area, especially the effects of sediment and nutrients from fertiliser (Arakel *et al.*, 1993). The concerns may be reflected in at least two areas:

1. Agricultural management to minimise runoff loss downstream from canelands. Ideally, activities in catchments should be managed to minimise the runoff to rivers of valuable agricultural resources (top soil, herbicides, fertiliser, petroleum products). This issue has been addressed by other presentations at this conference.
2. Agrochemical inputs to coastal waters and sediments and their effects on marine ecosystems.

To evaluate the sugar industry's ecological impact, the many factors in need of appraisal include:

- The physics and chemistry of the GBR receiving environment.
- Its marine habitats and their susceptibility to products resulting from sugarcane farming.
- The quantity and form of materials discharged from sugarcane areas to the GBR.
- The relative contribution of other land uses to runoff.

Such appraisals must consider:

- Knowledge of the changing spatial and temporal patterns of the agricultural practices.
- Understanding the scales of change and variability of coastal marine ecosystems.
- Knowledge of the diversity, extent and dynamics of marine communities especially in coastal waters.
- An inventory of sugarcane farming practices including the use of agrochemicals, tillage and harvesting regimes.
- Measurement of coastal sedimentation history, nutrient patterns and loadings to GBR waters.

Potential Versus Actual Impact in the GBR

The extent to which overall anthropogenic additions of nutrients and sediments can be detrimental to coastal systems has been discussed recently (Bell and Tomascik, 1994). For coral reefs, areas of concern are:

1. Coral growth inhibition by phosphorus. Phosphorus can weaken coral skeletons, making corals more susceptible to storm damage.
2. Nitrogen and phosphorus water concentrations increasing algal growth leading to death of live corals and algal invasions of areas where corals have been killed by some other disturbance.
3. Suspended sediments increasing turbidity and excluding light from corals, leading to their death.
4. Settling of sediments smothering and killing corals.
5. Accumulation of sediments on coral reef surfaces preventing settlement of microscopic coral larvae.

In many parts of the world, coral reefs have collapsed as productive and aesthetic ecosystems, a direct result of land-based human pressures and overharvesting of fishes and invertebrates (Ginsburg, 1994). While all the above outcomes can be seen within the GBR, we do not know the extent to which they are exacerbated above natural background levels, or the degree to which their geographic spread has been extended by agricultural activities.

The GBR-Receiving Environment

From the marine perspective, rivers discharging waters from sugar-producing catchments are point-sources of freshwater, sediments, nutrients and other contaminants. Exposure of the marine biota to riverine influence is seasonally variable and is not a function of the 'average year'. For six of Queensland's major catchments, for example, minimum recorded flows range from 4 to 25% of the mean, and maxima from 312 to 577% (Lough, 1994). The consequences of such variability for delivery of materials to the marine systems are being investigated by several research groups (Mitchell *et al.*, 1996). In addition to interannual and interdecadal variability in flows, there is great variability in the nature of the marine systems which come under riverine influence. There are two ways to view the ecological systems:

1. As individual units, whose responses are of intrinsic interest.
2. A 'reef-centred' view, in which mangroves, seagrass beds, pelagic and sea-floor systems act to modify water quality impinging on coral reefs.

Implicit in the latter view is that physical, chemical and biological processes in the mangrove, seagrass, planktonic and sea-floor communities improve the properties of water, sequentially and in parallel, before it reaches the coral reefs, the nearest of which range from 15-30km from the river mouths.

In the context of the Queensland catchments and coastal waters, the opportunities for such processing of the water differ according to differences in geographic and biological settings. For the Herbert River, the extensive mangrove system and large distances to the nearest reefs allow major trapping and stabilisation of riverine sediments and adsorbed nutrients. By comparison, the open coast discharge of the Johnstone River allows minimal processing. Seagrasses, like mangroves, also absorb nutrients through their root systems and a modest increase of sedimentation can benefit existing seagrass beds by providing increased nutrients for their growth.

Effects of Cane-growing on GBR Ecosystems

Physical effects

The sugar industry is responsible for several initiatives which have resulted in changes to sediments and nutrients reaching GBR waters. Damming of rivers (e.g. the Burdekin, Tully and Barron Rivers) has reduced sediment reaching the marine system. Dams have also reduced freshwater flows and extreme flood flows into the sea - i.e. the geographic pattern of river runoff has been permanently changed. Nutrient and sediment loads have generally increased from non-point sources originating from sugarcane-growing areas. Also, the draining of acid sulfate soils in coastal lands for sugar and other agricultural production (e.g. Tweed River) has had clear adverse impact on estuarine biota and coastal fishes (Bush, 1993).

Sediment and nutrient runoff

The fate and effects of materials discharged to the GBR from agriculture (Hunter, 1992) are influenced by rainfall patterns and the quantity of sediments in flood waters. Sediment discharge is a major concern, not only because of its potentially intrinsic effects on the marine environment, but because many chemical contaminants and nutrients are attached to the sediment particles. Generally, river flows from catchments are intermittent, because they are driven by episodic and intense rainfall associated with cyclones and tropical depressions. Hence, river discharges are aperiodic, summer events rather than chronic, but have the capacity to deliver large volumes of water-borne materials to the GBR (Mitchell *et al.*, 1996). River discharge plumes can be extensive, e.g. the Burdekin River plumes influence the nearshore areas north to Cairns (Wolanski, 1994), and briefly (for several days) extend to the outer reefs (Brodie, 1995; Furnas and

Brodie, 1996). While the significance of plumes to outer reefs is under examination, it is clear that freshwater, terrestrial sediments and nutrient inputs are concentrated in the shallow nearshore areas (about 10%) of the GBR (King and Wolanski, 1992; Furnas *et al.*, 1993). Anecdotal evidence points to sediment impacts on several coastal reef flats (Wachenfeld, 1996). Sediment inputs to the GBR are subsequently subject to resuspension and transport landward as a result of seasonal winds (Walker and O'Donnell, 1981) and cyclones (Gagan *et al.*, 1988).

Sediment

The products of natural soil erosion over the last 6000 years can be seen in the marine coastal zone as a narrow strip of terrestrial sediment trapped in mangroves and nearshore embayments (Belperio, 1983; Ward *et al.*, 1995). Under both natural and post-1900 land clearance conditions, 'coastal trapping' of riverine discharges by estuaries and the nearshore zone is probably very important to the health of the GBR (Bacon, 1988; Mantoura *et al.*, 1991). Land clearing has accelerated exponentially since the late 1800s (Wasson, 1994). High incidence of soil erosion and transport to the coastal zone occurred during the early years (1880-1930) of land clearance, before land conservation and management skills were developed (Kerr and Blyth, 1993). Lower catchments in north Queensland were rapidly transformed from natural forest / grassland / mangrove communities to a monoculture of sugarcane. Based on model and best-guess calculations, Moss *et al* (1992) estimated that the rate of sediment supply to the GBR has increased four-fold since European settlement due to grazing, crop land clearance, and other catchment developments. These estimates are under review (Neil and Yu, 1995).

The sugar industry has initiated cultivation and tillage practices which minimise sediment loss. In Queensland's wet tropics, the use of green manuring and no tillage has been shown to reduce soil erosion tenfold; from average annual soil erosion of 148t ha^{-1} y^{-1} to < 15t ha^{-1} y^{-1} (Prove *et al.*, 1995). These best management practices are important to farm economics, sustainable production and the well-being of the GBR. Grazing and other agricultural practices in catchments are being assessed for their influence on sediment loss and pollutant impact on the GBR.

Nutrients

Fertiliser use on agricultural lands is seen as a major source of N and P to the GBR lagoon. Sugarcane fertiliser use varies between catchments. Fertiliser use depends on soil type (for P) and the areal extent of sugar in proportion to other agriculture (Table 23.1). Land-derived nutrients can enter the GBR lagoon through groundwater aquifers and from rivers and floods. Elevated nitrate levels (of the order of 50-100mg nitrate l^{-1}) have been detected in groundwater associated with sugar production areas in Queensland (e.g. Bundaberg, Mackay, Burdekin) (Keating *et al.*, 1996). However, direct cause-effect relationships require further evaluation. While it is not possible to make any 'impact' link from these sources to the marine environment, freshwater 'seeps' have been observed in the nearshore waters of the GBR (J.E. Brodie, *pers. comm.*) Elsewhere, there is evidence that nitrate derived from sugarcane cultivation has resulted in local eutrophication of fringing reefs (D'Elia *et al.*, 1981; Crossland, 1983).

Table 23.1. Total nitrogen and phosphorus applied as fertiliser to some eastern Queensland basins in 1990, and proportion applied to sugarcane (from Pulsford, 1993).

	Phosphorus ($t\ y^{-1}$)			Nitrogen ($t\ y^{-1}$)			Mean annual runoff
River basin	Total	Sugar	% Sugar	Total	Sugar	% Sugar	('000Ml)
Daintree	100	85	85	340	320	94	3560
Mossman	240	230	96	820	812	99	687
Barron	625	6	1	1680	25	1.5	1153
Mulgrave-Russell	605	597	99	4720	4453	94	4193
Johnstone	1700	895	53	7300	4867	67	4698
Tully	530	409	77	2660	1890	71	3683
Murray	220	130	59	1290	640	50	1628
Herbert	1330	1025	77	9800	8848	90	4991
Black	0	0	0	5	0	0	509
Ross	0	0	0	21	0	0	372
Haughton	613	300	49	8805	7783	88	756
Burdekin	256	112	44	3180	2654	83	10100
Don	380	41	11	1445	851	59	689
Proserpine	459	376	82	3040	2786	92	1431
O'Connell	539	475	88	4390	4146	94	1668
Pioneer	648	421	65	5490	4626	84	994
Plane	995	937	94	7685	7468	97	1370
Styx	3	0	0	0	0	0	825
Shoalwater	27	0	0	0	0	0	832
Waterpark	73	0	0	475	0	0	700
Fitzroy	786	0	0	7290	0	0	7127
Calliope	20	0	0	62	0	0	340
Boyne	2	0	0	7	0	0	401
Baffle	105	85	81	405	360	89	750
Kolan	360	320	89	1690	1500	89	464
Burnett	1160	240	21	4545	844	19	1743
Burrum	650	580	89	2970	2655	89	718
Mary	1050	175	17	2800	1250	45	2309
TOTAL	**1347**	**7439**		**82915**	**58778**		

Discharge peaks in the first few days of a major flood event transport significant quantities of N and P to the GBR lagoon (Mitchell *et al.*, 1996). Study of a flood event in the Herbert River catchment showed that the particulate-associated nutrients comprised 50% of N and 80% of P transported. Similar relationships for the timing of peak discharges and ratio of dissolved:particulate N and P have been shown in the Tully and Johnstone Rivers (Furnas *et al.*, 1996).

Water quality measurements in the GBR lagoon indicate that dissolved and particulate N and P (and phytoplankton concentrations) are low throughout the GBR region (Furnas and Brodie, 1996). Coastal waters are characterised by slightly higher nutrient concentrations than the outer reef areas. However, the spatial and temporal patterns of nutrient concentration are considered to be natural (Furnas *et al.*, 1993). Episodic riverine and flood events elevate nutrients and sediment concentrations in the nearshore. The effects on the GBR nutrient budgets remain to be established. While there is potential for sediment-associated 'time bombs' of P and other contaminants to develop in the nearshore deltas of rivers, the open coast Johnstone River is free of phosphate accumulation (Pailles and Moody, 1996).

Pesticides/herbicides

Agrochemicals appear in north Queensland coastal sediments and marine organisms. Olafson (1978) showed lindane and DDT degradation products in corals, fish, and molluscs from the GBR. Brodie *et al.* (1984) identified the pesticides HCH and heptachlor in groundwater from the Burdekin Delta. Smith *et al.* (1984, 1985, 1987) found polycyclic aromatic hydrocarbons (PAHs, derived from combustion of petroleum products) in GBR sediment cores and molluscs. Gerretze *et al.* (1995) described DDT residues in estuarine sediments of the Herbert River. Some of these substances originate from global contaminant sources (Iwata *et al.*, 1994; Simonich and Hites, 1995) as well as local agricultural and urban sources. Kannan *et al.* (1994) have shown that foodstuffs from north Queensland have lower levels of most organochlorines compared with elsewhere in Australia

The last 100 years of land clearance and agricultural activities have resulted in contaminants entering the GBR lagoon. In determining the effects of agrochemical contaminants, consideration must be given to not only the present agricultural chemicals being added to the GBR lagoon, but also to the accumulated residues of agricultural products that have not been used for decades (e.g. DDT), both in agricultural soils (as a future source) and in coastal marine sediments. Hamilton and Haydon (1996), in auditing the Queensland sugar industry's pesticide use and environmental fate, have recommended monitoring priorities, particularly for soil-associated (e.g. atrazine, chlorpyrifos) and water-mobile and persistent pesticides (e.g. diuron, 2,4-D, asulam, aldicarb).

There are no published papers that show the effects of these contaminants on coastal ecosystems of the GBR. Studies on North American and European coastal regions predict that effects may first be seen in eggs and embryos of animals at the top of the food chain, such as dugongs, sea birds and carnivorous fish.

Trace metals

Studies of sediments accumulated on coastal landforms indicate an interesting history. Seismic profiles along the north Queensland coast reveal a wedge of terrestrial sediment within several kilometres seaward of the coastline. This results from (i) landward transport of alluvial sediment deposited during the low sea level stand, more than 12,000 years ago; and (ii) modern riverine sediment inputs (Carter *et al.*, 1993). Most terrestrial riverine sediments are stored in the estuarine mangrove zones, salt flat/mangrove tidal creeks, and in mudbanks in coastal embayments where eddies occur behind and northward of capes (Belperio, 1979; King and Wolanski, 1992). In specific locations, recent sediment accumulation rates are very rapid (5-15kg m^{-2} y^{-1}) with sediment mixing occurring to 200-500mm depth.

Sediments and their associated agrochemical contaminants are delivered to river estuaries where they are subjected to large changes in salinity, pH, redox potential, mixing energy, and particle concentrations. Natural and contaminant substances undergo complex transformations in this chemical and energy gradient between river and oceanic waters. Many hydrophobic organic contaminants appear to be sorbed into lipid-rich organic particles, which are aggregated into larger (sinkable) particles in the salinity gradient (Ko and Baker, 1995). These large organic flocculant particles are resuspended and deposited many times before being incorporated into the historical sediment pile. The sediment storage reservoir is the main source of organic contaminants to marine food chains (Shaw and Connell, 1982). Trace element behaviour in the estuarine salinity gradient is complex, and examples can be given of trace element release from particles (Ba, Cd, U), or uptake by particles (Cu, Hg, Pb).

There will be surprises ahead for us, as many of the supposedly short-lived agrochemical compounds have been found to be preserved in anaerobic sediments of the coastal zone (Fletcher *et al.*, 1995; Meakins *et al.*, 1995). Tesiram (1995) has found evidence for enhanced concentrations of Cd, As, U and Hg in Hinchinbrook Channel mangrove sediments. Bioaccumulation of Cd has been described in dugong livers (Kemper *et al.*, 1994), but cause and implications are unknown.

Impacts on the Marine Environment

The evidence for both eutrophication on a regional scale and a causal relationship with the demise of coral reefs is ambiguous, but the spectre of the Bell and Tomascik (1994) hypothesis remains an important motivation for ongoing research on water quality, benthic community structure and processes on coral reefs (Done 1995a,b). In investigating these issues, researchers have found the following:

1. Although added P can inhibit coral growth (Simkiss, 1964), weakening of coral skeletons has only been achieved at concentrations unlikely to occur in any scenario for agriculture (Rasmussen *et al.*, 1993).
2. Chronic high levels of N and P can promote algal growth leading to the death of live corals. The only experimental attempt in the GBR using episodic, high concentrations of P and N has failed to produce that outcome (A.W.D. Larkum, *pers. comm.*)
3. Around the world, algal overgrowth of coral reefs has been shown to be more often a result of reduced grazing due to changes in populations of fishes and invertebrates rather than elevated nutrient concentrations (Done, 1992). Moreover, high grazing rates can prevent damaging levels of algae (Wilkinson and Buddemeier, 1994; McCook, 1996).
4. Suspended sediments can exclude light, leading to coral death (Rogers, 1979). However, there is also an enormous capacity for inshore corals to exist in chronically turbid waters (Kleypas, 1996).
5. Settlement of sediments can smother and kill corals not normally exposed to turbid waters (Rogers, 1990). However, the corals on nearshore reefs most likely to be subjected to sediment are often well-adapted to surviving substantial sediment loads (Kleypas, 1996; Van Woesik and Done, 1996).
6. Accumulation of sediments on coral reef surfaces can prevent settlement of microscopic coral larvae (Rogers, 1990). However, many coral larvae settle on silt-free undersides of reef cavities or can occupy silted areas by asexual reproduction (fragmentation or budding).

The studies have thus demonstrated a capacity among corals to persist, but again, the key word is ‘can’. It is not sufficient for scientists or managers to demonstrate the existence of a capacity for elements of a marine system to resist impacts. Rather, it is necessary to demonstrate the extent to which that capacity can, or cannot, preserve ecosystem attributes and values.

On-Going Studies

Further studies need to separate out both the influence of different agricultural practices on the quality of discharges to the ocean, and the regional pattern of natural and anthropogenic influences. This is largely an agricultural and terrestrial ecological problem, but it is of great importance to the management of marine resources in the GBR. Information about catchment history is needed for models of past contaminant inputs, to compare with what we currently find in the coastal zone ecosystem. Quantitative histories (1900 to present) of land use change, crop productivity and profit, agrochemical usage, water usage and demographic change will be critical in evaluating past change and predicting future scenarios for the coastal zone of the GBR.

Much of what we understand about sugarcane production and the receiving marine environment is from inference rather than hard data, particularly in the case of the GBR. Major research initiatives are underway in a number of Australian research institutes (CRC Reef Research Centre (CRC Reef), CRC for Sustainable Sugar Production (CRC Sugar), Department of Primary Industries (DPI), Commonwealth Scientific Institute Research Organisation (CSIRO), Australian Institute Marine Science (AIMS)) to provide information for industry, catchment and Reef managers. For example, the CRC Reef Research Centre's Program (CRC Reef, 1995) is using two approaches:

1. Interpretation of the current status of marine systems (provided by rapid assessment and monitoring tasks in AIMS and the CRC Reef).
2. Prediction of likely outcomes given current projections for growth in the various sectors of activity in the GBR and adjacent coastal catchments.

An important issue is how long does it take for a 'cause' to be manifested as an 'effect', and how long does it take to see a remedial influence? Where there is good evidence by inference from other areas of research, managers and industry must proceed within the 'precautionary principle' in their work towards the goal of a sustainable sugarcane production industry in harmony with the nearby marine environments (CRC Sugar, 1996).

References

Arakel, A.V., Loder, T., McConchie, D. and Pailles, C. (1993) Environmental consequences of land degradation in coastal drainage basins of north Queensland, Australia: influence of farming practices. *Land Degradation & Rehabilitation* 4, 99-112.

Bacon, M.P. (1988) Tracers of chemical scavenging in the ocean: boundary effects and large-scale chemical fractionation. *Philosophical Transactions of the Royal Society, London* A 325, 147-160.

Bell, P.R.F. and Tomascik, T. (1994) The demise of the fringing coral reefs off Barbados and of regions in the Great Barrier Reef lagoon - impacts of eutrophication. In: Ginsburg, R.N. (ed.), *Global Aspects of Coral Reefs: Health, Hazards and History.* Case histories for the colloquium and forum, June 10-11 1993. University of Miami, Miami, USA, pp. 319-325.

Belperio, A.P. (1979) The combined use of wash load and bed material load rating curves for the calculation of total load: an example from the Burdekin River, Australia. *Catena* 6, 317-329.

Belperio. A.P. (1983) Terrigenous sedimentation in the Great Barrier Reef Lagoon: a model from the Burdekin region. *BMR Journal Australian Geology & Geophysics* 8,179-190.

Brodie, J.E. (1995) River flood plumes in the Great Barrier Reef. *Great Barrier Reef: Terrigenous Sediment Flux and Human Impacts Workshop*, 22-24 September, CRC

Reef Research Centre & Dept Earth Sciences, James Cook University, Townsville, Australia, p. 20.

Brodie, J.E., Hicks, W.S., Richards, G.S. and Thomas, F.G. (1984) Residues related to agricultural chemicals in the groundwaters of the Burdekin River Delta, North Queensland. *Environmental Pollution* (Series B) 8, 187-215.

Brown, B.E. (1986) *Human Induced Damage to Coral Reefs*. UNESCO Reports in Marine Science 40, Paris, France, 179pp.

Bush, R. (ed.) (1993) *Proceedings of the National Conference on Acid Sulphate Soils*, 24-25 June 1993, Coolangatta, Australia, 165pp.

Carter, R.M., Johnson, D.P. and Hooper, K.G. (1993) Episodic post-glacial sea-level rise and the sedimentary evolution of a tropical continental embayment (Cleveland Bay, Great Barrier Reef Shelf, Australia). *Australian Journal of Earth Sciences* 40, 229-255.

CRC Reef (1995) *Annual Report, 1994-5*. Cooperative Research Centre for the Ecologically Sustainable Development of the Great Barrier Reef, Townsville, Australia, 56pp.

CRC Sugar (1996) *Draft Strategic Plan, March 1996*. Cooperative Research Centre for Sustainable Sugar Production, Townsville, Australia, 25pp.

Crossland, C.J. (1983) Dissolved nutrients in reef waters of Sesoko Island, Okinawa: a preliminary study. *Galaxea* 1, 47-54.

D'Elia, C.F., Webb, K.L. and Porter, J.W. (1981) Nitrate-rich groundwater inputs to Discovery Bay, Jamaica: a significant source of N to local coral reefs? *Bulletin of Marine Science* 31, 903-910.

Done, T.J. (1992) Phase shifts in coral reef communities and their ecological significance. *Hydrobiologia* 247, 121-132.

Done, T.J. (1995a) Ecological criteria for evaluating coral reefs and their implications for managers and researchers. *Coral Reefs* 14, 183-192.

Done, T.J. (1995b) Remediation of degraded coral reefs: the need for broad focus. *Marine Pollution Bulletin* 30, 686-688.

Fletcher, C.A., Bubb, J.M. and Lester, J.N. (1995) Agricultural inputs of Mecoprop to a salt marsh system: its fate and distribution within the sediment profile. *Marine Pollution Bulletin* 30, 803-811.

Furnas, M.J. and Brodie, J.E. (1996) Current status of nutrient levels and other water quality parameters in the Great Barrier Reef. In: Hunter, H.M., Eyles, A.G. and Rayment, G.E. (eds), *Downstream Effects of Land Use*. Department of Natural Resources, Brisbane, Australia, pp. 9-22.

Furnas, M.J., Mitchell, A.W. and Skuza, M. (1993) *Nitrogen and Phosphorus Budgets for the Great Barrier Reef*. Great Barrier Reef Marine Park Authority Research Report 36, Townsville, Australia, 234pp.

Furnas, M.J., Mitchell, A.W. and Skuza, M. (1996) *Dissolved and Particulate Nutrients in Rivers Flowing into the Great Barrier Reef*. Great Barrier Reef Marine Park Authority, Draft Report, Townsville, Australia, 106pp.

Gagan, M.K., Johnson, D.P. and Carter, R.M. (1988) The Cyclone Winifred storm bed, central Great Barrier Reef, Australia. *Journal of Sedimentary Petrology* 58, 845-856.

Gerretze, R., Hernandez, F., Murray, A.S., Wallbrink, P.J. and Brunskill, G.J. (1995) Distribution of organochlorines, PAHs, phosphorus, and Cs-137 in sediment profiles from Ellen Brook in Western Australia. *Marine and Freshwater Research* 46, 843-851.

Ginsburg, R.N. (1994) *Global Aspects of Coral Reefs: Health, Hazards and History*. Case histories for the colloquium and forum. June 10-11 1993, University of Miami, Miami, FL, USA, 420 pp.

Hamilton, D. and Haydon, G. (1996) *Pesticides and Fertilizers in the Queensland Sugar Industry: Estimates of Usage and Likely Environmental Fate*. DPI/BSES Sugar Industry RD&E Program, Project ICM-R&D.94.01, DPI, Indooroopilly, Australia, 131pp.

Hatcher, B.G., Johannes, R.E. and Robertson, A.I. (1989) Review of research relevant to the conservation of shallow tropical marine ecosystems. *Oceanography and Marine Biology Annual Review* 27, 337-414.

Hunter, H.M. (1992) *Agricultural Contaminants in Aquatic Environments: A Review*. Department of Primary Industries, Brisbane, Australia, 53pp.

Iwata, H., Tanabe, S., Sakai, N., Nishimura, A. and Tatsukawa, R. (1994) Geographical distribution of persistent organochlorines in air, water and sediments from Asia and Oceania, and their implications for global redistribution from lower latitudes. *Environmental Pollution* 85, 15-33.

Kannan, K., Tanabe, R.J., Williams R.J. and Tatsukawa, R. (1994). Persistent organochlorine residues in foodstuffs from Australia, Papua New Guinea and the Solomon Islands: contamination levels and human dietary exposure. *The Science of the Total Environment* 154, 113-124.

Keating, B.A., Connell, D., Hillier, J., Ellis, R., Weier, K.L., Sunners, F. and Bauld, J. (1996) Leaching of nutrients and pesticides to Queensland groundwaters. In: Hunter, H.M., Eyles, A.G. and Rayment, G.E. (eds), *Downstream Effects of Land Use*. Department of Natural Resources, Brisbane, Australia, pp.151-164.

Kemper, C., Gibbs, P., Obendorf, D., Marvanek, S. and Lenghaus, C. (1994) A review of heavy metal and organochlorine levels in marine mammals in Australia. *The Science of the Total Environment* 154, 129-139.

Kerr, B. and Blyth, K. (1993) *They're All Half Crazy, 100 Years of Mechanical Cane Harvesting*. Canegrowers, Brisbane, Australia, 117pp.

King, B. and Wolanski, E. (1992) Coastal dynamics along a rugged coastline. In: D. Prandle (ed.), *Dynamics and Exchanges in Estuaries and the Coastal Zone*, American Geophysical Union, New York, USA, pp. 577-598.

Kleypus, J.A. (1996) Coral reef development under naturally turbid conditions: fringing reefs near Broad Sound, Australia. *Coral Reefs* 15, 153-168.

Ko, F-C. and Baker, J.E. (1995) Partitioning of hydrophobic organic contaminants to resuspended sediments and plankton in the mesohaline Chesapeake Bay. *Marine Chemistry* 49, 171-188.

Lough, J.E. (1994) Climate variation and El Nino-Southern Oscillation events on the Great Barrier Reef: 1958 to 1987. *Coral Reefs* 13, 181-195.

Mantoura, R.F.C., Martin, J-M. and Wallast, R. (eds) (1991) *Ocean Margin Processes in Global Change*. John Wiley & Sons, Brisbane, Australia, 469pp.

McCook, L. (1996) Effects of herbivores and water quality on the distribution of *Sargassum* in the central Great Barrier Reef: cross-shelf transplants. *Marine Ecology Progess Series* 139, 179-192.

Meakins, N.C., Bubb, J.M. and Lester, J.N. (1995) The mobility, partitioning and degradation of atrazine and simizine in the salt marsh environment. *Marine Pollution Bulletin* 30, 812-819.

Mitchell, A.W., Reghenzani, J., Hunter, H.M. and Bramley, R.G.V. (1996) Water quality and nutrient fluxes from river systems draining into the Great Barrier Reef Marine

Park. In: Hunter, H.M., Eyles, A.G. and Rayment, G.E. (eds), *Downstream Effects of Land Use.* Department of Natural Resources, Brisbane, Australia, pp. 23-33.

Moss, A.J., Rayment, G.E., Reilly, N. and Best, E.K. (1992) *A Preliminary Assessment of Sediment & Nutrient Exports from Queensland Coastal Catchments, A Desk Study.* Department of Environment and Heritage and Department of Primary Industries, Queensland Government, Brisbane, Australia, 27pp.

Neil, D.T. and Yu, B. (1995) Simple climate-driven models for estimating sediment input to the Great Barrier Reef lagoon. *Great Barrier Reef: Terrigenous Sediment Flux and Human Impacts Workshop*, 22-24 September, CRC Reef Research Centre & Dept Earth Sciences, James Cook University, Townsville, Australia, pp. 67-73.

Olafson, R.W. (1978) Effect of agricultural activity on levels of organochlorine pesticides in hard corals, fish, and molluscs from the Great Barrier Reef. *Marine Environmental Research* 1, 87-107.

Pailles, C. and Moody, P.W. (1996) Phosphorus dynamics of sediments from the Johnstone Rivers. In: Hunter, H.M., Eyles, A.G. and Rayment, G.E. (eds), *Downstream Effects of Land Use.* Department of Natural Resources, Brisbane, Australia, pp. 359-364.

Prove, B.G., Doogan, V.J. and Truong, P.N.V. (1995) Nature and magnitude of soil erosion in sugarcane land on the wet tropical coast of north-eastern Queensland. *Australian Journal of Experimental Agriculture* 35, 641-649.

Pulsford, J.S. (1993) *Historical Nutrient Usage in Coastal Queensland River Catchments Adjacent to the GBR.* Unpublished draft report to the Great Barrier Reef Marine Park Authority, Townsville, Australia.

Rasmussen, C.E., Cuff, C. and Hopley, D. (1993) Evidence of anthropogenic disturbances retained in the skeleton of massive corals from Australia's Great Barrier Reef. *Proceedings of the 7th International Coral Reef Symposium* 1, 201-210.

Rogers, C.S. (1979) The effects of shading on coral reef structure and function. *Journal of Experimental Marine Biology and Ecology* 41, 269-288.

Rogers, C.S. (1990) Responses of coral reefs and reef organisms to sedimentation. *Marine Ecology Progress Series* 62, 185-202.

Shaw, G.R. and Connell, D.W. (1982) Factors influencing concentrations of PCBs in organisms from an estuarine ecosystem. *Australian Journal of Marine and Freshwater Research* 33, 1057-1070.

Simkiss, P. (1964) Phosphates as natural poisons of calcification. *Biological Reviews* 39, 487-505.

Simonich, S.L. and Hites, R.A. (1995) Global distribution of persistent organochlorine compounds. *Science* 269, 1851-1854.

Smith, J.D., Bagg, J. and Bycroft, B.M. (1984) Polycyclic aromatic hydrocarbons in the clam *Tridacna maxima* from the Great Barrier Reef, Australia. *Environmental Science and Technology* 18(5), 353-358.

Smith, J.D., Bagg, J. and Sin, Y.O. (1987) Aromatic hydrocarbons in seawater, sediments and clams from Green Island, Great Barrier Reef, Australia. *Australian Journal of Marine and Freshwater Research* 38, 501-510.

Smith, J.D., Hauser, J.Y. and Bagg, J. (1985) Polycyclic aromatic hydrocarbons in sediments of the Great Barrier Reef region, Australia. *Marine Pollution Bulletin* 16, 110-114.

Tesiram, Y.A. (1995) The Use of Cadmium, Uranium, and Arsenic as Indicators of Anthropogenic Input into the Herbert River Estuary, Central Great Barrier Reef.

Honours Thesis, Department of Chemistry, James Cook University, Townsville, Australia, 57pp.

Van Woesik, R. and Done, T.J. (1997) Coral communities and reef growth in the southern Great Barrier Reef. *Coral Reefs* (in press).

Wachenfeld, D. (1996) Long-term reef-flat dynamics in the Great Barrier Reef: the historical photographs project. *Reef Encounter* 18, 7-10.

Walker, T.A. and O'Donnell, G. (1981) Observations on nitrate, phosphate and silicate in Cleveland Bay, North Queensland. *Australian Journal of Marine and Freshwater Research* 32, 877-887.

Ward, I.A.K., Larcombe, P. and Cuff, C. (1995) Stratigraphic control of the geochemistry of Holocene inner-shelf facies, Great Barrier Reef. *Marine Geology* 129, 47-62.

Wasson, R.J. (1994). Living with the past: uses of history for understanding landscape change and degradation. *Land Degradation & Rehabilitation* 5, 79-87.

Wilkinson, C.R. and Buddemeier, R.W. (1994) *Global Climate Change and Coral Reefs: Implications for People and Reefs.* Report UNEP-IOC-ASPEI-IUCN Global Task Team, IUCN, Gland, Switzerland, 124pp.

Wolanski, E. (1994) *Physical Oceanographic Processes of the Great Barrier Reef.* CRC Press, Boca Raton, LA, USA, 194pp.

Chapter twenty-four:

The Irrigation Experience in Australia - Lessons for the Sugar Industry

W.S. Meyer

Charles Sturt University, CSIRO Division of Water Resources, Private Mail Bag 3, Griffith, NSW 2680, Australia

Abstract

The Australian sugar industry is predisposed to maximise the benefits of irrigation. Its geographic and climatic location, its established research and extension record and a sound economic position are all favourable. To capitalise on this potential it must do what no other irrigation area has ever done completely - learn from history. In this paper, we examine the lessons from history in irrigation of Australian agriculture, but these lessons are broadly relevant to irrigation areas elsewhere in the world.

The single most outstanding issue is that irrigation must be accompanied by drainage. Without adequate provision for surface and subsurface drainage, without confronting the need to reuse, store or change the form of irrigation effluents, namely excess water, salt, nutrients and chemicals, irrigation will not be allowed or be economically viable in the long term.

Apart from the overriding issue of drainage management, irrigation provision must generate significantly more income, significantly lower crop failure risk and increase production opportunity over dryland agriculture to be warranted. If the focus of irrigation is lost by the fragmentation of control or the absence of irrigation farmers in system management, inefficient delivery and poor maintenance will result. Failure to support a strong and clear allocative policy has led to over-commitment and degradation of many river systems. The total available water resource needs to be defined and allocations made on crop need and not simply on an arbitrary value tied to land area. Delivery system management must have clearly defined performance criteria with losses being minimised. In this respect supply authorities need to avoid the conflict of interest

In: Keating, B.A. and Wilson, J.R. (eds), *Intensive Sugarcane Production: Meeting the Challenges Beyond 2000*. CAB International, Wallingford, UK, 1997.

of being supplier and regulator. Water prices must cover the longer term maintenance and improvement costs or the delivery systems will inevitably degrade. Pricing structures must separate fixed and variable costs and be in a form that irrigators can understand.

The Australian sugar industry is in a good position to direct further irrigation development for enhanced productivity and reduced resource impact. It can do this if it learns from existing irrigation areas and honestly confronts the need to irrigate for production and drain for effluent management.

Introduction

The largest irrigation developments in Australia are in the southern, and especially the south eastern, portion of the continent. The impetus for irrigation development largely resulted from the perceived need to mitigate the effects of recurring drought; a predominant feature of the Australian environment. Much thinking about the development and subsequent management of irrigation has come from this perspective. Management has been strongly influenced by the dryland experience, with irrigation being seen simply as providing more water. This mindset fails to appreciate the huge capital costs, the large running costs, and the management required for productive, long lasting irrigation areas.

To come close to justifying the investment in large scale irrigated areas they must be managed intensively to deliver a steady supply of highly saleable produce. Even with this, it is unlikely that any irrigated area will fully repay its investors in a strict financial sense (Davidson, 1969). Given this, it makes even less economic sense to have these areas producing fodder reserves in anticipation of the next drought in some remote location. Irrigation is not the panacea to give drought protection, rather it is an intensive production system which can reliably deliver a diverse range of high quality food and fibre produce. The continuing failure to appreciate this intensity has resulted in suboptimal productivity and resource protection (Smith *et al.,* 1982; Smith *et al.,* 1983) in most of the large scale irrigated areas of south eastern Australia.

Much of the documented information on irrigation in Australia is concerned with the larger scale, multi-user irrigated areas. It is from this context that the lessons for the sugar industry are drawn. While there is ample evidence for the arguments presented here, there is some extrapolation from existing trends and suggestions made to stimulate and predispose the irrigation industry to progress.

At its most basic and idealistic level, the purpose of irrigation is to promote plant growth and generate profit. However, the reality is that political considerations and not scientifically sound agronomy, soil science, hydrology, engineering or even rigorous economics have decided whether irrigation development will proceed. The legacy of these politically motivated decisions together with the failure of irrigation system managers to learn from history has resulted in many of the difficulties which are currently being addressed in developed irrigation areas of south eastern Australia.

While the focus of this paper is on the history of irrigation in Australian agriculture, the problems encountered and solutions developed are relevant to irrigation areas elsewhere in the world.

History of Development

The purpose of this section is to identify significant features of early development which have shaped the irrigated areas we see today. For more comprehensive historical treatments the reader is referred to Wadham *et al.* (1964), Davidson (1969), Smith *et al.* (1983), and Hallows and Thompson (1995).

Early optimism and failure

The earliest irrigation occurred along the coastal areas of New South Wales and in Tasmania. These schemes predominantly involved pumping water from rivers to irrigate small orchards, vegetable gardens, vineyards and some grass pastures (Hallows and Thompson, 1995).

European settlers moving to the inland areas of south eastern Australia in the first half of the 1800s quickly appreciated the importance of reliable water sources as they experienced the uncertainty of rainfall. Many early settlers, principally sheep owners, were lucky enough to claim land adjacent to streams and some advocated that irrigation was a way of mitigating the devastating effect of drought. Some built dams for irrigation supply, but found that the benefits were not enough to continue.

The uncertainties associated with land tenure and water rights tended to lead to extensive land acquisition rather than the development of more closely settled areas. At first, large scale irrigation faltered due to high establishment costs, poor technology and unreliable stream flows.

Tasmania

Irrigated fruit and vegetable growers in Tasmania were large enough to establish markets in other colonies, i.e. Melbourne, Sydney, Brisbane and Adelaide, until the introduction of barriers and custom duties on interstate trade in the 1850s. The colonial government of Tasmania took an interest in irrigation, an interest largely promoted by Arthur and Hugh Cotton, who were civil engineers with experience of irrigation in India. Arthur Cotton suggested the need to vest all rights to water in the Crown and to permanently reserve the bed and banks of all streams and water bodies. This suggestion was not taken up at the time nor was his advice that all irrigated areas must be designed with adequate drainage.

Victoria

Meanwhile in Victoria, a number of small private schemes were attempted and within a short time, failed. Factors such as unsuitable areas, poor administration, inadequate economic analysis, the inability to obtain reliable water sources, a failure to appreciate the huge capital investment involved and the long term nature of irrigation systems all contributed to their demise. The Victorian Colonial Government of 1883 was the first to pass legislation which provided for irrigation. There was public concern regarding the scale of expenditure required for irrigation schemes and concern about the failure of the private schemes. A Royal Commission on water supply chaired by Alfred Deakin was appointed in 1884. Deakin studied US irrigation development as he perceived the similarity in climatic and social settings. Deakin's inquiry led to the Victorian Government Irrigation Act of 1886. The Act vested in the Crown the right to the use of all water in any stream, lake or swamp, prohibited the issue of further riparian rights to water, authorised the construction of national works by the Government and enabled elected Trusts to carry out irrigation schemes with Government money. Public ownership of water is a major feature of water management in Australia.

By 1899 there were approximately 90 irrigation Trusts established in Victoria. Amongst these was the Chaffey brothers establishment near Mildura. Unfortunately the irrigation settlement at Mildura and a similar one near Renmark in South Australia initially failed due to high seepage losses from unlined supply channels, subsequent land salinisation, unsuitable crop varieties, a lack of transport and markets for produce, inexperienced and poorly skilled settlers and financial difficulties associated with the 1893 bank crash. The majority of these irrigation Trusts failed again for reasons similar to those above. The Relief Act of 1899 in Victoria wrote off three quarters of all Trust liabilities and the Water Act of 1905 abolished all of the Irrigation Trusts except the Mildura Irrigation Trust. The state placed the organisation of rural water supplies with the State Rivers and Water Supply Commission (Wadham *et al.,* 1964; Hallows and Thompson 1995) with an American, Dr Mead, as the founding Chairman. His fundamental belief that the more water that was applied, the greater the return, pervaded the way water was allocated and priced (Hodges, 1995). The end result was that settlers paid for a guaranteed quantity of water (often decided by the area of land) rather than charging for the actual amount used. Thus was set in train a pricing and allocation system which encouraged water use and failed to indicate in the price of water its true opportunity value. This policy has not been seriously challenged until the last decade.

New South Wales

During 1884, the same year Victoria set up the Deakin Royal Commission, the NSW Government appointed the Royal Commission into Water Conservation and related matters. Little eventuated from the Commission other than the establishment of a water conservation irrigation branch of the Department of Mines and Agriculture headed by Hugh McKinny. Following this, a number of small irrigation schemes were set up. It wasn't until 1896, ten years after Victoria passed similar legislation, that the NSW Government passed the Water Rights Act which vested in the Crown the right to use and control all surface waters. In 1902 the NSW Government passed the Water and Drainage Act which empowered the Government to construct works for water supply, water conservation, irrigation and drainage. In parallel with Victoria, the legislation also established the concept of a trust to administer the works.

Queensland

In Queensland the use of irrigation by individuals commenced in the 1880s. In 1884, John Drysdale started irrigating sugarcane in the Burdekin River delta, drawing groundwater from shallow aquifers (Connolly, 1964). Inspired by the irrigation developments at Mildura and Renmark the Queensland Government introduced the Water Authorities and the Irrigation Act of 1891. Unfortunately, the legislation proved to be of little value since it made no alteration to the common law relating to the diversion of water from water courses or underground sources.

Consequences of early development

These early experiences clearly showed that irrigation development was very expensive and large capital investments were involved. Rigorous economic analysis would have shown that despite the promise of improved productivity from irrigation, the early returns from production were insufficient to service the capital debts and in many cases insufficient to service the running and maintenance costs. However, with the strong political push for development and the sensitivity of interstate rivalries, investment in large irrigation schemes was seen to be a good thing. Fortunately good sense prevailed and the ownership of water was vested in Government. This overcame the very obvious disadvantages of the protracted litigious system which is a feature of US water ownership.

In line with the prevailing development ethic, early development was expansive. In terms of land and soil, there seemed to be unlimited resource and when large dams were constructed in the catchments, water on many of the major rivers also seemed unlimited. With Government now owning water, having invested heavily in infrastructure, and riding the politically popular water development sentiment, it was predisposed to kick start major and extravagant developments. One consequence was the institutionalisation of subsidisation and with it the expectation that this will continue for the immediate beneficiaries.

The excitement and needs of early development led to the formation of successful engineering-oriented government departments. The development push often resulted in little or no concern for longer term environmental impacts. The selection of many irrigation areas on the basis of sound soil and agronomic characteristics took a secondary place to the push for acquisition for land, for satisfying political interests such as 'making the desert bloom' and giving 'land for heroes' (associated with soldier settlement areas).

With this focused euphoria over development and the strong role that government was to play, a number of consequences ensued. There was little or no provision made for the continuing and increasing need for delivery system maintenance and upgrading. Many early developments failed to incorporate adequate drainage and many natural drainage lines were either trampled over or were used as effluent dumps. The perception of unlimited water once rivers were dammed led to a failure to measure and monitor where water was going and how much was productively used. With strong Government involvement, the institutionalisation of the single supplier and regulator was set in place.

The huge Government investment in the development of the Murrumbidgee Irrigation Area (MIA) in NSW is an example of successful engineering and social design. One reason for this success was the tight focus for development. One agency was responsible for the design, construction, settlement and, in effect, the productivity of this large undertaking. Subsequently, the responsibility for the irrigation areas was split between many government agencies: local government, Departments of Agriculture, Water Resources, Lands, Soil Conservation, Minerals (groundwater resources) and more recently Parks and Wildlife and Environment Protection. This fragmentation has caused a loss of focus on the real purpose of the irrigation areas, that is, to produce food and primary products that are marketable at competitive prices and produced with a minimum impact on the resource base.

At the time of these irrigation developments, areas of land and amounts of water for viable farms were allocated. These allocations were no doubt made on the best understanding at the time. However the limitations that this has posed as socio-economic conditions changed have been severe. For example, many horticultural farms in irrigated areas of NSW, Vic. and SA are currently not viable as single entities and only with amalgamation to achieve economies of scale are they likely to become economically viable (Bryant *et al.,* 1992; Meissner, 1995). For water, the almost arbitrary allocation of 4-10Ml ha^{-1} irrespective of region crop and soil type, irrigation method or agronomy, has perpetuated the inefficiency of not closely matching water supplied to plant need. Rather, access to this water allocation has become enshrined as the de facto 'water right'. This, combined with the expectation of continued cheap water, has made the introduction of more economic and flexible pricing and allocation policies extremely difficult. Only very recently has it been possible to sell water volumes separately from land. A consequence of tying water to land area has been that the opportunity value of the irrigation water is built into the land value. The whole basis of the original allocation has never been revisited despite improved knowledge of crop water requirements. Perhaps it is destined never to be revised because of extreme political sensitivity.

General Impacts of Irrigation

Large expenditure on Government irrigation schemes has considerable economic, social and environmental consequences. Public acceptance of the net benefit of them is presumably the reason why irrigated areas have proceeded. However, Davidson (1969) argued that on pure economic grounds more tangible benefits would have accrued from alternative investments, such as dryland agriculture. The pursuit of this argument leads one to conclude that the only justifiable reason for much politically motivated government expenditure is because it makes someone feel good. The balance is provided in the value of irrigated production, most recently estimated at nearly A$6B (AIC, 1994). This is about one quarter of all rural production. Sugar contributed nearly 4.5Mt (SRDC, 1995) at A$390 t^{-1} (QSC, 1995) of which some 40% by area (Ham, 1994) was produced with the assistance of irrigation. Further, Powell *et al.* (1985) estimated that there was a flow-on effect such that for every $1 generated at the farm gate another $5 to $7 was generated in processing and support services. It is this tangible productivity and the associated social structure which makes rural communities resolve to remain viable. Perceived and real benefits have accrued from the decentralisation of population, and the provision of constant and reliable supplies of fresh fruit, vegetables and other irrigation dependent products such as rice.

The impact of irrigation on the environment is summarised with the following excerpt from Meyer and Noble (1993):

'Irrigation development from its very beginnings in the late 1800s (Smith et al. 1983), largely followed a pattern of replacing vegetation, land forming, frequent cultivation and extensive application of water in vast distributed shallow floods. While immediate productivity, security and social benefits were eagerly anticipated, economic viability of most irrigation schemes in the Murray-Darling Basin has been shaky right from the start (Wadham et al., 1964). People have learnt skills and changed practices to secure continuance of irrigation by replacing inappropriate practices. At first the resources of land and water appeared boundless, especially in relation to the impact that the energy of a few people could exert in the course of a day. The accumulation of this exertion, however, has seen our landscape and rivers transformed with the ecological impact of this activity only recently becoming obvious. Three things have contributed to a more recently increasing awareness of the impact:

1. The lag and capacitance of soil and water systems to changed inputs and outputs has taken 40 to 50 years to become plainly different and distinct from the usual year to year variability.
2. The growing realisation, that the resources of arable land and available water are finite within the Murray-Darling Basin.
3. A growing sensitivity by people to caring for the food and fibre producing resource base as our collective social outlook moves from colonial exploiter to indefinite resident.'

Review of Older Irrigated Areas and Lessons to be Learnt

Economic inefficiencies in operation and management

Because of the high level of Government involvement in setting up all the major irrigation schemes in southern Australia there is a high level of institutionalised subsidisation. This extends not only to the initial capital costs which are now 'irretrievable investment' costs but more particularly to Government supported services such as distribution, regulation, drainage and farm management advice. There are many instances of cross subsidisation. Explicit recognition of this situation has only come about in the last few years as broad economic reform (e.g. Hilmer Report and COAG Water Policy) has moved through various sectors.

One effect of this subsidisation has been the capitalisation of the value of water in the cost of land. The dilemma then is that the real cost of water (quite apart from any consideration of charging for the resource) is not obvious to the end user. This leads to the difficulty that end users are not aware of the total costs, how they are derived, or why they need to change. Further, the performance criteria for effective water delivery are not set by the end users, leading to poor performance and inefficiency of the system. Fragmentation of the government agencies responsible for various components of water delivery, resource management, regulation and planning, has led to a loss of focus in the real purpose of the irrigation area, that is, to be highly productive with very efficient use of the available resources.

In most irrigation areas inadequate provision for maintenance, let alone improvement of the delivery system has led to deteriorated infrastructure which is poorly monitored. It is now difficult to decide where the best value for money will come in restoration. Most pricing structures for water delivery have not separated the fixed and variable costs so that revenue is completely dependent on the volume delivered - which encourages water delivery irrespective of seasonal conditions. The failure to update delivery systems and the financial stringency imposed by a poorly responding pricing structure have entrenched a system which lacks design and management flexibility for change. This entrenchment is further exacerbated by work practices which have little or no relation to optimising delivery system performance.

The failure to adequately monitor system performance and the active encouragement to use water has produced a system of water delivery and on-farm use which is inefficient (Cole, 1985). How inefficient is hard to quantify because of the lack of quality assured monitoring and measurement. This deficiency has worked discouragingly against the irrigation industry because it lacks a solid data base to show that there are some very efficient producers and that improvements are being made. The generally low efficiency has many contributing factors including fixed water delivery schedules, poor on-farm channel design, poor paddock layout, improperly matched pump and sprinkler systems, no farm storage or drainage retention, and a lack of understanding and skill to schedule and control water additions. Not only are irrigators not getting the best return for water added but a whole host of environmental consequences are set in train.

Environmental impacts

The biggest environmental consequences often arise from excessive withdrawal from the supply system - a situation exacerbated by over generous licensing which has occurred in most rivers of the Murray-Darling Basin. Most regulated rivers now have very different flow regimes and water more often moves across the landscape in man-made channels and shallow floods than down the natural drainage channel of the river. This changed hydrology almost inevitably causes an increase in the recharge to the groundwater. The rate at which water tables rise is dependent on the intensity of irrigation and the hydrogeological conditions in the area. The Murray Darling Basin Commission estimates that all irrigation areas in the Murray Basin will be affected by water tables less than two metres from the ground surface within the next 20 to 30 years. Shallow water tables reduce the storage capacity of the soil profile, slow internal drainage, predispose the soil to longer wet periods (waterlogging) and greatly decrease management flexibility. They also predispose the root zone to increasing salinisation. Salt accumulation is inevitable when irrigation is practised in arid and semi-arid environments. Generally, the more water and the more saline the water which is introduced the more rapid will be the rate of salt increase especially when soil profile drainage is inadequate. Shallow water tables restrict the leaching of salts below the root zone and capillary upflow from the watertable becomes a significant component of the water balance (Meyer *et al.*, 1990). This upflow can bring salt from the watertable, and with plant water uptake and soil evaporation, salt will be deposited in the upper profile or on the ground surface.

Waterlogging and salinity were estimated to cost A$65M in lost production in the Murray Basin around 1986 (MDBMC, 1988). People in the older irrigation areas are now all aware of the increasing need to manage salt. The obvious deleterious effect of increasing salinity is no better demonstrated than in the Kerang district of northern Victoria where dairy farmers are contracting their managed pastures back to the slightly higher portions of the landscape while the low lying areas are essentially abandoned. In the Wakool district of southern NSW an extensive subsurface drainage scheme has afforded protection to 47,000ha at a total capital cost of A$28M. However, there is considerable difficulty in managing the salt storage area because maintenance costs for the scheme are proving to be much more than originally estimated.

A lack of through profile drainage and shallow water tables leads to wetter soils, particularly in a climate with high rainfall variability. This in turn means that trafficking often occurs on wet, low strength soils, with considerable damage to soil structure. The reduction of physical fertility is often accompanied by reduced levels of organic matter. While the main chemical nutrients are generally replaced by fertilisers and in some cases by the use of legumes for nitrogen supplementation, there is increasing evidence that calcium and other cations are being displaced so that even well buffered soils are becoming acidic.

The increased awareness that catchments are interconnected has focused attention on the discharge of effluents into river systems. Irrigation-induced drainage is effluent, and excessive volumes are an expression of inefficient water use. Too often the provision of well designed drainage has been used to overcome grossly inefficient irrigation practices. Too much emphasis has been placed on fixing the symptom and not addressing the cause. One strategy which has been built into the MIA and is increasingly being encouraged in some Victorian irrigation areas is the use of drainage waters for irrigation. At first sight this seems eminently sensible. However, when a downstream irrigator becomes dependent on the drainage of those upstream the latter group must operate at a level of inefficiency sufficient to generate adequate drainage water. The problem is increasingly difficult if water quality specification, e.g. maximum levels of salt or absence of chemical contaminant because of human health or crop sensitivity, becomes an issue. The general principle of keeping supply and drainage waters separate should only be violated as a rare exception. Minimising and concentrating drainage as in any effluent management program will be the most cost effective practice and the one most likely to be workable in the long term.

Increasing evidence is now being gathered to show that some irrigation practices produce significant nutrient discharge (Harrison, 1994) and chemical contamination (Korth *et al.,* 1995). With increased environmental scrutiny and continued pressure from downstream users, irrigators are being pressed to contain and manage these sources of contamination.

Recent surveys (AITP, 1994) have shown a lack of skilled irrigators and equipment and service providers within the irrigation industry. This reflects a lack of commitment on the part of the industry to be involved with education and training. There has generally been slow adoption of new technologies associated with irrigation management, e.g. irrigation scheduling (Meyer, 1993) for a variety of social and economic reasons. Change in productivity and resource management are much more likely if those who can effect change are informed and appreciate the consequences of their actions.

The lessons to be learnt

In summary, the older irrigation areas of south eastern Australia are characterised by low water use efficiency, a result of poor management of both the delivery systems and on-farm application. The performance criteria for highly productive irrigation have either been lost after initial development, have not been explicitly set or, in the worst case, are wrong. There is clear evidence of a lack of skills and commitment to education and training, not only amongst irrigators, but amongst the service industry which supports irrigation. Within the cost structures of water supply and the taxation system there is a lack of incentive to change to more effective and efficient practices.

Audits on the land and water base indicate a high impact of irrigation. This impact runs to the excessive extraction and over-commitment of water from river sources and in a few cases over-use of groundwater reserves. Excessive drainage being generated and returned to rivers, declining soil structure and soil chemical imbalance all point to exploitive and non-sustainable ecological practices. In many areas adequate drainage provision has not accompanied irrigation supply provision, but more particularly, little or no provision has been made to identify where salt will be stored. The off-site and downstream impacts of nutrient and other chemical contaminants is only just entering management thinking. Irrigation has changed regional hydrology, and accessions to groundwaters from low efficiency delivery and application have caused water tables to rise, predisposing large areas to increasingly wet and saline soils, and inevitably declining productivity.

Current Developments in Southern Irrigated Areas

Thus far, this review of established irrigation areas has tended to concentrate on deficiencies and problems. The large social and economic benefits have been inferred rather than listed in detail. Concentrating on the difficulties has been done to analyse how developing irrigation areas might be predisposed to improved productivity, long term flexibility and viability.

All existing irrigation areas of the Murray Darling Basin are attempting to improve their economic and ecological sustainability. Change in water policy now allows water entitlement to be traded, independently of land, within catchment systems. Market forces are now active in setting the absolute and marginal value of water for irrigation and other purposes. The separation of water from land through allowing tradeable water transfers will cause shifts of water from lower producing areas to higher producing areas. Within the controlling Government agencies the separation of the regulator role from the supplier role is occurring through moves to 'corporatise' and in some cases 'privatise' management of irrigation areas. There are both positive and negative effects of this user ownership. Positive effects are that users can focus the suppliers' energies on providing an efficient, timely and cost effective service. However, some of the negative effects are associated with the privatised entities concentrating on immediate and local problems, the loss of development and specialist expertise that was part of the larger State Government system, a loss of appreciation of wider and longer term national and international leadership needs, and a lack of appreciation for training and skills acquisition.

The withdrawal of state agencies from directly managing regional water distribution networks means that regional authorities and local irrigators are now shaping more efficient and responsive supply systems. Costings, infrastructure needs and work practices are all being more closely scrutinised. Accompanying the withdrawal of direct government involvement has been the development of resource maintenance planning in the form of land and water management plans (LWMP) or 'salinity action' plans. These are seen as necessary for rehabilitating and maintaining basic land, soil and water resources and also providing a future operating and regulatory framework. While formulating and documenting these plans is well advanced, with many examples of excellent community involvement, the implementation and future organisational processes are not clear. Substantial progress has been made in integrating regional LWMPs into the integrated or total catchment management process and more recently the inter-connections with regional business and economic development are being appreciated. As an example of a LWMP, the MIA and Districts of southern NSW (Cummins, 1996) have taken 4 years to develop a draft plan with two main aims:

1. Achieving ecological sustainability by reducing accessions to the groundwater system.
2. Protecting downstream users and the environment by reducing drainage volumes, keeping drainage water quality within agreed standards, managing drainage treatment and storage and protecting natural resources.

To achieve these aims, performance criteria for the delivery system and farm water management have been negotiated. Seepage losses from channels will be identified and channels lined and repaired. Farm recommendations include whole farm planning for improved water distribution, bay layout, recirculation and storage structures. Incentives for farmer education and skill development will be introduced. Stricter controls on excessive water applications to rice crops, assistance with conversion of horticultural irrigation to fully piped systems and retention of major runoff and drainage volumes on farm are included. Tradeable drainage rights have been proposed (Bryant *et al.*, 1993).

All of these plans are highly dependent on measurement and monitoring. It is not possible to manage a complex system without accurate and consistent quantification. Thus the need to know how much water flows down channels, how much is lost, how much is delivered to each paddock and how much drainage of what quality is produced, is fundamental to improved management.

Sugar Irrigation

Brief overview of development and current status

Nearly 40% (182,720ha) of the Australian sugarcane area uses either supplementary or full irrigation (Ham, 1994). The approximate contributing value of this is calculated here. Annual irrigation additions reported range from 3.8Ml ha^{-1} at Bundaberg (Kingston, 1994) to 10Ml ha^{-1} in the Burdekin (Raine,

1995). Assuming an average addition of 5Ml ha^{-1} or 500mm, and using the relationship between total sugar yield and water use (Kingston, 1994), this represents an additional yield of about 3t sugar ha^{-1}. Over the total area and with a return to grower price of A$390 t^{-1} sugar, irrigation is contributing about A$214 million or 13% of the total gross value of production.

Irrigation has been part of sugarcane production particularly in the Ayr (Burdekin) and Bundaberg regions since their beginnings (Leverington and Ridge, 1975). At Mackay, it was initially used only for crop establishment but by 1979, Chapman and Chardon (1979) reported a 'dramatic increase' in irrigation by cane farmers in this region. The increased yields attributed to irrigation and the associated costs were examined early on by Ferguson and Hampson (1975), and Chapman and Chardon (1979). The latter authors concluded that irrigation may not be 'economically justified'. Despite this, it is generally true that the areas and interest in irrigation have increased, perhaps spurred on in recent years by the series of drier than average seasons, and an awareness of the high yields being achieved by irrigated cane in the Burdekin. More recent analyses (Wegener, 1990) show that although irrigation increases costs and thus the risk of more variable income, this risk is offset by the added income.

Production of sugarcane is highly dependent on water availability. Quantification of water use (Ham, 1985) and developing water use efficiency relationships have provided some good guidelines for comparative and extension purposes. The original rule of thumb of 10t cane per 100mm of evapotranspiration (Thompson, 1976) has been reworked (Kingston, 1994) to 8.55t cane per 100mm for Bundaberg as a commercial benchmark. However, as Robertson and Muchow (1994) point out, this linear interpolation is a major simplification of yield response to water, with the reported range, 5.8 to 15.8t per 100mm, indicating that there is much scope for moving beyond a statistically derived threshold value (see also Robertson *et al.*, 1997).

In common with many other irrigated crops in Australia there is plenty of scope to improve irrigation efficiency. This has been demonstrated recently (Raine, 1995; Shannon and Raine, 1996) by studies in the Burdekin River Irrigation Area, where application efficiencies for individual irrigation events ranged from 14 to 90% and seasonal efficiencies from 31 to 62%. Simply implemented management practices such as changing furrow length, shape, cultivation practice and farm irrigation scheduling greatly improve water use efficiency and economic return.

The high frequency of papers on irrigation and water use appearing in sugarcane related conference proceedings and the availability of the BSES and SRDC 'Irrigation of Sugarcane' publication indicates a high level of research and grower interest. Indeed, the early assessment of sub-surface drip for cane (Nicholson, 1977) would be one of the first applications in Australia. Many irrigation methods are used through the industry, from surface furrow in the Burdekin Delta, surge irrigation (Stewart, 1994), sub-surface trickle (Morgan,

1991; Sullivan and Ridge, 1991) to a range of sprinkler types (Churchward and Poulsen, 1991).

Until the last few years the main focus with irrigation has been to improve production levels. However, like irrigation areas in southern Australia, the wider dimension of water use and resource impact are now of concern. Evidence of over-exploitation of groundwaters, particularly in delta and coastal plain aquifers, has become painfully apparent in several areas including the Burdekin Delta and the Bundaberg region. The vulnerability to sea water intrusion means sustainable groundwater yields must be carefully evaluated and managed accordingly. Localised areas of salinity are becoming evident, even in reasonably high rainfall areas. While the off-site impact of saline drainage flows has not been an issue, the potential impact of nutrient and other chemical contaminants in drainage and runoff waters is. While runoff will occur with rainfall, some drainage and increased runoff will occur from irrigation areas. The losses of N and P (Mitchell and Furnas, 1994) represent significant financial loss (Bramley *et al.*, 1994) as well as a source of contaminant. Managing regional land, soil and water resources and minimising downstream and off-site impacts requires well planned and implemented catchment management. Such planning is being urged (Kapitzke *et al.*, 1994) and acted upon with the development of a comprehensive environmental management plan for the Burdekin River Irrigation Area (BRIA). Other areas must follow this lead.

Ensuring productivity from irrigation of sugarcane

Evidence from irrigation areas around the world suggests that separation of land ownership and water rights provides the potential for market forces to direct water to where its greatest economic return is possible. So long as this is done in a transfer process where it is physically feasible to shift water allocations (i.e. within a catchment) and where there is protection for maintenance of shared delivery infrastructure, as in a large community irrigation area, water will generally be used for the greatest net gain. In community schemes there must be clear and open accounting so that fixed and variable costs are evident, water pricing needs are evident and there is an explicit provision for continued maintenance and upgrading.

End users, e.g. cane growers, must be involved in setting up and monitoring performance criteria of the supply system. The irrigation water boards of the Burdekin River Irrigation Area are good examples of this. Once those involved get an agreed criterion such as ‘provision of acceptable quality water on time, in the right amount, with minimum loss and at a minimum cost’ they must then seek more specific detail. For example, ‘system losses should not exceed 10% of diverted volume’, or ‘90% of farm deliveries will begin within 48 hours of notification’.

In addition to irrigators, the involvement of industry, the service sector, state agencies and the local community is important in retaining the focus for a viable, productive irrigation area. These groups are fundamental to the process of land and water management planning. Without their commitment, no plan will be successfully implemented. Every region must develop this kind of plan and hopefully such resource use planning will become as accepted and committed an activity by Local Government as urban infrastructure planning is currently.

Measuring and monitoring of water diverted, conveyed, applied and drained is fundamental to better management. Measurements will give information on where cost effective improvements can be made, and enable comparison with agreed performance criteria. Non-performance must result in decisive action - either to correct the deficiencies or to develop more realistic criteria. This information, so long as it is quality assured, will provide the most credible evidence that improvements are being made.

The link between productivity improvement and resource maintenance (environmental or ecological responsibility) is efficient water use. This may be measured in terms of mass or $ produced per unit of water input. Setting performance criteria which have the long term aim of very closely matching plant water use with water applied (at minimal conveyance loss) will be the single most important factor in ensuring longevity of irrigation areas. Current irrigation practices will inevitably add water to the groundwater. Even in the few situations where direct groundwater pumping is used as the irrigation source, e.g. the Burdekin Delta, unnecessary water addition to the groundwater will leach nutrients, predispose the system to wet conditions and add to pumping costs. In most areas minimising net recharge is the key to prolonging the life of the irrigation area. This can be greatly assisted by encouraging the use of more controlled, well managed irrigation systems.

In all irrigation areas, and especially in hotter drier climates, salts will accumulate. The more water, and the more saline this water, the more rapid will be the salt increase. Salt must be managed, it cannot be ignored. Those few areas where subsurface drainage is high, rainfall leaching is high and which are located close to a salt storage area (e.g. the sea) are indeed fortunate. Most areas are not so fortunate and they must actively plan their salt storage strategies. Salt is but one, albeit the largest, effluent produced by irrigation areas. Nutrients and chemicals will enter the effluent stream. Minimising drainage and retaining drainage waters is not only good effluent management, but it is also good agronomic practice. Good chemical practice requires a high level of continued education and training. Industries such as sugar have a particular responsibility to ensure suppliers and growers are fully informed. Formation of regional chemical task forces, to oversee practices and to provide training and accreditation, are a necessary part of intensive irrigation practice.

The development of regional land and water management plans will focus definition and documentation of acceptable best management practices. Resources such as soils, water, chemicals and crop residues will all need to be examined. In doing so, the single commodity industry like sugar may find that encouraging growers into alternative crops is a sensible rotational strategy for long term productivity. This may also be true from a regional socio-economic perspective given the likely continued volatility of primary product prices.

Finally, the importance of continued research and development into all aspects of the industry and, in particular, farm production cannot be over emphasised. Without a balanced portfolio of basic, strategic and applied research, intimately connected to education and extension, no industry will continue to prosper in an ever-changing and competitive world.

Acknowledgements

The assistance of BSES staff, Mr Gary Ham and Dr Steven Raine in providing references and information is gratefully acknowledged.

References

AIC (1994) *Value of the Irrigation Industry.* Information paper. Australian Irrigation Council.

AITP (1994) *Australian Irrigation Training Project: Project Brief.* B. Breen and P. Foley, NSW Agriculture and Department Primary Industries and Energy, Canberra, Australia.

Bramley, R.G.V., Johnson, A.K.L. and Smettem, K.R.J. (1994) Quantifying the effects of different rural land use on water quality in the catchment of the Herbert River - preliminary results. *Proceedings of the Australian Society Sugarcane Technologists* 16, 21-29.

Bryant, M.J., Faulkner, R.J. and Drilon, M.L. (1992) *Report to the Murrumbidgee Irrigation Management Board on Horticultural Viability and Future Irrigation Supply Options in the MIA*, Dec. 1992. 31pp and Appendices.

Bryant, M.J., Greiner, R. and Prathapar, S.A. (1993) Evaluation of policies to reduce recharge to shallow water tables in irrigated areas. International Conference on *Environmentally Sound Water Resources Utilization.* Bangkok, Thailand Nov. 1993. Vol II, pp. 158-165.

Chapman, L.S. and Chardon, C.W. (1979) An evaluation of irrigation at Mackay. *Proceedings of the Australian Society Sugarcane Technologists* 1, 117-123.

Churchward, E.H. and Poulsen, N.J. (1991) Management strategies for Bundaberg canegrowers for the 1990s. *Proceedings of the Australian Society Sugarcane Technologists* 13, 131-137.

Cole, P.J. (1985) *The River Murray Irrigation and Salinity Investigation Programme.* Results and Future Directions. River Murray Irrigation Salinity Programme Report 69, 48 pp.

Connolly, R. (1964) *John Drysdale and the Burdekin.* Ure Smith, Sydney, Australia, 214 pp.

Cummins, T. (1996) *Murrumbidgee Irrigation Area and Districts. Land and Water Management Plan.* Revised draft MIA and Districts Land and Water Management Plan Working Group. Murrumbidgee Catchment Management Committee. Griffith, Australia, Jan. 1996, 33 pp.

Davidson, B.R. (1969) *Australia Wet or Dry? The Physical and Economic Limits to the Expansion of Irrigation.* Melbourne University Press. Melbourne, Australia, 264 pp.

Ferguson, G.A. and Hampson, R.J. (1975) Some factors influencing the decision to irrigate sugarcane. *Proceedings of the Queensland Society Sugarcane Technologists* 42, 99-103.

Hallows, P.J. and Thompson, (1995) *The History of Irrigation in Australia.* Australian National Committee on Irrigation and Drainage, Mildura, Australia.

Ham, G.J. (1985) Water consumption by sugarcane under commercial growing conditions. 5th Afro-Asian Region Conference International Committee on Irrigation and Drainage. *Planning and Management of Water for Agriculture in the Tropics.* Townsville, Australia, pp. 23-33.

Ham, G. (1994) *Irrigation and Drainage - a Queensland Perspective.* 41st Annual General Meeting, Australian National Committee on Irrigation and Drainage, 11-14 Sept., Kununurra, Australia, 10 pp.

Harrison, J. (1994) *Review of Nutrients in Irrigation Drainage in the Murray-Darling Basin.* CSIRO Division of Water Resources, Water Resources Series: No. 11, 149 pp.

Hodges, R. (1995) *Historic Land Oriented Socio-Political Decisions and Their Effects on Current Land Management.* CSIRO Division of Water Resources, Seminar Series, Griffith Laboratory, 8 Nov. 1995, Griffith, Australia, 19 pp.

Kapitzke, I.R., Johnson, A.T. and Pearson, R.G. (1994) Responsible stream management practices for the north Queensland sugar industry. *Proceedings of the Australian Society Sugarcane Technologists* 16, 7-14.

Kingston, G. (1994) Benchmarking yield of sugarcane from estimates of crop water use. *Proceedings of the Australian Society Sugarcane Technologists* 16, 201-209.

Korth, W., Bowmer, K.H., Thomas, M., Foster, S., Napier, G. and McCorkelle, G. (1995) *River Pollution with Agricultural Chemicals.* Final Report. Land and Water Resources Research Development Corporation Research Support Grant No. CWN5 (89/14), Canberra, Australia.

Leverington, K.C. and Ridge, D.R. (1975) A review of sugarcane irrigation. *Proceedings of the Queensland Society Sugarcane Technologists* 42, 37-43.

MDBMC (1988) *Draft Salinity and Drainage Strategy.* Murray-Darling Basin Ministerial Council Discussion Paper No. 1. Jan. 1988, Canberra, Australia, 12 pp.

Meissner, T. (1995) *The SIMPET Report. An Approach to Integrating Policy, Education and Technology for Sustainable Irrigation in the Mallee.* Report to Murray Darling Basin Commission, Natural Resources Management Strategy project M373, Canberra, Australia.

Meyer, W.S. (1993) Addressing the problem of poor adoption of irrigation scheduling techniques. In: *Irrigation 93 'Under Pressure'.* Proceedings of the National Conference of Irrigation Association of Australia and Australian National Committee on Irrigation and Drainage Launceston, Australia, pp. 17-20.

Meyer, W.S. and Noble, C.L. (1993) Assessing the impact of irrigation on resources: irrigation management to meet environmental constraints. In: *The Future of Irrigation in the Murray-Darling Basin*, CSIRO Division of Water Resources, Griffith, Australia.

Meyer, W.S., Prathapar, S.A. and Barrs, H.D. (1990) Water flux to and from water-tables on two irrigated soils. In: *Management of Soil Salinity in South East Australia.* Australian Society of Soil Science Incorporated, Riverina Branch Symposium, Albury, Australia, pp. 79-87.

Mitchell, A.W. and Furnas, M.J. (1994) River inputs of N and P to the central Great Barrier Reef (16°- 19°S). *Proceedings of the Australian Society Sugarcane Technologists* 16, 30-37.

Morgan, T.E. (1991) Trickle and furrow irrigation of sugarcane in the Burdekin. *Proceedings of the Australian Society Sugarcane Technologists* 13, 334.

Nicholson, W.J. (1977) The installation of sub-surface drip irrigation in the Mackay district. *Proceedings of the Queensland Society Sugarcane Technologists* 44, 139-142.

Powell, R.A., Jensen, R.C. and Gibson, A.L. (1985) The economic impact of irrigated agriculture in NSW. *A Report to the New South Wales Irrigators Council Limited.* 163pp.

Raine, S. (1995) Increasing the efficiency of furrow irrigation for sugarcane production in the Burdekin. *12th Irrigation Association of Australia Conference Proceedings* Tamworth, 1995, pp. 52-58.

Robertson, M.J., Inman-Bamber, N.G. and Muchow, R.C. (1977) Opportunities for improving the use of limited water by the sugarcane crop. In: Keating, B.A. and Wilson, J.R. (eds), *Intensive Sugarcane Production: Meeting the Challenges Beyond 2000.* CAB International, Wallingford, UK, pp. 287-304.

Robertson, M.J. and Muchow, R.C. (1994) Future research challenges for efficient crop water use in sugarcane production. *Proceedings of the Australian Society Sugarcane Technologists* 16, 193-200.

Shannon, E.L. and Raine, S.R. (1996) Improving the irrigation efficiency of Burdekin canegrowers. *Proceedings 8th Australian Agronomy Conference* Toowoomba, Australia, 1996. pp. 502-503.

Smith, R.C.G., Mason, W.K., Meyer, W.S. and Barrs, H.D. (1983) Irrigation in Australia. Development and prospects. *Advances in Irrigation* 28, 99-153.

Smith, R.C.G., Meyer, W.S., Blackwell, J. and Mason, W.K. (1982) Improving the productivity of irrigated agriculture. *Australian Society of Agronomy Conference* Wagga Wagga, Australia, July 1982.

SRDC (1995) *Research and Development Plan 1995-2000.* Sugar Research and Development Corporation, Brisbane, Australia, 46 pp.

Stewart, R.L. (1994) Some recommendations for surge irrigation in sugarcane. *Proceedings of the Australian Society Sugarcane Technologists* 16, 362.

Sullivan, G.T. and Ridge, D.R. (1991) Trickle irrigation - increasing irrigation options. *Proceedings of the Australian Society Sugarcane Technologists* 13, 331.

Thompson, G.D. (1976) Water use by sugarcane. *South African Sugar Journal* 60, 593-600, 627-635.

Wadham, S., Wilson, R.K. and Wood, J. (1964) *Land Utilization in Australia.* 4th Edition. Melbourne University Press. Melbourne, Australia, 295 pp.

Wegener, M.K. (1990) Analysis of risk in irrigated sugarcane at Mackay. *Proceedings of the Australian Society Sugarcane Technologists* 12, 45-51.

Chapter twenty-five:

The Economics of Resource Use and Environmental Impact in Intensive Agricultural Systems

J. Quiggin
James Cook University, Townsville, Qld 4811, Australia.

Abstract

Like most industries, particularly intensive agricultural industries, the sugar industry must address a range of environmental concerns. The main object of this paper is to describe the frameworks used by environmental economists to analyse environmental problems and propose policy responses. Three such frameworks are considered, based on the concepts of externality, sustainability and property rights, respectively. Because of the complexity of the problems raised by the sugar industry, it may be appropriate to use different analytical frameworks in different contexts rather than to attempt a single all-embracing approach. On the whole, a property rights analysis incorporating a mixture of private and common property concepts appears to be most applicable to the problems of intensive agricultural systems. However, both externality and sustainability concepts have an important role to play.

Introduction

Most industries, particularly intensive agricultural industries like the sugar industry, must address a range of environmental concerns. Many of the environmental issues faced by the sugar industry are shared with other intensive agricultural industries. Some environmental issues arise from practices specific to the sugarcane industry, most notably, the practice of burning fields prior to harvesting. Others, such as mill effluents, arise from the nature of sugar as an

In: Keating, B.A. and Wilson, J.R. (eds) *Intensive Sugarcane Production: Meeting the Challenges Beyond 2000*, CAB International, Wallingford, UK, 1997.

industrial crop and the fact that, because of the costs of transporting cane, processing must be performed close to the point of harvest.

The main object of this paper is to describe the frameworks used by environmental economists to analyse environmental problems and propose policy responses. Three such frameworks are considered, based on the concepts of externality, sustainability and property rights, respectively. These approaches are evaluated in terms of their strengths and weaknesses in the analysis of different types of environmental problems. Because of the complexity of the problems raised by the sugar industry it may be appropriate to use different analytical frameworks in different contexts rather than to attempt a single all-embracing approach. The final section of the paper is concerned with the environmental implications of policy changes affecting the sugar industry.

Environmental Issues Facing the Sugar Industry

Mary Maher and Associates (1996) identify a number of environmental issues relevant to the sugar industry. The most pressing issues arise from expansion of the area of assigned land. In the absence of careful planning, expansion can create problems including:

1. Altering the existing drainage regime, including wetlands, poorly drained coastal plains and coastal waterways.
2. Clearing of habitat critical for the continued viability of particular fauna or flora species.
3. Clearing significant vegetation communities.
4. Reduction in riparian vegetation which has an impact on aquatic life, water quality and fish breeding grounds.
5. Fragmentation of previously integral native habitat.

Species at risk from clearing include the southern cassowary, the mahogany glider, the spot-tail or tiger quoll and Lumholtz's tree kangaroo.

As with other intensive agricultural industries, soil erosion and land degradation are important issues, raising concerns about the long term sustainability of the industry. Soil erosion was identified by Dawson *et al.* (1983) as the main environmental problem facing the industry. Land used in the sugar industry was shown to have erosive indices (a measure of erosion damage) of between 500 and 1500. Levels of 200 or more are considered to indicate serious erosion problems. In addition to the loss of productive capacity, soil erosion in the sugar industry generates downstream environmental damage arising from siltation and deposition of soil along and off the coast. Considerable progress in controlling soil erosion has been made in recent years, combining traditional stabilisation measures such as contour banks with innovations in farming systems such as green cane harvesting and trash blanketing.

Freebairn *et al* (1996) suggest that these measures can reduce annual soil loss from 150t ha^{-1} to 5t ha^{-1}.

Additional environmental problems, also shared with other intensive agricultural industries, include problems arising from runoff of pesticides, fertilisers and mill effluents and problems associated with the demand for irrigation water.

The Externality Model of Environmental Problems

The approach most frequently used by economists in the analysis of environmental problems is based on the concept of externalities, first proposed by Pigou (1924). An action by a firm or individual is defined as having external effects if it directly affects either the productive capacity of other firms or the welfare of other individuals. The key idea is that the actions of one individual directly impinge on others, without any direct market interaction. Sometimes external effects are favourable. A classic example is that a person who maintains an attractive garden for their own pleasure also gives pleasure to passersby. But most interest is centred on negative externalities, such as pollution. Several different kinds of external effects have been considered in the literature. The most important are point source externalities, diffuse or nonpoint externalities and congestion externalities.

A point source externality problem arises when a single firm generates externalities affecting one or more other firms or individuals. An example of a point source pollution problem in the sugar industry is that of effluent from sugar mills. There will normally only be a single mill on a given stretch of river and it is therefore possible, at least in principle, to identify the pollution generated by a given mill and to evaluate the damage caused by that pollution. Even in this case, however, it is sometimes difficult to distinguish the effects of mill effluent from the effects of other more diffuse sources of pollution such as the runoff of fertiliser and pesticides from canefields and other farms (Pearson and Penridge, 1992).

Diffuse or nonpoint externalities arise when many firms or individuals contribute to an external effect on one or more others. The classic example is that of urban air pollution arising from motor vehicle use. Most environmental problems associated with the sugar industry are of this kind. For example, clearance of land by a single farmer has no significant effect on biodiversity. But clearance by many farmers reduces and fragments the habitat available to native animal and plant species resulting in local loss of biodiversity and, in some cases, species extinction.

Congestion externalities arise when members of a group generate negative externalities affecting each other. A standard example is that of an open access fishery. Each fishing boat reduces the catch available to the others. This results both in excessive effort to achieve a given catch and in depletion of fish stocks over time. Congestion externalities frequently arise in irrigation areas.

Application of irrigation water results in rising water tables, with consequences including waterlogging and salinisation.

Because the individual generating the externality does not bear the consequences, activities that generate negative externalities may yield net benefits to the person who causes the externality, while making society as a whole worse off. The standard remedy is to 'internalise the externality', for example by imposing on polluters a tax, called a Pigovian tax, equal to the marginal damage. Such a policy will not eliminate pollution. However, the Pigovian approach is based on the assumption that, in the absence of externalities, individuals will make socially optimal decisions. So, when externalities are internalised, the 'socially optimal' amount of pollution is generated.

The externality approach works well in the analysis of point source pollution problems, those where the source of pollution can easily be identified. In cases of this kind, Pigovian taxes can be applied to equate private and social returns. In the case of mill effluents, for example, a base standard expressed, say, in terms of concentration of pollutants could be set and the mill could be required to pay a penalty for each unit of pollution observed in excess of the standard.[1]

For most of the environmental problems associated with the sugar industry, the sources of the problem are diffuse rather than localised. These nonpoint source pollution problems have been analysed in the externality framework, but the results are not particularly satisfactory, especially when pollution damage is related in a nonlinear fashion to the activity generating the damage. For the analysis of diffuse environmental problems, analytical frameworks based on property rights ideas and on the concept of sustainability appear to offer more valuable insights than those of the externality approach.

Sustainability

In recent years, considerable attention has been focused on the notion of sustainability. So many different definitions of sustainability have been offered that discussion of the issue is inevitably confused. A summary and critique of the literature is given by Lélé (1991). I shall interpret sustainability very broadly to encompass two main concerns: (i) the interests of future generations should be given equal weight with our own in making decisions affecting the long term future (Norgaard 1991); (ii) it should not be assumed that capital, that is, technology embodied in produced goods, can be substituted indefinitely to compensate for land, taken broadly to include all the contributions of the natural environment to human welfare, and agricultural production in particular (Hartwick 1977).

[1] To make a Pigovian tax work perfectly, the mill should also receive a payment for reductions in pollution below the minimum standard. The issue of payments for pollution abatements is complicated however, because of the possibility that firms may establish themselves in order to obtain rights to compensation for pollution abatement (Baumol and Oates, 1975).

From these two main concerns we may summarise the core of the notion of sustainability as the idea that production systems should be required to maintain the stock of 'natural capital' on which they rely. The notion of sustainability may be developed fairly simply in the case of a fishery. A sustainable yield in a fishery is one which is consistent with a stable fish population. In a sustainably managed fishery, a constant average catch will be achieved with no change in fishing effort. Many fisheries are unsustainably managed with steady increases in fishing effort being required to maintain catch rates. Such policies almost invariably result in catastrophic collapse of the fish stock and commercial, if not biological, extinction of the target species. In simple cases such as that of the fishery, it is possible to incorporate the notion of sustainability into a management rule.

From this point, analysis has proceeded in two quite different directions. Attempts to derive general principles lead naturally to the conclusion that unsustainable practices must arise as the result of an inappropriate choice of discount rates (Quiggin, 1996). The resulting policy recommendation is to place greater weight on future environmental benefits in any formal or informal benefit-cost analysis used to decide policy.

One way of doing this, frequently advocated by environmentalists, is to reject discounting of environmental benefits. An alternative, more appealing to many economists, is to project a rising price of environmental benefits in the future. Over the medium term, say, from 10 to 50 years, the two approaches are equivalent, assuming that the rate of price increase for environmental goods is equal to the rate at which future benefits are discounted. Over the very long term, the zero discounting approach is equivalent to an assumption that environmental benefits become infinitely valuable. This does not seem reasonable given that the economy and population must ultimately enter a sustainable steady state.

Another direction of thinking about sustainability has been to seek to derive, for every form of agricultural enterprise, specific sustainability rules analogous to the sustainable catch rule for fisheries. Within this framework specific practices may be classed as 'sustainable' or 'unsustainable', according to whether they satisfy the specific sustainability rule. A variety of sustainability rules are discussed by Barbier (1987) and Barbier *et al.* (1990). The sustainability rule approach works best in the case of activities, like fisheries, that are dependent on a specific local stock of natural capital. Where activities draw on nonspecific stocks of natural capital attempts to apply specific sustainability rules encounter difficulties.

Consider, for example, the use of fertilisers in intensive agricultural industries such as sugar. In aggregate, it seems likely that current rates of fertiliser use are unsustainable, both in the sense that natural stocks of phosphate fertilisers will be exhausted and in the sense that the capacity of the environment to absorb runoff is limited. However, any one farm could carry on indefinitely with current rates of fertiliser use. More generally, there are many different patterns of fertiliser use that would be consistent with an aggregate constraint requiring total use to be reduced to some given level. In some of these patterns, the use of fertilisers in the sugar industry would be largely unaffected and other industries would bear the entire burden. In other patterns, the use of fertiliser in the sugar industry would be heavily reduced. It does not seem possible to deal with problems of this kind by *ex ante* specification of sustainable farming practices. Rather it is necessary to impose either a Pigovian tax or an implicit shadow price on the use of fertilisers, taking account of differences in environmental impact in different locations. As a scarce resource, rights to use fertiliser will be allocated to their highest value uses.

The sugar industry relies on a complex mix of land and water resources for its natural capital. The issues of sustainability are correspondingly more complex than in the case of fisheries. The most acute questions of sustainability arise in relation to water resources. The problems arising from runoff of fertiliser and pesticide have already been mentioned. One of the most important concerns in the context of the sugar industry arises from potential impacts on the Great Barrier Reef Marine Park. Further sustainability problems arise with the expansion of the sugar industry from its traditional base in high rainfall areas to dryland areas reliant on irrigation.

Intensive agricultural industries based on irrigation have encountered numerous sustainability problems relating to salinity, siltation and waterlogging. In these cases, the general approach to sustainability based on reducing the rate of discount of future environmental benefits appears appropriate. It is also necessary to deal with complex interactions between users of the water resource. For the analysis of problems of this kind, a property rights approach is most appropriate (Dahlman, 1979; Runge, 1981; Quiggin, 1986).

Property Rights Approaches - Private and Common

The most prominent critique of the Pigovian position is that developed by Coase (1960), who advanced an alternative analysis based on the concepts of 'property rights' and 'transactions costs'. In the absence of transactions costs, the famous 'Coase Theorem' states that the allocation of property rights, and in particular 'rights to pollute', makes no difference to resource allocation. If transactions costs preclude Coasian bargaining, the allocation of property rights should be that which would promote efficiency in any particular case.

The argument may be illustrated by an example. Suppose that initially there are no restrictions on land clearance by farmers, and that government is considering restricting clearances in order to protect endangered species. In Coase's analysis, the effect of restrictions on clearance is to transfer some of the property rights associated with a parcel of land from farmers to those members of the community who value species preservation. Coase observes that if there were no difficulties in reaching and enforcing bargains, groups concerned with species preservation could compensate farmers in return for an agreement not to clear land. Difficulties in reaching and enforcing bargains are referred to as transactions costs. An example of such a difficulty is that landholders with no intention of clearing land could threaten to do so in order to attract compensation. Assuming, as seems likely, that such transactions costs prevent the parties concerned from reaching direct agreements, governments or courts must decide what restrictions to impose on clearance rights. Under certain circumstances, a 'market' in clearance rights may emerge, achieving some of the efficiency gains associated with Coasian bargaining. During the 1960s and 1970s, Coasian ideas were developed by the private property rights school, notably represented by Furubotn and Pejovich (1974). However, the private property rights school found great difficulty in analysing changes in property rights. This is critical, since, within this framework, it is only through changes in property rights structures that environmental issues can be addressed.

Randall (1983) distinguishes between two post-Coasian traditions: a Coase-Posner tradition, following Posner (1972), which advocates flexibility in property rights when this would promote efficiency, and a Coase-Buchanan tradition which emphasises security and stability of rights. Randall points out the weaknesses of the Coase-Posner tradition. In addition to the reduction in the value of rights, and the associated disincentives for investment, Randall points out that instability in property rights encourages 'rent-seeking' behaviour aimed at securing a reassignment of rights. The Coase-Buchanan tradition is based on a contractarian theory of rights, of the type specified by Buchanan (1977). Following a hypothetical initial assignment of rights, all subsequent changes would be through voluntary exchange or other consensual processes. This approach meets Randall's objections to the Coase-Posner tradition, but involves significant limitations of its own.

The strengths and weaknesses of the private property rights approach may be seen by considering the case of the mahogany glider. There is a conflict between the clearing of land for sugar and other intensive agricultural industries and the preservation of habitat for animals such as the mahogany glider. In the absence of transactions costs, as Coase observed, this issue could be resolved by negotiations between farmers and environmentalists concerned with biodiversity. However, transactions costs are very high in cases of this kind. The critical issue, then, is how property rights are allocated and how they should change over time. Fifty or even twenty years ago, there would have been little doubt about the appropriate allocation of rights; the production of a valuable export commodity was regarded as being far more important than the habitat of an obscure marsupial. Today, however, community concern with species preservation is much greater, and this is reflected in restrictions on land clearance. In the terminology of the property rights school, these restrictions represent an attenuation of rights in land.

The private property rights framework provides some useful descriptive insights into environmental problems of this kind, but very little policy guidance. The Coase-Buchanan view, that property rights should be inviolate, is of appeal only to those who started out with plenty of property, and the initial allocation of rights gave almost nothing to the environment. The Coase-Posner view that property rights will evolve to take account of changing values seems to be correct. However, policy applications of this view assume the existence of an activist judiciary, willing to make the necessary changes through their interpretation of common and statute law. In the Australian context, this is not a realistic assumption, and the Coase-Posner view gives little guidance to legislators or bureaucrats.

One response to the difficulties associated with the private property rights analysis has been an increasing interest in the concept of common property. Economists in the private property rights tradition had long used the term 'common property' as a synonym for 'no property'. As long as this terminology was confined to the study of open access resources, as in the work of Scott (1955), it was merely confusing. However when economists in this tradition, such as Demsetz (1967), were confronted with actual common property institutions, they readily assumed that their open access model was applicable. This error was popularised by Hardin's (1968) 'tragedy of the commons' description of the medieval open field system[2] and the accompanying argument for private property rights solutions to modern environmental problems. The work of Ciriacy-Wantrup and Bishop (1975) and Dahlman (1980) refuted Hardin's description of the open field system. This was followed by work describing the actual operations of contemporary common property systems in less developed countries (LDCs), such as that of Jodha (1985) and Wade (1987). Much of this work has been focused on the analysis of common property in agricultural and grazing land. However, common property ideas have also been applied to more general environmental issues such as the management of airsheds and river systems. An example of the latter approach, applied to the problem of salinity in the Murray-Darling river system, is Quiggin (1988).

The problems of fertiliser and pesticide runoff in sugar production have many features in common with the salinity problems of the Murray-Darling river system. In both cases there are interactions among producers. In addition, the impact of effluent flows on the Great Barrier Reef Marine Park is paralleled by the impact of irrigation-related salinity on urban water users in Adelaide. In each case, it is necessary to establish institutions which permit both collective management of interactions among agricultural producers and resolution of interactions between agricultural producers as a group and downstream water users.

The common property approach extends naturally to problems of land and water management in the sugar industry. Land ownership involves a mixture of private property rights over the production of crops and common property rights arising from the fact that regional properties of the land resource, such as biodiversity, are determined by the interaction of many separate land use decisions. A wildlife corridor, for example, is a collectively produced good, and not merely the aggregate of separate decisions on what proportion of land should be cleared.

---

[2] Hardin based his tragedy of the commons story on a similar analysis of the commons by Lloyd (1833). It may be noted that Lloyd was writing just when the final wave of the enclosure movement had eliminated the possibility of an empirical test. Earlier critics of the open field system, such as Arthur Young (1801), were more circumspect.

Comparing Analytical Frameworks

The core concerns of the competing analytical frameworks used by environmental economists may be summed up by saying that the externality framework is concerned with prices, the sustainability framework with production processes and the property rights framework with social institutions.

Each of these approaches has its strengths and weaknesses in different situations. The externality approach is most useful in dealing with point source pollution problems and with cases where environmental damage is directly proportional to output or to the use of some input such as fertiliser or pesticide. The sustainability approach is most useful in evaluating policy constraints that may be imposed on specific farming practices. The property rights approach is most useful in examining non point resource problems such as degradation of waterways and biodiversity loss associated with land clearance. The ideal of an all-encompassing analytical framework capable of resolving all of the environmental problems associated with sugar production is not attainable, at least at present. It is necessary to draw on a variety of frameworks for economic analysis, as well as on insights derived from other social sciences and from the natural sciences.

Regulation in the Sugar Industry

Sugarcane has special characteristics as a crop that have led, in Australia, to a unique system of regulatory controls. Like many agricultural commodities, the final product is consumed far from the point of crop production. Sugarcane is grown almost exclusively in the tropics and consumed primarily in wealthy countries in the temperate zone. The international nature of the sugar market means that it is natural for large national and multinational enterprises to be engaged in the refining, transport and marketing of sugar. However, cane is so bulky that the initial stages of processing can only be economically undertaken close to the point where the cane is grown. Cane can only be grown profitably if a mill is established nearby and most growers have little choice but to send their cane to the nearest mill for processing. The result is a situation, referred to in the economics literature as monopsony, where growers must bargain with a single buyer, which may be one of many branches of a large enterprise. The inequality of bargaining power is likely to lead to a low price for cane, and to a transfer of wealth from growers to processing companies.

A complex system of regulation has built up around the canegrowing and processing industries. Initially, at least, the object of this system of regulation was to prevent exploitation of growers by processors, and to provide an equitable resolution of disputes about the appropriate price for cane. Regulations involved restrictions on where cane could be grown, where the cane grown on any given piece of land must be processed and on the terms and conditions under which growers and processors negotiated prices. Subject to these regulations, a single firm CSR Ltd handle the marketing of all sugar produced in Queensland and therefore, for practical purposes, all sugar produced in Australia. Although growers and processors are in conflict regarding the price of cane, both stand to benefit, at the expense of consumers, from an increase in the price of sugar. The regulatory structures designed to prevent the exploitation of growers by processors could also be turned to facilitate the exploitation of consumers by producers. With the effective exclusion of imports through a tariff, the price of sugar within Australia is determined by the quantity released for sale within Australia, and this quantity was controlled by negotiations between a monopoly enterprise and regulators concerned primarily with the interests of the sugar industry. Although the stated rationale for the maintenance of a fixed domestic price was stabilisation, many economists came to the conclusion that regulation was designed to facilitate the charging of monopoly prices to domestic consumers in order to increase returns to canegrowers and processors. Boston Consulting Group (1996) dismiss the stabilisation rationale for tariffs on sugar imports. They observe that it is questionable whether sugar producers should be insulated from market fluctuations and that, in any case, the stabilising effect of current arrangements is minimal.

Changes in Policy and Their Environmental Implications

Because of the complex economic relationship between mills and growers, the view that the regulatory structure is primarily designed to extract higher prices from consumers (Sieper, 1982) is simplistic. However, with the tide of opinion generally turning against intervention, the regulatory structure in the sugar industry has become increasingly difficult to sustain. Already, considerable relaxation of the regulatory structure has taken place. Most notably, the move to local control over land assignment has resulted in considerable expansion in the area of assigned land (Boston Consulting Group, 1996).

The 'Hilmer' competition policy, already enacted into law by State and Federal Parliaments, places the burden of proof on the defenders of existing regulatory arrangements to show why they should not be deemed anticompetitive. Although representatives of the sugar industry are vigorously resisting application of the Hilmer reforms to sugar, it seems unlikely that the current regulatory structure can be sustained. The Boston Consulting Group (1996) discusses a range of options for changes in the land assignment system, bargaining between mills and growers over prices, and controls on domestic and international marketing of sugar. Regulatory structures designed to achieve economic and social goals frequently have unforeseen environmental consequences. For example, by raising the profitability of mechanised agriculture, the European Common Agricultural Policy, designed to preserve village communities and protect farmers' incomes, has had the effect of transforming large areas of countryside from patchworks of fields and hedgerows into vast open fields. Similar unforeseen consequences may be seen in the case of the canegrowing industry.

The principal instrument used to regulate canegrowing was the system of assigned land, which restricted cane production to specific plots of land. The result of the system as a whole was to reduce the total area used for cane production and the result of land assignment was to restrict the reallocation of land from cane production to other agricultural activities and vice versa. However, the combination of high prices and restricted areas of land meant that more intensive production techniques were profitable under regulation than would have been the case otherwise. If farmers wished to increase output, they were constrained by the assignment system from doing so by expanding their crop area. Considerable improvements in sugar yields were achieved in response to the incentives provided by the assignment system. From 4.7t ha^{-1} in 1920, shortly after the introduction of the assignment system, yields rose to 13.9t ha^{-1} in 1994 (Boston Consulting Group, 1996), a rate of improvement considerably greater than in industries such as wheat, where output growth was achieved primarily through increases in farm area. However, the incentives for intensive production tend to increase the severity of problems such as soil erosion.

The gradual relaxation of regulation has yielded both environmental benefits and environmental costs. On the one hand, relaxation of restrictions on the use of land has made feasible a less intensive production technology with the adoption of innovations designed to reduce such negative effects as soil erosion. On the other hand, the expansion of cane growing into new areas, typically displacing relatively low-intensity beef cattle operations, has created new problems associated with tree-clearance and habitat loss.

Concluding Comments

A variety of economic frameworks have been used in the analysis of resource use and environmental impact in intensive agricultural systems. Each of these approaches has its strengths and weaknesses in different situations. On the whole, a property rights analysis incorporating a mixture of private and common property concepts appears to be most applicable to the problems of intensive agricultural systems. However, both externality and sustainability concepts have an important role to play. In the case of the sugar industry, there is a complex interaction between environmental impacts, market forces and regulatory policies. The removal of existing regulatory structures will have unforeseen impacts, some favourable and some unfavourable. As policy change proceeds, social and natural scientists should monitor its environmental impact and seek to design policy response that will mitigate unfavourable impacts and exploit favourable ones.

References

Barbier, E.B. (1987) The concept of sustainable economic development. *Environmental Conservation* 14, 101-110.

Barbier, E.B., Markandya, A. and Pearce, D.W. (1990) Environmental sustainability and cost-benefit analysis. *Environment and Planning A* 22, 1259-66.

Baumol, W. and Oates, W. (1975) *The Theory of Environmental Policy*, Prentice-Hall, New Jersey, USA.

Boston Consulting Group (1996) *Report to the Sugar Industry Review Working Party.* Boston Consulting Group, Sydney, Australia.

Buchanan, J. (1977) *Freedom in Constitutional Contract*, Texas A & M University Press, College Station, Texas, USA.

Ciriacy-Wantrup, S. and Bishop, R. (1975) 'Common property' as a concept in natural resource policy. *Natural Resource Journal* 15, 713-27.

Coase, R. (1960) The problem of social cost. *Journal of Law and Economics* 3, 1-44.

Dahlman, C. (1979) The problem of externality. *Journal of Law and Economics* 22, 141-62.

Dahlman, C. (1980) *The Open Field System and Beyond.* Cambridge University Press, Cambridge, UK.

Dawson, N., Bernt, R. and Venz, B. (1983) Land use studies - Queensland canelands. *Proceedings of the Australian Society of Sugarcane Technologists,* 5, 43-52.

Demsetz, H. (1967) Towards a theory of property rights. *American Economic Review* 57, 347-59.

Freebairn, D.M., Truong, P., Miles, R., Ciesiolka, C.C.A., Titmarsh, G. and Norrish, S. (1996) Practical measures to counter soil erosion. In: Hunter, H.M., Eyles, A.G. and Rayment, G.E. (eds), *Downstream Effects of Land Use.* Department of Natural Resources, Brisbane, Australia, pp.213-219

Furubotn, E. and Pejovich, S. (1974) *The Economics of Property Rights*, Ballinger, Cambridge, USA.

Hardin, G. (1968) The tragedy of the commons. *Science* 162, 1243-1248.

Hartwick, J.M. (1977) Intergenerational equity and the investing of rents from exhaustible resources. *American Economic Review* 67, 972-974.

Jodha, N.S. (1985) Common property resource in farming in arid and semi-arid tropical India. *Proceedings 29th Australian Agricultural Economics Society Conference*, February 1985, Armidale, Australia.

Lélé, S. (1991) Sustainable development: A critical review. *World Development* 19, 607-19.

Lloyd, W. (1833) Two Lectures on the Checks to Population, Oxford University Press, Oxford Reprinted in part in Hardin, G. (ed.) *Population, Evolution and Birth Control*, Freeman, San Francisco, USA 1964.

Mary Maher and Associates (1996) *Environmental Issues and the Assignment System.* Boston Consulting Group, Sydney, Australia.

Norgaard, R. (1991) *Sustainability as Intergenerational Equity: The Challenge to Economic Thought and Practice.* World Bank, Internal Discussion Paper no IDP-97, World Bank, Washington DC, USA.

Pearson, R. and Penridge, L. (1992) *An Ecological Survey of Selected Rivers in Queensland with Particular Reference to the Effects of Sugar Mill Effluents.* Australian Centre for Tropical Freshwater Research Report no 92/02, James Cook University, Townsville, Australia.

Pigou, A. (1924) *The Economics of Welfare,* Macmillan, London, UK.

Posner, R. (1972) *Economic Analysis of Law*, Little Brown, Boston, USA.

Quiggin, J. (1986) Common property, private property and regulation: the case of dryland salinity. *Australian Journal of Agricultural Economics* 30, 103-117.

Quiggin, J. (1988) Murray River salinity - an illustrative model. *American Journal of Agricultural Economics* 70, 635-45.

Quiggin, J. (1997) Discount rates and sustainability. *International Journal of Social Economics* (in press).

Randall, A. (1983) The problem of market failure. *Natural Resources* 23, 131-148.

Runge, C. (1981) Common property externalities: isolation, assurance and depletion in a traditional grazing context. *American Journal of Agricultural Economics* 63, 595-606.

Scott, A. (1955) The fishery: the objective of sole ownership. *Journal of Political Economy* 63, 116-124.

Sieper, E. (1982) *Rationalising Rustic Regulation*, Centre for Independent Studies, Sydney, Australia.

Tisdell, C. (1988) Sustainable Development: Differing Perspectives of Ecologists and Economists, and Relevance to LDCs. *World Development* 16, 373-84.

Wade, R. (1987) The management of a common property resource: Finding a co-operative solution. *The World Bank Research Observer* 2, 219-34.

Young, A. (1801) *Enquiry into the Propriety of Applying Wastes to the Better Support and Maintenance of the Poor*. London, UK.

Chapter twenty-six:

Doing Things Differently: the R,D&E Revolution?

E.J. Woods[1], P. Cox[2] and S. Norrish[3]

1. *Horticulture Industry Services, Department of Primary Industries, Brisbane, Qld 4000, Australia*
2. *Socioeconomics and Policy Division, ICRISAT, Patancheru, Andhra Pradesh 502 324, India*
3. *The Faculty of Agriculture & Rural Development, University of Western Sydney, Richmond, NSW 2753, Australia*

Abstract

Agricultural R,D&E practice is moving from the provision of novel technological components for production and resource management to include participatory processes based on adult learning which recognise human development (of farmers and scientists) as a legitimate aim. The traditional clients of R,D&E face increasingly complex issues; scientific 'solutions' developed in the absence of a real-world management context can represent poorly targeted and inefficient R,D&E investment. But the emerging perceptions of R,D&E context, processes and criteria provide significant opportunities. Participatory processes provide one opportunity of gaining value from different perspectives and a way of opening up novel possibilities by questioning underlying assumptions. The negotiation of different courses of action (technology in a broad sense) and how to choose between them or improve them then comes to the fore. This raises fundamental methodological and institutional issues for how we do science, what constitutes an outcome, who controls the agenda, and the accountability of the participants to others. Working through these issues will bring about substantial shifts in culture of RD&E professionals and other stakeholders including producers. Case studies from the grains and sugar industries, based on recent patterns of innovation in both production and information technologies, illustrate these issues within the Australian context.

In: Keating, B.A. and Wilson, J.R. (eds), *Intensive Sugarcane Production: Meeting the Challenges Beyond 2000*. CAB International, Wallingford, UK, 1997.

Introduction: The R,D&E Revolution

The potential benefits of investment in R,D&E are large, but they are not always captured, in agriculture and in other industry sectors. The scope for improvement in agricultural R,D&E was first clearly recognised in less developed economies. Discussion of the global problem has evolved over the last 25 years and forms the basis of a quiet revolution in agricultural R,D&E which is still going on. There are many threads to this revolution, which involves questioning the assumptions which underpin existing R,D&E practices. The revolution recognises the concept of human agency - simply defined as *the inherent capability of people to make a difference to the state of affairs or course of events* (Vanclay, 1994, p.111-2). It recognises the importance of choice, and that agriculture (and R,D&E) is as much about responsibility and accountability within a wider community as effectiveness and efficiency in production and marketing. It is a quiet revolution because to date it is better developed in concept than in practice. The emerging practice in R,D&E is about facilitating purposeful and beneficial change in diverse communities comprising a range of stakeholders. The thrust is not abrupt change, but learning and evolution (encouraging cycles of action and reflection to gestate change). Our conclusion is that we are not very far along this path. In this paper, we examine progress to date, why the revolution is important and what it may mean to the sugar industry.

Progressing New Concepts: the Global Perspective

Substantial progression in agricultural R&D can be linked to Farming Systems Research (FSR) (Collinson, 1987). FSR used systems analysis with a technology focus but recognised socio-economic constraints on production. Although social science approaches, mainly through economists but also anthropologists and sociologists linked with the international research centres, were first included during this phase, evaluations indicated the FSR model often failed to properly accommodate the socio-economic and cultural contexts (Davidson, 1987). The *Farmer First* model (Chambers *et al.*, 1989) grew from criticism of FSR as conceptually based on the linear 'transfer of technology' model, with experts delivering solutions to relatively passive farmers. *Farmer First* emphasised farmers' involvement in defining the problem as well as the importance of the social context, using methods such as Rapid Rural Appraisal and Participatory Rural Appraisals which evolved to include local people and non-government agencies as participants in data collection, interpretation and planning.

Farmer First was itself criticised as driven by outside experts and their assessment of a problem situation, for example, by Scoones and Thomson (1994). They explored the underpinning assumptions of the R,D&E system, asking questions such as:

1. Whose criteria and priorities count?
2. Whose knowledge is more important?
3. Whose modes of learning and analysis and whose tests, experiments, observations, assessments, are most critical to the direction of change?
4. Whose reality counts?

Their approach incorporated systems thinking, constructivist philosophy, and the reality of multiple perspectives. The feminist critique of science (Harding, 1991; Haraway, 1995; Rose, 1994; Jiggins, 1994), viewing science from outside, has also questioned the assumptions underpinning R,D&E. They noted the *situatedness* of knowledge in people's lives and asked how values such as care and a responsible rationality are handled by the R,D&E system.

The issues we face in agriculture are not unique. In the industrial/commercial world, TQM (total quality management) and reengineering (Kelada, 1996) formed new directions for change, displacing energy and resources from R,D&E as the primary driver of progress. In education, action-learning challenged the essential role of teaching and the expert teacher.

The conceptual revolution in R,D&E champions 'participation' of different groups of stakeholders. Participation is variously seen as solution, as context and as opportunity. Participation is a way of countering dependence, of capturing novel insights, and of building upon (rather than ignoring) the capacity of people to take action. The shift from product to also include process raises new issues for science - how it is done; how it is evaluated; how it is rewarded.

The situation in Australia

During the 1980s and 1990s there has been intense interest in Australia in capturing the benefits of agricultural and resource management R,D&E. Concern was heightened in an environment characterised by reducing government expenditures for the agricultural sector and a shift in focus to encompass resource management issues. The ensuing debate focused on research structures and scale but rarely questioned the intrinsic value of research. By comparison, the downward trend in available funds and the feasibility of private sector delivery of at least some extension services led to intense questioning of the value and future shape of extension services (eg. Russell *et al.*, 1989).

Challenged to demonstrate their worth and cope with declining operating budgets, extension workers moved to increase their professionalism. Government extension workers with backgrounds mostly in science and technology focused on increasing their awareness of social processes and skills in professional practice (craft skills). In the private sector, skills in business management and marketing became the increasing focus of professional development. An important characteristic of the extension debate was its relative isolation from research and development workers. Extension's increasing awareness of social science concepts and skills has resulted in an increasing philosophical separation of R&D and E. Some research and extension workers have rejected this as a civil war within R,D&E ranks! For others who have chosen to remain aloof from new perspectives, the perception may be 'What revolution? Did we miss something?'

We believe that some significant changes have occurred in Australia in response to this evolution. Examples include Bawden and colleagues at the University of Western Sydney who pioneered a different approach to R,D&E and agricultural education (Macadam and Wilson, 1993), and the rise of Landcare following recognition of the community as a possible focus for learning and action to improve resource management. Professional science has invested in models (for systems analysis) and the development of Decision Support Systems as a delivery tool to support improved management. There has been a shift in R,D&E focus from providing tangible technologies (e.g. new cultivars developed through plant breeding) to how to use technologies in management, and from research on research stations to increasingly on-farm research. Researchers and research funders have moved to support and train farmers in conducting their own research (e.g. the Producer Initiated R&D project supported by the Meat Research Corporation and the Do Our Own Research project in the nursery industry).

However, R,D&E faces continuing challenges. Technical developments in information technology and bioengineering and the increasing stress facing the agriculture sector (due to declining terms of trade, rising awareness of the environment and sustainability, and adverse climatic conditions) have raised questions about the ethics and values underpinning agricultural R,D&E and the willingness and capacity of R,D&E workers to contribute to improving the situation of rural people. Changes in funding and institutional arrangements for R,D&E have focused R,D&E workers on being accountable in relation to the interests of government and industry collectively (through the rural R&D Corporations), but not necessarily on the problems and issues facing individual farmers or groups of farmers, or in ways that are acceptable to them and to other client groups. There continues to be a strong perception that all the benefits of R,D&E are not being captured.

Clearly substantial discussion of R,D&E has occurred. But has Australian R,D&E been affected by these contributions? Institutional science has the inertia of any established organisation and it is possible that old concepts are reinforced by external groups such as funding bodies whose members share the same history. To what extent has R&D moved to recognise that its role is to contribute to learning, and to link with the tacit knowledge of agricultural practitioners and the situations facing rural people? Do we and other R,D&E workers want to do things differently and better? What do the changes entail?

Case Studies

We develop two major case studies: one relating to a major investment in information technology in the grains/cotton industries; the other to the development of a production technology in the sugar industry. There are parallels in these two experiences: a questioning of the assumptions underpinning R,D&E informed by the valuable tension between technological and social perspectives, with first a scientific imperative, then an overt questioning of the purpose of the process and then a progressive re-negotiation. In each case, resource management issues are both a source of questions and an alternative framework to work within; and participation is recognised as a useful process, but proves difficult to achieve.

Case study 1: APSRU

APSRU (the Agricultural Production Systems Research Unit) is a joint project established by the Queensland Department of Primary Industries and the CSIRO Division of Tropical Crops and Pastures in 1990 in recognition of the skills within the two organisations in constructing simulation models of crop and soil processes. Several examples (*SIRATAC*, *Wheatman*) of the use of these had been developed in farm extension. Although a diverse group (with peripheral interests in agroclimatology and the socio-economic aspects of agricultural R,D&E), much of the first five years was directed at the development of a modelling environment or shell called APSIM (McCown *et al.*, 1996). The intention was to use this as a vehicle for combining (in parallel and in sequence) a number of modules dealing with the growth of different crops (both summer and winter) in response to historical patterns of rainfall and temperature, and to describe the resulting distributions of yield and soil properties (water content, fertility). Use of a systems model in this way was proposed as an 'operations research' approach to obtaining improvements in farm management through quantification of the 'risk' (i.e. the level and variation in response) associated with different options.

By their nature, the context of R,D&E activities differs from the context in which the technologies they generate are intended to be used. This happens despite the best intentions of the participants to be relevant to clients. The inevitability that the contexts of R,D&E workers and their clients will differ leads both to their potential complementarity and to a need for both parties to be constantly questioning and negotiating underlying assumptions. It raises questions about the nature of our systems for generating and adapting technology, the social processes we use to engender novel technologies, and the patterns of innovation that we pursue. This case study illustrates some of the available choices faced when designing APSRU's R,D&E activities. It demonstrates the process of questioning not only what happens but also the underlying assumptions in our perspective on the situation. This is known as double loop learning (Argyris *et al.*, 1985), an approach which is integral to the concept of organisational learning. Double loop learning is suggested as one of the strategies by which organisations can learn new skills and shift their core business to match new needs.

APSRU, like most human activity systems, was set up on the basis of interlinked and mutually supporting assumptions reflecting values and perspectives about the nature of reality: about the limits to the performance of agricultural production systems; and about the way such limits can be overcome. These assumptions were not immediately apparent to all members at the time; again in common with most human activity systems, they became clearer and available for wide debate only later on reflection. Unlike many R,D&E groups, the APSRU group had included in its initial design the potential for several perspectives. The vigorous debate in which the group has engaged - both internally and in public arenas - provides a sentinel case study through which to examine the sorts of assumptions which underpin R,D&E. For example, the perceived severity of a child's fall at play is affected by whether it is your child, and by your recent experience of children's injuries. In the same way, the perspectives of progress by a research team differ depending on an individual's starting point and relationship with the team's processes. The interpretation presented here is from the perspective of an individual participant and therefore reflects both the participant and his place in the team activities. It does not purport to represent all the members of the team or to demonstrate good and bad. Its value lies in exposing some questions which can be used to negotiate the meaning (applicability) both of research results and of the overall research direction which need to be periodically addressed in research management and review processes.

Thus, in retrospect, one view of the series of cascading assumptions underpinning APSRU included the following:

1. That farm management is about decision-making and that, to the extent that this process can be made more rational, it can be improved.
2. That poor decision-making can result from either poor information inputs and/or poor information processing, i.e. poor decision-making behaviour.
3. That scientists can contribute where producers' decision-making is deficient by meeting information needs.
4. That researchers can identify where decision-making is deficient, i.e. the information needs of farmers.
5. That these information needs can be met by building a scientific description of the production system.
6. That the factors that limit physical yield are where the important gains in performance are to be found.
7. That improved analysis is equivalent to improved precision in prediction.
8. That prediction of the future is equivalent to prediction of the past.
9. That more precise specification of the system will lead to the identification and exploitation of novel opportunities.

In summary, this chain of reasoning concludes that the further development of the current generation of biophysical simulation models will lead to improved decision-making behaviour on the farm.

Questioning RD&E assumptions

In this case study, we apply learning organisation principles and raise questions about the assumptions perceived to underpin the APSRU experience. We know of few other cases of Australian agricultural R,D&E where this overt questioning has occurred. It certainly occurs in relation to scientific issues and informally between stakeholders, yet such a fundamental process to improving the scope and design of research activities is rarely documented or open to broader consideration. Let us deal then with the cascading assumptions that were perceived to underpin APSRU.

Assumption 1: That farm management is about decision-making
Looking at management from the outside and in the abstract, it is easy to accept that management is largely about making 'decisions'. Decisions are conceptualised as conscious choices between distinct alternatives each associated with well-defined cost and benefit streams. However, when farmers' decision-making is examined, it is no longer so clear that management is about decision-making. The choices are not obvious, well-defined, or with predictable consequences. It is even less obvious that the management of broadacre dryland cropping in the north-eastern grain belt of Australia (the focus of APSRU's work) has much to do with decision-making in this formal sense, because the variation in water supply is such that only a limited sample from the set of possible outcomes can ever be experienced. In considering the scope and design of research, the question to be answered is what evidence is there that simulation models will contribute to the evolution of a better-managed system within the narrow set of choices in any particular situation?

Assumption 2: That poor performance reflects poor decision-making
There is an assumption that farm performance can be improved, but we have little evidence that poor decisions are being made or about the nature of the constraints preventing better decisions. It is possible to make good decisions (on the basis of the evidence available when the decision has to be made) which turn out to have bad outcomes; the issue then is about how to manage unexpected outcomes. For example, there is a concern amongst researchers that too little nitrogen is being applied to dryland crops but it is a long jump from here to the assumption that a more precise estimate of the shape of the response surface is the most appropriate way to facilitate the desired change. The departure of decision-making practice from an established norm may be due to the operation of specific constraints not taken into account by the standard model. We don't know unless we are overt about the assumption and investigate its rigour.

Assumption 3: That poor decision-making results from either poor information inputs and/or poor information processing
From a scientific point of view, decisions based on more information, more scientific analysis and more quantitative data are better decisions. Poor decisions result from incomplete information or inadequate skill in calculation. From this point of view, there is clearly a case for the provision of more information and better tools to structure decisions (so-called decision support systems). On the other hand, if there is little evidence that poor decisions *are* being made, or about what constitutes a poor decision, or how it can most usefully be corrected, then the case is not clear. There is evidence that additional information can have little impact on behaviour if behaviour is constrained by practical (not informational) considerations, e.g. information about the shape of the fertiliser response surface at planting is not much help if no equipment is available to apply fertiliser at planting or if this is not a feasible option because it would slow down the planting operation.

Assumption 4: That researchers will be able to identify where producer decision-making is deficient, i.e. their 'needs'
This assumption views needs as objective entities which can be captured, described, experimented upon, then released back into the wild. In other words, that the social world (the world of needs) behaves like the natural world with which scientists are already familiar. Market research could be pressed into play, since clearly we are aiming to produce something (information, a DSS?) which has value and for which there will be a market. This fits with the current push towards commercialisation of extension (user pays). If farmers can be encouraged to express their needs, then we (scientists) can respond to those needs, or at least to those needs where there is a commonality of interests. On the other hand common sense suggests that separating needs from the context of their articulation carries the risk of misinterpretation. It encourages the view that science solves problems for farmers rather than farmers being involved in the solution of their own problems. There is also a tendency for both the needs expressed and the needs heard in answer to *scientists*' questions to be the ones scientists have a vested interest in addressing (Röling, 1988).

Assumption 5: That these needs can be alleviated by a scientific description of the production system
In order to make good (rational, informed) decisions, it is assumed that the starting point must be a description of the system which is best represented in scientific models grounded in crop physiology and soil physics. However, a focus on scientific models may exclude other ways of handling these issues. In R,D&E, farmers' models are often not recognised, let alone valued as a contribution. Yet farmers' models (captured as rule sets) are already in use, and there is some evidence that they are an elegant solution to the problem of managing open systems (Cox *et al.*, 1996a,b). Scientists' models are descriptive; farmers' models

are prescriptive. The two categories of model differ in purpose and in their use of fundamental constructs. Do we have constructive ways to move confidently between these models, to inform both R,D&E and industry practice?

Assumption 6: That the factors that limit yield are the ones where improved analysis is likely to be most beneficial
It is not clear that more precisely estimating the likely outcome of a decision (through improving scientists' models) is more likely to influence behaviour or that it should exert a greater influence on behaviour in all situations. Scientific models represent a managed system by using a set of often sparsely sampled parameters of the physical natural system. These models cannot capture all of the management of a production system rich with social, economic and psychological aspects. The collection of the inputs (daily rainfall, temperature, radiation, gravimetric soil water, soil N levels at different depths in the soil) is expensive, time-consuming and arduous. The models may exclude important limiting factors. (APSRU'S model, APSIM, includes water and N, but not mice, locusts, diseases or spray drift.) The decisions as to which factors to include are hotly debated by modelling groups (including APSRU) - and include assessing the best use of limited modelling skills. Do they also include researchers and extensionists in debate with farmers about which parts of their systems are problematic and perhaps most susceptible to improvement through interaction with the modelling process? By comparison, the decision points in farmers' models are usually quite restricted and often determined by exogenous events ('if it rains, I plant').

Assumption 7: That improved precision in prediction is a prerequisite for improved analysis
The validation of scientific models is based on their capacity to explain what happened by recreating the past. In choosing to seek increased precision, APSRU faced the risk that increased precision would be bought at a price - loss of transparency, increased complexity, less stability. The models become increasingly difficult to handle and interpret. This reduces the opportunities for other stakeholders to understand what is happening and why. (On the other hand, the inherent risk in not seeking high precision is scepticism about the accuracy of the conclusions.)

The optimal level of precision in description is itself a design criterion. It is possible that the optimal level of precision required to guide decision-making in broadacre dryland farming is quite low (Cox *et al.,* 1996a). There are several reasons for this: (i) response surfaces are often rather flat (or appear flat in view of the variable circumstances under which production is taking place); (ii) the biological feedbacks in the production system (carryover effects) mean that the outcome of a decision can be actively managed (e.g. if a farmer picked the wrong fertiliser rate in relation to the seasonal conditions that actually occurred, this can be corrected on the next cycle); and (iii) the detailed context of the decision can vary enormously according to the particular circumstances facing each farmer.

Assumption 8: That prediction of the future is equivalent to prediction of the past
The scientists' models are validated by retrodiction, i.e. successfully predicting the past. The point of view then rotates towards the future and the same model is applied. Of course, the future is uncertain so the prediction is now a probability distribution. But we have no way of knowing whether an actual outcome represents a movement along a probability distribution rather than a shift in the distribution. That experiment cannot be done. There are clues about the future to be found in the past (that we can learn from history and avoid past mistakes to some extent), but the past does not provide a complete picture of the future; the world is changing. Description of the past as a strategy for managing our way through the future risks excessive conservatism (attempting to fit significant change into previous experience) and to preoccupation with traditional technologies such as fertiliser use rather than alternative approaches to sustainable fertility management. We may be looking in precisely the wrong direction.

Assumption 9: That more precise specification of the system will lead to the identification and exploitation of novel opportunities
In some cases, this is of course the case. But how precise does one need to be in order to change behaviour? This question is discussed within the scientific community but often too quickly discounted. An increase in precision will only have value if the outcome (the action precipitated by the more precise information) is different. At some point, the decision-maker has sufficient information on which to make an informed and sensible decision. Once that point is reached, further increases in precision are of no value to that decision unless the situation changes. If the comparison of two options is sufficiently clear (large, significant), a comparatively coarse model will be sufficient to detect it. If models of greater precision are required to detect the difference, the difference between the two options is becoming increasingly marginal. Worse, the marginal cost of precision increases just as the marginal benefit of increased precision falls. Increases in precision are often obtained through increases in model complexity; understanding is lost and interpretation becomes more difficult. The marginal benefits of increasing precision can become negative: the greater the precision, the *less* valuable the model for certain purposes. These purposes include both routine farm management decision-making and the ability to spot emerging opportunities for doing things differently.

Constraints on new directions

Questioning the assumptions underpinning APSRU's work led to questioning the APRSU concept and to another set of questions that should be the subject of lively discussion within all agricultural R,D&E teams. There is a general risk that core research capabilities become core rigidities (Leonard-Barton, 1995). We have an inherent tendency to assume that the future will follow the same path as the past. APSRU was set up on the basis of an optimistic interpretation of the effectiveness of computer-based DSS at farm level. Reflections on SIRATAC (e.g. Cox, 1996) took place within APSRU, and the Wheatman experience was progressively re-interpreted as a process tool to facilitate discussion at farmer meetings rather than as a DSS to support routine decision-making. APSRU acknowledged and responded to these changes by a shift from a computer-based DSS as its primary product to recognition of the potential role of the model in learning contexts. The general question which arises is about the capacity of R,D&E groups to shift focus, and re-invent themselves, with enough speed to avoid wasting declining resources.

As Leonard-Barton points out, one of the simplest explanations for a rigid adherence to prior approaches for solving development problems is that alternatives don't occur to the people involved. The key is for research teams to continue to explore alternative ways of doing things and to cast off defensiveness as the only response to an implied criticism of the core capability. Yet many issues conspire to keep a research team preoccupied with existing approaches rather than on how a particular purpose can be achieved with or without that approach. For example, in APSRU, debates about the intellectual property rights of those contributing to the development of APSIM and fairness in licensing arrangements to use APSIM are perceived as having reduced the energy to explore bigger questions about the direction of the research.

Both technical and social scientists suffer from a tendency to continue with a research approach into which a significant investment has been made rather than incorporating new approaches. For example, at the same time as APSRU was building APSIM, new tools and methods were emerging from two directions. Off-the-shelf computer applications (such as @RISK™, DPL™, iTHINK™) might have been used either as process tools or as applications and may have better characteristics for useability, transparency and explanatory power for other stakeholders than APSIM. Farmers' own models (heuristics, rules of thumb, back-of-an-envelope calculations) could have formed a starting point for model development. At what point (if ever) should attention divert to consider these tools when APSRU already had world class simulation technology? How do the needs of researchers for completeness and elegance balance with the purpose of contributing to improving farmers' situations? What is the role of extension workers in facilitating the debate?

The boundaries to which a research team extend, will reflect the interests and abilities of the staff. Even research teams with deliberately diverse capacities face the risk that different perspectives offering scope for other ways of doing things will be marginalised. Researchers may listen better to those who confirm the value of their core capability, and not so carefully to those with alternative views. Typically R,D&E workers convene farmer reference groups for their purposes rather than building enduring relationships based on joint purposes.

Despite the unquestioned professional competence of its staff and APSRU's initial determination to do things differently, the APSRU case study highlights how difficult it is for a research group to operate effectively as a learning organisation. What are the characteristics of strategies which might assist in linking a research group to its end purpose and the other stakeholders in a co-learning relationship? They are strategies which recognise the 'situatedness' of knowledge (and knowledge generation) in people's lives and that therefore work with farmers and their existing knowledge wherever possible; strategies that emphasise interconnectedness rather than separateness of factors which affect outcomes; strategies which build around real action rather than describing; and strategies which build enduring relationships and care rather than isolated transactions driven by the immediate needs of the research team.

Cast Study 2: Green cane trash retention systems in the Queensland sugar industry

The development of green cane technologies

Crop production methods based on minimal soil disturbance and the retention of harvesting residues have been the focus of considerable agricultural research and development since the early 1980s (Cornish and Pratley, 1991). Retention of residues after unburnt harvesting has been discussed for most of the period in which sugar has been grown (Denman, 1900; King, 1934). In the mid 1970s improvements in harvester technology produced machines capable of harvesting larger green crops reasonably efficiently (Ridge et al., 1979). Coinciding with a period of wet harvesting conditions which caused substantial losses of crop and sucrose content in some regions, this development resulted in early proposals for green cane harvesting and conserving unburnt residues. At this early stage suggested benefits included an ability to harvest when conditions made it impractical or undesirable to burn, a lower rate of commercial cane sugar (CCS) loss between cutting and crushing, and a lower rate of dextran formation (Forster et al., 1977). Agronomic benefits included stabilising soils, recycling plant nutrients, improving soil structure, suppressing weeds during early ratooning, preserving soil moisture, and being a less labour intensive farming system (Prove et al., 1986; Wood, 1991).

By 1996, more than half of the Queensland crop was produced using unburnt harvesting and residue retention methods. However, despite substantial research and extension efforts over almost twenty years green cane trash retention techniques had been adopted inconsistently across the industry (Table 26.1), with a distinct dichotomy in adoption based on latitude. Green cane harvesting dominated crop production methods in areas north of Townsville with some mills receiving more than 98% of their cane harvested green. South of Townsville, the adoption of green cane trash retention systems was much slower.

Table 26.1. The percentage of the regional Australian sugarcane crop harvested green 1989-1994.

Region	1989	1990	1991	1992	1993	1994
North	68	73	77	79	86	88
Burdekin		5	7	4	3	3
Central	11	12	13	12	19	27
South	18	15	17	17	23	24
NSW	2	1	1	4	4	5

Issues impacting on the adoption process

While the previous case study focused on the assumptions driving research directions and designs, this case study illustrates the R,D&E and industry interface in practice. The disparity in adoption levels across the industry occurred despite significant efforts by many technologists to promote green cane methods and despite the environmental management context prevailing in the community. Research on the resistance to the technology was based on focus group discussions which drew on the experience and attitudes of industry members in central and southern regions, and four key issues emerged from the study.

Inherent risk and attitudes to a new technology

Attitudes to a technology result from continuing evaluation and decision processes where benefits are weighed against inferred costs. Vanclay (1994) argued that resistance to technological change in the farming sector was based on experience and rational consideration. Benefits and costs are often ill-defined, and encompass intangibles such as ease of process, amenity, and personal risk. Identifying barriers to adoption can be valuable in at least two ways. Firstly, precepts influencing attitudes can be validated. Secondly, the contextual setting of the problem can be enriched, eliciting personalised information, making links between attitude and circumstances, and creating a sound basis for designing future activity.

The over-riding influence producing negative attitudes to green cane trash retention was the perception of risk inherent in the technology. The risk was related to the influence of cooler temperatures and high moisture conditions created by the trash blanket. (Harvesting unburnt sugarcane generates residues in the order of 30-40% of the dry matter of harvested cane (Chapman *et al.*, 1994; Rozeff, 1994).) As a consequence, growers experienced slow ratoon emergence, impaired crop growth, and reduced yields. In less than ideal conditions, trash retention resulted in a less vigorous crop. As conditions deteriorated, invasive pathogens and the effects of impaired root function degraded the plant stool, to the point of crop failure in some cases. The perceived risk of reduced yields and shorter crop cycles grossly depressed the economic rationale for adopting the system.

Yield and profitability are important parameters influencing decisions for many growers (Squires *et al.*, 1991) and they are critical for economic viability on an industry scale. Although the importance of risk has been discussed in other agricultural industries, there has been little documented study of risk in the Australian sugar industry. In this study, soil type, time of harvesting, and adverse rainfall events consistently dominated experiences of the new technology yielding poor results, indicating limitations of the current technology. While acknowledged by technologists on a bio-physical level, the limitations had not been well understood. They were not clearly defined in terms of risk probabilities nor evaluated in terms of their potential impact on medium and long term profitability.

Risk issues can be approached in two ways. The first strategy is to minimise or negate risk through improving the management methods to overcome technical limitations. The amount of effort applied to this is dictated by the degree and likelihood of incremental gain. As has been discussed in the previous case study the marginality of the risk environment often warrants R&D resources being allocated elsewhere! (In making this decision it is important to note that for some risks which are not significant on an industry scale, externalities such as market or regulatory requirements can increase the risk for specific individuals to perilous levels.) The second strategy is to quantify risk to allow informed evaluation by individuals as opposed to simple awareness and semi-informed aversion.

An example relates to how well an individual grower understands the limitations of poorly drained soil types for green cane trash blanketing. Are decisions regarding investment in laser levelling, improved drainage and machinery suitable for green cane production best made in a framework of probability capable of analysing economic advantage over the medium to long term? This would require definition of the degree of limitation in the inherent soil properties, objective assessment of the probability of receiving adverse climatic conditions and extending the R,D&E framework to encompass sensitivity analyses of the contributing variables and objective determination of acceptable risk. Johnson (1995) argues for consideration of specific conditions present in the research domain to negate problems derived from standardised approaches at a broad spatial level.

Alternatively, in theory the adaptation/refinement process is expedited if a greater number of individuals contribute to the continuing development of the technology in the knowledge pool manner (Salmon, 1980). The focus group study identified that many growers felt current agronomic impediments within the technology would be overcome by progressive and committed growers, and favoured this approach over innovation developed through research organisations. This suggests that growers question the validity of existing research which was viewed as conducted in at least partial contextual isolation.

Poor understanding of interacting management factors

The apparent success of many growers using green cane methods south of Townsville generated considerable experimentation by farmers as well as research and extension activities. After experimenting on farm without suitable information many growers experienced poor results. They learnt that green trash retention required significant change to interacting farming practices (Fig. 26.1) and was not a simple change to the harvesting process. Some southern mill areas experienced a rapid uptake followed by rejection of trash blanketing techniques.

The initial formal research focused mainly on the soil stabilisation benefits of green cane trash blanketing (Sallaway, 1979; BSES & QDPI staff, 1984). Subsequently, instead of focusing on influencing and aiding decision processes (which farmers were showing was critical) much of the work was reduced (apparently by limited resources) to focusing on a single treatment - no pre-harvest burning - and a single measurement - yield. While some research was designed at least partially within the context in which the generated technologies were intended to be used (e.g. Ridge *et al.,* 1979), in other studies technologists selectively diminished the context, cognisant of but disregarding related farm resource and management issues.

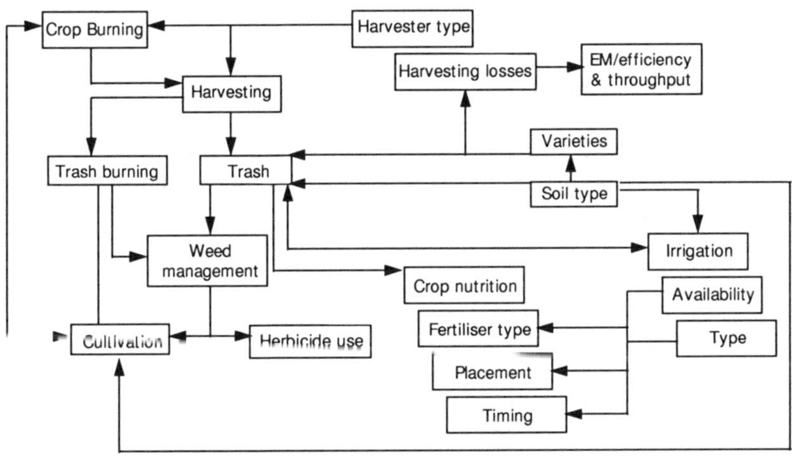

Fig. 26.1. Interaction of management components for green-cane trash blanketing (GCTB).

The narrow scope of the research approach continued to marginalise growers needing to manage the interactions between a complex set of management components. There were minimal trial data supporting insight into variables contributing to yield, such as seasonal conditions, soil type, and soil moisture characteristics. Growers expressed frustration that many technologists continued to promote green cane trash blanketing focused on a single management strategy - not burning. Subsequent promotion further alienated the target group and retarded future adoption. Many growers expressed frustration that they were perceived by the wider community and technologists as recalcitrant, when they had tried green cane trash blanketing and found the available management technology to be unsuitable. The processes linking farm management components were important, but they were poorly understood and not well documented.

A shift was required from thinking about the technology to thinking about the process in which the technology was to be used. Bawden and Macadam (1991) suggest variables influencing the success of technology must be considered in partnership with client groups. Recognition that farmers know and have some understanding of interactions within their farming system and are actively involved in their own experimentation and evaluation of farming system changes through a process of interaction and feedback, linking research, extension and farmers, may have reduced frustration and expedited technical improvements.

Inconsistent information

The relatively fragmented and *ad hoc* approach to green cane trash retention research exacerbated the lack of confidence in the technology and weakened extension efforts. Between 1978 and 1992 over 100 field trials comparing the conventional system against green cane trash blanketing were conducted in production regions throughout Queensland (Smith, 1993). Of these, over 60 trials were unreplicated and a significant number of the remainder were dual replicates. Importantly, no significant differences were detected at this level of rigour in many of the trials. On-farm yield results mirrored inconsistencies experienced in research trials. Growers concluded that yield gains from green cane trash blanketing were unstable in all situations apart from responses to conserved moisture on sandier soils in the absence of full irrigation and in low rainfall seasons.

Lack of economic analysis

Few participants in the study had any information on the influence of green cane trash retention on costs, margins and profitability. Apart from isolated studies (e.g. Hardman *et al.*, 1985) any input in this area was estimation based on empirical observation. The need to evaluate technological innovation on a sound economic basis is linked to risk probabilities and the requirement for consistent information as an essential to adaptation of the technology. Evaluating the impact of the technology on all sectors of the industry was also required, on a scale sufficient to accommodate their inter-dependence in the agro-industry structure of sugar production.

Lessons from the green cane experience

Science students are introduced early in their careers to the value of negative results, but in practice, as this case study illustrates, the admirable desire of technologists to assist industry to utilise improved technologies encouraged slow adoption to be interpreted as a consequence of inadequate demonstration and promotion. This led to redoubled efforts to 'push' the new technology. Researching the barriers to adoption occurred only when adoption fell short of expectation after considerable R,D&E effort and demonstrated that there were significant technical limitations. In the face of these problems, extra effort on promotion and demonstration was actually frustrating and angering rather than helping. Learning-based R,D&E approaches applied early and during the process (in an iterative manner) should have encouraged consideration of alternative explanations. Use of such processes was hampered by discipline and organisational boundaries which restrict access to potential collaborators with different perspectives, and by lack of skills and methods by which to strengthen understanding of the context within which stakeholders were operating and involve end-users in the design, review and refinement of an activity.

Bawden (1990) argued that future R,D&E must focus on enabling participants to cope with complexity using strategies based on action researching principles and collaborative enquiry. The strategy is not one of problem solving or technological innovation but of on-going situation improvement, with each participant seen as a co-researcher. The approach is adaptable and relies on development through the appreciative systems concept (Checkland and Casar, 1986). By comparison research and extension organisations are trying to implement change, but remain entrenched in the prevailing linear model of technology transfer. This case study illustrates an attempt to change, which was limited by a focus on symptomatic improvement to the R,D&E approach rather than fundamental review.

Implications: RD&E in the sugar industry

We began this discussion by noting the interest in better capturing the benefits of R,D&E in the face of declining funding. This interest persists despite R,D&E being a success by any standard and making substantial strides through generating technologies which people do use. It exists because something is wrong with existing processes - we are getting research products which are not adopted and opportunities for doing something different are being lost. The case studies presented illustrate both substantial gains achieved and opportunities to do better.

The focus for R,D&E in the sugar industry must be on delivering outcomes, not just in relation to yield and profitability but in relation to business viability and quality of life, maintaining the condition of resources, and meeting challenges to the Australian industry in the market place and in marketing arrangements. As the focus of R,D&E shifts to include improved management systems requiring higher levels of intervention as well as new products, the uptake time (which is the key determinant of the rate of return to R,D&E) is likely to rise. The question which is being asked of extension world-wide, 'What have you done for us recently?' (Patton, 1993), will also face R,D&E in the Australian sugar industry.

The quiet revolution that we identified was based conceptually on a search for approaches which fit R,D&E into the processes people naturally undertake to learn. Developing such approaches is the first critical requirement of R,D&E in its contribution to maintaining the competitiveness of the sugar industry and the viability and satisfaction of individuals within it. In contrast to the widespread scepticism of the capacity of science and technology to deliver benefits to people (e.g., see Goldman, 1989), it seeks to empower both farmers and R,D&E workers. Both need to be seen as learners in this process. Both have data, knowledge and agency.

In the case studies we have had the opportunity to view some of the challenges involved in learning not just to solve today's problems, but to transform R,D&E projects and programs by testing underlying assumptions as new evidence emerges. Hence the second requirement for R,D&E in the sugar industry is a willingness to adopt the concept of the learning organisation. The challenge inherent in this concept is for R,D&E not just to work to improve existing industry systems and processes using established R,D&E approaches, but to question and re-question the underlying assumptions which provide clues to the shifts in core business, both of industry and R,D&E, that will be future priorities.

Acceptance of a diversity of providers and the development of complementary rather than competitive relationships will be the necessary third requirement, to avoid the rigidity identified earlier as a threat to R,D&E organisations meeting new challenges. A recent review of extension in the cotton industry (in which private sector extension services are well developed) identified highly complementary roles between public and private sector services (J.A. Coutts, *pers. comm.*). The public sector services contributed on-farm validation of improved and more sustainable practices while the private sector provided focused individual advice on tactical management decisions. The sugar industry will need to strive to ensure similar diversity and complementarity is maintained between different service providers.

What characteristics will define suitable R,D&E strategies? They must be strategies which recognise the need for close examination of farmer practice as well as close examination of biophysical processes. Many researchers have already moved to more participatory processes, but additional skills will be needed to listen in a way that invites and truly values farmers' responses. Diversity of skills within R,D&E teams - in particular the inclusion of economic and social science perspectives - will be the essential fourth requirement, to design approaches to work with farmers and other client groups in developing new management strategies for sugar production. While the emphasis in the case studies has been on farmers and the production end of the industry, similar issues are likely to arise in R,D&E for the milling and processing sections of the industry.

Where this diversity of skills and perspectives does not exist, R,D&E organisations will face decisions about acquiring new people or encouraging existing staff to develop new capacities. Farmers in all agricultural industries are recognising the need to upgrade their own skills, and rising skill levels as well as partnerships with providers of education and training will also need to be considered in designing R,D&E approaches.

Fifth, the industry (including R&D) will need to look sideways not just in agriculture and in science but to other industry sectors adapting to change. Quality management and reengineering are two current approaches to changes in management initiated in industry outside agriculture, which may adapt well to improving sugar production. Future initiatives need to be assessed for their potential.

The sixth requirement is for strategies which emphasise interconnectedness rather than separateness. For example, the impact of new technologies on all industry sectors must be considered. Integrated systems rather than single discipline products will most commonly be the output of R,D&E activities. Electronic communication capacities will inevitably play an important part in future access to information in the sugar industry and may prove to be a real asset in allowing the maintenance of wide networks. There are indications that other intensive industries - pigs and cotton as well as horticulture - are already valuing this potential and planning communication and marketing strategies using email and the World Wide Web.

Strategies will increasingly need to build enduring relationships; this seventh requirement is one in which the sugar industry, with its close culture and strong organisations, has a potential advantage. The challenge is to maintain collaborative and cooperative relationships which build on this strength rather than allowing competitive, defensive and conservative stances to dominate.

Finally, if the purpose of R,D&E is changed behaviour in key actions, then strategies are needed which focus on action. Many R,D&E processes are based on description rather than action. R,D&E are interventions and contributions to the broad process of human development. They are characterised by a perception that they can make a difference to the future; their skills, methodologies and actions need to confirm that conviction in working with industry towards a promising future.

Concluding Remarks

Our R,D&E organisations have some way to go before they are well-developed ‘learning organisations’. There is much to learn but entrenched practices are hard to shift. The present disquiet with publicly funded research reflects this concern that we are not doing as well as we should or could.

There is a significant mismatch between ‘science as ballet’ and ‘science to make a difference’. We can do science as ballet, but the market for its product is restricted to an elite few. That is not our view of where the potential contribution of science is greatest. Science to make a difference requires new approaches based on greater tolerance of different points of view (especially when intended and expressed in a caring way). There are no single answers to the many important questions that we face in the way we manage our industries and our environment. Science, and professional researchers, can develop learning skills with clients: if anyone can learn how to learn, surely it is scientists. If 'shaking of the foundations' will result in a transformation which produces improved outcomes for many constituents (including scientists themselves) then *shakin' and stirrin'* is what science ought to be about.

Acknowledgments

We wish to thank Peter Ridge, Mike Grundy and Arnold Wissemann for helpful comments and suggestions.

References

Argyris, C., Putnam, R. and Smith, D.M. (1985) *Action Science. Concepts, Methods and Skills for Research and Intervention.* San Francisco, CA: Josey-Bass Publishers.

Bawden, R. (1990) Towards action researching systems. In: Zuber-Skerritt (ed.), *Action Research for Change and Development.* Griffith University Press, Queensland, Australia, pp. 21-50.

Bawden, R. and Macadam, R. (1991) Action Researching Systems. In: Tillman, H.J. (ed.), *Proceedings International Workshop Agricultural Knowledge Systems and the Role of Extension.* Institut fur Agrar Soziologie, Landwirtschaftliche Beratung und Angewandte Psycologie, Stuttgart, Germany, pp. 367-386.

BSES and QDPI staff (1984) A review of results of trials with trash management for soil conservation. *Proceedings of the Australian Society of Sugarcane Technologists* 6, 101-106.

Chambers, R., Pacey, A. and Thrupp, L.A. (1989) *Farmer First: Farmer Innovation and Agricultural Research.* Intermediate Technology Publications, London, UK, pp. 2-18.

Chapman, L.S., Haysom, M.B. and Saffigna, P.G. (1994) The recovery of N^{15} from labelled urea fertilizer in crop components of sugarcane in soil profiles. *Australian Journal Agricultural Research* 45, 1577-1585.

Checkland, P. and Casar, A. (1986) Vicker's concept of an appreciative system. *Journal of Applied Systems Analysis* 13, 3-17.

Collinson, M. (1987) Farming systems research: procedures for technology development. *Experimental Agriculture* 23, 365-386.

Cornish, P.S. and Pratley, J.E. (1991) Tillage practices in sustainable farming systems. In: V. Squires and P.G. Tow (eds) *Dryland Farming. A Systems Approach*, Sydney University Press, Sydney, Australia, pp. 76-101.

Cox, P.G. (1996) Some issues in the design of agricultural decision support systems. *Agricultural Systems* 52, 355-381.

Cox, P., Parton, K., Shulman, A. and Ridge, P. (1996a) On the articulation of simulation and heuristic models of agricultural production systems. In: Teng, P.S., Kropff, M.J., ten Berge, H.F.M., Dent, J.B., Lansigan, F.P. and van Laar, H.H. (eds), *Applications of Systems Approaches at the Farm and Regional Levels Volume* 1. Proceedings of the Second International Symposium on Systems Approaches for Agricultural Development, Los Baños, Philippines, pp. 213-228.

Cox, P.G., Shulman, A.D., Ridge, P.E., Foale, M.A. and Garside, A.L. (1996b) An interrogative approach to system diagnosis: an invitation to the dance. *Journal for Farming Systems Research-Extension* (in press).

Davidson, A.P. (1987) Does Farming Systems Research have a future? *Agricultural Administration and Extension* 24, 69-77.

Denman, E. (1900) The cultivation of sugarcane. *Queensland Agricultural Journal (Conference Supplement)* 7, 16-19.

Forster, D.H., Leverington, K.C., Price, R.A. and Melrose, I.K. (1977) The implications of green cane harvesting. Submission for Sugar Industry Review Committee. Bureau of Sugar Experiment Stations, Internal Technical Report, Brisbane, Australia.

Goldman, S.L. (1989) Images of technology in popular films: discussion and filmography *Science, Technology and Human Values* 14, 275-301.

Haraway, D. (1995) Situated knowledges. The science question in feminism and the privilege of partial perspective. In: A. Feenberg and A. Hannay (eds), *Technology and the Politics of Knowledge*. Indiana University Press, Bloomington, US, pp. 175-194.

Harding, S. (1991) *Whose Science? Whose Knowledge? Thinking from Women's Lives*. Open University Press, Buckingham, UK, 319pp.

Hardman J.R., Tilley, L.G.W. and Glanville, T. (1985) Agronomic and economic aspects of various farming systems for sugarcane in the Bundaberg district. *Proceedings of the Australian Society of Sugarcane Technologists* 7, 147-153

Jiggins, J. (1994) *Changing the boundaries. Women-centered Perspectives on Population and the Environment*. Island Press, Washington, US, 291pp.

Johnson, A.K.L. (1995) Risk perceptions and nutrient management responses in the Australian sugar industry: preliminary results from the Herbert River district. *Proceedings of the Australian Society of Sugarcane Technologists* 17, 172 178.

Kelada, J.N. (1996) *Integrating, Reengineering With Total Quality*. ASQC Quality Press, Wisconsin, USA.

King, N.J (1934) Trash conservation. *Cane Growers' Quarterly Bulletin* 1, 126-132.

Leonard-Barton, D. (1995) *Wellsprings of Knowledge. Building and Sustaining the Sources of Innovation*. Harvard Business School Press, Harvard, USA, 334pp.

Macadam, R.D. and Wilson, A.M. (1993) An historical perspective on the evolution of extension theory and practice as a basis for identifying future directions to meet emerging challenges. *Proceedings: Australia Pacific Extension Conference*, Queensland Government, Brisbane, Australia, pp. 605-611.

McCown R.L., Hammer, G.L., Hargreaves, J.N.G., Holzworth, D.P. and Freebairn, D.M. (1996) APSIM: a novel software system for model development, model testing, and simulation in agricultural systems research. *Agricultural Systems* 50, 255-271.

Patton, M.Q. (1993) World trends and challenges in extension. *Proceedings: Australia Pacific Extension Conference*, Queensland Government, Brisbane, Australia, pp. 644-654.

Prove, B.G., Truong, P.N. and Evans, D.S. (1986) Strategies for controlling caneland erosion in the wet tropical coast of Qld. *Proceedings of the Australian Society of Sugarcane Technologists* 8, 77-84.

Ridge, D.R., Hurney, A.P. and Chandler, K.J. (1979) Trash disposal after green cane harvesting. *Proceedings of the Australian Society of Sugarcane Technologists* 1, 89-93.

Röling, N. (1988) *Extension Science. Information Systems in Agricultural Development.* Cambridge University Press, Cambridge, UK, 233pp.

Rose, H. (1994) *Love, Power and Knowledge. Towards a Feminist Transformation of the Sciences.* Polity Press, Cambridge, UK, 326pp.

Rozeff, N. (1994) Sugarcane biomass and burning: an empirical scenario for the Lower Rio Grande Valley of Texas. *International Sugar Journal* 95(2), 2-5.

Russell, D.R., Ison, R.L., Gamble, D.R. and Williams, R.K. (1989) *A Critical Review of Rural Extension Theory and Practice*, University of Western Sydney Hawkesbury, Richmond, Australia, 67pp.

Sallaway, M.M. (1979) Trash retention as a soil conservation technique. *Proceedings of the Australian Society of Sugarcane Technologists* 1, 133-137.

Salmon, P. (1980) Personal psychology of change in management. *School of Agricultural and Forestry Monographs*, University of Melbourne, Melbourne, Australia, pp. 28-47.

Scoones, I. and Thompson, J. (1994) *Beyond Farmer First. Rural People's Knowledge, Agricultural Research and Extension Practice.* Intermediate Technology Publications, London, UK.

Smith, M.A. (1993) Is green cane trash blanketing as profitable as the burnt cane alternatives? *Bureau Sugar Experiment Stations Technical Report*, Brisbane, Australia.

Squires, V.R, Tow, P.G. and Boast, C.M. (1991) Future directions: Meeting the challenge. In: Squires, V.R. and Tow, P.G., (eds) *Dryland Farming: a Systems Approach.* Sydney University Press, Sydney, Australia, pp. 271-282.

Vanclay, F. (1994) The barriers to adoption have a rational basis. *Proceedings 7th International Soil Conservation Organisation Conference,* Sydney, Australia. International Soil Conservation Organisation, pp. 452-458.

Wood, A.W. (1991) Management of crop residues following green harvesting of sugarcane in North Queensland. *Soil & Tillage Research* 20, 69-85.

Chapter twenty-seven:

Doing Things Differently: The Policy Revolution

G.J. Syme[1], G.T. McDonald[2] and J.E. Butterworth[1]

1. *CSIRO, Australian Research Centre for Water in Society, Private Bag, PO Wembley, WA 6014, Australia*
2. *Department of Geographical Sciences and Planning, University of Queensland, St Lucia, Qld 4067, Australia*

Abstract

This paper provides a brief history of public policy and its administration in the primary industry sector in Australia. Major changes have occurred in the last decade, including the rise of Environmentally Sustainable Development as a focus for planning and policy development and greater emphasis on coordination in planning and implementation for Natural Resources Management. Increased community demand for input into policy and planning and decreased government resources devoted to policy implementation and institutional restructuring within government have been forces for change. These changes are demonstrated in the rise of the Landcare movement and community-oriented Integrated Catchment Management programmes. Discussion pertaining to the successful implementation of these programmes identifies specific challenges concerning the effective vertical and horizontal integration of decision making among community, industry and government stakeholders. The tension between long term aspirations and short term decision making is explored. We suggest that a conceptual model for evaluation of policy implementation in this area is urgently needed and a tentative one is offered with suggestions about how to make it work. We conclude that, despite the problems associated with policy changes, the goal of environmental sustainability and changed government roles provide a major opportunity for the sugar industry. The concept of a 'green' sugar industry should provide a core platform for its operation in the next century.

In: Keating, B.A. and Wilson, J.R. (eds), *Intensive Sugarcane Production: Meeting the Challenges Beyond 2000*. CAB International, Wallingford, UK, 1997.

The Policy Revolution

Sugar is a major export industry with a long and distinguished history in the development of northern Australia. It is highly vertically integrated industry with a history of government protection and intervention at both State and Commonwealth levels. It has, in the past, epitomised attempts at centralised and orderly planning in terms of markets, prices and production. As Herbert (1956, p. 116) noted at that stage, the three tiers of the industry, growers, millers, and refiners 'through a peculiar mixture of competition and cooperation had ensured progress and relative efficiency'. Everyone from the individual cane grower to the CSR executive and politician have 'known their place' in the industry.

In the last 10 years or so there has been a marked change in the way in which the Federal and State governments have looked at the sugar industry and indeed most of primary industry generally. Five major forces in government have challenged the old sugar 'order'. Firstly, the rise of economic rationalist thought and an ideological belief in the benefits of market forces in ensuring efficiency and prosperity. Secondly, the associated belief that infrastructure associated with industry is most appropriately paid for by the user or beneficiary rather than the general public. Thirdly, the unrelated, but equally significant, recognition of the importance and development of ecologically sustainable farming practices. This has been accompanied by a raised political profile of the value of conservation generally and the tropical ecologies in cane growing country in particular. Fourthly, there has been an erosion of the view that hierarchical or 'top-down' planning by professionals is appropriate for sustainable management. Finally, while north Queensland used to be self evidently a 'sugar community' there is now also a thriving tourism industry. The social complexities associated with the growth of this industry and others are increasing. Because of these last three changes, the sugar industry is gradually finding that it has to consider the views of others carefully when addressing its own future needs.

The less interventionist views of government in terms of markets and infrastructure have, perhaps surprisingly, also been reflected in government approaches to environmental management. Governments are allocating decreased centralised resources to assist landholders to change or modify farm management practices. Noted as the fourth force above, they are increasingly relying at the local and regional levels on rewarding voluntary participation. Resource issues are seen to be ideally 'community owned'. The new approach has also tended to integrate environmental agencies into larger departments to promote efficiency and coordination of government activities at a catchment level.

Indeed, the theory and practice of resources planning have changed markedly in recent years, adopting community consultative processes, stakeholder involvement and partnership between government and the local and regional communities. This contrasts with the old survey-analysis-plan paradigm or even the rational planning paradigm that focused on goals and systematic selection of alternatives. Now more than ever, the effectiveness of plans depends on the degree of stakeholder acceptance of plans. Programmes such as Landcare, Integrated Catchment Management (ICM), Coastcare, Waterwatch, Local Area Planning, and regional planning are clear evidence of these policy approaches. The effect has been that most planning agencies now involve a process of public consultation in preparing their plans. The challenge is to optimise these processes to use stakeholders' time efficiently, and to be transparent so that participants can see how (if at all) their preferences have been incorporated into the outcomes.

Unfortunately, many planners and resource managers have no training in appropriate consultative techniques. Indeed, the planning paradigm they adopt, such as scientific manager or design professional, has only a hazy role for community-based decision making. Many plans are defective because they fail the final implementation test or are never really implemented at all because the community was not adequately involved. Perhaps most importantly, conflicts and power sharing issues were not resolved at the planning stage. Furthermore, often the biggest difficulties may not be with community-based stakeholders but, rather, with bureaucratic stakeholders.

Clearly, as has been discovered with some genuine community planning programmes (e.g. ICM and Landcare), there is a vast difference between community-based planning and loose commitments to public involvement (e.g. Carman and Keith, 1994).

We have moved from an environment where government looked after the industry both collectively in terms of supporting its institutions and individually through the provision of local infrastructure or personalised extension advice. We now have a new environment in which the government and its public service have de-emphasised top-down leadership (either political or professional) and moved to a position where the industry and communities play a much greater role in dealing with their own problems. There is a greater emphasis on coordination and accountability between government activities, and governing by incentive and consultation.

Viewed positively, this new way of doing things does have an attractive face. Members of the community will own their own problems, have more control over their own destiny, and become more community oriented from their involvement in Landcare, Community Catchment and similar groups. They will find government easier to deal with because of its increasing coordination and efficiency. Finally, with these policies the landholder will have pride that he or she is commercially viable and does not need government intervention to guarantee economic wellbeing. Thus, producers will be confident of being able to hand down their farms to their children or the next generation in a sustainable state. Nevertheless, as one friendly farmer has observed, 'by this stage pigs may also be able to fly!'. The protectionist 'sugar senator' from the first half of this century, T.J. Crawford, must be spinning in his grave (see Pagani, 1989 for an account of the politics of protectionism in the first seventy years of the industry).

It is too early to evaluate how well this new government approach will work. While there are now literally thousands of Landcare Groups throughout Australia

(Campbell, 1994) and numerous examples of vigorous community-led ICM programmes (Syme *et al.*, 1994), there are also signs of frustration. There is the beginning of calls from government, professionals and landholders for systematic evaluation as to what outcomes are actually occurring and disagreement about what outcomes can be expected. The early part of this debate has occasionally been divisive as expectations about the effects of ICM or Landcare have varied from region to region and according to the individual's role in the process.

The performance of government in coordinating itself in the face of increasing social and biophysical complexities also remains to be seen. Often government coordination has been seen as a solution in the past. For example, McCreddin and Syme (1996), in examining government functioning in near urban land and water allocation in Western Australia, have suggested that this has been difficult to achieve. Better coordination in this regard was the major platform of a Perth mayoral candidate in 1900 and has been the recommendation of the last six consultancy reports on the issue, appearing over the last decade or so. Coordination is only a solution if it can be implemented and implementation is a problem in its own right (Webb, 1991).

Finally, in the more economic arena there are several new challenges ranging from competition in international markets to domestic government policies to changes in the institutions which serve the sugar industry. For example, many are concerned about the feasibility of the creation of 'level playing fields' which will ensure appropriate international market behaviour. Arguments about the nature of public versus private goods and who really is the beneficiary of infrastructure are also rife in many rural areas in the light of 'user pays' government philosophies. Through a series of reviews the industry is being forced away from protection and towards a more corporate stance. But even if these policies are uncomfortable, the sugar industry, at least in the medium term, will have to follow the economic rationalist trends, as will most primary industries.

It is important for the sugar industry to be aware that relevant policies and their implementation have changed radically over the past few years. How the industry should adjust its behaviour or its interactions with other industries, government and even the general community needs to be considered. A failure to notice radical changes in policy and to respond positively by searching for opportunities could well lead to major problems for the industry in the future. For example, a failure to respond to current environmental imperatives may mean that environmental standards may be imposed in lieu of appropriate standards being set or adopted by industry itself. The potential effects of nutrients on the Great Barrier Reef is possibly one example of where industry could, and is beginning to, show leadership.

The sugar industry, from its past performance in interacting with governments, has clearly had the skills with which to assume an assertive role in negotiations. It is reasonable that such negotiations will continue to be productive as the environmental goals of the sugar industry are likely to be compatible, in the long term, with those of other industries and the government itself. The tourism industry, for example, may provide increasing spin offs for the sugar community as ecological and cultural tourism expand their horizons.

Our Purpose

This paper deals mainly with environmental policies and their implications. Approaches to policy implementation in the environmental and economic fields are, of course, related. Unless the government wishes to look totally inconsistent, policy instruments are likely to reflect a similar style or philosophy. Morally speaking, for example, it is harder for a government to use compulsory regulatory means (at least initially) when the industry is being asked to 'stand on its own two feet' economically. It is more likely that the voluntary incentive-based approaches currently being used will be maintained, at least in the medium term.

We will outline the issues emerging from two voluntary and incentive driven initiatives, the development of Landcare and the promotion of ICM in its various guises (TCM or Total Catchment Management, WCM or Whole Catchment Management). Although early days, the issues already arising and their relationship to similar experiences in other previous planning and policy situations allow us to identify a series of challenges relevant to the sugar industry at local and regional levels. We conclude this paper with some suggestions as to how the industry can maximise its control over policy and implementation in the future.

The Policy Changes: What's Happened So Far?

The Landcare movement

The rise in number of Landcare groups from 200 in 1989 to > 2000 in 1994 (Campbell, 1994) has been a conspicuous development in the role of landholders in the sustainable management of their own properties and others in the local district. Such a movement reflects success of governments' persuasive and incentive driven policies for promoting community action to solve localised problems. Along with the rise in the numbers of groups has been a relatively new profession of Landcare coordinator/facilitator. Some moneys are provided to some groups to employ a group coordinator, often recruited from the community itself. This new profession of working with and coordinating groups contrasts with the more traditional concept of the agricultural extension officer from the 'department' providing individual advice to the landholder on the property. In the latter case, the extension officer would, among other things, be 'selling' departmental solutions to problems identified for the farmer by the professional agricultural planners and scientists.

Such a change has many obvious advantages. Ostensibly, locals can define their problems, and use the facilitators as well as Landcare technicians and project officers to assist in gaining government expertise and resources. This process can lead to sustainable practices which will become part of the local culture. Positive evaluations attesting to progress in 'group thinking' and action are already beginning to be reported (Curnow and Burnside, 1995; Curtis, 1995). However, some concerns have begun to emerge about future challenges that may face the Landcare movement. Vanclay and Lawrence (1995) warn us that we need to be clear about the background ideology of such a movement towards groups. According to these authors, 'it may be that the current rhetoric about a commitment to self-actualisation, farmer empowerment, and of farmers setting agendas and finding solutions to their own problems, masks an intention on the part of the state to reduce its funding and overall commitment to agriculture'. Vanclay and Lawrence go on to state that group extension may in fact increase individual demands for advice and that if it is not forthcoming because of lack of resources, the process would have failed. Finally, they note that environmental problems do not always lend themselves to easy recognition or solution. Local farmer groups cannot therefore be relied upon to provide answers in all situations. The 'non obvious' situation for environmental problem definition is certainly evident in the multiple use north Queensland catchments of such importance to the sugar industry. Priorities may differ between differing parties, and, physically speaking, environmental degradation is not as vivid as in other rural areas (e.g. those markedly affected by dryland salinity).

Martin and Woodhill (1995) suggest three other issues which may need to be confronted, and which may inhibit the usefulness of the Landcare approach: (i) the spatial and temporal scope of Landcare activities; (ii) the inconsistency between current economic rationalism and their effects on adoption patterns; and (iii) the apparent lack of commitment of resources for fostering active coordination and monitoring at a regional level.

The first concern is a central issue if community participation in this form is to be effective. Martin and Woodhill (1995), along with those undertaking wider community involvement in research, contend that those involved are more likely to be individuals who have the time and probably, financial resources. The advantage to them is that their personal properties are likely to be used for demonstration projects or research trials. Localities where there are greater resources and social cohesion will also be more competitive.

The second concern identifies the relationship between economic and environmental policy noted above. It is considered that the opening up of the market to external influences encourages a sense of uncertainty and thus long term investment in sustainable practices may be discouraged.

Finally, smallish cohesive groups such as Landcare groups tend to operate in terms of social boundaries often unrelated to the scale, significance or cause of environmental problems. Many of these environmental problems are best dealt with at the regional or catchment level. If this is to occur, more serious attention needs to be given to planning, monitoring and evaluation. Adequate resources to encourage and maintain coordination are also required.

The concerns outlined make an evocative contribution to our understanding of the processes and values underlying our currently evolving policies. They are echoed in policy development as the government has apparently begun to respond or think about some of these issues. For example, there has also been some public government discussion and analysis about the desirability of rural reconstruction to support sustainability via long term socioeconomic planning (e.g. within the development of the Murray Darling Basin Commission's Irrigation Management Strategy). The ICM policies discussed briefly below represent a movement towards appropriate environmental boundaries.

The potential problems, however, should not at this point outweigh the positive outcomes already identified. No programme is perfect in its early days and representative and more beneficial involvement could well be achieved once the Landcare movement has become more mature. From the perspective of some, part of this maturity will derive from increased skills within government and its agencies in using the detailed local knowledge of landholders. Nevertheless the issues raised by these critical reviews do provide significant insights for reference when planning. Above all, they emphasise the importance of the development of common expectations between governments and the community and agreed processes to meet them.

Integrated (total, whole) catchment management (ICM)

Each State in Australia now has a community or stakeholder oriented environmental programme based on catchment management principles. The names of the programmes differ and there are differing emphases in defining the government's role. The perceived need for legislative backing for the programmes also varies. Nevertheless, in all cases the emphasis is on government coordination and integrated community action. The philosophy of top-down planning and implementation is eschewed.

In recent years, this approach has become evident in northern Queensland and in sugar growing areas. The maturity of the process and its mode of operation have varied (e.g. between the Johnstone and the Herbert). Nevertheless, the introduction of such programmes has represented an opportunity for locals to gain more influence on overall government functioning and at a geographical level which is appropriate for producing some effect (Sale, 1991).

Some theorists may contend that such an approach has been forced on governments as traditional bureaucratic structures have inevitably failed when complex issues needing shared vision are addressed (see Grandori and Soda, 1995 for a review of such theories). An even more cynical landholder who was asked to clear property in the wheatbelt of Western Australia only twenty years ago may suggest that many in the bureaucracy can remember their tree clearing advice and do not want to be responsible for further mistakes!

What is motivating the move towards ICM is, however, probably not particularly significant. It is clear that our environmental problems are emerging on a catchment basis and that social and economic factors need to be considered in unison with biophysical considerations, if solutions are to be derived. Policies and programmes therefore must be delivered at this level and in these spheres. ICM is not optional.

The question becomes one of how ICM can be successfully implemented. Unfortunately, despite encouraging signs of 'success' in ICM there are anxieties in some areas that the process is not achieving what some participants expected. This may be because these implementation questions have not received due attention. Coordination of processes and integration of knowledge does not often happen spontaneously, even when peoples' objectives are similar and there are agreed methods of approach. Recent reviews of ICM methodologies and processes have identified the need for attention to details regarding processes within a social systems context (including the roles and responsibilities of differing parties) and the need for incorporation and integration of social, economic and environmental data in the planning, implementation and evaluation of ICM (Syme *et al.*, 1994; AACM, 1995). Further, it is difficult to see how the success of ICM will be evaluated in the future without formal models of policy implementation (Mazmanian and Sabatier, 1989).

The above discussion is not meant to reduce appreciation of the important practical activities already occurring under the banner of ICM to esoteric theoretical dimensions, nor to criticise current activities. It merely states the obvious. We need detailed understandings of where we want to go, *exactly* how we will try to get there and a rough idea (given current knowledge) about when we expect to arrive. The latter two are often not thought through and, as a consequence, realistic assessment of the resources required to implement programmes successfully is impossible. Failure to discuss expected rates of progress has led to many petty squabbles related to how things are going. This is inevitable if the parties had differing expectations in the first place.

Such comments are not novel. They could be made about the British regional planning exercises in the early to mid-1970s. By and large these programmes were considered less than successful (Friend *et al.*, 1974; Martens, 1981). Emphasis on regional planning certainly led to a lively debate among professionals about what went wrong. The outcome was a general lessening in confidence in regional planning by decision makers. Process variables such as coordination and power coalitions among government agencies were identified as issues which had been given too little attention. All these difficulties occurred without the extensive public involvement expected for ICM! It is interesting that the latest federal development in environmental management, overlaying ICM, is the revamping of regional planning.

Some Specific Challenges

This brief identification of issues emerging for both Landcare and ICM is relevant for primary industry generally and provides an agenda for discussion by the sugar industry. Such a discussion would, no doubt, be influential as the industry is an important cohesive force on northern Queensland catchments. There are, however, some fundamental issues to be addressed, and which have not yet been discussed in assessments of current practice.

Vertical integration

As noted earlier, there is a critical problem of spatial scale in dealing with resources planning. The ultimate hope of ICM is that the solution to this problem lies in creating institutional structures (rules and regulations defining powers and responsibilities) for the agencies involved in planning at the different scales required to address sustainable development; i.e. vertical integration. Nijkamp and colleagues refer to this as 'institutional assignment' (Nijkamp *et al.,* 1992). Difficulties in achieving policy coherence increase exponentially with the number of levels of government, making institutional assignment particularly difficult in federal systems such as the United States, Canada and Australia where there are three levels. In principle, higher levels of government provide a framework for implementation at lower levels and as long as lower levels act consistently with the guidance from levels above, many decisions can be taken at the local level or at least partially delegated.

The policy 'revolution' involves creating partnership between the different levels of government. *Integrated Development Approval Systems* are emerging in Australia, the best known being the Intergovernmental Agreement on the Environment (IGAE) which defines standards for environmental assessment and commits the States and Commonwealth to honour planning done by lead agencies at other levels. It is anticipated that the IGAE should, with the assistance of natural environmental standards prepared by CEPA, minimise political conflict and allow vertically coordinated planning to occur.

Horizontal integration

Irrespective of the level at which planning is done and decisions are made, environmental planning requires simultaneous consideration of ecological and socioeconomic factors. Unfortunately, government agencies are not created to do this, and throughout the country there remains a tension between the production and protection aspects of resource use. Inter-agency and community conflict is common because the environmental considerations have been separated out.

The policy revolution has resulted in massive and continuing governmental restructuring and in a snowstorm of legislative and administrative actions to achieve horizontal integration. The legislative and administrative relationship between production agriculture, land and water management, and nature conservation varies widely between states and changes from time to time as new governments are elected. Land and water conservation and environmental protection legislation proliferates and changes constantly in an attempt to create integration, but incoherent land management programmes persist.

Even within local governments, environmental management and expertise has been separated from planning and development control, reflecting the sectoral separation usually found in state governments. If strongly enforced policies or even statutory plans emerge in these areas, there is inevitably a problem of coordination, inefficient development approval and determining which plan shall predominate. The big policy challenge is to integrate this diversity of plans across government into a coherent package to guide development.

Long term aspirations and short term decision making

While there have been moves towards longer term government decision making (e.g. The Decade of Landcare) there is no doubt that, for sustainable environmental, social and economic objectives to be achieved, government processes require a longer term flexibility and commitment. Sustainability objectives will be achieved, though, through a series of many small gains, and innumerable 'blind alleys' associated with an occasional major breakthrough. At a community/government interaction level this will require frequent realignment of objectives and checking of mutual expectations. For all parties this will require patience. During this process the relative roles of governments and the community have the potential to change significantly and permanently. This will have particular visibility at the ICM and regional planning levels. Thus, in summary, we may be at the crossroads of being able to implement radical change in the service delivery of government programmes.

But despite the importance of thoroughly examining possibilities for long term administrative change there has been a tendency to rely on short term mechanisms. These have included smallish grants to Landcare groups or short term contracts to facilitators to carry the programme through. There are of course, exceptions, for example the Liverpool Plains in Northern NSW and Blackwood Catchment in WA, where community-based Steering Groups are beginning to define their longer term requirements for government services.

Similar roles to the short term coordinator have been tried in regional planning in the past (Martens, 1981). Generally they have disappeared and been reincorporated back to traditional structures within government planning departments. Apparently this has been largely because of initial lack of attention to the definition and role of a coordinator (as is still often the case) even when only interactions between government departments were being considered. Long term commitment to changing the way government programmes are devised and administered requires rigorous and ongoing evaluation of the relationship between the ongoing politics of the implementation and their relationship to long term environmental, economic and social goals of the programme. It should not be assumed that coordination means the same thing to all participants (Webb, 1991), nor assumed that the community and under-resourced agency staff will always agree or perform well when they are preoccupied with 'bushfires'. It should not be assumed that an industry or individual facing short term economic uncertainty will be motivated by long term survival. It should not be assumed that self interest will magically disappear because we are all concerned about the environment. Finally, the ability of locals to contribute to the long term benefit of their environment should not be doubted if the ability of government to make a longer contribution has not been demonstrated. In short, worthy long term goals or decision making processes should not be abandoned if they have been tried once with a faulty implementation process. The assumptions of the implementation process need to be constantly identified and tested.

Dovers (1995) summarises this position when he states that to achieve sustainability we need 'better institutional and policy abilities to handle uncertainty; redoubled efforts and greater independence and permanence in basic environmental monitoring; more continuity in policy settings'. In simple terms, old style extension workers often said that it would take at least a decade for a new technology to be adopted by farmers. The improvement of the new system over the old needs to be demonstrated. In addition, patience may be severely tested as we search for workable new forms of public administration which reflect a commitment to social learning processes (Mintzberg, 1994).

A Conceptual Model for the Implementation of Policy

Commonwealth and State government policy as it relates to Landcare is clearly explained in government documentation. The modes of implementation of the policy in terms of general roles and responsibilities are also described. A simple analysis of how well we are going may therefore be to provide counts of obvious indices. For example, how much of their own money sugar farmers and others on a catchment have charged, donated or invested in environmental protection, number of Landcare groups, or change in phosphorus level in the local river could all be measured and reported. But the real issue for the long term is not only whether indices have changed but why. Alternatively, if they have not, why not? There is little general use in knowing if we succeeded on one catchment if we do not know why. There is also little value in throwing out all apparently unsuccessful programmes if we do not know why they 'failed'. In short, we need a model to systematically compare performances in different localities.

Mazmanian and Sabatier (1989) provide one of the surprisingly few models for evaluating policy implementation. It is basically a top-down model. They assume that government and its agencies have clear policies which are then 'dispensed' from the department to the community who respond to the implementation of the policy. On the face of it this seems inappropriate for the participative policies of Landcare and ICM. However, in an addendum to their model the authors acknowledge that, in reality, policy implementation should incorporate bottom-up perceptions and that more of a partnership approach be adopted. Because of successful application to other problems this model was adapted by Bellamy (1995) and colleagues after consultations with landholders and others on the Herbert River catchment in northern Queensland (Fig. 27.1). There is more emphasis in the revised model on process issues (i.e. who does what and when) and clear links are made to ongoing evaluation activities. The model is relatively crude and self-evident. Nevertheless, it clearly demonstrates the variables that should be measured or monitored in all facets of policy implementation. This allows some predictability to data collection and early identification of 'missing links' and missing data. In this way strong and testable hypotheses as to why events happened can be formulated.

The model could be used by individual stakeholder groups so that they can view the system from their own point of view. Encouraging this activity will assist in avoiding misunderstandings and therefore unproductive conflict (Martin, 1995). It is not contended that all conflict is unhealthy in this arena.

Making the model work

The proposed model represents an organised approach to conceptualising what we are doing. Nevertheless it has to be reasserted that it is very much a top-down as opposed to bottom-up approach. Some authors (Hjern and Hull, 1982) suggest that policies and their implementation are best evaluated from a bottom-up perspective. This would certainly be a compatible view with many who see ICM and its subsidiaries as community-driven exercises. But perhaps, as Mazmanian and Sabatier and other authors (Hollick and Mitchell, 1991) have suggested, the implementation of policy needs to be seen in a social systems perspective. That is, if the processes of implementation fail to capture everyone's perspectives, the coordination between parties will be difficult to sustain in the long term. This is because there will be a lack of understanding of the differences between parties and therefore inadequate policy formulation in the first place.

In their review of ICM processes, Syme *et al.* (1994) attempted to provide a model which allowed for flexibility in policy formulation and its implementation, and also made specific suggestions as to the roles and responsibilities of all stakeholders in the ICM system. These roles and responsibilities were linked within the ICM social system as a whole. That is, the issue of bottom-up or top-down views of policy implementation were replaced by a tentative description of how the system incorporating government, institutions, industry and landholders would function as a coherent whole. This exercise needs to be repeated in the context of the modified Mazmanian and Sabatier model and with more reflection on the sugar growing areas. In this way all players will be in agreement about what should happen and the difficulties needing to be faced to get there.

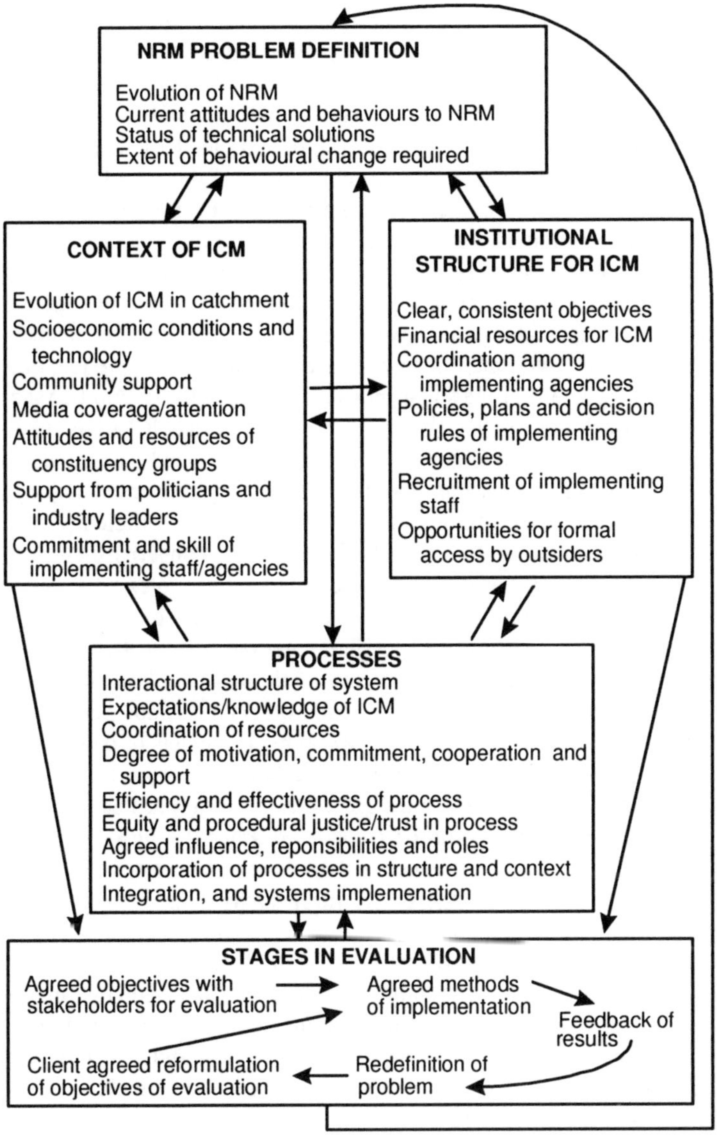

Fig 27.1. Integrated Catchment Management evaluation framework (adapted from Mazmanian and Sabatier, 1989).

What Does All This Mean for the Sugar Industry?

This paper has identified policy trends, and issues relating to the long term significance of these trends. Clearly there are many challenges ahead for the sugar industry. We have responded in the usual academic fashion by suggesting a model of policy implementation hypothesised to be relevant to the Herbert catchment, and hopefully others. Other models are possible and should be developed for comparative evaluation. We contend that model creation is important because, otherwise, all developments in ICM will end up as 'one-off' activities. This will result in much reinventing of the wheel and confusion about why things work. We contend that the model will only be useful if we have an accompanying model of the social processes which will need to occur. People need to be clear about their roles and responsibilities if ICM is to succeed.

ICM and regional planning are an opportunity for the industry to create refocused partnerships with government and newer players in the field. Given current lack of government resources and a 'hands off' approach, due recognition and encouragement of the sugar industry's needs will have to be made within this new way of doing things. The reason for this is that ICM has taken us back to basics in terms of governments and their role. Newly negotiated government roles will include discussions about priorities for government action, definition of fair community contributions and issues relating to efficiency of operation of policy implementation. Policy and administrative concerns such as the definition of public and private goods, the leadership roles of government and professionals, and how to manage the expectations of competing stakeholders will all become part of the ongoing negotiations of ICM. This means that sustainability will become much more of an everyday business concept for industry. Because sustainability depends on relationships between environment, economics and society, it will impinge on many aspects of daily life and differing functions of government. It will therefore become a core focus for industry and its lobbyists. If ICM works, it is a great forum for all participants!

In many ways this could be considered a healthy turning back of the clock. The agreements struck, catchment by catchment, will resemble those of previous times when common property in rural areas was managed by community negotiation. These negotiations included rules of fairness and methods for ensuring compliance (Ostrom, 1990). Many negotiated rules for managing the commons have lasted over centuries and are still in sustainable management. The sceptics of the role of the community in managing their own policies, as in earlier times, suggest that society is now more complex and that old rules cannot be made by the community in modern contexts. Nevertheless, authors such as Gardner and Stern (1996) have identified the prerequisites for deriving such rules in modern, socially complex situations. They also show how rules made in relation to water allocation can be derived at a regional level by organising decision making appropriately through representative organisations.

We suggest that given certain conditions, which include clarity of purpose of competing interests and clarity of representation, the community part of the ICM system can be vibrant. Novel and lasting deals can be struck with government institutions and these can be maintained over the medium to long term. This will not only help the industry maintain control of its destiny but also provide guidance for government institutions in fine tuning their ICM operations. Such trends are beginning to occur in the rice industry (and other industries depending on irrigation) with community operation of irrigation areas (e.g. in the Coleambally Irrigation Area, NSW and the South West Irrigation Area, WA).

Conclusion

Over the past two decades it has become obvious that primary industry has lost political clout. One needs only to look at the relative number of votes 'in the bush'. Economic deregulation of rural industries has exacerbated increasing feelings of powerlessness. The advent of ICM and regional planning, however, provide an important opportunity for the sugar industry and its peers to regain some of the influence it once had, but in a different context. Old government institutions and their mechanisms have been found wanting. ICM requires local knowledge, wisdom and commitment. By offering this, the sugar industry will, over the next ten to twenty years, be rewarded with ICM policies that fit its needs and those of its neighbours. ICM can serve an enlightened self interest. If sugar industry participation is not offered, those who insist that top-down institutional control is better will reassert their authority. The authors do not care if this happens - they are not driven by organisational ideology - but do have grave fears whether sustainability will occur under these conditions. There is some theory and several recent memories to support their concerns.

It is widely known that the sugar industry received protection and support from the government in the early part of this century because of astute association with the 'White Australia' policy (Pagani, 1989). We suggest that in the early part of the twenty-first century the sugar industry's remarkable contribution can be continued by a change of colour. The 'Green Australia' policy may well be the path to influence and prosperity in the next century.

References

AACM (1995) *Enhancing the Effectiveness of Catchment Management Planning.* AACM International, Adelaide, Australia.

Bellamy, J.A. (1995) *Evaluation of Integrated Catchment Management in a Wet Tropical Environment.* Progress Report. CSIRO, Division of Tropical Crops and Pastures, Brisbane, Australia.

Campbell, A. (1994) *Landcare: Communities Shaping the Land and the Future.* Allan and Unwin, Melbourne, Australia.

Carman, K. and Keith, K. (1994) *Community Consultation Techniques: Purposes, Processes and Pitfalls.* Department of Primary Industries, Brisbane, Australia.

Curnow, M. and Burnside, D. (1995) *Technical Support to Land Conservation Districts (1989-1995): An Evaluation.* Agriculture Western Australia, Miscellaneous Publication 14/95.

Curtis, A. (1995) *Landcare in Victoria: The State of Play.* Report No. 24. Johnstone Centre of Parks Recreation and Heritage, Charles Sturt University, Albury, Australia.

Dovers, S. (1995) Information, sustainability and policy. *Australian Journal of Environmental Management* 2, 142-156.

Friend, J.K., Power, J.M. and Yewlett, C. (1974) *Public Planning: The Inter-Corporate Dimension*, Tavistock Publications, London, UK.

Gardner, G.T. and Stern, P.C. (1996) *Environmental Problems and Human Behaviour*, Allyn and Bacon, Boston, USA.

Grandori, A. and Soda, G. (1995) Inter-firm networks: Antecedents, mechanisms and forms. *Organization Studies* 16, 183-214.

Herbert, H.W. (1956) The Australian Sugar Industry. In: *South Pacific Enterprise: The Colonial Sugar Refining Company Limited.* Angus and Robertson, Sydney, Australia, pp. 91-117.

Hjern, B. and Hull, C. (1982) Implementation research as empirical constitutionalism. *European Journal of Political Research* 10, 105-116.

Hollick, M. and Mitchell, B. (1991) *Integrated Catchment Management in Western Australia: Background and Alternative Approaches.* Centre for Water Research, UWA Publication ED 606 MH, Perth, Australia.

Martens, P. (1981) *An Organisational Approach to Regional Planning.* Gower Press, Aldershot, UK.

Martin, P. and Woodhill, J. (1995) 'Landcare in the balance': Government roles and policy issues in sustaining rural environments. *Australian Journal of Environmental Management* 2, 173-183.

Martin, S. (1995) Partnerships for local environmental action: Observations on the first two years of rural action for the environment. *Journal of Environmental Management and Planning* 38, 149-165.

Mazmanian, D.A. and Sabatier, P.A. (1989) *Implementation and Public Policy.* University Press of America, Lanham, USA.

McCreddin, J.A. and Syme, G.J. (1996) *Developing Fair and Equitable Land and Water Allocation: Principles, Processes and Decision Making.* CSIRO, Australian Research Centre for Water in Society, Working Paper, Perth, Australia.

Mintzberg, H. (1994) *The Rise and Fall of Strategic Planning: Reconceiving Roles for Planning, Plans, Planners.* Free Press, New York, USA.

Nijkamp, P., Lasschuit, P. and Soeteman, F. (1992) Sustainable development in a regional system. In: Breheny, M.J. (ed.), *Sustainable Development and Urban Form*, Pion, London, UK.

Ostrom, E. (1990) *Governing the Commons: The Evolution of Institutions for Collective Action.* Cambridge University Press, Cambridge, UK.

Pagani, M. (1989) *Politics and the Queensland Sugar Industry.* James Cook University, Townsville, Australia.

Sale, K. (1991) *Dwellers on the Land: The Bioregional Vision* (2nd Ed.). New Society Publishers, Philadelphia, USA.

Syme, G.J., Butterworth, J.E. and Nancarrow, B.E. (1994) *National Whole Catchment Management: A Review and Analysis of Processes,* Land and Water Resources R&D Corporation, Occasional Paper No. 01/94, Canberra, Australia.

Vanclay, F. and Lawrence, G. (1995) Agricultural extension as social welfare. *Rural Society* 5, 20-33.

Webb, A. (1991) Coordination: A problem in public sector management. *Policy and Politics* 19, 229-241.

Chapter twenty-eight:

Distilling the Spirit: A Vision for the Sugar Industry in 2020

R.J. Clements

Australian Centre for International Agricultural Research, GPO Box 1571, Canberra, ACT 2601, Australia

A 2020 Vision

'It is the year 2020, and there are about 8 billion people on earth - 2 billion more than in 1996. Greater Brisbane, Australia's second biggest city, has a population of more than three million people, and extends for 150km along the south-east Queensland coast. The small sugarcane industry that was previously situated between Brisbane and the Gold Coast has long since disappeared, swallowed up by a sea of houses and hobby farms. The Australian sugar industry has largely moved northwards, beyond the Tropic of Capricorn.

The sweetener industry world-wide has changed dramatically. World sugar production has risen to 190Mt y^{-1}, but sucrose from sugarcane and sugar beet now provides only 70% of the world demand for sweeteners. Sugar beet's share of world sweetener output has fallen to 25%, and cane sugar holds a 45% share.

It is 2020, and the world sugar production scene has changed dramatically. More than half the world's sugar is produced and consumed in Asia. India and Thailand are dominant in that region. Production from South America has grown rapidly, and Brazil is the world's largest single producer of sugarcane and of sugar, with a swing away from ethanol production. Cuba and Mexico are no longer major producers, but southern Africa has lifted production significantly.

In 2020, Australia dominates international trade in sugar, with an annual production close to 10Mt, 90% of which is exported. Among the other major exporters are Brazil, India, southern Africa and Thailand. Since the liberalisation of international trade, the removal of trade sanctions by the USA against Cuba, and the decline of communism, Cuba has become a major tourist

In: Keating, B.A. and Wilson, J.R. (eds) *Intensive Sugarcane Production: Meeting the Challenges Beyond 2000*, CAB International, Wallingford, UK, 1997.

destination for USA citizens and its sugar industry has declined, but it has retained a useful niche market in the USA, particularly in Florida where more than 50 million people now live, and where environmental concerns forced the local sugarcane industry to close down early in the 21st century.

It is 2020, and the Australian sugar industry has changed dramatically. More sugar is produced by fewer farmers, but the family farm is still the key production unit. The industry is concentrated in north and central Queensland, particularly in the Burdekin Valley and on the Atherton Plateau - but there is a flourishing regional industry in north-western Australia. Average cane yields are almost 120t ha^{-1} y^{-1} (fresh weight), with CCS levels averaging 15.5 in most years, with genetically engineered cane. Every new cane variety is transgenic, and Australia has a significant equity in the world's genetically engineered sugarcane, following a farsighted decision by Australia's National Agricultural Research System to invest in sugarcane biotechnology during the 1990s. Using genetic engineering technologies, Australia's cane breeders have modified the process of sugar accumulation, improved sugar quality, provided durable resistance to major pests and diseases, modified the patterns of uptake of plant nutrients, and improved resistance to low temperatures and drought. But even more importantly, they have transferred to sugarcane several genes to produce high value pharmaceutical chemicals, which are extracted at the mill by joint venture partners. They have thus turned sugarcane into a biological factory for pharmaceuticals. Around the world, other cane producers have expanded the range of by-products from sugarcane.

Australia's cane farmers have led the way in developing and utilising environmentally-friendly technologies. Computerised equipment is used routinely in farming operations. At harvest time, yields are monitored continuously, and the data are fed into a computer linked to global positioning system (GPS) equipment fitted to the tractor. Subsequently, when fertiliser is applied, the application rates are adjusted continuously by equipment linked to the yield database. The maximum application rate is within limits set by the cane farmers in consultation with the local community. Herbicides and pesticides are rarely used. Technical innovations to cane harvesters and improved planting arrangements have reduced cane losses during harvesting to less than 2%. Irrigation is used effectively, and, driven by realistic water prices, the efficiency of application averages 90%. Soil acidification is minimised, and is reversed routinely by applying lime and gypsum. Stream banks are well-protected by a 30-50m riparian zone. The sugar industry is widely recognised and respected as a careful user of natural resources.

It is 2020, and a new generation of sugarcane mills handles the Australian crop. They are twice as energy-efficient per tonne of sugar as the mills of the 1990s. State-of-the-art cane cleaning technology is in place. Dirt is recovered and returned to the farm. Highly sophisticated membrane technologies and other liquid separation, ultra-filtration and flocculation processes remove unwanted contaminants and isolate the high-value pharmaceuticals. Sugarcane wax and other valuable by-products are separated for further value-adding.

In 2020, the Australian sugar researchers are concentrated in north Queensland, where they work closely with industry and the community. A loose coordination of R&D activities keeps the researchers focused but allows them a great deal of freedom to pursue truly novel solutions to industry challenges. They interact on a daily basis with collaborators world-wide in a global network of sugar researchers. Although they keep an eye on the value of intellectual property, they recognise that most of the time it is more helpful to the Australian industry for them to exchange information than to hoard it. A wide diversity of R&D providers ensures that core research capacities do not become core rigidities. Industry leaders and researchers recognise that in the year 2020, applications of information technology and biotechnology are the key to industry innovation and technological competitiveness.'

This is, of course, a fanciful view of the future; but almost every one of those outcomes can be imagined from a careful analysis of the information presented at this conference. All of those outcomes - and more - are possible. Whether or not they occur depends on decisions that are made in many countries during the next few years.

At an early time in the organisation of this conference, this final session was referred to as 'Distilling the spirit'. It follows other sessions referred to as 'Sugar and spice', 'Sweetness and light', 'One lump or two', 'Sweet and sour' and 'Shaken and stirred' and I have retained this thematic idea in the title of this address and in my approach to summarising this remarkable conference. However, I will also attempt to capture some of the flavour of the presentations while I concentrate the essence. Perhaps this will help my summary to age gracefully.

Looking back over the four days of this conference, two issues have clearly dominated the proceedings. The first is complexity and the second is opportunity.

Complexity

Let me deal first with complexity. Speaker after speaker mentioned the complexity of the sugar industry and its R&D challenges at every scale. Let me give you some examples.

Although I did not hear either James Fry or Ross Chapman use the word 'complexity', it is quite clear that the sugar industry they described at the global and Australian levels is extremely complex, and is changing rapidly. Indeed, James Fry said that the world sweetener industry is changing so rapidly that one can lose sight of the underlying trends; and managing the Australian sugar industry at the downstream end of the value chain is all about understanding and managing complexity at the level of global supply and demand both for sweeteners and sugar. At the other end of the 'big picture', the sugarcane production end, Andrew Johnson noted that the level of political complexity has increased. Geoff Syme said that integrated catchment management involves complex issues requiring shared vision, and that the social complexities associated with the growth of the sugar industry are increasing. Mike Robertson said that decisions by policy makers and farmers on the use of limited water are complex, and Martin Kropff noted the complexity of dealing with multiple objectives, and the need for clearly stated goals from people with different perspectives. Of course, the complexities at each end of the business system are not independent, and John Quiggin mentioned the complexity of the interaction between environmental impacts, market forces and regulatory policies. Indeed, my organisation is just commencing a project in Indonesia to examine some of these interactions (but not specifically for the sugar industry).

On the farm, the complexities abound. Alan Garside described the complex causes of the sugarcane yield plateau in Australia - indeed, a major outcome of the research was to demonstrate the complexity of the production environment. Mal Wegener and Ian Gray described the complexity of farm decision-making, and the ways in which farmers cope with change. Beth Woods showed how complexity in the local implementation of green cane harvesting slowed down the adoption rate in many districts, and talked about the difficulty of getting complex management practices adopted by farmers. This is true of every agricultural industry, in my experience. Lisa Brennan talked about the complex cane harvesting system in Australia, and said that it required 'coordination of inter-related innovations in harvesting'.

At the research bench, complexity is still the order of the day. Russell Muchow described the complexity in assessing the climatic limits to sugarcane and sucrose production. Brian Keating mentioned the complex interactions that determine the fate of nitrogen fertiliser applied to sugarcane crops. Tom Sinclair, I think, did not talk about complexity, although it was implicit in his 'mountain model' of crop yield, but he showed that simple equations can reduce the complexity of crop yield to a manageable level. Those equations form the basis for simulation models, and I will say more about them presently. Tom also introduced the concept of a socio-economic plateau, and suggested that a possible plateau in grain yields in several countries was associated with a socially induced downturn in rates of fertiliser application.

Interestingly, the molecular biologists, who deal with genetic manipulation of plants at the most complex level - a level of complexity that most of us can barely understand - talked the least about complexity! John Manners noted the complexity of the sugarcane genome, and Robert Birch introduced us to the complexity of IPR (intellectual property rights). Paul Moore described the complexity of the sucrose metabolism pathway, and noted that any one or more of the steps (or none) could be rate-limiting. Complexity occurs at every level. In the face of complex challenges, it is possible to be frozen into inaction, like a rat in the spotlight of a cane farmer. We can analyse the situation interminably, and end up in paralysis by analysis. But at this conference we have seen many examples of how to move forward in the face of complexity, and there are at least two approaches we can use. They are not mutually exclusive.

The first approach is to use information technology to explore options. One of the big messages from this conference is the power of information technology to provide solutions to intractable problems when many things vary simultaneously. There were nine poster papers on applications of information technology at this conference. Speaker after speaker described the use of information technologies or indicated where they could be used: Martin Kropff; Russell Muchow; Tom Sinclair; Brian Keating; Andrew Wood; Michael Robertson; Andrew Johnson; Angela Arthington. Anyone who may have doubted the value of simulation modelling, geographic information systems, or remote sensing, must surely have emerged from this conference with a heightened awareness of their future application to the sugarcane industry. I suggest to you that this is perhaps the clearest message from this conference.

The second approach is the use of effective group processes to negotiate compromises and trade-offs on resource use. Of course, trade-offs at a number of levels in the Australian sugar industry are already commonplace. Lisa Brennan described trade-offs in the harvesting operation. Chris Allen described them in the milling and processing operations, and Mal Wegener described them in relation to farm inputs and outputs. But we need to extend the process. Andrew Johnson pointed out that we need to negotiate acceptable trade-offs between successive generations of people, for example. Many speakers at the 'environmental' end of the spectrum mentioned the need to achieve ecologically sustainable development, and ultimately this will require compromises and trade-offs. As Geoff Syme and Beth Woods pointed out, the immediate challenge is to develop effective group processes to achieve these compromises. It is important to recognise that, as in all conflicts about resource use, there will be winners and losers. It will not be all 'sweetness and light'. There may be a lot of 'shaking and stirring'. We have some useful models to try - the Landcare and Integrated Catchment Management models, for example - and we must push ahead. But there will be many failures and disappointments along the way. Just as for any other scientific experiment, there will be blind alleys.

Opportunity

Let me turn to the second big issue from this conference: opportunity. At the end of four days, most of us were impressed by the possibilities for technical innovation in the sugar industry. Let me give you some examples:

1. Russell Muchow described the opportunity in Australia to increase sugar production by expanding the sown areas on land most suitable for cane, and to improve yield forecasting and crop scheduling practices.

2. Wayne Meyer noted the opportunity to improve the efficiency of irrigation by modifying water prices, and Michael Robertson highlighted the opportunity to develop best-bet irrigation strategies using long-term climate and crop models.

3. Andrew Wood and Brian Keating described the opportunity that individual cane farmers and the industry as a whole could grasp to re-evaluate and improve their approach to nutrient management, including the use of computer-based decision support systems. Both the industry and the wider community would benefit.

4. On the achievement of ecologically sustainable development, several speakers highlighted significant opportunities. Andrew Johnson urged us to develop incentives to promote desirable resource use, and to empower communities and industries to ensure ESD. Angela Arthington noted the opportunity to develop indicators of the condition of aquatic ecosystems and riparian vegetation, and Chris Crossland outlined the opportunity to obtain hard data on the impact of land management practices on the offshore environment. Without such information we are defenceless in the face of well-organised critics.

5. In relation to the application of biotechnology, the opportunities are stunning - but the investment risks are significant. Robert Birch encouraged the industry to choose its targets and invest in the research. His group has shown what can be achieved; they have recently transferred a gene from a humble bacterium into sugarcane to protect it against leaf scald. Frank Smith stated that 'the opportunity for improved efficiency of nutrient use by our major agricultural plants has never been greater', and made the first public announcement of the structure of the phosphate transporter APT 1. This remarkable genetic achievement was not made using sugarcane as the model plant, but it has great relevance to cane farmers. Paul Moore described the opportunities to modify the sucrose accumulation pathway, and a flurry of related posters described the research in progress in several countries. Indeed, there were 19 posters on biotechnology at this conference. The second big take-home message from the conference is that after a late start, sugarcane has entered the age of biotechnology and there is a growing team of molecular biologists on the job.

6. In a strike from left field, Colin Adam has raised the tantalising opportunity of value-adding sugarcane in ways that were almost unimaginable a few years ago, and changing the nature of the vendor pyramid for the sugar industry.

With so many opportunities, which ones do we select? Again we have some useful models in Australia for prioritising research, and some alternative sources of funding. Of course, everyone will see the priorities slightly differently, and we will benefit from a diverse group of lightly coordinated R&D providers. In the pressure to achieve the desired goal, enunciated by Harry Bonanno in his opening address, of an R&D system which is better coordinated, better directed, more accountable and has less duplication, we need some of those group processes described by Beth Woods and Geoff Syme to achieve some acceptable trade-offs between industry pull and technology push. Having seen the evidence during these four days of the benefits of globalisation of research, a range of research providers, and collaboration between groups of scientists in different organisations, who would retreat to the sugar research world of ten years ago?

Flavour

The organisers of this conference developed an ambitious game plan. Part of their purpose was to review the contribution of R&D to the technical evolution of the sugar industry along the entire value chain or business system. This in itself was a significant undertaking. But they also set out to analyse the contemporary challenges facing the industry, to illustrate some of the current exciting (but highly specialised) research that might provide solutions to these challenges, and to do all this in language that could be understood by an audience of people of diverse backgrounds. In the written papers, but more particularly in the spoken presentations, the contributors did not hesitate to use colourful language to communicate their ideas. Not surprisingly, there were some memorable and thought-provoking statements. My summary would be incomplete if I did not capture some of these for you. My personal selection of written or spoken quotations is as follows:

- 'There is a large chasm between what is known and what is done' (Malcolm Sumner)
- 'There has been relentless and continuous improvement in production aspects of the industry' (Andrew Johnson)
- 'In no case shall nutrient concentrations of a body of water be altered so as to cause an imbalance in natural populations of aquatic flora or fauna' (David Anderson, quoting a Florida State Law of 1996)
- 'The new reality in crop production may be one of yield plateaus defined by socio-economic constraints' (Tom Sinclair)
- 'There is a general risk that core research capacities become core rigidities' (Peter Cox, in the presentation by Beth Woods)
- 'It is difficult to be innovative if you're wearing a straitjacket' (Nils Berding)
- 'The focus for future R&D is to enable participants to cope with complexity using strategies based on action researching principles and collaborative enquiry' (Beth Woods).

Every one of these statements - and they are only a sample - has the capacity to shock the listener. They are memorable because in one way or another they challenge us to reconsider our views on particular issues: technology transfer, environmental degradation and research management. Collectively they illustrate the vigour of the presentations and the strength of the viewpoints.

A Parable for Our Times

Let me conclude by returning to my themes of complexity and opportunity, and the need to move forward.

'I once heard a story about a religious teacher - a preacher. Each week he did his best to interpret for his people the mysteries and complexities of life. He was happily married, but unfortunately his wife died and he became very lonely. His congregation, who loved and admired their preacher, were greatly concerned by his loneliness, and after much discussion they purchased a cat for him. The preacher and the cat became good friends. One day, as he entered the church to talk to the congregation, the cat followed him. The congregation thought this was harmless enough, and at least their preacher was happy. The following week it happened again, and it soon became a ritual. Each week, the preacher and the cat would enter the church together. The congregation liked the cat. Besides, it distracted them from the effort of trying to understand the complexities of life.

The years went by, and the cat grew old and finally died. The preacher, too, was old, and his eyes were no longer sharp. When the congregation found another cat and taught it to follow him down the aisle, he hardly noticed that it was not his old friend. Each week, the old preacher and the new cat would enter the church in solemn procession.

After many more years, the preacher also died, and a new preacher arrived to interpret the mysteries and complexities of life to the congregation. As he entered the church, he was surprised to find that a cat appeared beside him and accompanied him down the aisle. The congregation seemed quite unconcerned, and the preacher saw that this was an expected and regular part of the ceremony, and was clearly of great religious and cultural importance. The oldest and most senior members of the congregation did seem to understand its mystical significance, but they did not easily reveal the mysteries of their ancient rituals to outsiders such as him. They grew still older, and eventually died, and the cat grew old and was replaced, and in the end nobody could remember the origins of this most important ritual. But the congregation was comforted by it, and they knew that it must be important in understanding the meaning of life.'

Ladies and gentlemen, let us not be like this imaginary congregation. Let us not avoid complexity or confront it with ritual. If there are ritualistic cats in our thinking about R&D for the sugar industry, let us dispose of them. Let us instead embrace the opportunities revealed at this conference. These opportunities and technologies will take us beyond our present level of understanding, and will lead to industrial innovation beyond the year 2000.

Index

In: Keating, B.A. and Wilson, J.R. (eds) *Intensive Sugarcane Production: Meeting the Challenges Beyond 2000*. CAB International, Wallingford, UK, 1997.